ELEMENTS OF
MODERN
MATHEMATICS

Mathematics is the majestic structure conceived
by man to grant him comprehension of the universe.

LE CORBUSIER

ELEMENTS OF
MODERN
MATHEMATICS

Kenneth O. May

DOVER PUBLICATIONS, INC.
Mineola, New York

Bibliographical Note

This Dover edition, first published in 2019, is an unabridged republication of the 1962 Addison-Wesley Publishing Company, Reading, Massachusetts, second printing of the work originally published in 1959.

Library of Congress Cataloging-in-Publication Data

Names: May, Kenneth O. (Kenneth Ownsworth), 1915-1977, author.
Title: Elements of modern mathematics / Kenneth O. May.
Other titles: Modern mathematics
Description: Dover edition. | Mineola, New York : Dover Publications, Inc.,
   2019. | Unabridged republication of: Reading, Massachusetts :
   Addison-Wesley Publishing Company, 1962; second printing of the work
   originally published in 1959. | Includes index.
Identifiers: LCCN 2019015959| ISBN 9780486836577 | ISBN 0486836576
Subjects: LCSH: Mathematics.
Classification: LCC QA37 .M39 2019 | DDC 510—dc23
LC record available at https://lccn.loc.gov/2019015959

Manufactured in the United States by LSC Communications
83657601
www.doverpublications.com

2 4 6 8 10 9 7 5 3 1

2019

CONTENTS

(Starred Sections and Chapters are optional)

PREFACE

This book is written for the student who is just beginning a serious study of mathematics but who wishes to become acquainted as quickly as possible with the interesting and powerful ideas that have flowered in this century. The main prerequisites are a strong desire to learn and a willingness to think energetically. No more mathematical background than standard high-school courses in plane geometry and elementary algebra is assumed. Additional experience, mathematical or not, is helpful primarily because of its contribution to maturity and motivation. The purpose of the book is, briefly, to help the student obtain a modest mathematical literacy. By energetic study he may expect to acquire from it a minimal orientation in modern mathematics, a basic vocabulary of mathematical terms, and some facility with the use of mathematical concepts and symbols. He should then be ready to study mathematical statistics and to take further work in mathematics, and he should be equipped with the mathematical tools most essential in the physical sciences, engineering, the biological sciences, and the social sciences.

The beginner in mathematics may be compared with someone about to take up a foreign language. He may wish merely to learn to speak, read, and understand the simplest phrases, in short, just enough "to order breakfast." There are several books directed toward the student who wants only this so-called "practical" acquaintance with elementary mathematics. On the other hand, the student of a language may wish to be able to carry on an intelligent conversation, to read significant material, to write, and even perhaps to think in the language. He must be interested in grammar, syntax, and, above all, in acquiring an understanding and a "feel" for the language. Such a student of mathematics acquires manipulative skill as a by-product of striving for insight and thorough knowledge. It is for this kind of student that the present book has been written.

The traditional college curriculum in mathematics is largely an ossification of the program designed to meet the needs of engineers before World War I. Since that time mathematics has developed as never before in its history. Altogether new branches have grown up, and new light has been shed on old ones. Mathematical methods have been applied to new fields, and unexpected developments in science and technology have required the use of novel mathematical theories. The old curriculum no longer meets our needs in the training of scientists, engineers, and mathematicians.

The inadequacy of the traditional mathematics curriculum is most obvious in relation to the needs of the social scientist. Today there is hardly a branch of mathematics that does not find application in economics and psychology. Since 1930, when the Social Science Research Council appointed a committee to report on college mathematics, there has been considerable study, experimentation, and discussion. There seems to be fairly general agreement that social scientists need a mathematical curriculum that brings them quickly into contact with modern mathematics with the omission of traditional topics that no longer contribute either to applications or understanding.

It might seem that the increasing needs of the social scientist would make it desirable to plan different introductory courses for social and natural scientists. However, it turns out that the topics considered desirable by the social scientist are exactly those that are fundamental for all users of mathematics, for the educated citizen, and for the prospective mathematician.* The proposals for curricular change made by social scientists and psychologists are very similar to those proposed by committees of mathematicians who have recently considered the high-school and elementary college programs for all students. This is not really surprising, because the basic characteristic of mathematics is its universality of application. Therefore it is natural that ideas of fundamental importance in one field of application and in mathematics itself should turn out to be also of fundamental importance elsewhere.

This book has developed out of a number of years of experimentation with elementary mathematics courses at Carleton College. Experience proved that students of all degrees of ability and interest were able to understand, enjoy, and apply supposedly very "abstract" and "difficult" ideas of modern mathematics. A preliminary edition has been used at Carleton and by a number of professors in different parts of the country. The present book is a completely rewritten version incorporating major and minor revisions based on classroom experience and a number of detailed critiques.

To achieve the aim of moving rapidly to significant modern material without getting bogged down in traditional detail, many innovations have been necessary. Among these are: (a) The symbolism of logic and set theory is presented early and *used* throughout the book. (b) A simple notation indicating substitution for variables in formulas is adopted. (c) Exercises are inserted in the body of the sections in order to require reader participation. Answers are given at the end of each section. (d) Applications are to a very wide field in the humanities, arts, biology, and social sciences, as well as to the physical sciences and engineering. (e) There is an intentional and carefully explained variation in rigor, the

---

* See, for example, "Mathematics for Social Scientists," by R. R. Bush, W. G. Madow, Howard Raiffa, and R. M. Thrall, *American Mathematical Monthly*, October, 1954.

greatest rigor being applied to simpler topics. (f) There are numerous departures from traditional order where this makes for greater brevity and clarity. I am convinced that there remain many unexplored possibilities for further simplifying elementary mathematics.

The mathematician who casually picks up this book and glances at the first pages may get the impression that it is a textbook on foundations of mathematics. *This is not the case.* The material on foundations, logic, and sets has mathematical importance in itself, but it is presented at the beginning of the book as a pedagogical device to speed up and make more efficient the student's acquisition of mathematical understanding and skill. In the first three chapters, of the twenty-nine sections not used for introducing a chapter and not marked for possible complete omission, seventeen are devoted entirely to review of algebra. In the others, much of the content consists of traditional elementary mathematics, and many of the details relating to logic and set theory may be omitted by the teacher who wishes to move forward quickly to other topics in algebra and analysis. This is not a textbook on logic. It is a textbook that *uses* logic to attack elementary mathematics. In doing so it follows the tradition founded by Euclid, whose treatise on geometry used logic as a pedagogical and organizational device.

In the preliminary edition, applications were concentrated in the social sciences. The present book still contains social science applications, but there is now a balance between applications in the physical and social sciences. A careful check has been made to be sure that the student will be supplied with every technique needed in elementary courses in physics, chemistry, and engineering. For example, every mathematical technique required for the study of *University Physics* by Sears and Zemansky has been included, often by requiring the student to work out the very manipulations and results he will meet when reading that text.

Perhaps the title, *Elements of Modern Mathematics*, calls for some explanation. In the 20th century mathematics has been characterized by an explosive creativity, rapid expansion of its fields of application, an increasing use of abstract theories, and a tendency toward unification in terms of the concepts of set theory. "Modern" has come into use as a catchword to describe mathematical teaching that reflects these features, and it is used in this sense in the title. It does not signify that everything of old vintage has been rejected, but rather the belief that everything, of whatever age, should be reconsidered in the light of our current knowledge of mathematics and its applications. I have no doubt that this book represents only one step along the path of fully modernizing the mathematics curriculum.

The help of many colleagues and students was acknowledged in the preliminary edition. However, I wish to acknowledge again the influence and encouragement of the pioneering texts of Moses Richardson, the College Mathematics Staff of the University of Chicago, Carl Menger, E. Begle, and F. L. Griffin. The following have made helpful suggestions for revision of the preliminary edition: W. J. Baumol, E. J. Cogan, C. H. Dowker, S. Dreyer, John Dyer-Bennet, W. L. Garrison, H. Gulliksen, F. Harary, H. Howden, S. P. Hughart, N. L. Jacobson, M. Kline, R. P. May, H. S. Moredock, J. Nystuen, M. W. Oliphant, E. Pimsler, C. Poggenburg, J. Roth, D. Schier, J. Shear, R. Sloan, R. Stewart, D. M. Stowe, G. R. Trimble, A. W. Tucker, S. Weintraub, and F. L. Wolf. I am also indebted to several anonymous reviewers and to the staff of Addison-Wesley Publishing Company. I hope that readers of this book will not hesitate to make corrections, criticisms, and suggestions.

K. O. M.

Northfield, Minnesota
January, 1959

PREFACE TO SECOND PRINTING

No major changes have been made, but along with the correction of errors there are numerous minor revisions suggested by classroom experience. It would be impractical to list the nearly one hundred individuals who called attention to errors or suggested improvements. However, I must thank particularly W. R. Scott, L. O'Rear, A. M. Glicksam, H. Dickson, A. Broman, and S. Birnbaum.

K. O. M.

Northfield, Minnesota
September, 1962

NOTE TO THE STUDENT

On the first page of his *How to Play the Chess Openings*, Znosko-Borovsky writes, "It is not, then, by memorizing variations that we shall become proficient in playing openings, but by understanding their meaning, their purpose, and the general ideas and principles which are their foundations." And on the very last page, he repeats: "Do not trust to your memory or learn variations by heart. Learn to pick out the directive ideas of an opening for yourself, and above all remember that you alone are the creator of your game, and your task starts at the very first move." This excellent advice is as appropriate to mathematics as it is to chess. If you wish to study mathematics efficiently, you must center your work on understanding fundamentals and on doing your own thinking. Of course, you have to memorize some definitions and rules of procedure (just as you must learn the moves in chess), but forgotten facts can easily be looked up and those that are used frequently will naturally come to be remembered. Understanding, however, can neither be looked up nor forgotten.

According to the dictionary, to understand something is to be thoroughly familiar with it, to visualize it clearly, to be able to explain, justify, and use it. How, then, can you come to understand the mathematics in this book? *Only by working with it!* Mathematics cannot be mastered passively, by merely reading. Reading is only the first step, to be followed immediately by your own activity. Rephrase what you read in your own words. Illustrate it by examples. Draw a picture or diagram suggested by it. Try to find an example that contradicts it. Relate it to previous discussion. Try to justify it. Try to prove it wrong. Experiment with variations. Try to generalize and specialize its meaning. Analyze its form and parts. Explain it to yourself. As Lewis Carroll wrote in the introduction to his *Symbolic Logic*, "One can explain things so *clearly* to one's self! And then, you know, one is so patient with one's self: one *never* gets irritated at one's own stupidity!" Of course, some of what you read will not deserve such detailed treatment, but many parts of this book, especially the formal theory, compress a tremendous amount of information into a very small space. These can be understood only by extensive activity on your part. Your most important tools in this work are a pencil, plenty of paper, and a large wastebasket. Exercises are inserted in the body of the text to help you take an active part. *These exercises must be done as you read, since they are essential to the discussion.*

Unfortunately, mathematics is often taught as a mere matter of learning instructions and following them. You may have the habit of not trying anything in mathematics until you already "know what to do," until you feel that you "understand" the problem. If this is your habit, try to overcome it as quickly as possible, for it will be the main obstacle to your progress. In order to *learn* to understand, in order to develop the ability to *find out* what to do, you must go boldly ahead. If you are not sure what to do with a problem, *experiment and see what happens!* If the results are not satisfactory, try something else. Experiment, record your work, think about your results, and experiment again. In this way, and only in this way, will you develop the confidence to think independently in mathematics.

If in your reading or your efforts to solve problems you are unable to get satisfactory results in spite of numerous experiments and hard work, review previous parts of the book, or do something else with the intention of returning to the troublesome question later. Often a change of line of thought, or even a complete rest, will accomplish more than hard work. You cannot understand everything at once, and there is no need to give up because some items are obscure. In the words of one of the best-known living mathematicians, Norbert Wiener, "It is not essential for the value of an education that every idea be understood at the time of its accession. Any person with a genuine intellectual interest and a wealth of intellectual content acquires much that he only gradually comes to understand fully in the light of its correlation with other related ideas. . . . Scholarship is a progressive process, and it is the art of so connecting and recombining individual items of learning by the force of one's whole character and experience that nothing is left in isolation, and each idea becomes a commentary on many others."

K. O. M.

NOTE TO THE TEACHER

Although this book is tightly structured, it may be adapted to a wide variety of courses, provided only that the teacher wishes to make use of the simplest ideas of logic and sets. Many sections (marked with a star) may be omitted altogether without disturbing the continuity. Others may be skipped if minor adjustments are made later. Considerable variation is possible by choice of topics and problems within the sections. Problems are graded from the easiest exercises in the body of the sections to really difficult problems toward the end of the list following each section. Answers to exercises and to selected problems are given at the end of each section. Many problems are paired, one that is answered being followed by a similar one without an answer. There are more problems than can be done in the time usually available, so that students and teachers can choose those appropriate to their interests.

For those who wish to reach as quickly as possible the material on analysis beginning in Chapter 4, the number of days spent on logic and sets may be reduced to a total of six. Probably, even for the students with strongest preparation in algebra, another seven class periods should be spent on algebra review in the first three chapters. For students with weak backgrounds in algebra, the time spent on review should perhaps be doubled. This suggests that Chapter 4 could be reached in a minimum of eight assignments spent on logic, sets, and other basic ideas, that a feasible schedule for well-prepared students calls for thirteen days, and that a total of twenty-one days or more would be required for those with weak preparation in algebra. The following are schedules for thirteen and twenty-one days.

1. 1–1 through 1–4	5. 2–1, 2–2, 2–3	9. 3–1, 3–2
2. 1–5 through 1–10	6. 2–5, 2–6, 2–7	10. 3–3, 3–4
3. 1–11 through 1–15	7. 2–8, 2–9	11. 3–5
4. 1–16, 1–17	8. 2–10	12. 3–6
		13. 3–9

1. 1–1, 1–2, 1–3	8. 1–14	15. 2–9
2. 1–4	9. 1–15	16. 2–10
3. 1–5, 1–6, 1–7	10. 1–16, 1–17	17. 3–1, 3–2
4. 1–8, 1–9	11. 2–1, 2–2	18. 3–3, 3–4
5. 1–10, 1–11	12. 2–3	19. 3–5
6. 1–12	13. 2–5, 2–6, 2–7	20. 3–6
7. 1–13	14. 2–8	21. 3–9

For engineering courses in which it may be desirable to reach some calculus (Chapter 7) early in the first semester, either of the two plans above could be followed by a schedule such as this:

1. 4–1, 4–2	9. 7–3
2. 4–3	10. 7–4
3. 4–5	11. 7–5
4. 5–1, 5–2, 5–3	12. 7–6
5. 5–4	13. 5–9
6. 5–7	14. 7–7
7. 5–8	15. 5–10
8. 7–1, 7–2	16. 7–8

Before continuing with Chapter 7, the class should cover the omitted sections in Chapters 4, 5, and 6 (except the starred optional sections). Variations are possible. Such accelerated schedules require concentration on essentials and occasional class references to omitted material. If slighted topics can be taken up again later in the year, advantages can be gained from double exposure and increased student maturity.

For a year course stressing foundations and meeting three hours a week, the following is a possible schedule. (Parts of Chapter 7 could be included by cutting shorter the work in Chapters 6, 8, 9, or 10.)

Chapter 1: 15 assignments	Chapter 6: 12 assignments
Chapter 2: 11 "	Chapter 8: 6 "
Chapter 3: 9 "	Chapter 9: 3 "
Chapter 4: 10 "	Chapter 10: 7 "
Chapter 5: 11 "	

A year course concentrating on analysis can be based on very thorough coverage of the first seven chapters. If the students are strong in algebra, time not needed for review can be used for selections from Chapters 6, 8, 9, and 10. Chapter 7 can occupy from fifteen to thirty class hours, depending on the emphasis and thoroughness of coverage.

For a one-semester course to be followed by a course in analytics and calculus, Chapters 1, 2, 3, 4, and 5, and selections from Chapters 6, 8, 9, and 10 would be appropriate.

Any of these courses could, by choice of problems and topics within sections, be adapted to terminal or introductory courses, to special courses for teachers, social scientists, or engineers, and to varying levels of student ability.

K. O. M.

CHAPTER 1

ELEMENTARY ALGEBRA

1-1 The purpose of this chapter. Fruitful discussion requires some
common knowledge and understanding. Unless we agree on the meaning
of words and other symbols, we shall be like players in a game without
rules, and just as likely to be confused.

The task of creating a basis for mathematical discussion is simplified
by the fact that mathematics can be grasped in terms of a very few pro-
foundly simple ideas. This results in a beautiful unity of logical structure,
and also makes mathematical knowledge much easier to acquire, organize,
understand, and recall. It underlies the cumulative character of mathematics courses, where each step is based on those before, and the whole
develops in a logical way.

We take for granted an understanding of everyday language, a background of experience such as may be assumed common to the reader and
the author, and some memories of elementary geometry, algebra, and
arithmetic. The purposes of this chapter are (1) to remind the reader of
certain simple ideas associated with numbers, (2) to discuss informally the
fundamental ideas used throughout the book, (3) to explain the meanings
that we assign to certain words, and (4) to show how some laws of elementary algebra may be derived from others.

There is an old saying that "a hard beginning makes a good ending."
The reader will find the proverb confirmed if he attacks this chapter with
energy. Of course, he must not expect to master immediately the new
ideas presented here. They are among the most generative man has
produced and can be appreciated fully only by meeting and using them
in various forms.

1-2 Numbers and points. Numbers may be visualized and applied in
many ways. For example, a *positive integer*,* such as 1, 2, 3, and 37, may
be thought of as the result of counting the members of some set of objects. A ratio of positive integers, such as 1/3, 4/3, 2/12, and 117/29,
may be visualized as indicating the number of fractional parts in some
collection of parts of objects. Zero may be conceived as the number of
members in a club from which all members have resigned. A negative

---

* A word or phrase that is to be used in this book in a special technical sense
is printed in italics in the sentence that defines it. Other words are used with
meanings that are familiar to the reader or may be found from a dictionary.

number, such as -2, $-4/3$, -2.75, and -100, may be thought of as a temperature below zero or as the balance in an overdrawn bank account. The variety of interpretations and uses is endless. We concentrate here on one that gives us a simple geometric picture of any number.

We imagine a straight line, as pictured in Fig. 1–1. On it we choose a point, call it the *origin*, and let it serve as the geometrical image of the number 0. We select another point (customarily to the right of the origin), call it the *unit point*, and let it correspond to the number 1. The direction from the origin to the unit point is the *positive direction;* the opposite direction is the *negative direction;* and the distance between the origin and the unit point is the *unit distance*. We call such a line an *axis*. The scale on a thermometer is a familiar example of a portion of an axis.

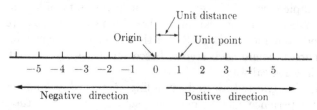

FIGURE 1–1

Every point on an axis serves as the geometric image of a unique number. We call the number the *coordinate* of the corresponding point, and the point the *graph* of the number. If the point is in the positive direction from the origin, this number is just the distance between the origin and point, measured in terms of the unit distance. If the point is in the negative direction, the corresponding number is the negative of the distance between the point and the origin. For example, the number corresponding to the point 10 units to the left of the origin is -10. We now look more carefully at this correspondence between points and different kinds of numbers.

Positive integers. To find the point corresponding to a positive integer we lay out the unit distance to the right a number of times equal to the positive integer.

Integers. The positive integers, zero, and the negative integers are called *integers*. Every negative integer is the negative of some positive integer. For example, -6 is the negative of 6. And every positive integer is the negative of a negative integer. For example, 6 is the negative of -6, since $-(-6)$ is the same as 6. To find the point corresponding to a negative integer, we lay out the unit distance to the left a number of times equal to the negative of the number. For example, to find the point corresponding to -6 we lay out the unit distance $-(-6)$, or 6, times to the left. It is much easier to describe this by using a letter, say "x," to stand for any number. Then we can say that if x is a positive integer, we find the point corresponding to $-x$ by laying out the unit distance x times to the left.

*(a) Draw an axis and show the origin, the unit point, and the points corresponding to 6, -6, -2, and 7. Complete the following: (b) The points corresponding to x and $-x$ are _____ from the origin but in opposite directions. (c) If x is a positive integer, the point x is to the _____ of the origin and the point $-x$ is to the _____, but if x is a negative integer, the point x is to the _____ of the origin and the point $-x$ is to the _____.

Rational numbers. A number that is the ratio of two integers is called *rational.* Examples are 27/19, $-1/2$, 1.25, and 3. The last two are rational because 1.25 is the same as 125/100 and 3 is the same as 6/2. To find the point corresponding to 1/3 we simply divide the unit distance into three equal parts and lay out a distance equal to one of them. To locate 7/3 we lay out the segment of length 1/3 seven times. More generally, if x and y are positive integers, we locate the point x/y by dividing the unit segment into y equal parts and laying out one of these parts x times to the right of the origin. If x and y are positive, we locate $-x/y$ by performing the same operation toward the left (see Fig. 1-2). Since an expression such as "5/0" is meaningless, a fraction with zero denominator does not represent a number. (This is explained in Section 1-14.)

FIGURE 1-2

(d) Draw an axis and show the points corresponding to 4/3, $-5/2$, $-10/3$, 1.25, -0.3.

We now have a method for locating the unique point corresponding to any given rational number, but suppose instead that we are given a point. Can we always determine a corresponding rational number? Of course we can if the point lies at a whole number of unit lengths or fractional parts of the unit length from the origin. But is this so for all points? The answer is no. Some points (indeed, countless points) do not correspond to any rational number. Suppose we lay out from the origin a segment equal in length to the circumference of a circle of diameter 1. The number cor-

---

* The exercises inserted in the body of the text are an essential part of the discussion and must be read and carried through (not necessarily in writing) if the material is to be understood. Answers are given at the ends of the sections.

responding to its endpoint is π, the ratio of the circumference to the diameter of a circle. But π is not a rational number, although rational numbers such as 22/7, 3.14, and 3.1415926536 are used as approximations to it.

Other examples can be obtained in the following way. Let x be the length of the diagonal of a square whose sides have length 1. We have by the Pythagorean theorem (see Fig. 1-3) that $x^2 = 1^2 + 1^2 = 2$, and $x = \sqrt{2}$. We prove in Section 6-8 that there is no rational number equal to the square root of 2. Hence if we lay out the diagonal from the origin as in the figure, we determine a point that does not correspond to any rational number.

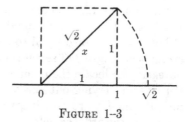

FIGURE 1-3

A number that corresponds to a point on an axis but is not a rational number is called an *irrational number*. For the time being, we shall always replace an irrational number by some rational number that is approximately equal to it. Thus we replace $\sqrt{2}$ by 1.414 even though $(1.414)^2 = 1.999396$. Similarly we replace π by 3.142 even though neither this rational number nor any other is exactly equal to the ratio of the circumference to the diameter. Table I in the Appendix gives rational approximations of the square and cube roots of positive integers up to 100.

Real numbers. The rational and irrational numbers are called *real numbers*. Every real number has a unique corresponding point on the axis, and every point corresponds to a unique real number. The whole axis is a picture of the real numbers, each point being the image of just one real number. Because of this correspondence, we visualize and talk about numbers as points whenever that is convenient. For example, we may speak of the point 6 (meaning actually the point corresponding to 6) or of the number 6, as convenient.

Decimals. When a real number is given in decimal form, the corresponding point may be located by rewriting the decimal as the quotient of two integers and using the method described above for rational numbers. For example, $3.142 = 3142/1000$. However, it is easier to think of the decimal as the sum of units, tenths, hundredths, and so on. Thus $3.142 = 3 + 1/10 + 4/100 + 2/1000$. Then we locate the point by going to 3, then 1/10 more, then 4/100 more, and finally 2/1000 more. If the decimal is negative, we do the same but in the negative direction.

Sketch each of the following, drawing an axis for each, and using approximations from Table I in the Appendix where necessary:	(e) 2.1,	(f) 3.14, (g) 22/7,	(h) −1.5,	(i) −3.85,	(j) $\sqrt{3}$,	(k) $\sqrt[3]{5}$.

Imaginary numbers. Not all numbers are real. For example, there is no real number whose square is −2. Hence $\sqrt{-2}$ is not a real number. The same remark applies to $\sqrt{-6}$, $\sqrt{-9}$, $1 + \sqrt{-3}$, and so on. Nevertheless, such numbers can be defined so that they are as useful and meaningful as real numbers. (See Section 6–13.) They are called *imaginary numbers*, although they are no more imaginary in the usual sense than the reals. The real and imaginary numbers are called *complex numbers*. For the time being we shall limit our discussion to real numbers.

PROBLEMS

1. Which of the following numbers are rational? real? irrational? 2, −3/2, $\sqrt{4}$, $\sqrt{5}$, $\sqrt{2} - 1$, $\sqrt{-4}$, 1.79, −22/7, 3.14.

2. Answer the same questions for these numbers: −43.08, $\sqrt{345}$, $\sqrt{169}$, −85/9, $\sqrt{2} - \sqrt{-3}$, 1, 0, −1, $1 + 2\sqrt{13}$.

★3. Explain and justify the method of dividing a segment into several equal parts by straightedge and compass construction.*

ANSWERS TO EXERCISES†

(a) In order from left to right: −6, −2, 6, 7.	(b) equidistant.	(c) right, left, left, right.	(d) In order from left to right: −10/3, −5/2, −0.3, 1.25, 4/3.	(g) Note that 22/7 is not the same as 3.14, though they are approximately equal.

ANSWERS TO PROBLEMS†

1. All but $\sqrt{5}$, $\sqrt{2} - 1$, and $\sqrt{-4}$ are rational. All but $\sqrt{-4}$ are real. $\sqrt{5}$ and $\sqrt{2} - 1$ are irrational.	3. See a geometry textbook.

1–3 Constants and variables. Written communication is accomplished by means of various marks on paper—letters, punctuation marks, conventional signs, words, abbreviations, phrases, sentences, pictures, dia-

---

* Problems marked with a star are optional. They are interesting, sometimes difficult, but not essential to the main argument of the book.

† Answers are given to most exercises and to selected problems. Abbreviated answers are sometimes given, but the reader should give answers in full, of course.

grams, and so on. We call any intentional mark a *symbol*. Examples are the heading of this section, the previous sentence, the word "two," the sign "+," the numeral "18," and the question mark. A symbol may consist of just a single mark, or it may be made up of several marks, each of which is a symbol. Examples are, respectively, "8" and "753." We often call a symbol an *expression*, especially if it is made up of several symbols. "Symbol" and "expression" are synonymous.

The significance of a symbol depends upon the way it is used. For example, the sound of "a" is different in "hat" and in "hate." The *context* of a symbol is the collection of symbols among which it appears and the other circumstances of its use. The meaning of an expression depends upon its context. For example, "2" stands for 20 in the expression "327" and for 200 in "271."

(a) Compare the meaning of "The Constitution of the United States" in 1859 and 1954. (b) Why does a word-for-word translation from one language to another seldom preserve meaning?

Some symbols serve in particular contexts as names of specific things—persons, places, tangible objects, or ideas. Examples are "the Pacific Ocean," "the Eiffel Tower," "the English alphabet," "the President of the United States," and "New York City." A symbol that in a particular context is the name of just one specific thing is called a *constant*. Grammarians call constants proper nouns. The thing that a constant names is called its *value*. A constant stands for, or names, its value. The value of a constant is its meaning. The symbol "F.D.R." is a constant standing for the late Franklin D. Roosevelt. The man F.D.R. is the value of the symbol "F.D.R." A symbol that stands for a number is called a *numeral*. For example, the numeral "2" stands for the number 2. We say also that "2" has the value 2.

The distinction between a constant and its value is simply the distinction between a name and the thing it names. A constant always names its value. If we wish to name a constant or other symbol, we do so by enclosing it in quotation marks (as illustrated in the above paragraphs*). The quotation marks are sometimes omitted when no confusion is likely. However, the distinction between a symbol and its value is important if language is to be understood. When we communicate we do so not by using the objects about which we are talking, but by using their names. For example, to talk about the number 7 we use "7," "VII," "seven," or some other constant standing for it.

---

* As the reader will note, it is customary to place the closing quotation mark outside a comma or period immediately following the expression, even though the comma or period is not part of the expression.

(c) Distinguish between "27" and 27.   (d) Why is the Pacific Ocean not a constant?   (e) Describe exactly what you see below:

A Live Cobra

(f) Why is "a chair" not a constant?   (g) The English alphabet consists of 26 letters, yet we call "the English alphabet" a constant, i.e., the name of one specific thing. Why?

Many symbols are not constants. Obvious examples are punctuation marks, which certainly do not name anything. But many nouns do not name one specific thing. For example, in "A chair has four legs," the word "chair" does not name any particular chair. In mathematics when we wish to refer to an unspecified object of a certain kind, we usually use some arbitrary letter or other special symbol, such as "x." For example, the statement that any integer equals itself divided by one could be written $x/1 = x$, where we use "x" to refer to an unspecified number. Here "x" is not a numeral, but it stands for any number. By this we mean that any constant standing for a number (that is, any numeral) may be substituted for it. For example, we may write $2/1 = 2$ or $3/1 = 3$ or any other expression obtained by substituting a numeral for x. Similarly, in "A chair has four legs," "a chair" stands for any chair in the sense that we may substitute for it the name of any particular chair.

A symbol that, in a particular context, is not a constant but for which any one of certain constants may be substituted is called a *variable*. For example, "x" is a variable in the context $x/1 = x$, and "a chair" is a variable in the context "A chair has four legs." The constants that may be substituted for a variable in a particular context are called *significant substitutes* for it. The value of a significant substitute is called a *value* of the variable in a particular context. In our examples, "2" is a significant substitute for "x" and "the first chair in the first row" is a significant substitute for "a chair." The number 2 and the first chair in the first row are the corresponding values.

Note that a constant has just one value, whereas a variable has more than one value. This means that a single object is associated with a constant, whereas a collection of several objects is associated with a variable. The variable may be thought of as standing for some unspecified object in this collection.

Cite some values of the variables in: (h) "the point x and the point $-x$ are equidistant from the origin";   (i) "Mr. X";   (j) "the price of y"; (k) "the response to r."

A handy method of creating variables is to place an additional symbol (called a *subscript*) to the right and below another. Thus we may create an unlimited number of different variables from x by writing x_1 (read

"x-sub-one"), x_2, x_3, x_k, x_n, and so on. Often the same letter with different subscripts is used to indicate different variables having the same values. Thus we might speak of the men M_1 and M_2, the women W_1 and W_2, and the numbers x_1 and x_2.

Most of the confusion concerning variables and constants is avoided if a careful distinction is made between symbols and their values. Since a variable is a placeholder for which constants are to be substituted, it appears in its context in the same way as if it were a constant. But a variable is not the name of any particular thing and has no definite value. It is merely a symbol used to indicate a place in which a constant may be substituted. It is a blank, or hole, to be filled in a certain way. Thus in "$3 + x$," "x" appears in the same way as does "2" in "$3 + 2$," but "x" is not the name of any particular number.

Problems

Describe the values of the variables in Problems 1 through 4.

1. "$2 + x$."

2. "A better tennis player than x."

3. "Thus, if the demand for commodity A increased relative to the demand for commodity B, this would cause more of A to be produced."

4. "Economic theory can tell us absolutely nothing more than that for the attainment of a given technical end x, y is the sole appropriate measure or is such together with y_1 and y_2 respectively, and that their application and thus the attainment of the end x requires that the 'subsidiary consequences' z, z_1, and z_2 be taken into account." (Max Weber, *The Methodology of the Social Sciences*, 1949, p. 37)

5. Distinguish carefully between the meaning and role of "chair" in "Bring a chair" and "Bring that chair in the corner."

6. Does "Mr. X" name anyone? When "X" is replaced by a man's name, what happens to "Mr. X"? Does it make sense to ask what happens to Mr. X?

7. Is it true that the value of a constant never changes?

8. Does a variable change? How can variables be used to indicate change?

9. There is only one value of x for which $2x = 4$. Does this mean that x is a constant in $2x = 4$?

★10. "In language, which is the most amazing symbolic system humanity has invented, separate words are assigned separately conceived items in experience on a basis of simple one-to-one correlation." (S. K. Langer, *A Theory of Art*, 1953, p. 30) Comment.

★11. Write a brief essay on the different dictionary definitions of "variable."

★12. In a magazine advertisement we read, "Everything you need for baking a cake is right on this page." Comment.

Answers to Exercises

(a) Amendments have been added. (b) Meaning depends on context. (c) The first is a numeral, the second a number. (d) It is a body of water, not a symbol. "The Pacific Ocean" is a constant. (e) Not a live cobra! (f)

It names no particular chair. (g) It names a single object, which is a collection, or set, of letters. (h) 4, −4, −1.7, $\sqrt{3}$, −$\sqrt{7}$. (i) Jones, Smith, etc. (j) Butter, sugar, etc. (k) A pin prick, a letter, etc.

Answers to Problems

1. Numbers. 2. People. 3. Commodities. 5. The first refers to any unspecified chair, the second to a particular chair. 6. No. It becomes a constant. 7. It changes with the context. 8. No. By substituting different constants for them. 9. No. It has many values in the equation, even though only one of them yields a true statement.

1–4 Vectors. Suppose an aircraft flies due east 75 miles. We may picture its flight by an arrow as in Fig. 1–4, where an axis has been chosen so that the beginning of the flight is at the point −10. If the aircraft had flown 35 miles due west, the arrow would point in the other direction, as in the figure. We have labeled this arrow −35 instead of 35 in order to indicate its direction. Clearly, any motion in the east-west line can be indicated by an arrow whose length is the distance moved and whose direction is the direction of motion. Corresponding to each such arrow there is a unique real number equal to the length of the arrow if it points to the right and the negative of this length if it points to the left.

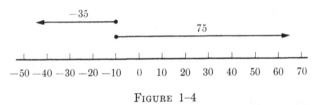

FIGURE 1–4

Arrows such as those pictured in Fig. 1–4 are sometimes described as *directed distances* and called *vectors*. We see that every real number determines a unique vector and, conversely, that every horizontal vector determines a unique real number.

Complete the following: (a) The vector corresponding to the real number 7 is a vector of length _____ pointing to the _____. (b) The vector corresponding to −7 is of length _____ pointing to the _____. (c) The vector corresponding to the number x is the vector of length x pointing to the right if x is _____, and is the vector of length −x pointing to the _____ if x is _____.

Wherever a vector is placed, we may think of it as the directed distance from its *initial point* to its *terminal point* (see Fig. 1–5). For example, 3 is a directed distance of three units toward the right and may be visualized as a vector of length three pointing toward the right. It is the directed distance from the origin to the point 3, the directed distance from the point

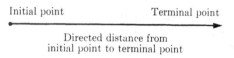

FIGURE 1-5

1 to the point 4, and the directed distance from any point to a point 3 units to the right of it. On the other hand, −3 is a directed distance of three toward the left and may be visualized as a vector of length three pointing toward the left. It is the directed distance from the origin to the point −3 and from any point to another point 3 units to its left (see Fig. 1-6).

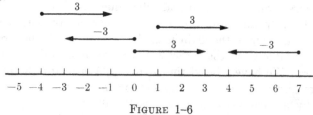

FIGURE 1-6

From now on we shall think of numerals as standing for numbers, points on a real axis, or vectors, according to convenience. For example, we use "−1" to stand for the number −1, the point −1 (the point one unit to the left of the origin), or the vector −1 (the vector of length 1 pointing to the left).

Draw an axis for each of the following and the vector given by the number with its initial point at the origin and in two other positions: (d) 5,   (e) −5, (f) −1.75,   (g) $1 + \sqrt{2}$.

Complete the following.   (h) If the vector x is placed with its initial point at the origin, its terminal point will be at the point _____.

Addition. Among the many uses of vectors is the vector interpretation of the operations of arithmetic. Suppose an aircraft flies 75 miles east, then 35 miles more in the same direction. Its total flight is given by adding the numbers or by adding the vectors, as in Fig. 1-7. The addition, $75 + 35 = 110$, and its vector picture can be given many interpretations. For example, if two forces of 75 and 35 pounds are pulling in the same direction, their resultant is the force of 110 pounds.

We define the *sum of two vectors* as follows: To find the vector $a + b$,

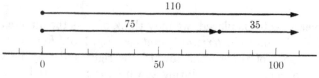

FIGURE 1-7

lay the vector b with its initial point on the terminal point of a. Then $a + b$ is the vector from the initial point of a to the terminal point of b (see Fig. 1–8). In this definition the values of the variables "a" and "b" are all real numbers, i.e., all vectors of the kind we are considering. With this definition, vector addition corresponds exactly to numerical addition. For example, $7 + (-3) = 4$ by the rules of arithmetic; and the same result is obtained by adding the vector 7 and -3 according to the rule (see Fig. 1–9).

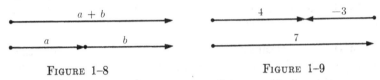

FIGURE 1–8 FIGURE 1–9

In every case we may visualize $a + b$ as the motion given by the vector a followed by the motion given by the vector b. For example, as illustrated in Fig. 1–9, going 7 units to the right then 3 to the left is the same as going 4 to the right. This interpretation of addition in terms of motion is very old, as illustrated by the fact that the ancient Egyptian symbols for "plus" and "minus" were pairs of legs walking in different directions. It is worth while to cultivate the habit of visualizing addition in this way, since it gives a check on accuracy, a quick way of finding sums by "just looking," and an important tool for dealing with distance.

Draw vector diagrams for the following: (i) $3.2 + 4.5$, (j) $8 + (-3)$, (k) $-4 + (-2)$, (l) $-4.1 + 6$.

Complete the following: (m) If a is placed with its initial point at the origin, then the terminal point of the vector $a + b$ is at the point _____.

We now give vector interpretations of other arithmetical concepts.

Negation. The expression "$-a$" stands for the number obtained from a by the operation of negation. If a is a positive number, say 17, then $-a$ is the corresponding negative number, -17. If a is a negative number, say -7, then $-a$ is the corresponding positive number, 7. The points a and $-a$ are at the same undirected distance from the origin, and the vectors a and $-a$ have the same length. If a is a vector pointing toward the right, its length is just a. If a is a vector pointing toward the left, its length is $-a$. For example, the length of -7 is $-(-7)$ or 7. We designate the *length* of the vector a by $|a|$, which is also called the *absolute value* of the number a. It is the *undirected distance* between the initial and terminal points of the vector. The reader should note that in this discussion both negative and positive numbers are values of a. Thus a may be positive or negative, so that $-a$ may be either positive or negative. Indeed if a is negative, $-a$ is positive. *The length of a is therefore a if a is positive and $-a$ if a is negative.*

Sketch x and $-x$ for each of the following values of x. Find $|x|$ in each case.
(n) 5,   (o) -5,   (p) $\sqrt{2}$,   (q) -10,   (r) -7.1.

Subtraction. The *difference* $a - b$ is the number that yields a when it is added to b, that is, $b + (a - b) = a$. For example, $7 - 5 = 2$ because $5 + 2 = 7$. We may find the vector $a - b$ by placing the initial points of a and b together and then drawing the vector from the terminal point of b to the terminal point of a, as suggested in Fig. 1–10.

Sketch the following as in Fig. 1–10:   (s) $5 - 2$,   (t) $7 - 10$,   (u) $-3 - 4$,   (v) $-4 - (-3)$.

From the above examples it is evident that $a - b$ is just the vector from the point b to the point a, that is, the directed distance from b to a. Hence we may always visualize a subtraction as the directed distance, or vector, extending from the subtracted point to the other point. *The directed distance from one point to a second point is found by subtracting the first (beginning) point from the second.* In Fig. 1–11 we use subscripts to emphasize this idea. *The absolute value of the difference gives the undirected distance between the points.*

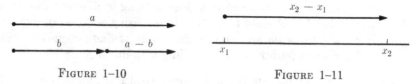

FIGURE 1–10   FIGURE 1–11

Multiplication. The *product* of a and b, symbolized by "ab," "$a \cdot b$," or "$a \times b$," is the number obtained from the *factors* a and b by multiplication. When a is a positive integer, $a \cdot b$ is obtained by adding b to itself a times. In terms of vectors, $a \cdot b$ is found by laying b out a times. If a is negative we interpret this to mean that b is laid out a times in the negative direction or, what comes to the same thing, $-b$ is laid out a number of times equal to $-a$. (Note that $-a$ is a positive integer here, since a is a negative integer.) This is illustrated in Fig. 1–12.

If a is not an integer but is rational, $a \cdot b$ is found by laying out b a certain integral number of times and then laying out some fractional part

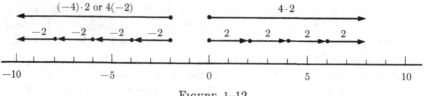

FIGURE 1–12

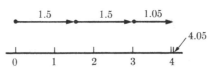

FIGURE 1–13

of it. Thus $(2.7) \cdot (1.5)$ is found by laying out 1.5 twice, then laying out 7/10 of it (see Fig. 1–13). This seven-tenths can be found by arithmetic, since $(7/10)(1.5) = (7)(1.5)/10 = (10.5)/10 = 1.05$. Or it may be found by dividing the segment 1.5 into 10 equal parts and laying out 7 of them. If a is irrational, we shall replace it by a rational approximation. (A more satisfactory procedure is suggested in Section 6–9.)

Sketch: (w) $2 \cdot (-7)$,   (x) $\sqrt{2} \cdot (2.9)$.

Division. The expression "$a \div b$" (or "a/b") means the number that results from dividing a by b. It is the number that yields a when it is multiplied by b. Thus $6/3 = 2$ because $6 = 3 \cdot 2$. If b is a positive integer, we can find a/b by dividing a into b equal parts. In other cases, the geometric interpretation is not so simple, but in every case if the vector a/b is multiplied by b according to the vector interpretation of multiplication, the result is a.

Sketch: (y) $5/3$,   (z) $-18/5$.

PROBLEMS

Sketch the following with vectors.

1. $-2.35 + 1.28$.   2. $-10 - (-3.9)$.   3. $(-24.1)(-1.9)$.
4. $(5.2)(-3.1)$.   5. $-7.1 - 8.01$.   6. $-5/2$.

For the following draw vectors as in Fig. 1–11 with initial and terminal points as given (in that order) and verify that the appropriate difference gives the vector.

7. 3 and 8.   8. 8 and 3.   9. -1 and 4.
10. -3 and -9.   11. 1.1 and -1.1.   12. 7 and -1.
13. -14.2 and 14.2.

14. If the initial point of the vector x is placed at the point a, where will its terminal point be?

15. What are the values of the variables in this section?

16. What are the significant substitutes for them?

17. Is "0" a significant substitute for "b" in the expression "a/b?" (*Hint:* What number would be represented by "$a/0$" for any particular value of "a"?)

18. The expression (Deaths $\cdot$ 1000)/(Population) gives the death rate in deaths per thousand inhabitants. (a) What are the variables? (b) What are the constants? (c) Cite some values of the variables. (d) Calculate *roughly* the value of the expression corresponding to deaths of 1,340,000 in a population of 119,000,000 (figures for the U. S. in 1930).

Answers to Exercises

(a) 7, right.   (b) 7, left.   (c) Positive, left, negative.   (g) Use approximation from Table I in the Appendix.   (h) x.   (l) Fig. 1–14.   (m) $a + b$.
(o) Fig. 1–15.   (p) $|\sqrt{2}| = \sqrt{2}$.   (q) $|-10| = -(-10) = 10$.   (∴) $|-7.1| = -(-7.1) = 7.1$.   (s) See Fig. 1–16.   (z) See Fig. 1–17.

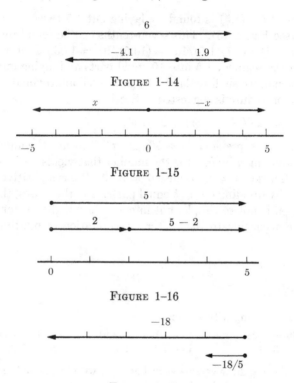

Figure 1–14

Figure 1–15

Figure 1–16

Figure 1–17

Answers to Problems

2. See Fig. 1–18.   7. Note that the initial point is to be subtracted; in this case $8 - 3 = 5$.   8. $3 - 8 = -5$.   14. $x + a$.   15. Numbers.   16. Numerals.
17. No.   See Section 1–14.   18. (a) "Deaths," "Population."   (b) "1000."
(c) Say 25 and 1000 for "D" and "P."   (d) $(1{,}340{,}000{,}000)/(119{,}000{,}000) = 1340/119 \doteq 11.3$.   (We use $\doteq$ to mean approximate equality.)

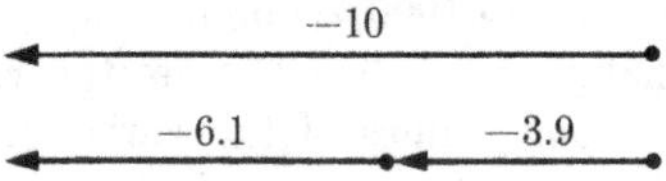

-10

-6.1 -3.9

Figure 1–18

```
┌─────────────────────────────────────────────────────────────────┐
│                                                                   │
│   NORTHFIELD, Minn.   _____ 19 ____ No. ____            │
│                                                                   │
│   THE  FIRST  NATIONAL  BANK            75-147/912                 │
│   PAY  TO                                                         │
│   THE  ORDER  OF _____  $ _____           │
│                                                                   │
│   _____ Dollars       │
│                                                                   │
│                      _____                 │
│                                                                   │
└─────────────────────────────────────────────────────────────────┘
```

FIGURE 1–19

1–5 Formulas. Everyone has had some experience with forms in which blanks are to be filled in. For example, the form pictured in Fig. 1–19 becomes a bank check when correct entries are made. Evidently the blanks in the form are variables, since constants are to be substituted for them.

The check form is convenient for filling in, but it is easier to discuss if we write it in one of the following ways.

(1) Northfield, Minn. *Month, Day* 19*year* No. *Number* The First National Bank 75–147/912. Pay to the order of *payee* $ *Amount* (in a decimal) *Amount* (written out) Dollars. *Payer* (signed)

(2) Northfield, Minn. *m, d* 19*y* No. *n* The First National Bank 75–147/912. Pay to the order of *x* $ *a*, *A* Dollars. *z*.

Form (1) makes clear the nature of the significant substitutes for the variables. Form (2) is more compact and easier to talk about. For example, we may say that the value of "*a*" must be the same as the value of "*A*" if the check is to be valid.

(a) Describe the values of each variable in Form (2). (b) Why are letters more convenient variables than blanks? (c) What happens to the check if *a* is larger than the balance in *z*'s account? (d) Restate the question in (c) using words for variables. (e) What do we call "*z*" if it is written by someone other than *z* or an authorized agent? (f) Name some constants in the check form.

Some expressions in the check form are neither constants nor variables. Among these we observe that certain expressions, such as 19__, become constants if appropriate constants are substituted for their variables.

For example, 19__ becomes 1955 if 55 is substituted for the blank. An expression that contains at least one variable and that becomes a constant when significant substitutions are made for all its variables is called a *formula*. The entire check form is a formula, because when significant substitutions are made for its variables it becomes a valid check, i.e., a name of an order to pay. In the previous sections we have used such formulas as x, $-x$, $-a$, $a + b$, $a - b$, ab, and a/b. A single variable is a formula according to this definition.

When significant substitutes displace the variables in a formula, the formula becomes a constant. The value of any such constant is called a *value* of the formula. A formula, like a variable, has no uniquely determined value. Rather, it has many values, each one determined by substituting constants for the variables that appear in the formula. Significant substitutes for the variables in a formula are those that yield values of the formula, i.e., those that transform the formula into a constant. We say that formulas stand for their values and that they take certain values when constants are substituted for their variables.

In each of the following formulas indicate the variables, some values of the variables, and the corresponding values of the formula. (g) "_____ 19__" in Fig. 1-19. (h) "Mrs. A, this is Mr. B." (i) "Area equals width times length." (j) "Profit equals gross income minus costs." (k) "_____a, _____am, _____ae, _____a; _____ae, _____as, _____arum, _____is, _____is." (l) "I _____ take thee _____ to be my wedded wife." (m) $a - b$, $-a$, a/b.

We call constants, variables, and formulas *terms*. Terms are symbols that have values. Not all symbols are terms. For example, punctuation marks and individual letters in words are not terms. A term is able to stand alone in the sense that it either names something (if it is a constant) or is a formula that names something when constants are substituted for its variable or variables.

Whether or not a symbol is a term depends on the context. For example, "moi" is a term in French but not in English. Most of the time, in mathematical as well as in other writing, the reader is supposed to be able to tell from the context what the significant substitutes and values of the variables are. This is easy when words are used as variables, since they usually name their own values. When letters are used as variables, the values must be determined from the context, unless specific indications are given. In formal mathematical theory all possible confusion is avoided by spelling out exactly which symbols are terms and what substitutions are allowed.

Which of the following are terms? Justify your answers. (n) "$2x$," (o) "Mr. $X + 1$," (p) "$ax =$," (q) "$x - (y - (z + 2))$," (r) "$a/$."

We say that an expression is *in the form* of a certain formula if it can be obtained from the formula by some substitutions for the variables without any other operations. For example, "2^2" is in the form of "a^2" because it can be obtained from "a^2" by substituting "2" for "a." However, 4 is not in the form a^2, because to get 4 from a^2 we must substitute 2 and then replace 2^2 by 4. We say that two expressions are in the *same form* if they are in the form of the same formula. For example, 2^2 and 3^2 are in the same form since they are both in the form a^2.

(s) Show that b^3 and 4^3 are in the form a^3.   (t) Show that 2^2, x^2, and 5^y are in the form x^y.   (u) Why is 2/3 not in the form a/a?

PROBLEMS

1. Why is a variable a formula?

2. Illustrate how the values of a variable may be determined from the context by discussing "Mrs. A" and "$x - 3y$."

3. When two trees have almost the same circumference, the following formula is sometimes used to determine which is bigger: the number of inches of circumference at 4.5 feet above the ground is added to the height in feet plus one-fourth of the crown-spread in feet. (a) State this formula in an equation, using words for variables. (b) Do the same, using letters. (c) In 1946 a white pine near Newald, Wisconsin, was reported to have a height of 140 feet, a crown-spread of 56 feet, and a circumference of 17 feet 2 inches at 4.5 feet above ground. Calculate the number of points by substituting in this formula.

4. R. C. Angel devised the following formula for measuring the "welfare effort" of a city in a community chest campaign: effort = (amount raised)/(quota) + (pledgers)/(number of families in area) + (amount raised)/(0.0033 × yearly retail sales). What is the significance of each term? In what numerical range would you expect the scores to lie? (Communities on the average contribute about 1/3 of 1% of retail sales.)

5. The distance traveled by a body moving at constant speed is equal to its speed multiplied by the time elapsed. Rewrite this sentence, using letters for variables.

★6. Show that an expression may be in the form of several different formulas.

★7. Argue that the meaning attached in this section to "in the form of" is not inconsistent with its use in such expressions as "The church was constructed in the form of a cross" and "The sonata is in the form A, B, A."

ANSWERS TO EXERCISES

(a) m: months, 12 values; d: days, 31 values; y: years, 100 values; n: natural numbers; x: people; a: amounts of money.   (b) Difficult to distinguish different blanks.   (c) It bounces!   (d) Replace a by "the amount" and z by "the payer."   (e) A forgery.   (f) "Northfield," "Minn.," "The First National Bank."   (g) For each substitution the formula becomes a name of a day, and its value is that day.   (h) A: women; B: men; formula: introductions.   (i) Area: numbers; width: numbers; length: numbers; formula: statements of equality.   (j) Like (i).   (k) This formula is a paradigm in which the blank is to be filled in with the stem of a Latin noun. Values of the formula are declensions.   (l) Blanks: men, women; formula: pledges.   (m) a, b; values are numbers.   (n) Yes.   (o) No.   (p) No.   (q) Yes.   (r) No.   (s) b for a and 4 for a.   (t) 2 and 2, x and 2, and 5 and y for x and y, respectively. (u) Any substitution yields same numerator and denominator.

Answers to Problems

1. A variable is an expression containing at least one variable and becoming a constant when its variables (just itself) are displaced by appropriate constants. 3. (a) Size = circumference + height + (1/4)(crown-spread).   (b) $C + H + S/4$.   (c) 360.

1–6 Equations. A constant has only one value in a given context, but different constants may have the same value. For example, "Minnesota," "the land of ten thousand lakes," and "the thirty-first state" all have the same value.

When two constants have the same value (are names of the same thing) we call them *synonymous*. When we wish to indicate that two constants are synonyms, we write an *equals sign* between them. By $1/2 = 4/8$ (1/2 *equals* 4/8) we mean that "1/2" and "4/8" are names of the same thing; that is, 1/2 and 4/8 are one and the same thing.

Cite several constants synonymous with   (a) "New York State,"   (b) "zero," (c) "−1."

We call an expression of the form $x = y$ an *equation*, in which x and y are the *sides* or *members*. The statement $x = y$ is true if and only if x is identical with y. The relation of equality, represented by the equals sign, is the relation that a thing has with itself and with nothing else We write $x \neq y$ for "x is *not* equal to y."

(d) Restate $a = b$ in words in as many ways as you can.   (e) Why is $2 = $ II true even though "2" and "II" are obviously different?   (f) If $a \neq b$, what do we know about a and b?   ★(g) One of the axioms of Euclid was, "If two things are equal to the same thing, they are equal to each other." Restate this so as to avoid any confusion as to the meaning of equality.   (h) Write ten expressions having the value 10.

Sometimes "equals" is used loosely to mean merely that some aspect of two things is the same. Thus it is customary in high-school geometry to say that two triangles are equal when their areas are the same. According to the meaning we have given "equals," $\triangle ABC = \triangle DEF$ if and only if the two symbols ("$\triangle ABC$" and "$\triangle DEF$") stand for the *same* triangle.

which can be so only if "D," "E," and "F" are just different names for the vertices of $\triangle ABC$. Hence if we wish to indicate that two triangles have the same area, we write (Area $\triangle ABC$) = (Area $\triangle DEF$), which is true when each side of this equation stands for the same number. Similarly, in The Declaration of Independence, "all men are created equal" means that the rights of all men are (or should be) the same. This could be stated more precisely (though less effectively from a literary point of view) by writing "all men should have equal rights."

We can always avoid using "equals" loosely by indicating the particular aspect of the two things that is actually identical, as we have done in the previous paragraph, or else by adopting a special symbol. For example, in geometry we write $\triangle ABC \sim \triangle DEF$ to mean "(The shape of $\triangle ABC$) = (The shape of $\triangle DEF$)," and write $x \cong y$ to mean that x and y have the same shape and size. *In this book, the equals sign always means identity, and it is not used at all in the loose sense.*

(i) What aspect of the two lines a and b is the same when a is parallel to b? (j) What precisely is meant in plane geometry when it is said that two segments are equal?   (k) Can two different numbers be equal? How then can we say that $2 = 10/5$?

An equation involving no variables is either true or false. For example, $1.5 = 3/2$ is true, and $1/2 + 1/3 = 1/5$ is false. However, if an equation involves variables, we cannot say that it is true or false until constants are substituted for the variables. For example, $2x = 6$ is neither true nor false. If we substitute "3" for "x" we get the true equation $2 \cdot 3 = 6$, and if we substitute "4" we get the false equation $2 \cdot 4 = 6$. Equations that are true for some values of their variables and false for others are called *conditional equations*. Equations that are true for all values of their variables, such as $x + x = 2x$, are called *identities*.

Which of the following are conditional equations and which identities? (l) $a + b = b + a$,   (m) $1 - 2x = 3$,   (n) $a^2 = a \cdot a$.

A value of a variable in a conditional equation that makes the equation true is called a *root*. For example, 3 is a root of $2x = 6$.

Show that the following are roots of the indicated equations: (o) 2 of $6 = 3x$,   (p) -3 of $x^2 = 9$,   (q) -4 of $7 + x = 3$,   (r) 1 of $(x - 1)x = 0$, (s) 2 of $1 - 3x = 2x - 9$.

The roots of an equation whose members stand for numbers are not altered by adding the same term to or subtracting the same term from both members, or by multiplying or dividing both members by the same non-zero term. Hence in solving equations we may add, subtract, multiply, or divide provided we perform the same operation on both members and take care not to multiply or divide by zero. (The justification for this is given in Sections 2–6 and 2–8.) For example, to solve $6 = 3x$ we may divide both members by 3 to get $2 = x$.

Solve the following by using the ideas in the previous paragraph. (t) $x - 3 = 2$,   (u) $5x - 10 = 0$.

The members of an identity may be constants or formulas. In either case we call them *synonymous* and we say that either is a *synonym* of the other. This extends the use of "synonymous," which was defined at the beginning of this section with reference only to constants. Synonymous expressions always have the same values. This means that if they are formulas, any significant substitution of constants yields constants with the same value.

PROBLEMS

Which of the following are identities? (*Suggestion:* Try values of the variables.) Solve those that are conditional, if you can.

1. $1/a + 1/b = 1/(a + b)$.   2. $x^2 = 2x$.
3. $a - 1 = 2a - (a + 1)$.   4. $2b + c = 2(b + c)$.
5. $ab = ba$.   6. $a(b + c) = ab + ac$.
7. $3x = 12$.   8. $2x - 1 = x + 4$.
9. $1.5x + 3.1 = 0.4x - 8$.   10. $2.3x - 10.9 = 3.4 - 1.9x$.
11. $(x - 1)^2 = 9$.

★12. Find two values of x for which $x^2 + 1/x^2 = a^2 + 1/a^2$.

13. Show that -3 is a root of $x^3 - 37x = 84$.

★14. A certain company advertises that its coupons may be redeemed as follows: a teaspoon for 5 cents and 34 coupons or for 20 cents and 2; a solid-handle knife for 5 cents and 89 or 50 cents and 3; a hollow-handle knife for 5 cents and 129 or 85 cents and 3; a tablespoon for 5 cents and 69 or 40 cents and 3; a butter knife for 5 cents and 69 or 40 cents and 3, a cold-meat fork for 5 cents and 109 or 75 cents and 7.   (a) Do the coupons have the same value for each premium?   (b) What amount do you think was used by the company as the value of a coupon?   (c) Using this value, which alternative is cheaper? (The first method in each case is called "The Thrift Plan.") (*Suggestion:* Let x be the value of a coupon. Then the two ways of getting a teaspoon cost $5 + 34x$ and $20 + 2x$. Equating them, we get an estimate of x.)

★15. According to Wittgenstein (*Tractatus Logico-Philosophicus*, p. 139), " . . . to say of *two* things that they are identical is nonsense, and to say of *one* thing that it is identical with itself is to say nothing." Comment on this in relation to the preceding discussion of equality.

ANSWERS TO EXERCISES

(a) "The Empire State," "the state whose capital is Albany," etc.   (b) 0, $2 - 2, 0/5$, etc.   (c) $5 - 6, 1.3 - 2.3, -8 + 7$.   (d) "a" has the same value as "b," a is the same as b, etc.   (e) The constants are different, but they have the same value.   (f) a and b are different in some respect.   (g) If two symbols

are each synonymous with a third, they are synonymous with each other.
(h) $2 \cdot 5$, $1 \cdot 10$, X, $20/2$, $6 + 4$, $11 - 1$, ten, $15 - 5$, $9 + 1$, 10^1.   (i) Direction or orientation.   (j) That their lengths are the same.   (k) No, two different things cannot be the same. Although "2" and "10/5" are different constants, they have the same value.   (l) Identity.   (m) Conditional.   (n) Identity.   (o) through (s) By substituting and simplifying.   (t) 5.   (u) 2.

Answers to Problems

3, 5, and 6 are identities. 7. 4. 9. Adding $-0.4x$ and -3.1 to both sides, we get $1.1x = -11.1$. Dividing both sides by 1.1, we get $x = -11.1/1.1 = -111/11$. 10. 143/42. 11. $x - 1 = 3$ or $x - 1 = -3$. Hence $x = 4$ or $x = -2$. 14. (a) No. (b) They appear to have taken $\frac{1}{2}$ cent as the value of the coupon. (c) The first method is cheaper except for teaspoons.

1–7 Sentences. The reader undoubtedly identifies

(1)   "2/3 is the reciprocal of 3/2"

as a complete declaratory sentence. It is a constant because it expresses (is a name of) an idea. It makes sense to say that (1) is true or to say that (1) is false. We say that it is a statement whose value is a proposition.

On the other hand, consider

(2)   "$x/3$ is the reciprocal of 3/2."

This is not a statement, although it does have the same form as (1). It makes no sense to say that it is true or to say that it is false. However, (2) does become a statement when we substitute properly for x. Evidently (2) is a formula whose values are propositions.

(a) What values of x make (2) true?

An idea that can be characterized as either true or false (but not both) we call a *proposition*. A constant whose value is a proposition is called a *statement*. A formula whose values are propositions we call a *propositional formula*. We call both statements and propositional formulas *sentences*. Sentences are simply terms whose values are propositions.

In formal mathematical theories it is customary to indicate explicitly what terms are sentences and how truth is to be established. When such rules are not stated in connection with a discussion, a good common-sense way of deciding whether an expression is a statement is to ask "Does it *make sense* (whether or not it is correct) to call the alleged statement true?" To test an alleged propositional formula one sees whether statements can be obtained from it by substitution.

Which of the following are sentences? statements? propositional formulas?
(b) $2 = 4$, (c) $x - y = -(y - x)$, (d) $x = -x$, (e) $a(b + c = 2$.
(f) Are all equations sentences?

A sentence that involves no variables must be a statement, and is either true or false. A sentence involving one or more variables is neither true nor false, since it is a propositional formula. But some substitutions may yield true statements, and others may yield false statements. Values of the variables that yield true statements are called *solutions*. Solutions are said to *satisfy* the sentence. As examples, 2 is a solution of "$x + 2 = 4$" and Herbert Hoover is a solution of "x was a mining engineer and x was President of the United States."

Finding all the solutions of a sentence is called *solving* the sentence. Variables whose values are sought are often called *unknowns*, and other variables that may appear in the same sentence are called *parameters*. For example, if we wish to solve the sentence "$x + a = 3$" for x, x is the unknown and a is a parameter. Similarly, if we wish to solve the sentence "y is a son of x, and x is a male" for x, x is the unknown and y is a parameter. The solution is evidently (the father of y). This can be tested by substituting "the father of y" for x to get "y is a son of the father of y, and the father of y is a male," which is certainly true for any appropriate value of y.

The oldest, and quite legitimate, way of finding solutions is to guess and experiment. Use this method to find a solution to each of the following. Check your guesses. (g) $3x = 12$. (h) $x - 4 = 3$. (i) $x = x$. (j) Dwight Eisenhower followed x in office. (k) $2x$ is the largest integer less than 5.

In the previous section we discussed sentences of the form $x = y$. Many mathematical ideas are expressed in such sentences, but by no means all of them. As a simple example, when the numbers x and y are unequal, one of them is always smaller than the other. When the point x lies to the left of the point y on the real axis, we say that x *is less than y*. Symbolically we express this idea by writing $x < y$ (x is less than y) or $y > x$ (y *is greater than x*). Since all negative numbers are less than zero, and all positive numbers greater than zero, we use "$x < 0$" and "$x > 0$" as abbreviations for "x is negative" and "x is positive," respectively. We refer to a sentence of the form $x < y$ or $y > x$ as an *inequality* and to x and y as the *members*.

For what values of x is (l) $x < 0$ false, (m) $x > 0$ false, (n) $-x < 0$ false, (o) $-x > 0$ true?

The solutions of an inequality often are infinite in number. For example, $2x < 12$ is satisfied by any number less than 6. The solutions make up all the axis to the left of 6, as indicated in Fig. 1–20.

Sketch the solutions of the following sentences: (p) $x < -1$, (q) $2x > 12$, (r) $3x < -9$.

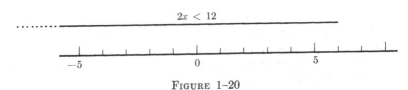

$$2x < 12$$

FIGURE 1–20

In solving inequalities, the same rules apply as those given for equations in Section 1–6 (addition of, subtraction of, multiplication by, division by the same number) with *one very important exception*. When both members of an inequality are multiplied or divided by a negative number, $>$ must be changed to $<$ or $<$ to $>$, i.e., the direction of the inequality must be changed. For example, $2 < 4$, but $(-3)2 > (-3)4$, since $-6 > -12$. To solve $-3x < 18$, we divide both members by -3 to get $x > -6$. (For a proof, see Section 3–5.)

Solve the following sentences and sketch the solutions:　(s) $-3x > 18$, (t) $2x < -4$,　(u) $-x < 2$,　(v) $3 + x < 2x - 1$.

PROBLEMS

Solve the following and sketch the solutions:

1. $3x < x + 1$.
2. $1 - 2x > 4x + 31$.
3. $8.25x < 5x + 3$.
4. $-x > 2x - 14$.
5. $2x + 3 - 1.5x < 1.7x + 8.9$.
★6. $-\sqrt{2x} + \sqrt{2} < -x + 1$.

The following intentional nonsense was written by the mathematician Charles Dodgson (Lewis Carroll), whose writings contain many delightful illustrations of mathematical ideas.

(3)
> *And as in uffish thought he stood,*
> *The Jabberwock, with eyes of flame,*
> *Came whiffling through the tulgey wood,*
> *And burbled as it came!*

As Alice said, "Somehow, it seems to fill my head with ideas—only I don't know exactly what they are!"

★7. Replace nonsense words in (3) by variables, and decide whether the result is a sentence.

★8. Make significant substitutions for the variables so as to form a statement.

★9. Is (3) a sentence from Alice's point of view?

★10. Is (3) a sentence in the context of the Jabberwock's fictional world?

Solve Problems 11 through 14 for x, indicating any parameters.

11. A is the father of the father of x.
12. A is a son of a son of x.
13. $x - a = 2$.
14. X served more than three terms as President of the United States.

★15. C. C. Fries in his *Structure of English* (Chapter II) refers to over one hundred definitions of "sentence." The word most often means any unit of discourse, and sentences are classified as declaratory, exclamatory, interrogatory, or imperative. To which of these kinds does "sentence" as used in this book refer? Give examples of each kind.

ANSWERS TO EXERCISES

(a) 2 is the only one. (b), (c), and (d) are sentences. Note that a false statement is still a sentence. (b) is a statement. (c) and (d) are propositional formulas. (f) Yes, if their members are terms. (i) Anything is a solution. (k) 2. (l) Positive or zero values. (m) Negative or zero values. (n) Negative or zero values. (o) Negative values. (p) All points to the left of -1. (q) All points to the right of 6. (r) All points to the left of -3. (s) All points to the left of -6. (t) All points to the left of -2. (u) All points to the right of -2. (v) All points to the right of 4.

ANSWERS TO PROBLEMS

1. $x < 0.5$. 2. $x < -5$. 3. $x < 12/13$. 4. $x < 14/3$. 5. $x > -59/12$. 7. And as in x thought he stood, the y, with eyes of flame, came z-ing through the w wood, and burbled as it came. Yes. 8. troubled, chimpanzee, rush-, tangled. 9. No. 10. Yes. 11. x is a paternal grandchild of A. 12. x is a paternal grandparent of A. 13. $x = a + 2$. 14. F.D.R. 15. Declaratory.

1–8 Laws. Sentences may be true for some values of their variables and not for others, as we have seen in the last two sections. Some sentences are false for all values of their variables. For example, $x/x = 2$ is never true. Still other sentences are true for all significant values of their variables. An obvious example is $x = x$.

A sentence all of whose values are true we call a *law*. A true statement is a law, since its one value is a true proposition. A propositional formula is a law if and only if *every* one of its values is a true proposition. This means that *any significant* substitution for its variables yields a true statement. A law has the convenient property that it can be applied to any situation in which the things under discussion are values of its variables.

(a) What do we call an equation that is a law? (b) In order to show that a sentence is not a law, it is sufficient to cite one false value. Why? (c) Show that $2x = x^2$ is not a law. (d) Show that $2x \neq x^2$ is not a law. (e) What do we mean by applying a law? (f) Why would it be incorrect to say that a law is a sentence that becomes true no matter what is substituted for its variables?

The main goal of scientific activity is the formulation of useful laws. For example, Boring writes, "The whole value of science is that it reduces the complexities of the world to general rules, which, once established, enable you to explain or predict many, many individual cases." (*Foundations of Psychology*, p. 14)

Discuss the following, restating them if possible so as to bring out the nature of their variables and their status as laws.   (g) "Any two bodies are attracted to each other by a force proportional to the product of their masses and inversely proportional to the square of the distance between them." (Newton's Law of Universal Gravitation)   (h) Every change in nature is produced by some cause. (The so-called law of causality)   (i) In every given state of the arts an increase of labor or capital beyond a certain point, as in the cultivation of land, causes a less than proportionate increase in the production from the unit to which the additional labor or capital is applied. (The law of diminishing returns)   (j) "The following formula may be postulated.   Laughter occurs when a total situation causes surprise, shock, or alarm, and at the same time induces an antagonistic attitude of playfulness or indifference." (John M. Willmann, "An analysis of humor and laughter," *American Journal of Psychology*, 1953, p. 70)   (k) "No peoples seem to lack magic formulas and magicians . . . ." (C. Grant Loomis, *White Magic*, 1948, p. 5)   (l) ". . . to ask for the cause of an event is always to ask for a general law which applies to the particular event." (R. B. Braithwaite, *Scientific Explanation*, 1953, p. 2)   (m) The negative of the negative of a number is the number itself.

Whether or not a sentence is a law depends upon the context. In the elementary algebra of numbers, "$ab = ba$" is a law. However, if we think of the variables as standing for acts instead of for numbers and think of ab as the act consisting of doing a and then doing b, "$ab = ba$" is not a law. For example, let $a =$ putting on one's shoes, and $b =$ putting on one's socks. The act of putting on one's shoes and then one's socks is not the same as the act of putting on one's socks and then one's shoes. This illustrates the fact that a law is a sentence that is true in a particular context and for particular substitutes for its variables, namely the significant substitutes for the variables in that context.

A sentence involving variables is a propositional formula, not a statement. Hence a law involving variables is not a statement; it is a propositional formula that has only true values. For example, "$ab = ba$" is not a statement. However, "'$ab = ba$' is a law" is a statement because it makes an assertion about the sentence "$ab = ba$." To avoid having to make awkward statements such as "'$ab = ba$' is a law," it is customary in mathematics to write laws as though they were statements. When this is done, the quotation marks and the phrase "is a law" are implicit. For example, when we write "$ab = ba$" we mean to claim that this sentence is a law. In the same way, when we say "A man has a soul," we mean that "x has a soul" is true for all human values of x.

Since sentences that are not laws often appear in mathematical writing, the above convention might lead to confusion. However, the danger is minimized by the convention that mathematical laws are displayed and labeled explicitly in some way. When this convention is not followed, the meaning has to be guessed from the context or by inspection of the sentence itself. For example, "Prove p" or "Show p" suggests that p is a law. On the other hand, "Solve p" suggests that p is not a law (though it may be), and "Consider p" leaves the matter open. In this book, a sentence displayed with a number in bold-face type is a law.

For each of the following, specify significant values of the variables and decide whether the sentence is a law: (n) $x^2 = x \cdot x$.   (o) If x is taller than y, and y is taller than z, then x is taller than z.   (p) $2 = 1 + 1$.   (q) If x is congruent to y, then x is similar to y.

We have seen that "p is a law" is a statement whenever p is a sentence, even if p is a propositional formula. We can always convert a propositional formula into a statement by making a statement *about* it. Whenever p is a sentence, "p is a law," "p is not a law," "p has at least one true value," "p is always false," etc., are statements. The operation of making a statement about the truth of the values of a propositional formula is called *quantification*. We discuss it further with the aid of special symbols in Section 2–10.

Is each of the following true for all, some, or none of the significant values of its variables?   (r) $x = 1$.   (s) $3x = x + x$.   (t) $(a + b)^2 = a^2 + b^2$.   (u) If p is a law, then p has at least one true value.

PROBLEMS

1. Give an example of a law, indicating the variables and their values.

In Problems 2 through 8, the values of the variables are numbers. Which are laws?

2. $x < x + 1$.

3. $a(1 + b) = a + ab$.

4. $|x| > 0$.

5. $\sqrt{a + b} = \sqrt{a} + \sqrt{b}$.

6. $1/a + 1/b = (a + b)/ab$.

7. $ca/cb = a/b$.

8. $a + (b + c) = (a + b) + c$.

In Problems 9 and 10, what are the values of the variables? For these values, is the sentence a law?

9. If x is parallel to y, then y is parallel to x.

10. If x is parallel to y, and y is parallel to z, then x is parallel to z.

★11. To what extent does our use of "law" correspond with its use to indicate a statute, i.e., a law in the legal sense?

ANSWERS TO EXERCISES

(a) An identity.   (b) Since a law has only true values.   (c) Let $x = 5$. (d) Let $x = 2$ or $x = 0$.   (e) Substituting for its variables.   (f) It is alleged to be a true statement only for significant values of its variables. For some substitutes it may have no meaning.   (g) "Any two bodies" is the key expression here. The law could be stated "If x and y are bodies, then x and y are attracted...." The sentence is claimed to be true whenever x and y are displaced by the names of bodies.   (h) "If x is a change in nature, then x has a cause."   (i) The variables are "state of the arts," "increase of labor or capital," and "increase in production."   (j) Variable is "a total situation."   (k) "If x is a people, x seems to have magic formulas...."   (l) Variable is "an event." (m) Variable is "a number."   (n) Numbers. Law.   (o) Persons, buildings, or other things that have heights. Law.   (p) No variables. Law.   (q) Geometrical figures. Law.   (r) Some (one).   (s) Some.   (t) Some.   (u) All.

Answers to Problems

2, 3, 6, 7, and 8 are laws. 4 is not a law since it is false for $x = 0$. 5 is not a law since it is false for $a = b = 1$. In 7, "0" is not a significant substitute for "c," so 7 is a law even though it is meaningless for $c = 0$. See the answer to Exercise (f). 9. Lines; a law. 10. Lines; not a law since it is false when $x = z$.

1–9 Simple arithmetic identities. The simple addition

$$(1) \qquad \begin{array}{r} 37 \\ 45 \\ \hline 82 \end{array}$$

is done by following rules about adding columns and carrying, with which the reader has been familiar for many years. But why does the procedure give the right answer? To see the situation more clearly, we write

$$(2) \qquad 37 + 45 = (30 + 7) + (40 + 5).$$

Now in (1) we began by adding the 7 and 5; that is,

$$(3) \qquad (30 + 7) + (40 + 5) = (30 + 40) + (7 + 5)$$

$$(4) \qquad \qquad\qquad\qquad = (30 + 40) + 12.$$

Then we carried; that is,

$$(5) \qquad (30 + 40) + 12 = (30 + 40) + (10 + 2)$$

$$(6) \qquad \qquad\qquad\qquad = (30 + 40 + 10) + 2.$$

Finally, we added the tens to get $80 + 2$ or 82. Such was our procedure, although considerably abbreviated in (1). Why are all these rearrangements of the terms justified?

Let us look at (3), for example. From the left to the right side we have interchanged the 40 and the 7. The shift can be justified by using the following laws:

(**7**) $a + b = b + a$ (Commutative law of addition),

(**8**) $a + (b + c) = (a + b) + c$ (Associative law of addition).

We apply (8) first by substituting in it $(30 + 7)$ for a, 40 for b, and 5 for c. We get

(9) $(30 + 7) + (40 + 5) = [(30 + 7) + 40] + 5.$

This begins the process of getting the 40 and 30 together and the 7 and 5 together. We continue as follows, where each right member is synonymous with the right member above it, and so with the left member of (9) and (3).

(10) $[(30 + 7) + 40] + 5 = [30 + (7 + 40)] + 5$

(11) $= [30 + (40 + 7)] + 5$

(12) $= [(30 + 40) + 7] + 5$

(13) $= (30 + 40) + (7 + 5).$

The steps are justified as follows: step (10) by substituting 30 for a, 7 for b, and 40 for c in (8); step (11) by substituting 7 for a and 40 for b in (7); step (12) by substituting 30 for a, 40 for b, and 7 for c in (8); and step (13) by substituting $(30 + 40)$ for a, 7 for b, and 5 for c in (8).

The above steps are tedious, and we should not wish to go through them when doing additions. However, they show how the commutative and associative laws underlie even the simplest arithmetic. They indicate that the procedures in (1) may be justified by reference to simple laws that can be applied repeatedly to rearrange terms in a sum according to convenience.

(a) Actually write out the results of making the substitutions indicated above in justifying steps (10) through (13).

We are inclined to think of (7) and (8) as obviously true, because our experience with numbers has made us acquainted with many instances in which they apply. In terms of the geometric interpretation of addition given in Section 1–4, we know by experience that to go a directed distance a and then b puts us at the same point as going a directed distance b and then a (see Fig. 1–21). This lends plausibility to (7).

(b) Give a similar argument for (8).   (c) Illustrate (7) and (8) by substituting numerals for the variables and evaluating the two sides.

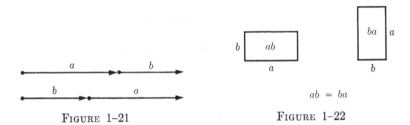

FIGURE 1–21   FIGURE 1–22

Two laws that for multiplication play a role similar to that played by (7) and (8) for addition are

(14) $ab = ba$ (Commutative law of multiplication),

(15) $a(bc) = (ab)c$ (Associative law of multiplication).

The vector interpretation of multiplication does not by any means make (14) evident. But (14) seems plausible if we recall that ab may be interpreted also as the area of a rectangle of base a and height b, as indicated in Fig. 1–22. But then ba is the area of a rectangle of base b and height a. The obvious equality of the two areas suggests (14).

(d) Give a similar argument for (15) by considering volumes.   (e) Illustrate them for particular numerical values of their variables.

The arguments above are arguments for the plausibility of (7), (8), (14), *and* (15). They certainly suggest their truth, *but they do not prove them* beyond any doubt, because they depend on diagrams and interpretations that may or may not always hold. (For example, what about negative numbers?) In elementary algebra (7), (8), (14), and (15) are usually assumed without proof. We shall discuss this matter and the nature of proof later.

We list below some other laws used in simple calculations of arithmetic and elementary algebra. They hold for all numerical values of their variables, except that no denominator can be zero. (We explain the reason for this in Section 1–14.)

(16) $$b + (a - b) = a.$$

This identity expresses the idea that $a - b$ is the number that yields a when it is added to b (see Fig. 1–10).

(17) $$b \cdot (a/b) = a.$$

(f) What idea does (17) express?

(18) $$a - b = a + (-b),$$

(19) $$a/b = a \cdot (1/b).$$

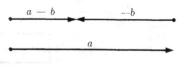

FIGURE 1-23

Fill in the following: (g) (18) expresses the rule "to subtract, _____ and _____."   (h) (19) expresses the rule "to divide, _____ and _____." (i) Explain Fig. 1-23, describe a method of doing vector subtraction based upon it, and compare it with that given in Section 1-4.   (j) Suggest an alternative method, based on (19), for finding a/b in vector terms.

$$(20) \qquad\qquad a - a = 0,$$

$$(21) \qquad\qquad a/a = 1.$$

(k) Express (20) and (21) in words.   (l) Draw a vector diagram for (20).

$$(22) \qquad\qquad a + 0 = a,$$

$$(23) \qquad\qquad 0 + a = a,$$

$$(24) \qquad\qquad 1 \cdot a = a,$$

$$(25) \qquad\qquad a \cdot 1 = a.$$

(m) How would you describe the vector corresponding to 0 in (22) and (23)? (n) Why do (22) and (23) seem plausible in terms of vectors?   (o) Why do (24) and (25) seem plausible in terms of vectors?

$$(26) \qquad\qquad a + (-a) = 0,$$

$$(27) \qquad\qquad a \cdot (1/a) = 1,$$

$$(28) \qquad\qquad a - 0 = a,$$

$$(29) \qquad\qquad a/1 = a,$$

$$(30) \qquad\qquad a \cdot 0 = 0,$$

$$(31) \qquad\qquad 0 \cdot a = 0,$$

$$(32) \qquad\qquad 0/a = 0.$$

The following identities are used for dealing with expressions involving minus signs.

$$(33) \qquad\qquad -(-a) = a,$$

$$(34) \qquad\qquad (-1)a = -a,$$

$$(35) \qquad\qquad (-a)b = -(ab),$$

$$(36) \qquad\qquad a(-b) = -(ab),$$

$$(37) \qquad\qquad (-a)(-b) = ab.$$

(p) Substitute numerals for the variables in (16) through (37), calculating each side separately and verifying that the two sides are equal. Do this for several different substitutions, for negative and nonintegral numbers as well as for natural numbers. Note that these examples show that the laws hold in some cases, *but examples cannot prove them conclusively*, because we might just have happened to choose solutions of the equations.

What is the importance of identities of the kind written above? Simply that they can be applied in order to simplify expressions or to rewrite them in some desired way. For example, if we see the expression $5 + (9 - 5)$, we know immediately that it has the same value as 9, because by substituting 9 and 5 for a and b in (16), we get $5 + (9 - 5) = 9$. Hence we may write the equation in this way, and we may also replace $5 + (9 - 5)$ by 9 or 9 by $5 + (9 - 5)$ whenever we wish, since they stand for exactly the same thing.

Because of (7), (8), (14), and (15) we can rearrange the terms in a sum or the factors in a product in any order and in any grouping. When an expression is written in a different form by using these commutative and associative laws, we call the process *rearrangement*. When laws are used to effect a reduction in size or complexity of an expression we speak of *simplifying*.

Simplify the following, indicating the identity used: (q) $2/1$, (r) $3 \cdot (1/2)$, (s) $-27/(-27)$, (t) $0/100$, (u) $(5000)(2/2)$, (v) $0 \cdot (-10)$, (w) $(-2)(-5)$, (x) $14 + (-14)$, (y) $(-1)(45)$, (z) $1000 \cdot (85/1000)$.

Problems

1. Justify the synonymity of the right members of (5) and (6) by reference to (8).

2. Which is easier, $2 \cdot 5 \cdot 37$ or $2 \cdot 37 \cdot 5$? What laws justify doing the easier calculation?

3. What laws justify writing $2(-3)(-2) = (2)(-2)(-3)$?

4. Which is easier to evaluate, $84 + (16 - 59)$ or $(84 + 16) - 59$? What law assures the equality of the two?

5. Rearrange the following so as to facilitate calculation: (a) $12 + 115 + 88$, (b) $5 \cdot 37 \cdot 2$, (c) $150 - 60 + 150$, (d) $-5 + 73 + 105$.

6. Justify (21) in terms of the vector interpretation of division.

7. Argue for (26) and (27) in vector terms.

8. Argue for (28) through (32) in vector terms.

9. Sometimes in doing calculations one "cancels" two terms. What law or laws justifies this? What remains after cancellation?

★10. When we do additions mentally, it is easier to rearrange in a way different from (1). Thus, we may think $37 + 45 = 37 + 40 + 5 = 77 + 5 = 82$. Describe this procedure in vector terms and draw a diagram.

Answers to Exercises

(a) (10): $30 + (7 + 40) = (30 + 7) + 40$; (11): $7 + 40 = 40 + 7$; (12): $30 + (40 + 7) = (30 + 40) + 7$; (13): $[(30 + 40) + 7] + 5 = (30 + 40) + (7 + 5)$.   (b) One goes a distance a, then a distance $b + c$; or one goes a distance $a + b$, then c.   (d) $a(bc)$ is the volume of a rectangular cylinder of base bc and height a.   (f) a/b is the number that yields a when it is multiplied by b.   (g) change the sign; add.   (h) invert; multiply.   (i) Rule: add the vectors a and $-b$.   (j) Divide the unit interval into b parts, then lay out a of them.

(k) Any number less itself is zero. Any number divided by itself is one. (m) Of zero length and indeterminate direction.   (n) Adding a zero vector produces no change.   (o) A vector laid out just once yields itself.   (q) 2 by (29).   (r) 3/2 by (19).   (s) 1 by (21).   (t) 0 by (32).   (u) 5000 by (21) and (25).   (v) 0 by (31).   (w) 10 by (37).   (x) 0 by (18) and (20), or by (26). (y) -45 by (35) and (24), or by (34).   (z) 85 by (19), (14), (15), and (27), or by (17).

Answers to Problems

1. By substitution of $(30 + 40)$ for a, 10 for b, and 2 for c.   2. The first. (14) and (15).   9. (20) and (21). When (20) is used, 0 remains and can then be dealt with by (22), (23), (30), or (31). When (21) is used, 1 remains and can be dealt with by (24), (25), or (29). Since "cancellation" is so ambiguous, it is better to think in terms of identities and avoid crossing out symbols carelessly.

1–10 Parentheses. Consider the following alleged sentence, based on one appearing in a book about the stock market.

(1)   This policy involves avoiding diversification and holding one's capital uninvested for long periods of time.

What did the writer mean? Did he mean to avoid both diversification and holding, or did he mean to avoid diversification but to hold capital uninvested? In everyday English, the meaning can be clarified by punctuation marks, additional words, or other devices. In mathematics, parentheses are used. One meaning of (1) may be expressed as follows:

(2)   This policy involves avoiding not only diversification but also holding one's capital uninvested for long periods of time.

(2′)   This policy involves avoiding (diversification and holding one's capital uninvested for long periods of time).

The other meaning may be expressed as follows:

(3)   This policy involves not only avoiding diversification but also holding one's capital uninvested for long periods of time.

(3′)   This policy involves [(avoiding diversification) and (holding one's capital uninvested for long periods of time)].

(a) Discuss the two meanings of the following sentence and rewrite it as we did (1): I liked Mary and John and Bill liked me.

A pair of parentheses in mathematics, as in (2′) and (3′), indicates that the expression within is to be treated as a single term with respect to any indicated operations. This means that the value of the entire expression within parentheses is to be determined *before* its parts are involved in evaluating the context in which it appears. Thus $(a + b)^2$ means that we evaluate $a + b$ and *then* square the result.

We have already seen in Section 1–9 the way in which parentheses affect the meaning. Consider the formulas $a + (b + c)$ and $(a + b) + c$ in the associative law of addition (1–9–8).* The first means to find $(b + c)$ and add the result to a. The second means to find $a + b$ and add c to the result. For $a = 2, b = 3$, and $c = 5$, the first becomes $2 + 8$, the second $5 + 5$. The associative law states that these two different procedures yield the same result. Because of this law, it is customary to omit parentheses when only the operation of addition is involved. Similarly, because of (1–9–15), parentheses are omitted when only the operation of multiplication is involved.

(b) Indicate two ways of evaluating $(4) \cdot (5) \cdot (6)$ by using parentheses.

When both addition and multiplication are involved, ambiguity is possible unless parentheses are used. Thus $(2 \cdot 3) + 4 \neq 2 \cdot (3 + 4)$, so that "$2 \cdot 3 + 4$" might be ambiguous. To reduce the number of parentheses required, the convention has been adopted to always do multiplications before additions. Then $2 \cdot 3 + 4 = (2 \cdot 3) + 4$ by this agreement, and parentheses have to be used only if we wish to indicate $2 \cdot (3 + 4)$. The omission of the dot for multiplication suggests this convention. Thus we usually write $(2)(3) + 4$ in place of $2 \cdot 3 + 4$, bringing closer the factors to be multiplied before the addition is performed.

A convention that covers all four operations is: perform multiplications first and then divisions in the order in which they appear from left to right;

---

* (1–9–8) means formula (8) in Section 1–9.

then additions and subtractions in order from left to right. As examples, $3 \cdot 4/7 \cdot 5 = (3 \cdot 4)/(7 \cdot 5)$ and not $3 \cdot (4/7) \cdot 5$; $3/4 + 2 = (3/4) + 2$ and not $3/(4 + 2)$; and $5 - 1 + 2 = (5 - 1) + 2$ and not $5 - (1 + 2)$. There is one operation that comes even before multiplication, and that is finding a power of a number. Thus $2 \cdot 3^2 = 2(3^2)$ and not $(2 \cdot 3)^2$. These conventions can be summed up by saying that the operations are performed in the following order: taking powers, multiplications in order, divisions in order, and finally additions and subtractions in order. Where any doubt is possible, parentheses should be used, but these conventions cut parentheses to a minimum.

Evaluate: (c) $1 + 2 \cdot 8$, (d) $3 \cdot 4 - 2 \cdot 17$, (e) $2 \cdot 5/3 \cdot 4 - 4$, (f) $5/5 - 1$, (g) $2 + 3/6$, (h) $(2 + 3)/6$, (i) $(1 + 2)^2$, (j) $2 \cdot 3^2$, (k) -2^2, (l) $(-2)^2$, (m) $2 \div 3 + 4 - 5 \cdot 7 + 3/6 \cdot 5$, (n) $(3/6) \cdot 5$, (o) $2 \cdot 4/3 - 1$.

Rewrite the following in as simple and compact a way as possible: (p) $a \cdot 2 - 1 + a$, (q) $5 \div (3 \div 4)$, (r) $8/2/4$, (s) $1 \div a - 2$, (t) $x \div y^2 \cdot z$, (u) $2 \cdot 4 \div 5$.

A useful concept for talking about the meaning of operation symbols is the concept of scope. The *scope* of an operation symbol (such as the plus sign or multiplication dot) or a relation symbol (such as the equals sign) is the expression to which it applies. Thus the scope of 2 in a^2 is a, but the scope of 2 in $(1 + a)^2$ is $1 + a$. Also, in $2 \cdot 3 + 4 \div 1$, the scope of the multiplication dot is 2 on the left and 3 on the right. However, the scope of the plus sign is $2 \cdot 3$ on the left and $4 \div 1$ on the right, since it is $2 \cdot 3$ and $4 \div 1$ that are to be added, not 3 and 4.

Indicate the scope of each operation or relation symbol in the following: (v) $(2 + 4 \cdot 3 - 5^6)/7 = a - 2c$, (w) $a = b$, and $3x = 4y - 2/3^2$, (x) $1 - (3 - 1/2)/4$.

Often it is necessary to have parentheses within parentheses. When parentheses are nested, reading is made easier if we use different kinds of enclosures. The most common are the ordinary parentheses ( ), brackets [ ], braces { }, and the bar ———. Braces are used in this book for the special purpose of designating sets, as explained in Chapter 3. The bar is used primarily for radicals and fractions. For example,

(4) $$\sqrt{a + b} = \sqrt{(a + b)},$$

(5) $$\frac{c + d}{a + b} = (c + d)/(a + b).$$

In (5) the bar serves as a parentheses for both numerator and denominator and also as a sign of division.

(y) What is the scope of $\sqrt{}$ in (4) and of the bar in (5)?

The use of different kinds of enclosures is illustrated by the following:

(6) $(a + b)((1 - x)(2 + (c - 1))) = (a + b)[(1 - x)(2 + [c - 1])]$.

The left side is a little harder to read, but the corresponding pairs of parentheses can be identified in it. Indeed, corresponding pairs are separated by none or an even number of parenthetical marks.

(z) Identify corresponding parentheses by assigning them the same number:
(( ))(( ) ( )( )).

Within reason it is better to have too many parentheses than too few. In science verbosity is better than ambiguity. Be as explicit as possible; then abbreviate if this can be done without causing doubt as to the meaning.

PROBLEMS

1. Discuss "Save rags and waste paper," and show how to use parentheses to give two different meanings.

Tell whether the removal of all parentheses from the left member of each of the following formulas would yield a synonymous expression:

2. (1–9–8).   3. (1–9–15).   4. (1–9–37).   5. (1–9–36).

6. Why do we use parentheses in 2(3.08) and not in ab?

Show that 7 through 10 are *not* laws.

7. $(a + b)^2 = a^2 + b^2$ (?).
8. $(a + b)/(c + d) = (a + b)/c + d$ (?).
9. $-a + b = -(a + b)$ (?).
10. $\sqrt{a + b} = \sqrt{a} + \sqrt{b}$ (?).

11. Multiply your age by 2, add 5, and multiply the result by 50. Add the amount of change in your pocket or purse if it is less than one dollar. Subtract the number of days in the year (365) and add 115. Let A be your age, C the amount of change if less than one dollar, and $C = 0$ otherwise. Represent this rule in a formula.

Do 12 and 13 as in Exercise (z).

12. ((( )( )( ))( )).   13. ((( ))( )).

14. The degree of spottedness of the solar disk is commonly expressed by a system of relative numbers known as Wolf's sunspot numbers. To find the relative number we first count one for each individual spot then ten more for each spot or cluster of spots. Moreover, in order to compare observations made at various observatories by different astronomers using a variety of instruments, Wolf reduced them to a common basis by means of multiplying by properly calculated factors. Write the formula for Wolf's number r in terms of the number of groups and single spots g, the total number of spots observed f, and the reduction factor for different observing conditions k.

Answers to Exercises

(a) I liked Mary and John, and Bill liked me = [I liked (Mary and John)] and (Bill liked me). I liked Mary, and John and Bill liked me = (I liked Mary) and (John and Bill liked me).   (b) $(4 \cdot 5) \cdot 6$ and $4 \cdot (5 \cdot 6)$.   (c) 17.   (d) -22.   (e) $-19/6$.   (f) 0.   (g) $5/2$.   (h) $5/6$.   (i) 9.   (j) 18.   (k) -4. (l) 4.   (m) $-907/30$.   (n) $5/2$.   (o) $5/3$.   (p) $3a - 1$.   (q) $20/3$. (r) 1.   (s) $1/a - 2$.   (t) x/zy^2.   (u) $8/5$.

(v) $(2 + 4 \cdot 3 - 5^6)/7 = a - 2 \cdot c$ (w) $a = b$, and $3 \cdot x = 4 \cdot y - 2/3^2$

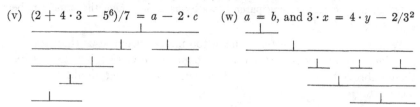

(x) $1 - (3 - 1/2)/4$

(y) $a + b$; $c + d$ on the left and $a + b$ on the right.

(z)   ((   ))((   )   ( )( ))
    1   2   2134   4   5   56   63

Answers to Problems

2. Yes.   3. Yes.   4. No. "$-a - b$" is not synonymous with "$(-a)(-b)$." 5. No.   6. $23.08 \neq 2(3.08)$.   7. Let $a = b = 1$. Similarly in 8 through 10. 11. $50[2A + 5] + C - 365 + 115$.

12. ((( )( )( ))( ))
     123   34   45   526   61

13. (((   ))(   ))
     12   3   324   41

14. $r = k(10g + f)$.

1–11 Replacement and substitution. If two expressions are synonymous, it is obvious that either can be used in place of the other without changing the meaning of any context in which they appear. This is true because synonymous expressions, whether constants or formulas, always have the same value. We call the process of putting one of two synonymous terms in place of another *replacement*. For example, in Section 1–9 we said that "$5 + (9 - 5)$" and "9" could replace each other because $5 + (9 - 5) = 9$.

The operation of replacement is a very useful one, even though it appears at first sight to introduce nothing new. For example, the calculation in (1–9–9) through (1–9–13) consists of replacing the right member of (1–9–2) by an expression synonymous with it and then continuing to make replacements until the final desired result is obtained.

Replacement of terms by synonyms is often helpful in deriving laws. For example, we can use this method to derive $a + (a - a) = a$ from the laws in Section 1–9. By (1–9–22), $a + 0 = a$, and by (1–9–20), $a - a = 0$. Hence in the first of these we may replace 0 by $a - a$ to get the desired result.

(a) Cite examples of replacement in previous sections and construct other examples.   (b) Derive $a(a/a) = a$ from the identities in Section 1–9.

We shall use replacement so frequently that it is worth while to have an explicit rule to justify it.

(1)   RULE OF REPLACEMENT: *If a term in an expression is replaced by a synonym, the resulting expression is synonymous with the original. If the original is a law, so is the result.*

Throughout this book we use "replacement" *only* in this sense. To illustrate the first part of the rule, $2(-a)b = 2(-ab)$, since the right member is obtained from the left by replacing "$(-a)b$" by "$(-ab)$", which is synonymous by (1–9–35). To illustrate the second part of the rule, $0/(a + 0) = 0$ is a law because it results from (1–9–32) by replacing "a" by "$a + 0$," which is a synonym by (1–9–22).

Replacement is useful, but it alone will not carry us very far. To make good use of laws we must take advantage of the fact that they are true for all values of their variables. To do this we make significant substitutions for variables in them to get the expressions we wish. For example, if we substitute "2" for "a" in (1–9–31) we get the law $0 \cdot 2 = 0$.

We use substitution so much that it is convenient to have a special notation to indicate it briefly. To indicate that an expression is to be substituted for a variable, we write the variable followed by a colon and the substitute. Thus "$(a:2)$" means the substitution of "2" for "a." Also,

(2) $$a^2(a:b) = b^2,$$

(3) $$(\text{hall})(a:i) = \text{hill},$$

(4) $$(a + x^2)(x:a) = a + a^2,$$

(5) $$(x = a^2)(x:x + y) = (x + y = a^2),$$

(6) $$(x + xy)(x:2a) = 2a + 2ay,$$

(7) $$(ab = ba)(a:3) = (3b = b3).$$

The brevity of the notation is emphasized by comparing (2) with "When b is substituted for a, a^2 becomes b^2" or with "The value of a^2 when the value of a is b is b^2." (Note here the omission of some quotation marks, whose inclusion would make these expressions even more unwieldly.)

Find: (c) $x(x{:}a)$,   (d) $y(y{:}a)$,   (e) $x^2(x{:}a^2)$,   (f) $(2a)(a{:}x^2)$, (g) $(b^2 + b)(b{:}a)$,   (h) (the successor of x)$(x{:}$Henry IV$)$,   (i) $(1{-}9{-}26)(a{:}3)$.*

It should be noted that $(a{:}b)$ is the substitution of "b" for "a" *everywhere* in the expression involved. This is essential, because we use this notation to indicate substitution for variables, and it is understood that a variable cannot take different values at the same time.

The notation is easily extended to take care of simultaneous substitutions for as many variables as we wish. To indicate that a and b are to be substituted for x and y, we write $(x{:}a, y{:}b)$. As examples, $(x - y)$ $(x{:}a, y{:}b) = a - b$; $(x - y)(x{:}a, y{:}a) = a - a$; and ($x$ is a sibling of y) $(x{:}$John, $y{:}$Mary$) = $ John is a sibling of Mary. Similarly, $(1{-}9{-}17)$ $(a{:}2, b{:}4) = [4(2/4) = 2]$.

Find: (j) (x never repeats exactly)$(x{:}$a physical system$)$;   (k) $(1{-}9{-}36)$ $(a{:}2, b{:}3)$;   (l) $(x^2 + x - 12)(x{:}3)$;   (m) $(x^2 + xy + 12y^2)(x{:}1, y{:}1)$.

Since a law is true for all values of its variables, we get a true statement by substituting for the variables in any law constants that are significant substitutes. (Compare Section 1–8.) Such a substitution amounts to applying the law to a particular case. It is a matter of specialization.

Write the following statements: (n) $(1{-}9{-}7)(a{:}2, b{:}7)$;   (o) $(1{-}9{-}18)$ $(a{:}0, b{:}{-}1)$;   (p) $(1{-}9{-}14)(a{:}3, b{:}{-}4)$;   (q) $(1{-}9{-}18)(b{:}1)$;   (r) $(1{-}9{-}15)$ $(a{:}2, b{:}5, c{:}{-}8)$.

What happens when we substitute other variables for the variables in a law? Clearly, the result is also a law, since this just amounts to writing the law in terms of other symbols. Thus $(1{-}9{-}8)(a{:}x, b{:}y, c{:}z)$ is another way of expressing $(1{-}9{-}8)$ itself. Hence we may substitute variables for variables in laws in order to get laws.

What may we say about the result of substituting a formula for a variable in a law? Clearly, if every value of the substituted formula is also a value of the variable, the result is a law, for any substitution in the resulting sentence yields a value of the original sentence and hence a true statement. For example, in $1 \cdot a = a$ any number is a value of "a." Now every value of "$2x$" is a number. Hence we may substitute "$2x$" for "a" in order to obtain the law, $1 \cdot 2x = 2x$.

Write the following laws: (s) $(x = x)(x{:}a + b)$;   (t) $(1{-}9{-}19)$ $(a{:}z, b{:}x + y)$;   (u) $(1{-}9{-}33)(a{:}a + b + c)$.

---

* Here $(1{-}9{-}26)$ is an abbreviation for formula (26) in Section 1–9.

Since substitution in known laws is the most common device used in deriving new laws, we formulate the following rule.

(8) RULE OF SUBSTITUTION: *If a sentence is obtained by making significant substitutions for the variables in a law, the result is a law.*

Note that significant substitutes include both constants and formulas. What the significant substitutes are must be determined from the context if no explicit indication is given. We use "substitution" *only* in this sense throughout this book.

Almost all the work of so-called elementary algebra in the schools consists of making replacements and substitutions in known laws. When an expression is inserted by either rule, care must be taken that the expression as a whole is subjected to the same operations as the term whose place it takes. This may be assured by enclosing the insertion in parentheses. When any doubt is possible, parentheses should be used and then removed if that would cause no change in meaning. Thus (1–9–33) $(x:a + 2) = [-[-(a + 2)] = a + 2]$. However, if we did not enclose $a + 2$ in parentheses before substituting, we should get $-(-a + 2)$, which is not equal to $a + 2$. Similarly, $a^2(a:1 + x) = (1 + x)^2$. Without parentheses we should get $1 + x^2$.

(v) Show that $a + 2 = -(-a + 2)$ and $(1 + x)^2 = 1 + x^2$ are not laws.

The way in which the rules of substitution and replacement are used in calculations is illustrated in (1–9–9) through (1–9–13). There we made replacements, each one of which was justified by an identity resulting from substitution in a law. For example, we went from (1–9–10) to (1–9–11) by replacing "$7 + 40$" by "$40 + 7$," which are synonyms because $7 + 40 = 40 + 7$, and we know this last is true because it is obtained by the substitution $(a:7, b:40)$ in the commutative law of addition, (1–9–7).

As a second example, we derive $a(1/a) = 1$ from other identities in Section 1–9. First we write $a/a = a(1/a)$ by the substitution of $(b:a)$ in (1–9–19). But $a/a = 1$ by (1–9–21). Hence we replace a/a by 1 to get $1 = a(1/a)$, which is what we wished to prove.

(w) Why are "$x = y$" and "$y = x$" synonymous?

It is important to understand both replacement and substitution, and to distinguish carefully between them. We list some of their differences.

In replacement the inserted term must be synonymous with the original. In substitution the inserted term must be a significant substitute, but that is all.

Replacement of a synonym may be for any term. Substitution may be made only for variables.

A symbol may be replaced by a synonym in just one of its appearances in an expression. A substitute for a variable must be inserted for it in every place where it appears.

Because of these differences, the substitution notation should not be used for replacement.

(x) Find $(a \cdot 0 = 0)(a{:}2a)$. Why is this way of writing preferable to a form such as "Find $a \cdot 0 = 0$ for $a = 2a$"? (y) What is wrong with $(x + y{:}y + x)$? (z) Comment on the following: We know that $x + y = y + y$, since it is obtained by substituting y for x in $x + y = y + x$.

Problems

Write the substitution by which the first expression is obtained from the second.

1. $1/1 = 1; a/a = 1$.
2. $a + b + c = c + a + b; a + b = b + a$.
3. $c(a + b) = (a + b)c; ab = ba$.
4. $2(x + y) = 2x + 2y; a(b + c) = ab + ac$.
5. $-2(1/3) = -2/3; a/b = a(1/b)$.
6. $0/(-1) = 0; 0/a = 0$.

Write the result of making the indicated substitutions:

7. $(-(-a))(a{:}x + y)$.
8. $((-1)a)(a{:}-a)$.
9. $((-a)b = -(ab))(a{:}x + y, b{:}x - y)$.
10. $((-a)(-b) = ab)(a{:}x^2, b{:}y)$.
11. $(a(b + c) = ab + ac)(a{:}a + b, b{:}a, c{:}-b)$.
12. $(a(b + c) = ab + ac)(a{:}a + b, b{:}a, c{:}b)$.

What substitution in what law in previous sections yields each of 13 through 18?

13. $-(-0) = 0$.
14. $2/3 - 2/3 = 0$.
15. $18 + 0 = 18$.
16. $0 \cdot (a - b) = 0$.
17. $(x + y)/(x - y) = (x + y)(1/(x - y))$.
18. $(a + b + c)/(a + b + c) = 1$.
19. Find $\dfrac{ar^n - a}{r - 1}$ $(a{:}R, r{:}1 + i)$.
20. May we substitute variables for constants?

Answers to Exercises

(b) By (1-9-21) replace 1 by a/a in (1-9-25). (c) a. (d) a. (e) $(a^2)^2$. (f) $2x^2$. (g) $(a^2 + a)$. (h) The successor of Henry IV. (i) $3 + (-3) = 0$. (j) A physical system never repeats exactly. (k) $2(-3) = -6$. (l) 0.

(m) 14.   (n) $2 + 7 = 7 + 2$.   (o) $0 - (-1) = 0 + (-(-1))$.
(p) $3(-4) = (-4)3$.   (q) $a - 1 = a + (-1)$.   (r) $2(5(-8)) =$
$(2(5))(-8)$.   (s) $a + b = a + b$.   (t) $z/(x + y) = z(1/(x + y))$.
(u) $-(-(a + b + c)) = a + b + c$.   (v) $(a:1)$ and $(x:1)$.   (w) In $x = y$
we can replace x by y and y by x by the rule of replacement to get $y = x$.
(x) $a = 2a$ only for $a = 0$, which is not what we have in mind.   (y) We substitute *only* for variables!   (z) Substitution has not been made throughout.

ANSWERS TO PROBLEMS

1. $(a:1)$. 3. $(a:c, b:a + b)$. 5. $(a:-2, b:3)$. 7. $-(-(x + y))$. 9. $(-(x + y))$
$(x - y) = -((x + y)(x - y))$.   11. $(a + b)(a + (- b)) = (a + b)a +$
$(a + b)(-b)$.   13. $(-(-a) = a)(a:0)$.   15. $(a + 0 = a)(a:18)$.
17. $(a/b = a(1/b))(a:x + y, b:x - y)$. 20. There is no logical basis for doing
so. Scientists try to find propositional formulas that become true statements
when substitutions are made for their variables. They do this by inserting
variables for constants in true sentences, but this is not substitution in the
sense in which it is used in mathematics.

1-12 The distributive law and its consequences. One of the most useful laws of arithmetic is *the distributive law of multiplication.*

(1) $a(b + c) = ab + ac.$

(a) Verify (1) for $(a:3, b:7, c:2)$ and $(a:2, b:3, c:-1)$.

The word "distributive" is used because we may think of (1) as asserting
that multiplication may be "distributed" over addition. That is, we may
perform the operation of multiplication either on the result of adding or
on the two terms before adding.

Show by substitution in (1) that: (b) $x \cdot 5 + x \cdot 2 = x \cdot 7$, (c) $x(x + 1) =$
$x^2 + x$.

The distributive law (1) is by no means obvious. It does appear plausible, however, if we think of the products as representing areas, as indicated
in Fig. 1-24. Of course this example does not prove (1), because, for one
thing, it does not apply to negative values of the variables. Actually the
property expressed by (1) is a very special one. If we interchange the

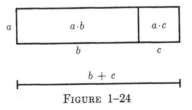

FIGURE 1-24

operations of multiplication and addition in (1), we get $a + (b \cdot c) = (a + b) \cdot (a + c)$, which is not a law. Hence addition is not distributive over multiplication.

(d) Show that the equation in the above paragraph is not a law.   (e) Show that $a \cdot (b \cdot c) = (a \cdot b) \cdot (a \cdot c)$ is not a law, so that multiplication is not distributive over multiplication. (f) Similarly show that $a + (b + c) = (a + b) + (a + c)$ is not a law, and interpret the result in words.

By applying the rules of replacement and substitution to (1) and previously stated identities, we can derive a large number of useful algebraic laws. For example, we can easily show that

$$(2) \qquad\qquad (b + c)a = ba + ca.$$

To derive (2) we begin by applying $(a{:}b + c, b{:}a)$ to the commutative law of multiplication to get

$$(3) \qquad\qquad (b + c)a = a(b + c).$$

Because of (1) we may replace the right member of this by $ab + ac$ to get

$$(4) \qquad\qquad (b + c)a = ab + ac.$$

By the commutative law of multiplication we may replace ab by ba and ac by ca in (4) in order to get (2) as desired.

(g) What substitution in the commutative law of multiplication justifies replacing ac by ca?

The above argument for (2) can be abbreviated conveniently as follows:

(3′)	$(b + c)a = a(b + c)$	$(ab = ba)(a{:}b + c, b{:}a)$
(4′)	$= ab + ac$	(1)
(5)	$= ba + ca$	(1–9–14) and (1–9–14)$(b{:}c)$.*

Note that each step was obtained from that above by a replacement, and that each replacement was justified by a law previously stated or obtained by substitution in such a law. The laws (1) and (2) are often called respectively the *left* and *right distributive laws*.

(h) Justify $2x + 7x = 9x$ by substitution in the right distributive law. (i) Derive (6) from (1).

$$(6) \qquad\qquad a(b + c + d) = ab + ac + ad.$$

---

* Although we sometimes use numerals as abbreviations for laws, the student should always write the laws themselves as we did in step (3′).

Among the many consequences of the distributive law, we list the following useful identities.

(7) $(a + b)(c + d) = ac + ad + bc + bd,$

(8) $-(a + b) = (-a) + (-b),$

(9) $(a + b)^2 = a^2 + 2ab + b^2,$

(10) $(a - b)^2 = a^2 - 2ab + b^2,$

(11) $(a + b)(a - b) = a^2 - b^2,$

(12) $(a + b)^3 = a^3 + 3a^2b + 3ab^2 + b^3,$

(13) $(a + b)(a^2 - ab + b^2) = a^3 + b^3,$

(14) $(a - b)(a^2 + ab + b^2) = a^3 - b^3.$

Proofs are similar to that given for (2) above. For example, we may derive (8) as follows:

(15) $-(a + b) = (-1)(a + b)$ $(-a = (-1)a)(a{:}a + b)$

(16) $= (-1)a + (-1)b$ (Distributive law)

(17) $= (-a) + (-b)$ $(-a = (-1)a)(a{:}b).$

Similarly, (11) is obtained as follows:

(18) $(a + b)(a - b) = (a + b)(a + (-b))$ (1–9–18)

(19) $= aa + ba + a(-b) + b(-b)$ (7)

(20) $= a^2 + ba + (-ab) + (-b^2)$ (1–9–36)

(21) $= a^2 - b^2$ (1–9–14), (1–9–20), (1–9–22).

(j) Explain the reasons in (18) through (21), indicating the substitutions. (k) Complete the following: "(8) expresses the rule that parentheses preceded by a minus sign may be removed provided_____." (l) Use (11) to easily find $1251^2 - 1250^2$.   (m) Check (10) for $(a{:}10, b{:}9)$.   (n) Show that (14) is not an identity if the first minus sign is displaced by a plus sign.

When an expression in the form of the right member of one of (1), (2), (6) through (14), or similar identities, is rewritten in the form of the left member, it is said to be *factored*. When it is originally in the form of a left member and is rewritten in the form of a right member, it is said to be *expanded*. Problems in factoring or expanding consist simply of making the proper substitution in some law so as to get the given expression on one side of an equation and the desired form on the other. For example, to factor $x^2 - 4a^2$ we note that it is in the form of the right member of (11). Hence we apply $(a:x, b:2a)$ to (11) to get $x^2 - 4a^2 = (x + 2a)(x - 2a)$. On the other hand, to expand $(3 - x)^2$ we note that it is in the form of the left member of (10). Hence we apply $(a:3, b:x)$ to (10) to get $(3 - x)^2 = 9 - 6x + x^2$.

Write out: (o) $(9)(a:3x, b:1)$, (p) $(6)(a:x, b:1, c:x, d:2)$, (q) (7) $(a:x + y, b:x, c:y, d:1)$.

Show by substitution: (r) in (11) that $(x - 2y)(x + 2y) = x^2 - 4y^2$, (s) in (14) that $9^3 - 7^3 = (9 - 7)(81 + 63 + 49)$.

Expand, indicating identities and substitutions used: (t) $(x - y)^2$, (u) $(2x + 1)(4x^2 - 2x + 1)$, (v) $(3 - a)(3 + a)$.

Factor, indicating identities and substitutions used: (w) $36 - b^2$, (x) $1 - 27x^3$, (y) $x^2 - 2x + 1$, (z) $9x^2 - y^2$

The distributive law and related identities are the basis of the rules of multiplication. For example,

$$(22) \qquad (25)(37) = (20 + 5)(30 + 7) = 600 + 140 + 150 + 35$$

by $(7)(a:20, b:5, c:30, d:7)$. The terms on the right are simply the numbers obtained if we carry out the multiplication in the usual way,

	37	or	37
	25		25
	35		185
(23)	150		740
	140		925
	600		
	925		

The familiar right-hand form in (23) is simply an abbreviated version of the left, and this in turn is just a different arrangement of the terms in the right member of (22).

PROBLEMS

Expand, indicating laws and substitutions:

1. $2(x - 3)$.
2. $x(x + 1)$.
3. $(4a - b)b$.
4. $-(x - y)$.
5. $(2a + 3b)^2$.
6. $(x + 3)(x - 4)$.
7. $(x - 1)^2$.
8. $(x + 1)(x - 1)$.

Factor Problems 9 through 16, indicating laws and substitutions:

9. $5x - x^2$.
10. $2x + 5x$.
11. $x^2 + 2x + 1$.
12. $x^2 - 4x + 4$.
13. $4 - x^2$.
14. $x^3 - 1$.
15. $x^2 - 6xy + 9y^2$.
16. $x^2 + 3x + 2$.

17. Simplify $([(3x - 4)x + 2]x - 2)x + 5$. *Note:* In proving an identity, work with one side by making replacements until you transform it to the other side, as we did, for example, in (15) through (17). Or, you may work with first one side and then the other until you have shown each side synonymous with the same third expression. However, do not state as a step in your argument what you are trying to establish until the end. Your *last* step should be the identity you are proving.

18. Show that $4ab + (a - b)^2 = (a + b)^2$. (For a geometric interpretation of this and other identities, see *An Introduction to the History of Mathematics*, by Howard Eves, p. 63 ff.)

19. Show that $(mq - sp)^2 - m(p - sq)^2 = (s^2 - m)(p^2 - mq^2)$. (This identity may be used to show that $\sqrt{m}$ is irrational unless m is the square of an integer.)

An *odd number* is one that can be expressed in the form $2x - 1$ where x is a positive integer, and an *even number* is one that can be expressed in the form $2x$ where x is a positive integer. In Problems 20 through 24 show that:

20. The sum of two odd numbers is even.

21. $n^2 + (n - 1)^2$ and $n^2 - (n - 1)^2$ are always odd, where n is a positive integer.

22. The product of two consecutive integers is always even.

★23. The sum of three consecutive integers is always divisible by 3.

★24. $n^3 - n$ is always divisible by 6, where n is a positive integer.

★25. Suppose a rope is stretched tight around the circumference of Mars and then 10 feet is added to its length. How much would the radius of Mars have to increase in order for the rope to fit tightly? (*Suggestion:* Let r = the radius of Mars. Then $r + x$ = the new required radius. Then the difference of old and new circumferences is 10. Express this and solve for x.)

26. A sociologist constructed an integration index by combining a crime index C and a welfare-effort index E as follows. The crime scores were "reversed" by subtracting them from 20, in order to make high values indicate strong integration. These reversed crime values were multiplied by 2 to give them double weight. The values of the welfare-effort index were transformed by the formula $E' = M_1 + (s_1/s_2)(E - M_2)$, where M_1, M_2, s_1, s_2 were determined in a specified way. The two modified scores were then added and divided by 3. (*The Language of Social Research*, by P. F. Lazarsfeld and M. Rosenberg, p. 61.) Write a formula for the integration index in terms of E and C.

27. Show that $(a^2 + b^2)(x^2 + y^2) = (ax + by)^2 + (bx - ay)^2$.

★28. Show that $(a^2 + b^2 + c^2 + d^2)(x^2 + y^2 + z^2 + t^2)$ can be written as a sum of squares in more than one way.

29. Show that (a) $(2uv)^2 + (u^2 - v^2)^2 = (u^2 + v^2)^2$; (b) $(2m)^2 + (m^2 - 1)^2 = (m^2 + 1)^2$. (*Note:* For any substitution, these formulas yield the Pythagorean relation between the squares of the sides of a right triangle. Hence they can be used to find such triangles. The second formula is credited to Plato. See the reference given in Problem 18.)

Multiplication can be done mentally by thinking directly in terms of (7) and holding in the memory only the totals. Thus one might think of $(25)(37)$

as $(20 + 5) \cdot (30 + 7)$, then think $20 \cdot 30$ is 600, plus $20 \cdot 7$ (or 140) is 740, plus $5 \cdot 30$ (or 150) is 890, plus $5 \cdot 7$ (or 35) is 925. It is methods of this sort that enable people to calculate rapidly without pencil and paper. Calculate the following mentally.

★30. $100^2 - 97^2$.   ★31. $(24)(36)$.   ★32. $25^2 - 28^2$.

★33. 32^2.   ★34. $(15)(139)$.   ★35. $(99)(32)$.

★36. During the performance of a "lightning calculator" a member of the audience decided to embarrass the performer by asking him to multiply 5362 by 9,999,999,999, but the speaker gave the answer almost instantaneously. How did he do it?

Answers to Exercises

(a) $3(7 + 2) = 3 \cdot 9 = 27$; $3 \cdot 7 + 3 \cdot 2 = 21 + 6 = 27$. $2(3 - 1) = 2 \cdot 2 = 4$; $2 \cdot 3 + 2(-1) = 6 - 2 = 4$.   (b) $(a{:}x, b{:}5, c{:}2)$.
(c) $(a{:}x, b{:}x, c{:}1)$.   (d) through (f) Substitute constants to get a false sentence.   (g) $(b{:}c)$.   (h) $(a{:}x, b{:}2, c{:}7)$.   (i) $a(b + c + d) = ab + a(c + d) = ab + ac + ad$ by $[(1)(c{:}c + d)]$ and $[(1)(b{:}c, c{:}d)]$.   (j) (18): No sub.; (19): $(c{:}a, d{:}{-}b)$; (20): $(a{:}b)$; (21): ba replaced by ab, then $(1{-}9{-}20)(a{:}ab)$, $(1{-}9{-}22)$ $(a{:}a^2)$.   (k) The sign of each term within is changed.   (l) $(1251 + 1250)$ $(1251 - 1250) = (2501)(1)$.   (m) Find the value of each side separately.
(n) By substituting constants.   (o) $(3x + 1)^2 = 9x^2 + 6x + 1$.
(p) $x(1 + x + 2) = x + x^2 + 2x$.   (q) $(x + y + x)(y + 1) = (x + y)y + (x + y) + xy + x$.   (r) $(a{:}x, b{:}2y)$.   (s) $(a{:}9, b{:}7)$.   (t) $(10)(a{:}x, b{:}y)$.
(u) $(13)(a{:}2x, b{:}1)$.   (v) $(11)(a{:}3, b{:}a)$.   (w) $(11)(a{:}6)$.   (x) $(14)(a{:}1, b{:}3x)$.
(y) $(10)(a{:}x, b{:}1)$.   (z) $(11)(a{:}3x, b{:}y)$.

Answers to Problems

1. $2x - 6$.   3. $4ab - b^2$.   5. $4a^2 + 12ab + 9b^2$.   7. $x^2 - 2x + 1$.
9. $x(5 - x)$.   11. $(x + 1)^2$.   13. $(2 - x)(2 + x)$.   15. $(x - 3y)^2$.   17. $3x^4 - 4x^3 + 2x^2 - 2x + 5$.

1–13 Definitions. As pointed out in Section 1–1, we must agree on what our terms mean if we are to communicate effectively. Hence we should like to define all our terms. But is this possible? To define a word we must use other words. If we try to define all our terms, we must define the terms we use in defining *them*. Then *these* terms have to be defined in turn. We shall go on endlessly in this fashion, or else we shall come around in a full circle by directly or indirectly defining one symbol in terms of a second and then the second in terms of the first. Thus, if we wish to define all our terms, circularity is inevitable.

Not long ago a reviewer wrote that the editors of a well-known dictionary ". . . held to one laudable ambition, to put an end to definitions couched in words that require reference to another definition, and another, until you land right back where you started."   (a) In view of the fact that no word is undefined in a dictionary, comment on this ambition.   (b) Suppose that a language contains just five words and that at least one different word must be used in defining a word. In a dictionary of this language how many words at most could be looked up without "landing right back where you started," that is, without finding in a definition a word already looked up? (c) Would the same sort of argument hold for a language of 100,000 words? (d) Experiment with an English dictionary by looking up a word, then all the words in its definition, and so on.

If it is impossible to define all our terms, how do we ever find out what terms mean? The answer is, of course, that we learn the meaning of a basic vocabulary by seeing how terms are used and by having objects pointed out as the values of certain terms. Knowing the meanings of these basic terms, we learn others from the dictionary and from the contexts in which they appear. If everyone had the same basic vocabulary, it could be omitted from dictionaries, and then circularities could be avoided. In mathematics such a procedure is possible, because we are constructing a language rather than describing one that has developed spontaneously. Accordingly, we can choose certain terms as undefined and define others in terms of these.

Terms that are not defined in a mathematical theory are called *basic terms* of that theory. Basic terms are used to define other terms, but are not defined within the theory itself. It does not follow, however, that we have no idea what our basic terms mean. On the contrary, we usually have some more or less definite idea of their significance. It is just that we must start with *some* terms as undefined if we are to avoid circularity or an endless series of definitions. For example, in this chapter we have been defining various special terms by using ordinary words of everyday language. These ordinary words are undefined in this book.

In formal mathematical theories we specify what terms are to be basic. They may include constants, variables, and formulas. Other terms are defined by stating that they are synonyms of expressions that are either basic terms or previously defined terms. The new term that is defined is called the *defined term*. The term that is used to define it is called the *defining term*. The definition is made by stating that the defined term is a synonym of the defining term, i.e., in the form "defined term = defining term." For example, if we take "$a \cdot b$" as a basic term, we may define "a^2" by writing $a^2 = a \cdot a$. To indicate that this is a law by definition, we may write $=_d$ in place of $=$ or precede the definition by "Let" or "Def." The following all mean that "a^2" is defined by "$a \cdot a$."

(1)   Let $a^2 = a \cdot a$.

(2) $a^2 =_d a \cdot a$.

(3) **Def.** $a^2 = a \cdot a$.

The first form is used for temporary definitions.

What is the meaning of this definition of a^2, and how did it arise? Does a^2 equal $a \cdot a$ and nothing else? Could a^2 have been defined differently? The meaning of (1) through (3) is simply that we agree to write a^2 as an abbreviation for $a \cdot a$. We could have chosen, instead, to write a sq, ^2a, or 2^a. In fact, the symbol a^2 is of relatively recent invention, and many other symbols were used before a^2 came to be generally adopted. If we used some other symbol for $a \cdot a$, we could then use a^2 with a different meaning. For example, we could let $a^2 = a + a$. Then it would turn out that $a^2 = 2a$ is a law. We do not do this because the use of a^2 as a synonym for $a \cdot a$ turns out to be more convenient. The fact is that words and other symbols are man-made and can be used as we choose. Of course, common words have well-accepted meanings, and some confusion would be caused by using them with unusual meanings. But even common words are used with different meanings in different contexts and change their meaning with time. So long as we make our meaning clear, we are quite free to use words and other symbols as we wish.

(e) At the conclusion of an astronomy lecture, one of the audience asked: "I can understand how astronomers found out the distances, sizes, and compositions of the stars, but how could they possibly discover their names?" What fallacy is involved?   (f) According to *Our Language*, by S. Potter, King James II described the great St. Paul's Cathedral built by Sir Christopher Wren as "amusing," "awful," and "artificial." Was the King lacking in aesthetic appreciation?

In mathematics a *definition* is simply an agreement as to how a symbol is to be used. It is a sentence that gives to a symbol, previously without meaning, the same meaning as a term already part of the theory. When the definition is given in the form "defined term = defining term," it is called a *formal definition*. When it is given in some other way it is called *informal*. The definitions in (1) through (3) are formal, whereas the definitions we have been giving of words printed in italics are informal. We could express (1) informally by saying, "We call the product of a number by itself the *square* of that number and write a^2 for $a \cdot a$." Formal or informal definitions are both used in mathematics, depending upon which is convenient in a given context.

(g) Write a formal definition of a^3, assuming that the meaning of a^2 is known. (h) Why is it foolish to ask for a proof of $a^2 = a \cdot a$?   (i) Why are all definitions laws?   (j) Why are defined terms usually shorter than defining terms? (k) In a mathematical theory we could dispense with all defined terms and express everything by using only basic terms. Why? Why is this not done? (l) In the formal definition "triangle $=_d$ three points not in a straight line and the segments joining them," what geometrical terms are assumed basic or previously defined?

A formal definition is easily recognized, but most definitions are informal. Often they are stated in a way that does not make clear which term is defined and which defining. They may even be asserted as though they were statements of fact instead of merely agreements about the use of symbols. Consider, for example, the following formal definition of "x is parallel to y."

(4) $x \parallel y =_d x$ and y are lines in the same plane and x does not meet y.

Here "$x \parallel y$" is obviously the defined term, and the right member is the defining term. Every time we see "$x \parallel y$" we can mentally replace it by "x and y are lines in the same plane and x does not meet y." However, this definition may appear in many forms, of which we give a few.

(5)   We define parallel lines as those that do not meet.

This has the disadvantage of seeming to define the objects, parallel lines, whereas we are really defining the term "parallel lines."

(6)   If two lines do not meet we call them parallel.

Here the phrase "we call" is the indication that the sentence is a definition, but the statement appears to leave open the possibility that we might still call them parallel even if they did meet! It also leaves implicit the idea that the lines must lie in the same plane, as did (5).

(7)   If lines do not meet, they are parallel.

This has the same drawbacks as (6) and, in addition, sounds like a statement of fact (that might conceivably be wrong), whereas it is actually intended to be true by definition.

In reading scientific literature, it is usually necessary for us to guess from the context whether or not a statement is a definition and, if so, what is the defined term. Alertness on this score saves much time that would otherwise be wasted by trying to see why a supposed claim is true, when it is merely a definition of one of the writer's terms.

In each of the following, identify the defined term, criticize, reformulate as a formal definition, and give, if possible, a better informal definition: (m) A triangle with three equal sides is equilateral.   (n) A political plan that cannot secure the results it envisages is utopian.   (o) A leaf with an even margin is said to be entire.

Often the meaning of a term is indicated intentionally or unintentionally only by the context, without any direct statement of its meaning. For example, in Section 1-4 we indicated the meaning of "factor" by the way in which we used it to describe multiplication. Of course, guessing the meaning of a word from its context may be risky, as illustrated by the schoolboy who said that an average was a "rack on which chickens lay eggs" because he had read somewhere that chickens lay one egg a day on an average!

Give a formal and an informal definition of the italicized term in each of the following: (p) *Factor*. (q) The *magician* seeks to control the forces of nature in an unnatural or in an unusual way. (r) Avoid *overcooking*, since cooking food too long spoils it.

In formulating a definition we must decide on both the defined and defining terms. Even if we know precisely the expression for which we want to adopt an abbreviation, the problem of picking the appropriate defined term is usually by no means easy. Some points to consider are the following:

(8)   *The defined term should not have been previously defined in the same context.*

(9)   *The defined term should be either lacking in previous meaning in the context or have a previous meaning and associations not inconsistent with the defining term.*

(10)   *The defined term should, if possible, seem to be a natural choice.*

(11)   *The defined term should be convenient to use.*

When it comes to choosing words as defined terms, a compromise among these desirable features is usually necessary. A newly coined word always satisfies (8) and (9), but often fails to satisfy (10) and (11), while old words seldom satisfy (9) completely. For example, in choosing "symbol" as a synonym for "any intentional mark" in Section 1-3 we considered other words, such as "sign," and came to the conclusion that "symbol" satisfied these conditions better than any other existing word, even though the exact meaning we gave to it was somewhat broader than its dictionary definition.

When we choose defined terms of nonverbal mathematics, we are not handicapped by previous meanings and can often invent an entirely new symbol that satisfies (8) and (9) and seems to have a good likelihood of satisfying (10) and (11). As to (11), we cannot tell whether a term will be convenient to use until we try it. For this reason it frequently happens that various terms are used with the same meaning in mathematics. Sometimes one finally emerges as the best, and sometimes several remain in use permanently, each in a context where it is most convenient.

(s) Give an example of two mathematical symbols that designate the same arithmetical operation.   (t) Why is "a^3" superior to "a cubed"?

There is no problem of choosing a defining term if we already have an expression for which we wish to adopt an abbreviation. Such is often the case in mathematical theories. Usually, however, we have only a more or less clear idea of something for which we wish to adopt a term. In such cases the problem is to formulate clearly a description of the thing to be named, then use this as the defining term. For example, in defining "symbol" in Section 1–3 we inserted "intentional," since otherwise accidental marks would be included as symbols, and we wished to include only marks made intentionally by man. On the other hand we did not specify "marks on paper" or "printed marks" because we did not wish to limit the meaning of the term in this way. The following may be helpful in formulating defining terms.

(12)   *The defining term should contain only undefined terms and previously defined terms.*

(13)   *The defining term should have a clear meaning.*

(14)   *The defining term should name precisely what is intended, neither including unwanted items nor excluding ones that are desired.*

(15)   *The defining term for a formula must make clear how its values may be found.*

These suggestions are intended to guarantee that the defining term is actually a term in the sense of Section 1–5. Definitions that conform to (15) are called *operational*. For example, $a^2 =_d a \cdot a$ is operational, because the right member tells us how to compute a^2 for every value of a.

(u) Criticize "$\sqrt{x} =$ the square root of x" as a definition of $\sqrt{x}$.   (v) Which of the following is operational?   "Economic progress is any improvement in economic welfare."   "Economic progress is the increase over time of per capita goods and services."   (w) Discuss the following two definitions: "National welfare = gross national product as determined by federal statistics."   "National welfare = the happiness of the people."   (x) Which criterion is not satisfied by the following: "A square is a quadrilateral with right angles at all its vertices."?   (y) Criticize: "A regular polygon is one with equal sides."

We close this section with two definitions that will illustrate some of the ideas discussed above.

You are already familiar with the symbol for square root, $\sqrt{}$. How can we formulate a precise definition? Why is $\sqrt{9} = 3$? Because $9 = 3^2$. This suggests the definition $(y = \sqrt{x}) =_d (y^2 = x)$. But this is unsatisfactory because both $(-3)^2 = 9$ and $(3)^2 = 9$, so that "$\sqrt{9}$" might mean either 3 or -3. Hence with this definition "$\sqrt{x}$" would not be a formula. To eliminate the ambiguity, we define "$\sqrt{x}$" to mean the non-negative number whose square is x; that is,

(16) Def. $(y = \sqrt{x}) = (y^2 = x$ and $(y = 0$ or $y > 0))$.

With definition (16) we can say that any positive number x has two square roots, $\sqrt{x}$ and $-\sqrt{x}$. We call $\sqrt{x}$ the *principal square root*. For the present we exclude negative values of x in $\sqrt{x}$.

In Section 1–4 we gave a very informal definition of $|a|$ in geometric terms, saying that it represented the length of the vector a. This was not very satisfactory, since there was no precise specification of what we meant by length! A more satisfactory operational definition is the following.

(17) Def. $|a| = a$ when $a > 0$ or $a = 0$.

$|a| = -a$ when $a < 0$.

Note that the definition has two parts, one specifying how to find $|a|$ when $a > 0$ or $a = 0$, the other when $a < 0$. This definition conforms to the remarks in Section 1–4 at the end of the paragraph dealing with absolute value.

(z) Illustrate (17) for $a = -10$.

There are many other definitions in elementary algebra. Our purpose here was to include a few in order to illustrate the following points:

(18)

> (a) *Definitions are arbitrary, but they are chosen in order to facilitate calculation and communication.*
>
> (b) *Any expression is meaningless and not even a term until it is introduced as basic or defined.*
>
> (c) *A formula is defined only for certain values of its variables, and is meaningless for other values.*

Problems

1. Why are there no circularities in an English-French dictionary?
2. Which is correct usage: "Define an elephant" or "Define 'an elephant'"?
3. Discuss from the point of view of this section one or more of the definitions of words given in the book so far.

Give precise definitions of the italicized terms in Problems 4 through 7 on the basis of the context.

★4. "To refer to the whole political organization of a particular people inhabiting a definite territory and ruled by a distinct governmental establishment a broader term *the state* is employed." (G. L. Field, "Toward a More Objective Definition of Political Concepts," *The Southwestern Social Science Quarterly,* July, 1946)

★5. "To adopt a term as *primitive* is to introduce it into a system without defining it." (N. Goodman, *The Structure of Appearance,* 1951, p. 56)

★6. The classical writers distinguished between *chronicle* and *history.* The former records what was done and in what year it happened; history must exhibit also the reason and cause of events.

★7. "... the technique of measuring *group morale* designed by L. D. Zeleny. In its simplest form the technique consists in determining the number of attractions or 'likes' that are found in a group and expressing it as a ratio of the total number of attractions which would be theoretically possible in that group, i.e., if each member were attracted to each other member." (P. F. Lazarsfeld and M. Rosenberg, *The Language of Social Research,* p. 24)

Answer Problems 8 and 9 with reference to definition (16).

8. What is $\sqrt{0}$ and why?

9. Why is $\sqrt{16} \neq -4$?

10. If $a > 0$, what is $\sqrt{a^2}$? If $a < 0$, what is $\sqrt{a^2}$? Show that

(19) $$(\sqrt{a})^2 = a,$$

(20) $$\sqrt{a^2} = |a|,$$

(21) $$\sqrt{a} \geq 0.$$

11. Show that

(22) $$\sqrt{a}\sqrt{b} = \sqrt{ab} \quad \text{and} \quad \sqrt{a}/\sqrt{b} = \sqrt{a/b}.$$

12. Use (22) to simplify (a) $\sqrt{2 \cdot 9}$, (b) $\sqrt{18}$, (c) $\sqrt{8}$, (d) $\sqrt{4a^3}$, (e) $\sqrt{8a^2}$, (f) $\sqrt{a^4}$.

13. Simplify the following: (a) $(a + \sqrt{b})(a - \sqrt{b})$, (b) $(a + \sqrt{b})^2$, (c) $(\sqrt{3} - \sqrt{2})^2$, (d) $(\sqrt{x} - \sqrt{2x})^2$.

14. Is there any real number whose square is -4? Is there any real number whose square is x, if $x < 0$? Why?

Answer Problems 15 through 17 with reference to definition (17).

15. What is $|0|$ and why?

16. Does "$|a|$" mean that a is positive?

17. Why is $|a^2| = a^2$ a law? Why is $|x| = x$ not a law?

18. Show that

(23) $$|a||b| = |ab|,$$

(24) $$|a| = |-a|,$$

(25) $$|a| \geq 0.$$

19. Define "$z\%$" and "x is $z\%$ of y."

Answers to Exercises

(a) Presumably their ambition was to use common words in defining uncommon ones, a very laudable procedure. However, in defining the common ones they would then have to "land you right back where you started." (b) Suppose the words are a, b, c, d, and e. Then b has to be used to define a. If we are not to "land right back," a different word must be used to define b, say c. Then still another must be used to define c. In this way d is used and e must be used to define it. Now we cannot avoid using a word again. So when we look up the fifth word, we get a circularity. Of course, we might get one sooner. (c) Yes.   (d) Circularity usually occurs in only two or three steps.   (e) Astronomers *assigned* the names.   (f) In his time, amusing = pleasing, awful = inspiring awe, and artificial = well constructed.

(g) $a^3 =_d a \cdot a^2$; Let $a^3 = a \cdot a^2$; or Def. $a^3 = a \cdot a^2$.   (h) Because it is true by our decision as to the meaning of the symbols.   (i) Same as (h). (j) They are introduced for brevity and convenience.   (k) Because we could replace all defined terms by their synonyms. It would make for awkward communication.   (l) "Point," "straight line," "segments," and "joining." (m) Equilateral triangle $=_d$ triangle with equal sides.   (n) Utopian plan $=_d$ plan that cannot secure the results it envisages.   (o) Entire leaf $=_d$ leaf with an even margin. Note that in some contexts "entire leaf" would mean simply a whole leaf, but this is no serious drawback, since the context would indicate the meaning.

(p) Factor $=_d$ the value of a term that appears as does "a" or "b" in "$a \cdot b$." (q) A magician $=_d$ a person who seeks to.... (r) Overcooking $=_d$ spoiling food by cooking it too long.   (s) $\cdot$ and $\times$; or $/$ and $\div$.   (t) Easier to manipulate; shorter.   (u) Does not help in that one does not know what "square root of x" means. Gives no means of finding $\sqrt{x}$.   (v) Second. It is also perhaps too narrow.   (w) First may be too narrow, but is operational. Second is nonoperational.   (x) Too broad, since it includes rectangles that are not square. (y) Too broad, since it includes polygons with unequal interior angles. (z) $-10 < 0$. Hence we substitute $(a:-10)$ in the second part of the definition to get $|-10| = -(-10) = 10$, as expected.

Answers to Problems

1. English terms are all undefined. 2. Second. 4. A state $=_d$ the whole political organization . . . by a distinct governmental establishment. 5. Primitive term $=_d$ a term that is undefined. 8. 0 since $0^2 = 0$ and $0 = 0$. 9. -4 is not > 0 or $= 0$. 10. $\sqrt{a^2} = a$ when $a > 0$ or $a = 0$; $\sqrt{a^2} = -a$ when $a < 0$. (20) is true since the right members of these two equations are the same as in (17). 11. $(\sqrt{a}\sqrt{b})^2 = \sqrt{a}\sqrt{b}\sqrt{a}\sqrt{b} = (\sqrt{a})^2(\sqrt{b})^2 = ab$ and $\sqrt{a}\sqrt{b} > 0$ or $= 0$. 12. (a) $3\sqrt{2}$, (b) $3\sqrt{2}$, (c) $2\sqrt{2}$, (d) $2|a|\sqrt{a}$, (e) $2|a|\sqrt{2}$, (f) a^2. 13. (a) $a^2 - b$, (b) $a^2 + 2a\sqrt{b} + b$, (c) $5 - 2\sqrt{6}$, (d) $x(3 - 2\sqrt{2})$. 19. $z\% =_d z/100$; x is $z\%$ of $y =_d (x/y = z/100)$.

1–14 Fractions. A *fraction* is an expression of the form a/b. The properties of fractions are consequences of the following definition.

(1) Def. $(a/b = c) = (a = bc).$

Is this an adequate definition in terms of the criteria of the previous section? Does it enable us to find a/b for any numbers a and b? We know by experience that it does in many cases, since the methods of division we learned in school consist of taking trial divisors and "multiplying back." For example, we know that $63/7 = 9$ because $63 = 7 \cdot 9$.

It is not hard to see that (1) does not define a/b for $b = 0$, for the substitution $(b:0)$ in (1) yields $[a/0 = c] = [a = 0 \cdot c]$. Now $0 \cdot c = 0$. Hence if $b = 0$ and $a \neq 0$, there is *no value* of c that makes the right member of (1) true, and hence no value of a/b is determined by (1). On the other hand, if $a = b = 0$, the right side of (1) becomes $0 = 0 \cdot c$. But this is satisfied by *any value* of c. Hence in this case (1) fails to define a unique value of a/b. We have shown that a/b is undefined for $b = 0$. Hence we say that division by zero is undefined, and we exclude 0 as a value of any denominator. Does (1) define a/b for all nonzero values of b? The answer is yes, but we postpone a proof until Section 1–16.

From (1) we can easily derive the laws in Section 1–9 that dealt with fractions. For example, $(1)(c:a/b)$ is $[a/b = a/b] = [a = b(a/b)]$. The left member is obvious; the right is just (1–9–17). Now (1–9–27) follows by the substitution $(a:1, b:a)$ in (1–9–17). Returning to (1), the substitution $(c:a(1/b))$ yields $[a/b = a(1/b)] = [a = b(a(1/b))]$. But the right member is just $a = a$, since $b(a(1/b)) = b(1/b)a = a$ by rearrangement and (1–9–17). Since the left member is (1–9–19), that law is proved.

(a) Prove (1–9–21) by substitution in (1). (You may, of course, assume the identities having to do with multiplication.)   (b) Justify $(-8)/(-2) = 4$ from (1).

(2) $ca/cb = a/b.$

Identity (2) says that the value of a fraction is unchanged if its numerator and denominator are multiplied (or divided) by the same term. It follows from $(1)(a:ca, b:cb, c:a/b)$ since $(cb)(a/b) = c(b(a/b)) = ca$ by rearrangement and (1–9–17). *It can be used to great advantage in simplifying fractions.* For example, it is the law that permits us to write $9/24 = 3/8$, since $9/24 = 3 \cdot 3/3 \cdot 8 = 3/8$. The last equation is obtained from (2) by $(a:3, b:8, c:3)$.

(3) $a/d + b/d = (a + b)/d.$

Law (3) tells us how to add fractions with the same denominator. We do not need a separate law for fractions with different denominators, because we may always write fractions so that they have any desired denominators by using (2). For example, to add $1/2$ and $5/6$, we cannot use (3) directly since the denominators are not the same. However, by (2) $1/2 = 3/6$. Hence

(4) $$1/2 + 5/6 = 3/6 + 5/6 \qquad (2)$$

(5) $$= (3 + 5)/6 \qquad (3)$$

(6) $$= 8/6 \qquad 3 + 5 = 8,$$

(7) $$= 4/3 \qquad (2).$$

Of course, we would not think of writing out all the details of such a simple calculation except to show how each step depends on the application of a law.

(c) Use (2) to simplify $1/(1/2)$.   (d) Indicate the substitutions used in the laws indicated in steps (4), (5), and (7).   (e) How was the rule of replacement used in (4) through (7)?

Perform the following as above, indicating the laws and substitutions used: (f) $2/3 + 5/6$.   (g) $1/3 + 4/5$.   (h) $4/5 + 1/6 + 2/3$.   (i) Prove (3) by (1) $(a{:}a + b,\ b{:}d,\ c{:}a/d + b/d)$.   (j) Show that $a/b + c/d = (ad + bc)/bd$.

The following laws cover multiplication and division of fractions.

(8) $$(a/b)(c/d) = ac/bd,$$

(9) $$(a/b)(b/a) = 1,$$

(10) $$1/(a/b) = b/a,$$

(11) $$(a/b)/(c/d) = ad/bc.$$

The first follows from $(1)(a{:}ac,\ b{:}bd,\ c{:}(a/b)(c/d))$, since $bd(a/b)(c/d) = b(a/b)d(c/d) = ac$ by rearrangement and a double application of (1–9–17). Then (9) follows by applying (8) to get $(a/b)(b/a) = ab/ba = 1$, the last equality being justified by rearrangement and $(a/a = 1)(a{:}ab)$. Then (10) follows from $(1)(a{:}1,\ b{:}a/b,\ c{:}b/a)$ and (9), and (11) follows from $(1)(a{:}a/b,\ b{:}c/d,\ c{:}ad/bc)$ and (2).

Simplify, indicating laws and substitutions: (k) $27/30$,   (l) $(3/4)(2/3)$, (m) $(3/4)/(2/3)$,   (n) $1/(1/3)$,   (o) $(-2/3)(3/(-2))$.

For fractions involving minus signs, the following are useful.

(12) $$a/b = (-a)/(-b),$$

(13) $-(a/b) = (-a)/b$

(14) $= a/(-b).$

Simplify, indicating laws and substitutions: (p) $(-2)/(-3)$,   (q) $(-7)/2$,
(r) $8/(-4)$,   (s) $0/(-3)$,   (t) $(-2)(-3)(-27)/(-9)$.

By using the identities of this and previous sections we can perform all
the manipulations needed to rewrite algebraic expressions in any desired
form (provided, of course, that the desired form is synonymous with the
given expression!).   In particular, by using (2) and the distributive law
we can easily manipulate fractions.   For example,

(15) $\dfrac{x - 1/3}{x + 3/2} = \dfrac{6(x - 1/3)}{6(x + 3/2)}$ $(2)(a{:}x - 1/3, b{:}x + 3/2, c{:}6)$

(16) $= \dfrac{6x - 2}{6x + 9}$ (Distributive law)

(17) $= \dfrac{2(3x - 1)}{3(2x + 3)}$ (Distributive law)

(18) $= (2/3)((3x - 1)/(2x + 3))$ (8).

Which one of these synonymous expressions is considered the simplest
depends on the situation.

Simplify: (u) $(x + 2y)/(2y + x)$,   (v) $(3x - 2)/(2 - 3x)$,
(w) $1/(a - b) + 1/2$,   (x) $(3/4 - 1/x)/(x + 1/x)$,   (y) $1/(1/x)$.

Just as the laws having to do with division follow from (1), so those
having to do with subtraction follow from the next definition.

(19) Def. $[a - b = c] = [a = b + c].$

The corresponding laws are paired in Section 1–9.   Of course, 0 plays the
role for subtraction that 1 does for division.   We leave the development
of this idea to the reader (see Problems 21 and 22).

PROBLEMS

Show that 1, 2, and 3 *are not laws.*

1. $1/a + 1/b = 1/(a + b)$.
2. $(1/a + 1/b)^2 = 1/a^2 + 1/b^2$.
3. $1/(1/a + 1/b) = a + b$.

4. Simplify $(1/a - 1/b)/(1/a + 1/b)$ without adding the terms in numerator
and denominator.

5. Prove (1–9–21), (1–9–29), and (1–9–32) by substitutions in (1).

Show that

6. $\dfrac{a+b}{2} + \dfrac{a-b}{2} = a.$

7. $\dfrac{a+b}{2} - \dfrac{a-b}{2} = b.$

8. $a + b/c = (ac + b)/c.$

9. $a/b - 1 = (a - b)/b.$

10. $a/(a+b) + 1 = (2a + b)/(a + b).$

11. Prove (3) by using the distributive law and $a/b = a(1/b).$

12. Show that $\dfrac{p-w}{w}(w-a) = p - w - ap/w + a.$ (*History of Economic Analysis*, by J. A. Schumpeter, p. 467)

13. Complete the following: The argument following (1) shows that "0" or any term synonymous to it is not a _____ for "b" in the formula "a/b." In fact "$a/0$" is not a formula, because _____ .

14. "The formula given earlier for the coefficient of determination is: $1 -$ (Error variance)/(Total variance). . . . It follows that an equivalent expression . . . is as follows: (Total variance $-$ Error variance)/(Total variance)." (*A Primer of Political Statistics*, by V. O. Key, p. 115) Derive the second formula from the first.

15. "Assign to 0_1 the value $R_1/(R_1 + R_2 + R_3 + R_4)$, to 0_2 the value $R_2/(R_1 + R_2 + R_3 + R_4)$, to 0_3 the value $R_3/(R_1 + R_2 + R_3 + R_4)$, and to 0_4 the value $R_4/(R_1 + R_2 + R_3 + R_4)$. The sum of these values should be equal to 1." (*The Design of Social Research*, by R. L. Ackoff, p. 25) Justify this statement.

16. Show that $p/(1 - (1 - p))^2 = 1/p$ and $\dfrac{n}{1 - (L_0 - n)/L_0} = L_0.$ (*Readings in Learning*, by L. M. Stolurow, p. 53)

17. Show that $(a + a')/(b + b') = a(1 + a'/a)/b(1 + b'/b).$ (*A Geometry of International Trade*, by J. E. Meade, p. 13)

18. Solve for T: $1/T = 1/S + 1/E.$ (*Manual of Astronomy*, by R. W. Shaw and S. L. Boothroyd, p. 98)

★19. In music an interval between two tones is measured by the ratio of the frequencies as follows: octave, 2/1; fifth, 3/2; fourth, 4/3; major third, 5/4; minor third, 6/5; major sixth, 5/3; minor sixth, 8/5; second, 9/8. To find the sum of two intervals we multiply the ratios; to find the difference we divide. (a) The sum of 2 fifths is called a ninth. What is its ratio? (b) Show that a fifth less a fourth is a second. (c) Show that the difference between 12 fifths and 7 octaves is 531441/524288. (d) Show that the major third taken three times differs from the octave by 125/128.

★20. The number π may be approximated by the equation

$$4/\pi = 1 + \cfrac{1^2}{2 + \cfrac{3^2}{2 + \cfrac{5^2}{2}}}.$$

Solve for π and express as a decimal. Improve the approximation.

★21. From (19), derive formulas (16), (18), (20), (26), and (28) of Section 1–9.

★22. What are the analogs for subtraction of (2) and (8) through (11)?

ANSWERS TO EXERCISES

(a) (1)$(b{:}a, c{:}1)$ is $[a/a = 1] = [a = a \cdot 1]$, and the right member is true by (1–9–25).   (b) $[(-8)/(-2) = 4] = [-8 = (-2)4]$.   (c) $(a{:}1, b{:}1/2, c{:}2)$.
(d) (2)$(a{:}1, b{:}2, c{:}3)$, (3)$(a{:}3, b{:}5, d{:}6)$, (2)$(a{:}4, b{:}3, c{:}2)$.   (e) Each right member was obtained by a replacement in the previous one.   (f) 3/2.
(g) 17/15.   (h) 49/30.   (i) Use distributive law and (1–9–17).   (j) $a/b +$ $c/d = ad/bd + cb/bd = \ldots$ (k) 9/10.   (l) 1/2.   (m) 9/8.   (n) 3.
(o) 1.   (p) 2/3 by (12).   (q) $-7/2$ by (13).   (r) -2 by (14).   (s) 0 by (1–9–32).   (t) 18.   (u) 1.   (v) -1.   (w) $(2 + a - b)/2(a - b)$.
(x) $(3x - 4)/4(x^2 + 1)$.   (y) x.

ANSWERS TO PROBLEMS

1, 2, 3. $(a{:}1, b{:}1)$.   4. (2)$(a{:}1/a - 1/b, b{:}1/a + 1/b, c{:}ab)$.   6 through 10. See note in Problem 17 in Section 1–12.   11. $a/d + b/d = (1/d)a + (1/d)b = (1/d)(a + b) = (a + b)/d$.   12. $\dfrac{p - w}{w} (w - a) = (p - w)(w -- a)/w = (pw - pa - w^2 + wa)/w = p - w - pa/w + a$.   13. significant substitute; it does not become a constant for any substitution for a.

1–15 Decimals. Which is larger, 11/12 or 23/25? The easiest way to find out is to express both numbers in decimal form. By long division we find $11/12 = 0.91666\ldots$ and $23/25 = 0.92$. (The dots in $0.91666\ldots$ are used to suggest that the 6 is repeated indefinitely.) Evidently $23/25 > 11/12$.

(a) Check the above by carrying through the long division.   (b) Which is larger, 5/7 or 44/61?

For purposes of such comparisons and for convenience in calculation, real numbers are often expressed by decimals. As we see from the above example, it is not always possible to write a terminating decimal, such as 0.92, to represent a number. An infinite decimal, such as $0.91666\ldots$, is often required. Since a terminating decimal may be thought of as an infinite decimal with only zeros after a certain point, we may think of all decimals as infinite decimals. More precisely we think of a *decimal* as an expression of the form

(1) $N.d_1d_2d_3d_4 \ldots d_n \ldots$ or $-N.d_1d_2d_3 \ldots d_n \ldots$,

where N is a non-negative integer and the d's are digits. We call d_n the digit in the nth decimal place. The dots suggest the endless sequence of digits, some or all of which may be zero.

(c) What is a digit?   (d) What is the digit in the fourth decimal place in -27.391204?

When the decimal ends in zeros after a certain point, we call it *terminating*. When after a certain point a decimal consists of the repetition of a digit or group of digits, we call the decimal *repeating*. We indicate this by writing dots over the repeated group of digits. For example, we write $11/12 = 0.91\dot{6}$. A terminating decimal is also a repeating decimal, since after a certain point it consists of a repetition of the digit 0.

Any rational number can be represented by a repeating decimal, which can be found by long division. For example, $2 = 2.\dot{0}$, $-31/10 = -3.1\dot{0}$, $1/3 = 0.\dot{3}$, and $12/7 = 1.\dot{7}1428\dot{5}$. Let us verify this last equation.

$$
\begin{array}{r}
1.7142857\ldots \\
7\overline{)12.0000000\ldots} \\
\underline{7} \\
5\,0 \\
\underline{4\,9} \\
10 \\
\underline{7} \\
30 \\
\underline{28} \\
20 \\
\underline{14} \\
60 \\
\underline{56} \\
40 \\
\underline{35} \\
50
\end{array}
$$

(2)

The calculation suggests the proof of our claim that any rational number can be represented by a repeating decimal. Suppose we try to find m/n by long division, where m and n are integers. Since the decimal representing m consists of zeros after the decimal point, there comes a time in the division after which we are "bringing down" only zeros. If after this point the same remainder turns up twice, as 5 does in (2), the group of digits since its last appearance will be repeated from there on. But when dividing by n there are at most n different remainders. Hence a remainder is bound to reappear in at most n steps.

(e) If zero appears as a remainder, does this spoil the argument? (f) Argue that in the repeating decimal representing a rational number r/s in lowest terms, there are at most s digits in the smallest repeating cycle of digits. (g) Find the decimal representation of 15/7, (h) 4/9, (i) 15/13.

We have seen that every rational number can be represented by a repeating decimal. Is the converse true? Is every repeating decimal a rational number? Consider, for example, $x = 3.\dot{9}\dot{1} = 3.9191\ldots.$ Since multiplication by 100 may be indicated by shifting the decimal point two places, we have $100x = 391.91\ldots = 391.\dot{9}\dot{1}.$ Then $100x - x = 391.\dot{9}\dot{1} - 3.\dot{9}\dot{1} = 388.$ Hence $99x = 388$ and $x = 388/99.$ The calculation may be written as follows:

$$
\begin{aligned}
100x &= 391.919191\ldots \\
-x &= -3.919191\ldots \\
\hline
99x &= 388.000000\ldots \\
x &= 388/99
\end{aligned}
$$

(3)

This example suggests the following rule: To find a ratio of natural numbers equal to a repeating decimal x, multiply by 10^n, where n is the number of digits in the repeated group of digits, subtract x, and divide by $10^n - 1$. (If this gives the ratio of two terminating decimals, multiply numerator and denominator by some power of 10 to get a ratio of integers.)

(j) Prove that $(10^n x - x)/(10^n - 1) = x.$ Use the rule to find the ratio of integers equal to   (k) $0.\dot{3}$,   (l) $0.3\dot{5}$,   (m) $2.3\dot{1}$,   (n) $1.\dot{9}.$

It is evident that the above process applies to any repeating decimal, but the reader may ask whether we are justified in manipulating infinite decimals, as we did in (3), as though they were terminating decimals. The answer is that it can be justified rigorously in various ways, all of which are too time-consuming for our purposes here. Moreover, in each case the above calculation can be checked in reverse by long division.

In Exercise (n) the reader found that $1.\dot{9} = 2 = 2.\dot{0}.$ Apparently, there may be more than one repeating decimal representation of a rational. Let us consider a repeating decimal ending in nines, $x = n.a_1a_2\ldots a_i\dot{9}.$ Then $x = n.a_1a_2\ldots a_i + 0.00\ldots\dot{9},$ where there are i zeros preceding the repeated part. But by a calculation similar to (3), $0.00\ldots\dot{9} = 0.00\ldots1,$ where there are $i - 1$ zeros preceding the 1. Hence $x = n.a_1a_2\ldots(a_i + 1)\dot{0}.$ Thus we see that any decimal ending in nines is equal to the decimal found by increasing by one the last digit preceding the nines and changing the repeated nines to zeros.

(o) Use the procedure of (3) to find $0.\dot{9}$, $0.0\dot{9}$, $0.00\dot{9}$. $0.000000\dot{9}$, and $32.185\dot{9}.$

To avoid this double representation of some rationals, *we agree to exclude decimals ending in nines.* Accordingly, from now on "repeating decimal" refers to repeating decimals other than those ending in nines. With this understanding, there is just one repeating decimal corresponding to each rational and one rational corresponding to each repeating decimal.

What about irrational numbers? Clearly we must use nonrepeating infinite decimals to represent them. It is easy to give examples of nonrepeating infinite decimals. For example, 0.1010010001 . . . , where it is understood that after each "1" we put one more "0" than previously. The decimal expansion of $\sqrt{2}$, π, or any other irrational number does not repeat. Nevertheless, the digit in every decimal place is definitely determined by the number.

To see why this is so, we recall from Section 1–2 that a real number is one that represents a point on an axis. Suppose that we have a real number x, which we assume positive to simplify the argument. We shall show that regardless of the position of x, there exists an infinite decimal representing it. If x is an integer, we have its decimal equivalent immediately. If not, it lies between two integers of which the smaller gives us the integral part of our decimal. Divide the segment between these two integers into tenths. Again if the point lies on one of these points of division, we have a terminating decimal. If not, it lies in the interval between two of them, and the smaller gives us the first place in our decimal. Now we divide this last interval into ten equal parts (of length 0.01) and repeat the process. Evidently, if the position of the point is exactly known (and this is what we mean by saying that we have a definite real number given), we can determine the digit in any decimal place in its decimal expansion, i.e., the decimal is uniquely determined. Hence, to every real number there corresponds a unique infinite decimal.

(p) Why is there no difficulty here with decimals ending in 9's?

Conversely, suppose we are given an infinite decimal, $N.d_1d_2d_3 \ldots d_n \ldots$. To find the corresponding point we first go to the integral point N, then $d_1/10$, then $d_2/100$, and so on as indicated in Section 1–2, except that now we never come to the last step unless the decimal happens to end in zeros. Nevertheless, it seems plausible that there is a unique corresponding point. For when we reach the point $N.d_1$, we know that the corresponding point (provided it exists) lies between $N.d_1$ and $N.(d_1 + 1)$. When we reach $N.d_1d_2$, we know that it lies between this and $N.d_1(d_2 + 1)$. At the next step we know our point lies between $N.d_1d_2d_3$ and $N.d_1d_2(d_3 + 1)$, and at the nth step between $N.d_1d_2d_3 \ldots d_n$ and $N.d_1d_2d_3 \ldots (d_n + 1)$. We have suggested these intervals in Fig. 1–25. Actually each of the segments is 1/10 the length of the previous, but we have distorted their size in the sketch.

We see that the process generates a sequence of segments each within the previous one and each 1/10 the length of the previous one. This sequence is without end. (If the decimal ends in zeros, after a certain point all segments have their left ends coincident.) Since their lengths approach zero, it seems plausible that there is one and only one point common to them all. Certainly there can be no more than one. For if there were two different points, however close, there would come a segment too small to

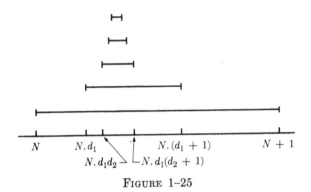

$$\text{Figure } 1\text{--}25$$

contain them both. If we assume that an infinite sequence of intervals, each contained in the previous and approaching zero in length, contains a common point, then this argument shows that each infinite decimal determines a unique point, i.e., a unique real number.

(q) To illustrate the above argument consider the decimal expansion of π, $3.14159 \ldots$. Complete the following: From the first two digits we know that π lies between 3.1 and _____. Using two decimal places, we know that it lies between 3.14 and _____. Using three, between _____ and 3.142. Using four, between _____ and _____. Using five, between _____ and _____. The length of this last interval is _____. (r) We have excluded decimals ending in 9's. However, if we applied the process of the previous paragraph to such a decimal, what would be the nature of the intervals and the corresponding point?

The above arguments are intended to make more reasonable the identification of real numbers and points that we mentioned in Section 1–2. We shall henceforth identify infinite decimals with real numbers. Then we may sum up the above discussion by saying that there is a one-to-one correspondence between real numbers and points on an axis.

Problems

Arrange in order of increasing size.

1. $8/7$, $28/25$.

2. 3.1416, π, $22/7$.

3. $\sqrt{31}$, 5.61, 5.5678.

4. $23/11$, $\sqrt[3]{9}$, 2.081.

Express as ratios of integers.

5. $35.\dot{2}\dot{7}$.

6. $35.0\dot{2}\dot{7}$.

7. $2.1\dot{4}2\dot{8}5\dot{7}$.

8. $8.1\dot{1}\dot{3}$.

Express as decimals.

9. $1/11$.   10. $3/11$.   11. $23/51$.   12. $1/17$.

★13. In the arguments showing the one-to-one correspondence between points and decimals, we assumed positive decimals. Give an argument for negative decimals.

★14. The following ratios are of interest in the theory of music: 81/80, 125/128, 531441/524288. Range them in order of size. (*Introduction to Musicology*, by Glen Haydon, p. 33)

★15. For another elementary discussion of decimals and real numbers, see *The Theory of Numbers*, by B. W. Jones, pp. 31 ff.

ANSWERS TO EXERCISES

(b) 44/61.　(c) One of the symbols 0, 1, 2, 3, 4, 5, 6, 7, 8, 9,　(d) 2. (e) No, since zero is then repeated.　(g) $2.1\dot{4}2\dot{8}5\dot{7}$.　(h) $0.\dot{4}$.　(i) $1.\dot{1}5\dot{3}8\dot{4}\dot{6}$. (j) $10^n x - x = x(10^n - 1)$.　(k) 1/3.　(l) 35/99.　(m) 104/45.　(n) 2. (o) 1, 0.1, 0.01, 0.000001, 32.186.　(p) The procedure would lead us to a terminating decimal equivalent to one ending in nines.　(q) 3.2; 3.15; 3.141; 3.1415, 3.1416; 3.14159, 3.14160; 0.00001.　(r) After a certain interval, all intervals would have the same right endpoint.

ANSWERS TO PROBLEMS

1. 28/25, 8/7.　3. $\sqrt{31}$, 5.5678, 5.61.　5. Check by long division.　9. Check by reverse procedure.

1–16 Axioms for the real numbers.

If we wish to convince a person of the correctness of a statement in such a way that our success does not depend on his weaknesses or our cleverness and so that our argument would be equally convincing to others, we must adopt a method of proof that meets with general agreement. Whatever this method is, it must certainly involve making statements, since it is hard to conceive of convincing anyone of anything without communicating with him. So we must carry on the proof by means of statements. But suppose that the listener objects to some of our statements. Then we must convince him of their validity before we can continue. We can do this only by making other statements. Again, if our listener does not agree with these, we should have to try to justify them—of course by making still other statements. Evidently we cannot convince anyone of anything unless we can get him to agree to something! Apparently we cannot prove all statements any more than we could define all terms. We must begin by assuming some statements without proof.

In mathematics, sentences that are assumed to be laws without proof are called *axioms*. Synonyms of "axiom" are "postulate," "basic assumption," and (outside mathematics) "hypothesis." Sentences that are proved to be laws are called *theorems*. The previous paragraph is intended to convince the reader that if we wish to prove any theorems we must begin by assuming some axioms.

In previous sections we have proved some laws of real numbers by applying the rules of replacement and substitution to other laws. To avoid confusion and possible circularity, it is best to pick out a few laws as axioms and derive all others from them, just as is customarily done in high-school plane geometry. In the following axioms we may think of real numbers as infinite decimals, points on an axis, vectors on a line, or in any other way not inconsistent with the axioms.

(1) Ax. *If a and b are any real numbers, there exists a unique real number $a + b$, called the sum of a and b, and a unique real number $a \cdot b$, called the product of a and b.*

(2) Ax. $a + b = b + a.$ $\left.\right\}$ (Commutative laws)

(3) Ax. $a \cdot b = b \cdot a.$

(4) Ax. $a + (b + c) = (a + b) + c.$ $\left.\right\}$ (Associative laws)

(5) Ax. $a \cdot (b \cdot c) = (a \cdot b) \cdot c.$

(6) Ax. $a \cdot (b + c) = (a \cdot b) + (a \cdot c).$ (Distributive law)

(7) Ax. *There is a real number 0 such that $a + 0 = a$.*

(8) Ax. *There is a real number 1 such that $a \cdot 1 = a$.*

(9) Ax. *If a is a real number, there exists a unique real number $-a$, called the negative of a, such that $a + (-a) = 0$.*

(10) Ax. *If a is a real number not equal to zero, there exists a unique real number a^{-1}, called the reciprocal of a, such that $a \cdot a^{-1} = 1$.*

(a) List the undefined constants that appear in these axioms.   (b) List the undefined formulas.   (c) Prove that $(-a) + a = 0$ and $a^{-1}a = 1$.

We now wish to prove that all the identities of algebra stated in previous sections can be proved from these axioms by using in each proof only axioms, definitions, and previously proved laws. Since some of these identities have already been derived from others, it will be sufficient to tie in these axioms with our previous derivations.

First we observe the importance of (1), obvious as it seems. Without it we could not even write $a + b = a + b$, for if the sum were not unique the two members might be different! From it we know that if $a = b$ and $c = d$, then $a + c = b + d$ and $ac = bd$. This justifies the rules given in Section 1-6 for manipulating equations. Axiom (1) also justifies the manipulations of infinite decimals in Section 1-15.

Next we derive the identities in Section 1-9 having to do with addition and multiplication. We note that (1-9-22) is just (7), and (1-9-23) follows from (7) and (2). Similarly, (1–9–24) and (1–9–25) follow from (8) and (3). (1–9–26) is just (9), but in order to get (1–9–27), we need to show that $a^{-1} = 1/a$. For this we need the definition of division (1–14–1). According to this definition, $[1/a = a^{-1}] = [1 = a(a^{-1})]$. But the right member is true by (10). Hence $1/a = a^{-1}$. Now (1–9–27) is seen to be the same as (10).

(d) Why did we not use $1/a$ in the axioms instead of a^{-1}?

To prove (1–9–30) we write

(11) $a \cdot 0 = a \cdot 0 + 0$ $(a = a + 0)(a{:}a \cdot 0)$

(12) $= a \cdot 0 + aa + (-aa)$ $(a + (-a) = 0)(a{:}aa)$

(13) $= a(0 + a) + (-aa)$ $(a(b + c) = ab + ac)(b{:}0, c{:}a)$

(14) $= aa + (-aa)$ $(0 + a = a)$

(15) $= 0$ $(a + (-a) = 0)(a{:}aa)$.

Here we have given reasons in full, showing the laws and substitutions. Below we shall usually omit part or all of the reasons.

(e) Prove (1–9–31) immediately from the above and (3).   (f) Prove (1–9–31) as in (11) through (15), using the right distributive law, which was derived in Section 1–12 solely from our axioms.

Now we can easily derive (1–9–33) through (1–9–37). First we prove (1–9–34) by

(16) $(-1)a = (-1)a + 0$

(17) $= (-1)a + a + (-a)$

(18) $= ((-1) + 1)a + (-a)$

(19) $= 0 \cdot a + (-a)$

(20) $= 0 + (-a)$

(21) $= -a.$

(g) Fill in complete reasons for the steps in (16) through (21).

We now have, similarly, $(-1)(-1) = (-1)(-1) + 0 = (-1)(-1) + (-1) + 1 = (-1)(-1) + (-1)1 + 1 = (-1)((-1) + 1) + 1 = (-1) \cdot 0 + 1 = 0 + 1 = 1$. With $(-1)(-1) = 1$ we can quickly derive the other identities concerned with negatives. For example, $(-a)(-b) = (-1)a(-1)b = (-1)(-1)ab = 1 \cdot ab = ab$. Also, $(-a)b = (-1)ab = -ab$.

(h) Similarly prove (1–9–33) and (1–9–36).

As shown in Section 1–14 we can derive the identities concerning fractions from the definition (1–14–1). However, we need still to show that (1–14–1) does define a unique number a/b for all real a and b, provided $b \neq 0$. First we see that it does determine at least one such real number. For if $b \neq 0$, we have $b(ab^{-1}) = abb^{-1} = a \cdot 1 = a$ by rearrangement, (10), and (8). Hence ab^{-1} is a value of c that satisfies the right member of (1–14–1) and $a/b = ab^{-1}$. Now suppose that there were two values c and c' that satisfied the right member of (1–14–1), that is, $a = bc$ and $a = bc'$. By the rule of replacement $bc = bc'$. Now we multiply both members by b^{-1}, which is possible if $b \neq 0$, to get $b^{-1}bc = b^{-1}bc'$. But $b^{-1}b = 1$, so this becomes $c = c'$, and we have proved that the quotient c is uniquely determined.

(i) Similarly show that (1–14–19) defines a unique number c.

We have now established all the identities in Section 1–9. Since those appearing in later sections were proved from them with the addition of the distributive law in Section 1–12 and the definitions of Section 1–14, we have completed our task of showing that the axioms are sufficient to derive the identities of previous sections. Actually, for the purposes of elementary algebra we need no further axioms about the real numbers.

PROBLEMS

1. We adopted a special symbol, a^{-1}, for the reciprocal for reasons given in the answer to Exercise (d). Why did we not have to adopt a new symbol for the negative?

2. How have we used (4) and (5)?

3. Write a detailed proof of $(-1)(-1) = 1$ along the lines indicated.

4. Similarly, write a detailed proof with reasons for $(-a)(-b) = ab$.

★5. Suggest an alternative system of axioms that could serve in place of (1) through (10).

★6. Show that 0 is the only real number with the property (7). (The number 0 is sometimes called the *identity element for addition*.)

★7. Similarly, show that 1 is the only real number with the property (8). (The number 1 is sometimes called the *identity element for multiplication*.)

★8. It is not true that any definition will yield a uniquely defined object. To illustrate this consider the following definition: $(a/b) \, \theta \, (c/d) =_{\text{d}} (a+c)/(b+d)$, where θ is a new operation consisting in "adding" rational numbers by adding numerators and denominators. Show that this operation does not yield a unique result by showing that if $a/b = a'/b'$ and $c/d = c'/d'$, then it does not follow that $(a/b) \, \theta \, (c/d) = (a'/b') \, \theta \, (c'/d')$!

★9. Suppose in (1) through (10) we inserted "rational" for "real." Would all the axioms hold?

★10. Answer the same question for "integer" in place of "real number."

Answers to Exercises

(a) "0," "1."   (b) "$a + b$," "$a \cdot b$," "$-a$," and "a^{-1}."   (c) (9), (2); and (10), (3).   (d) If we had used $1/a$ we would be in the position of defining a previously introduced undefined term in our definition of division (1–14–1), contrary to the rules of Section 1–13.   (f) $0 \cdot a = 0 \cdot a + 0 = 0 \cdot a + aa + (-aa) = (0 + a)a + (-aa) = aa + (-aa) = 0$.   (g) $(a + 0 = a)(a{:}(-1)a)$; $a + (-a) = 0$; $((b + c)a = ba + ca)(b{:}{-}1, c{:}1)$; $((-a) + a = 0)(a{:}1)$; $(0 \cdot a = 0; 0 + a = a)(a{:}{-}a)$.

(h) $-(-a) = (-1)(-1)a = 1 \cdot a = a$; $a(-b) = a(-1)b = (-1)ab = -ab$.   (i) $a + (-b)$ satisfies the definition by substitution in the right member of (1–14–19), rearrangement, (7), and (9). If $a = b + c$ and $a = b + c'$, then $b + c = b + c'$. Adding $(-b)$ to both members, we have $c = c'$.

Answers to Problems

1. $-a$ is already a symbol not previously appearing nor defined in (1–14–19), since the latter defines only the formula $a - b$ and not $-b$ standing alone. The formula $0 - a$ for the negative would have the same disadvantage in the axioms as would $1/a$ for the reciprocal. 2. In rearrangements. 5. For example, $0 + a = a$ in place of (7).

★1–17 **Check list for reading mathematics.** In this chapter we have been at some pains to explain everything in detail. Usually in mathematical discourse this is not possible. As a rule, explanations are incomplete. Moreover, important parts of calculations and manipulations may be omitted. For these reasons the reader of mathematical discourse must take an active part. He must use his mind (and also plenty of paper) to fill in the more or less bare outline that he reads. The following check list suggests some activities that may be helpful.

1. Give several interpretations in words.

2. If possible give a geometric interpretation, visualize, and make a drawing.

3. Experiment by substituting constants and formulas for variables that appear, and interpret the results.

4. Decide on the values and significant substitutes for any variables that appear.

5. Think of applications if you can.

6. Construct informal arguments, making full use of the meaning of terms, to establish the plausibility of axioms, definitions, and theorems.

7. Make slight variations in laws and see whether the results are still laws. This will help you see why the laws are stated in a particular way.

8. Check every step in proofs, actually writing out every reason in full and performing the indicated substitutions.

9. Rewrite incomplete proofs by inserting omitted steps.

10. Rewrite complete proofs in briefer form.

11. Note the key ideas involved in proofs.

12. Where no proof is given, try to construct one if time permits.

Note that the check list concentrates on experimental activities and intuitive thinking. It is such activities that lead to an appreciation of ideas and lay the basis for precise logical formulations and proof. As George Sarton writes in his *The Study of the History of Mathematics* (p. 19), "The ways of discovery must necessarily be very different from the shortest way, indirect and circuitous, with many windings and retreats. It's only at a later stage of knowledge, when a new domain has been sufficiently explored, that it becomes possible to reconstruct the whole theory on a logical basis, and show how it might have been discovered by an omniscient being, that is, if there had been no need for discovering it!" Mathematics is usually presented in this final logical form, but the learner must energetically experiment and think in order to understand its significance.

It is no accident that the construction of a proof is listed last. Proof should be the last step. As Polya writes in *How to Solve It*, "If you have to prove a theorem, do not rush. First of all, understand fully what it means. Then check the theorem; it could be false. Examine its consequences, verify as many particular instances as are needed to convince yourself of its truth. When you have satisfied yourself that the theorem is true, you start proving it."

(a) Reread *Note to the Student* in the prefatory material.

CHAPTER 2

ELEMENTARY LOGIC

2-1 Introduction. Logic may be described roughly as the theory of systematic reasoning. Symbolic logic is the formal theory of logic. There are many kinds of logic, but we shall consider only the logic that is most commonly used in mathematics and other sciences.

Symbolic logic has important applications in science and industry. In the *New York Times* of November 25, 1956, a well-known producer of electric products advertised for "men with ideas" to work in the field of electronics. The advertisement called for students of mathematics who had done "creative and original work in all fields of mathematics," and who had "an interest in the theory of numbers, theory of groups, Boolean algebra and symbolic logic...." In the following chapters the reader will get some inkling of why a great corporation is interested in such matters.

We introduce symbolic logic now for three reasons: (1) We can utilize its symbols and laws to simplify later work. (2) The axioms and proofs of elementary symbolic logic are simple and serve to illustrate the nature of a formal mathematical theory. (3) The laws and methods of logic will be useful to the reader in all his thinking in mathematics and other areas.

The purposes of this chapter are (1) to familiarize the student with the most important concepts and notations of symbolic logic, (2) to supply him with logical laws of wide applicability, (3) to develop his skill in reading formal mathematics, and (4) to apply logic to the algebra of real numbers.

2-2 Some simple logical formulas. Logical formulas are sentences whose variables stand for propositions. The purpose of this section is to familiarize the student with the following logical formulas.

	Logical formula	*Informal verbal synonym*
(1)	$\sim p$	It is false that p.
(2)	$p \wedge q$	p and q
(3)	$p \vee q$	p and/or q
(4)	$p \veebar q$	p or else q

We do not take the space constantly to suggest that the student carry out the operations listed in Section 1-17. Occasionally we ask questions, but the reader should still refer to the check list and work over the material on his own. Item 3 on the list indicates that sentences should be substituted for p and q in (1) through (4). (a) Carry out (p: I shall buy a car, q: I shall sell my old car) in (1) through (4). (b) What is the difference between (3) and (4)?

Negation. We call $\sim p$ the *negation* of p. It is the sentence that denies p. We read $\sim p$ as "It is false that p," "p is false," or "not-p." The negation of a sentence is the sentence that is false when the original is true, and true when the original is false. We indicate this in table (5), in which 1 means truth and 0 means falsity.

(5)

p	1	0
$\sim p$	0	1

When a proposition is true we say that it has the *truth value* truth (represented here by 1), and when it is false we say that its truth value is falsity (represented here by 0). Then table (5) indicates in its first row the possible truth values of p and in the second row the corresponding truth values of $\sim p$. It shows that p and $\sim p$ always have opposite truth values.

To find the negation of a sentence, we must find a sentence that conforms to *both* columns of (5). For example, "He is a good hunter" and "He is not a good hunter" are each the negation of the other, because if one is true the other is false. However, "He is a bad hunter" is not the negation of "He is a good hunter," because both might be false (if he is not a hunter at all!).

(c) Why, in view of the meaning we attach to "proposition," does (5) represent all possibilities? (d) Why is "It is white" not the negation of "It is black"? (e) Express $\sim p$ in several ways if $p = (x$ is happy). (f) Why is "~ 3" nonsense? (g) How do we usually express $\sim(a = b)$?

Conjunction. We call $p \wedge q$ the *conjunction* of p and q. It is the sentence that asserts *both* p and q. We read it as "p and q," "Both p and q are true," or "p is true and q is true." The conjunction of p and q is true when both p and q have truth values 1, and it is false otherwise. This is indicated in table (6), which gives the possible combinations of truth values of p and q in the first two rows and the corresponding truth values of $p \wedge q$ in the last row.

(6)

p	1	1	0	0
q	1	0	1	0
$p \wedge q$	1	0	0	0

(h) True or false? $(2 + 2 = 4) \wedge$ ("3" is a name of 3).   (i) What is the scope of $\wedge$ in (h)?   (j) Why is $(2 = 4/2) \wedge (0 = 1)$ false?   (k) Why is "$2 \wedge 5$" nonsense?

Disjunction, the inclusive "or." The word "or" in English has two distinct meanings. The exclusive "or" means "or else," as illustrated by "Either the Repubocrats will win or (else) I'll eat my hat!" The inclusive "or" means "and/or," as illustrated by "I intend to study French or (and/or) German." The inclusive "or" means one or both of the two possibilities; the exclusive "or" means one but not both. It turns out that the inclusive "or" is more frequently used in mathematics, and we adopt for it the symbol $\vee$.

We call $p \vee q$ the *disjunction* of p and q. It is a sentence that asserts that at least one of p and q is true. We read it as "p or q," "p and/or q," or "p or else q or else p and q," according to convenience. Table (7) shows how the truth value of $p \vee q$ depends on the truth values of p and of q.

(7)

p	1	1	0	0
q	1	0	1	0
$p \vee q$	1	1	1	0

(l) True or false? $(2 = 10/5) \vee (2 = 5)$.   (m) True or false? (George Washington was our first president) $\vee$ (John Hancock signed The Declaration of Independence).   (n) Why is "George $\vee$ Mary" nonsense?

Often we wish to state that one number, a, is less than or equal to another, b. We write $a \leq b$, defined as follows.

(8) Def. $[a \leq b] = (a < b) \vee (a = b)$.

(o) Why is $2 \leq 3$?   (p) Why is $2 \leq 2$?   (q) Define $a \geq b$

Exclusive disjunction. The exclusive "or" is not used as much as $\vee$, but it is occasionally convenient, and for this reason we include it here. We call $p \veebar q$ the *exclusive disjunction* of p and q. It is the sentence claiming that one and only one of p and q is true. We read it "p or else q," "One and only one of the following is true: p or q," or "p or q but not both." Table (9) indicates the way in which the truth value of $p \veebar q$ depends on the truth values of p and of q.

(9)

p	1	1	0	0
q	1	0	1	0
$p \underline{\vee} q$	0	1	1	0

(r) Why is "up $\underline{\vee}$ down" nonsense?   (s) Under what conditions is $p \vee q$ true but $p \underline{\vee} q$ false?   (t) Can $p \underline{\vee} q$ be true and $p \vee q$ false for the same truth values of p and q?   (u) Compare the meaning of "I will study mathematics $\vee$ I will study physics" with the same sentence with $(\vee : \underline{\vee})$.

Definitions. The logical symbols introduced above are not independent. We can take some of them as undefined and use these to define the others. For example, we note from (7) and (6) that $p \vee q$ is false only when p and q are both false, that is, when and only when $\sim p \wedge \sim q$ is true. Hence $p \vee q$ is true when and only when $\sim p \wedge \sim q$ is false, that is, when $\sim(\sim p \wedge \sim q)$ is true. This suggests the definition

(10) Def. $[p \vee q] = \sim(\sim p \wedge \sim q).$

Similarly (9) and (7) suggest

(11) Def. $[p \underline{\vee} q] = (p \vee q) \wedge \sim(p \wedge q).$

(v) In (10) and (11) what symbols are taken as undefined?   (w) Justify (11).

PROBLEMS

Translate 1 through 6 into words.

1. $\sim(2 > 3)$.
2. $(3 < 4) \wedge (4 < 5)$.
3. $(x = 3) \underline{\vee} (x = 4)$.
4. $(x = 3) \underline{\vee} (x < 0)$.
5. $(x < 0) \underline{\vee} (x = 0) \underline{\vee} (x > 0)$.
6. $\sim[x < 0 \wedge x > 0]$.

Translate 7 through 14 into symbols, using $\sim$, $\vee$, $\wedge$, $\underline{\vee}$.

7. 3 does not satisfy $x^2 = 10$.
8. 2 is less than 10, and 2 is a digit.
9. (1–13–16).
10. $2^2 = 2 \cdot 2$, but $3^2 \neq 2 \cdot 3$.
11. One and only one of the following holds: $a < b$, $a = b$, $a > b$.
12. $|5| = 5$, yet $|-5| \neq -5$.
13. $3 < 4$, however $-4 < -3$.
14. 0 is neither positive nor negative.

Suppose that p is true and q is false. Determine the truth value of 15 through 22.

15. $\sim q$.
16. $\sim p \lor q$.
17. $\sim(p \lor \sim q)$.
18. $p \land \sim q$.
19. $\sim p \lor \sim q$.
20. $\sim(p \lor q)$.
21. $p \underline{\lor} q$.
22. $\sim p \underline{\lor} q$.

23. Let $\bar{p}$ be the truth value of p. Then show from the truth tables that

(12)
$$\overline{\sim p} = 1 - \bar{p},$$

(13)
$$\overline{p \land q} = \bar{p} \cdot \bar{q},$$

(14)
$$\overline{p \lor q} = \bar{p} + \bar{q} - \bar{p} \cdot \bar{q},$$

(15)
$$\overline{p \underline{\lor} q} = |\bar{p} - \bar{q}|.$$

★24. Suppose we take $\sim$ and $\lor$ as undefined. Define $\land$.

★25. Define $\lor$ and $\land$ in terms of $\sim$ and $\underline{\lor}$.

Answers to Exercises

(a) (1) It is false that I shall buy a car. I shall not buy a car. (2) I shall buy a car and sell my old car. (3) I shall buy a car and/or sell my old car. (4) I shall buy a car or else I shall sell my old car. (b) The first means either or both, the second one but not both. (c) A proposition is either true or false but not both, by definition of "proposition." (d) Both might be false. (e) x is not happy; it is false that x is happy; not-(x is happy); (x is happy) is false. (f) Because "3" is not a sentence and hence not a significant substitute in (1). (g) $a \neq b$.

(h) 1. (i) "$2 + 2 = 4$" on left, "'3' is a name of 3" on right (j) See (6). (k) "2" and "5" are not sentences. (l) 1. (m) 1. (n) "George" and "Mary" are not sentences. (o) $2 < 3$. (p) $2 = 2$. (q) $[a \geq b] = \sim[a < b]$ or $[a \geq b] = (a > b) \lor (a = b)$. (r) "Up" and "down" are not sentences. (s) When p and q are both true. (t) No. (u) First means either or both, second either but not both. (v) $\sim$ and $\land$. (w) It says that $p \underline{\lor} q$ means that p is true or q is true but not both, which is what (9) indicates.

Answers to Problems

1. 2 is not greater than 3. 3. x is 3 or 4. 5. One and only one of the following holds: $x < 0$, $x = 0$, $x > 0$. 7. $\sim(3^2 = 10)$. 9. $[y = \sqrt{x}] = [y^2 = x \land y \geq 0]$. 12. $|5| = 5 \land |-5| \neq -5$. 15. 1. 17. 0. 19. 1. 21. 1. 23. From truth tables by considering all cases; or show (12) and (13), then use (10) and (11).

2–3 Implication. Sentences of the form "If p, then q" are very common in scientific discourse. The if-then idea is expressed in many ways, of which the following are synonymous examples.

(1) If p, then q.

(1′) q if p.

(2) p implies q.
(2′) q is implied by p.
(3) p only if q.
(4)   Hyp: p, Con: q.
(5) p is a sufficient condition that q.
(6) q is a necessary condition that p.

In everyday discourse such expressions are used with various meanings and connotations. In scientific discourse these sentences are synonymous and have a precise technical meaning. In mathematics a special symbol is usually used, the most common being an arrow pointing from the hypothesis to the conclusion. We shall adopt this notation and write

(7) $$p \rightarrow q$$

as a synonym for sentences (1) through (6).

The meaning we attach to $p \rightarrow q$ is indicated by the formal definition (12) below. To prepare for the definition we indicate the significant substitutes in (7), the conditions under which it is true or false, and the nature of the information that it may convey.

(8)   *The expression "$p \rightarrow q$" is a propositional formula in which significant substitutes for the variables are sentences and only sentences.*

(a) What does $p \rightarrow q$ become if for the variables we substitute statements? (b) numerals?   (c) names of people?   (d) Why is every value of $p \rightarrow q$ either true or false?

A statement of the form $p \rightarrow q$ is considered true under any one of the following three conditions:

(9)
p is true, and q is true.

p is false, and q is true.

p is false, and q is false.

It is considered false *only* in the following case:

(10) p is true, and q is false.

A statement of the form $p \rightarrow q$ makes the claim that one of the three possibilities listed in (9) is the case, but *it makes no other claim*. In particular, it says *nothing* as to whether p is true or false, as to whether q is true or false, as to the meanings of p and q, or as to the relation between these meanings.

We may summarize the above in terms of a table for $p \to q$.

(11)

p	1	1	0	0
q	1	0	1	0
$p \to q$	1	0	1	1

Which of the following are true?   (e) $(2 = 2) \to (3 = 3)$,   (f) $(2 = 3) \to$ $(4 < 1)$,   (g) $(2 = 3) \to (4 > 1)$,   (h) $(2 = 2) \to (2 = 3)$.

We note from (11) that $p \to q$ is false just in the one case when p is true and q false.   In other words, $p \to q$ is true just when $p \wedge {\sim}q$ is false, that is, when and only when ${\sim}(p \wedge {\sim}q)$ is true.   This suggests

(12) Def. $(p \to q) = {\sim}(p \wedge {\sim}q).$

We call $p \to q$ the *conditional* of p and q.   The special technical meaning assigned to $p \to q$ [and to the synonymous expressions (1) through (6)] by (12) may not seem entirely natural to the reader.   He may think that we are not following the criteria of (1–13–9).   It turns out that this idea of implication is entirely satisfactory for scientific purposes, is more in keeping with ordinary usage than first appears, and is more convenient than any alternative yet proposed.   However, other kinds of implication are considered by logicians, and the one defined by (12) is called *material implication* to distinguish it from the rest.

For each of the following, first decide whether it is true or false on the basis of the everyday meaning, then decide the same question on the basis of (11): (i) If the ocean is mostly water, then it contains about twice as many hydrogen atoms as oxygen atoms. (*Note:* We assume that the formula for water is H_2O, i.e., each molecule of water contains 2 hydrogen atoms and one oxygen atom.) (j) If the ocean is entirely grade-A milk, then it contains about twice as many atoms of hydrogen as atoms of oxygen.   (k) If the ocean is entirely grade-A milk, then ocean water is a nourishing beverage.   (l) If the ocean is mostly water, then ocean water is a nourishing beverage.

The previous exercises serve as examples to indicate that unless we permit $p \to q$ to be true under any one of the three conditions of (9), we shall find a marked contradiction between our technical meaning and ordinary usage.

A still further illustration of the advantages of (9) is the way in which it facilitates the statement of laws.   Consider, for example, the following law of elementary algebra (to be proved in Section 2–6):

(13) $(a = b) \to (ca = cb).$

We certainly are inclined to agree that this is a law, i.e., that if $a = b$, then $ca = bc$. Indeed, this is the law suggested by "If equals be multiplied by equals, the results are equal."

Now we recall that a law is a sentence that is true for *all* significant values of its variables. Since in (13) any number is a significant value of any of the variables, we are free to consider the following cases:

(14)　　$[2 = (6 - 4)] \rightarrow [3 \cdot 2 = 3 \cdot (6 - 4)]$　　$(13)(a{:}2, b{:}6 - 4, c{:}3),$

(15)　　$(2 = 3) \rightarrow (0 \cdot 2 = 0 \cdot 3)$　　　　　　$(13)(a{:}2, b{:}3, c{:}0),$

(16)　　$(2 = 3) \rightarrow (5 \cdot 2 = 5 \cdot 3)$　　　　　　$(13)(a{:}2, b{:}3, c{:}5).$

If (13) is a law, then (14) through (16) must each be true. In (14) both hypothesis and conclusion are true, and hence it is true by the first case in (9). In (15) the hypothesis is false and the conclusion true, and hence it is true by the second case. In (16) the hypothesis and conclusion are both false, and hence it is true by the third case in (9). We cannot find an example for which the hypothesis is true and the conclusion false, because (13) is a law. We see that unless we agree that $p \rightarrow q$ is true in all three cases in (9), we shall be unable to say that (13) is a law! The same is true of many other laws of the form $p \rightarrow q$.

(m) Discuss the law $(a = b) \rightarrow (a^2 = b^2)$ as we did (13).

In discussing and using the implication concept and the symbol $\rightarrow$, we shall make use of the different synonyms listed in (1) through (6). For this reason and because these synonyms appear very frequently in scientific discourse, it is important to be able to translate from any one form into any other, and particularly to and from the form $p \rightarrow q$. The essential thing is to think of the meaning and to recall that a hypothesis or sufficient condition is always at the heel of an arrow, whereas a conclusion or necessary condition is always at the point of an arrow, as indicated in Fig. 2-1. This reflects the fact that one argues from hypotheses (sufficient conditions) to conclusions (necessary conditions).

Sufficient condition　　　　　　　Necessary condition

FIGURE 2-1

Translate into each of the forms (1) through (6):　(n) (13).　(o) The sentence of Exercise (i).　(p) If two triangles are congruent, they are similar.

We call $q \rightarrow p$ the *converse* of $p \rightarrow q$. It is the sentence obtained by interchanging hypothesis and conclusion in $p \rightarrow q$. We call $p \leftrightarrow q$ the *biconditional* of p and q. It is the sentence that claims that $p \rightarrow q$ and $q \rightarrow p$; that is, $p \rightarrow q$ and its converse are both true.

(17) Def. $[p \leftrightarrow q] = [(p \rightarrow q) \wedge (q \rightarrow p)]$.

From this definition and (11) we easily get table (18):

(18)

p	1	1	0	0
q	1	0	1	0
$p \leftrightarrow q$	1	0	0	1

There are many ways to read $p \leftrightarrow q$. One is "$p \rightarrow q$ and conversely," and in this we may replace $p \rightarrow q$ by any one of the readings indicated in (1) through (6). Other forms are "p if and only if q," "q if and only if p," "p is a necessary and sufficient condition for q," "p has the same truth value as q," and "p is logically equivalent to q." The last form refers to the equality of the truth values of p and q when $p \leftrightarrow q$.

(q) True or false? $(2 = 4) \leftrightarrow (3^2 = 2 + 2)$, $(2 = 2) \leftrightarrow (2 = 3)$, $(24 = 2 \cdot 12) \leftrightarrow (2 = 2)$, $(1 = 2) \leftrightarrow (3 = 3)$.   (r) Why is "$2 \leftrightarrow 2$" nonsense? (s) Give an example from geometry of a law whose converse is not a law. (t) What conclusion can be drawn if $(p \leftrightarrow q)$ is true and q is true?   (u) and p is true?   (v) and p is false?   (w) and q is false?

PROBLEMS

In Problems 1 through 4 cite cases corresponding to the possibilities in (9).

1. If ABC is an equilateral triangle, then ABC is isosceles. (Draw figures for each case.)
2. $(x = y) \rightarrow (-x = -y)$.
3. $(AB \parallel A'B' \wedge AC \parallel A'C') \rightarrow (\angle BAC \cong \angle B'A'C')$.
4. $(x^2 = y^2) \rightarrow [(x = y) \vee (x = -y)]$.

In Problems 5 and 6 both the theorem and its converse are true. Cite cases to illustrate the possibilities. Why do you find only two?

5. $(ABC = 90°) \rightarrow (\overline{AB}^2 + \overline{BC}^2 = \overline{AC}^2)$ (The Pythagorean theorem).
6. $(a = b) \rightarrow (a + c = b + c)$.

★7. How are the following consistent with our definition of implication? "If to do were as easy as to know what were good to do, chapels had been churches and poor men's cottages princes' palaces." (*Merchant of Venice*, Act 1) "If all the year were playing holidays, to sport would be as tedious as to work." (*King Henry IV, Part 1*, Act 1)

Translate 8 through 23 into symbolic language.

8. There can be no great smoke arise, but there must be some fire.

9. There is no fire without smoke.

★10. "The existing economic order would inevitably be destroyed through lawless plunder if it were not secured by force." (*The Law as a Fact*, by K. Olivecrona, p. 137)

11. 3 is a root of $x^2 = 9$.

★12. In a good group you can't tell who the leader is.

★13. "In order for a country that imports capital goods to have a high rate of investment, it must have a large export industry." (*A Survey of Contemporary Economics*, Vol. II, p. 156)

14. "When an organism is conditioned to respond to one stimulus, it will respond in the same way to certain others." (*Principles of Psychology*, by F. S. Keller and W. N. Schoenfeld, p. 115)

★15. ". . . an adequate command of modern statistical methods is a necessary (but not sufficient) condition for preventing the modern economist from producing nonsense. . . ."

★16. "He is well paid that is well satisfied."

★17. I mean what I say.

★18. I say what I mean.

★19. "The surface of the leaf must be coated so as to prevent evaporation of the water that has been so laboriously gathered by the root system." (H. E. Stork, *Studies in Plant Life*, p. 36)

★20. You will make 6% provided the dividend is paid.

★21. He'll win as long as he's better.

22. "It has been shown . . . that the equality is necessary if the third and fourth marginal conditions are not to be violated."

23. It is known that the totally blind are able to detect objects at a distance. Although the blind often think they possess "facial vision" based on skin sensations, experiments have shown that such clues are neither necessary nor sufficient for the blind's perception.

★24. Restate each of Problems 8 through 23 in terms of necessary conditions.

★25. Restate 8 through 23 in terms of sufficient conditions.

★26. Explain the following: "The argument that no clear and present danger to American democracy now exists inside the country should not be taken to mean that no group would constitute a danger if it were powerful. For that would be a confusion of necessary and sufficient conditions." (R. G. Ross, "Democracy, Party, and Politics," *Ethics*, January 1954)

★27. One of the symptoms of tuberculosis is persistent coughing. Is coughing a necessary or a sufficient condition for tuberculosis? What can you say about medical symptoms generally?

★28. Read "What does 'if' mean?" in the *Mathematics Teacher* for January 1955.

★29. Does $p \rightarrow q$ mean that q follows p in time?

★30. Speaking of gamblers, Cardano (1501–1576) wrote, "If a man is victorious, he wastes the money won in gambling, whereas if he suffers defeat, then either he is reduced to poverty, when he is honest and without resources, or else to robbery, if he is powerful and dishonest, or again to the gallows, if he is poor and dishonest." Formalize and draw a conclusion. (Oystein Ore, *Cardano, The Gambling Scholar*, p. 187)

ANSWERS TO EXERCISES

(a) A statement.   (b) Nonsense.   (c) Same.   (d) By (8), and see Section 1–7.   (e) T.   (f) T.   (g) T.   (h) F.   (i) True, as is evident by arguing from the hypothesis to the conclusion; here p and q are true.   (j) True, as above, since milk is mostly water; here p is false, and q is true.   (k) True, since milk is nourishing; here p and q are false.   (l) False. In each of the first three cases the conclusion follows from the hypothesis by everyday reasoning. It does not in the last, since many liquids containing mostly water are not nourishing. Here p is true and q is false.   (m) $(a{:}2, b{:}2)$, $(a{:}2, b{:}{-}2)$, $(a{:}2, b{:}3)$.

(n) If $a = b$, then $ac = bc$. $a = b$ implies $ac = bc$. $a = b$ only if $ac = bc$. From $a = b$, it follows that $ac = bc$. Hyp: $a = b$, Con: $ac = bc$. $a = b$ is a sufficient condition that $ac = bc$. $ac = bc$ is a necessary condition that $a = b$. (o) The ocean is mostly water $\rightarrow$ it contains . . . . The ocean is mostly water implies it contains . . . . The ocean is mostly water only if it contains . . . . From the ocean is mostly water it follows that it contains . . . . Hyp: The ocean is mostly water. Con: It contains . . . . The ocean is mostly water is a sufficient condition that it contains . . . . (*Note:* The best procedure in performing translations of this kind is to first write the expression in the form $p \rightarrow q$, *being careful that p and q are sentences.* Then write in other forms and make whatever adjustments are required to conform to good English usage.)   (p) A sufficient condition that two triangles be similar is that they be congruent. A necessary condition that two triangles be congruent is that they be similar.   (q) T, F, T, F.   (r) "2" is not a sentence.   (t) p.   (u) q.   (v) $\sim q$.   (w) $\sim p$.

ANSWERS TO PROBLEMS

1. Cite an equilateral triangle, one that is isosceles but not equilateral, and a scalene triangle.   3. Cite congruent angles, placed with their sides ∥, congruent angles placed otherwise, and noncongruent angles with sides not ∥.

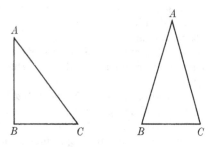

FIGURE 2–2

5. When the converse is also true we have only the possibilities of both hypothesis and conclusion true or both false by (18). Figure 2–2 shows two triangles ABC. In the first, the hypothesis and conclusion are both true; in the second they are both false. Hence, unless we wish to deny that the Pythagorean theorem is a law, we must agree that $p \rightarrow q$ is true when both p and q are false as well as when both p and q are true. This is an example of the convenience of the property of $\rightarrow$ embodied in the last column of (11). Of course, we do not care to apply a theorem when its hypothesis is false, but the meaning assigned to $\rightarrow$ and its various verbalizations makes the stating of laws much simpler.
9. There is fire $\rightarrow$ There is smoke. Fire is a sufficient condition for smoke. (Note here a violation of the agreement that only sentences be substituted for variables in $p \rightarrow q$. However, this is really an abbreviation for "A sufficient condition that there is smoke is that there is fire.") 11. $(x = 3) \rightarrow (x^2 = 9)$. 13. A country importing capital goods has a high investment rate $\rightarrow$ It has a large export industry. 27. Some are necessary but not sufficient (always present with the disease but also present at other times), some are sufficient but not necessary (their presence always indicates the disease but they may not always accompany it), some are both, and others are neither.

★2–4 Truth tables. A *truth table* of a logical formula is a table that shows the truth value of the formula that corresponds to each combination of truth values of its variables. Truth tables are not essential to the development of logic, but they are a convenient tool for investigating logical formulas and provide a procedure for testing any logical formula to see whether it is a law. If we set up truth tables as our criteria for establishing laws in logic, we could use them as a device for proof. We prefer to work with axioms instead and to use tables as an informal device outside the theory.

For example, is $p \rightarrow (p \wedge q)$ a law? To answer the question we make table (1):

(1)

p	1	1	0	0
q	1	0	1	0
$p \wedge q$	1	0	0	0
$p \rightarrow (p \wedge q)$	1	0	1	1

The third row comes from (2–2–6). The fourth row is obtained from the first and third by reference to (2–3–11). We see that the formula is not a law, since it is false when p is true and q false.

To be able to compare the truth tables of different formulas it is essential to adopt a standard form of construction. To this end we always write the first rows in the following way. If one variable is present:

(2)

p	1	0

If two variables are present:

(3)

p	1	1	0	0
q	1	0	1	0

If there are three variables:

(4)

p	1	1	1	1	0	0	0	0
q	1	1	0	0	1	1	0	0
r	1	0	1	0	1	0	1	0

Note that each of these tables is formed from the previous one by re-writing each column of the previous one twice, the first time with a 1 below it and the second time with a 0 below it. For example, the first column of (3) reads 1, 1; and the first two columns of (4) read 1, 1, 1 and 1, 1, 0. The second column of (3) reads 1, 0; and the third and fourth columns of (4) read 1, 0, 1 and 1, 0, 0.

(a) Using the indicated procedure, write the first four rows in a truth table involving four variables.

In Section 2–5 we assert a number of identities, each of which claims that two logical formulas are synonymous. If two logical formulas are synonymous, certainly both should be true or both false in any given instance. In other words, their truth tables should have the same entries in the last row. In symbols, $(p = q) \rightarrow (p \leftrightarrow q)$. However, the converse is not true, since $p \leftrightarrow q$ whenever p and q have the same truth value, and this may happen without p and q being equal.

(b) Give an example of the observation in the preceding sentence.

From the above discussion, we see that we can test an alleged logical identity by making truth tables for its members. If their last rows are different, we know the equation is not an identity. If the last rows are the same, we have verification but not proof of the identity.

Verify the following by truth tables: (c) (2–5–2),   (d) (2–5–4),
(e) (2–5–7),   (f) (2–5–8),   (g) (2–5–23),   (h) (2–5–28).

The truth table of a law of logic ought to contain only 1's in its last row. Consider, for example, the formula $p \lor \sim p$. Its table is (5).

(5)

p	1	0
$\sim p$	0	1
$p \lor \sim p$	1	1

Evidently this formula is a law. If we read it verbally, we have "p or not-p," "p is true or not-p is true," "p is true or false," or "any proposition is either true or false." We prove this in Section 2–7.

Make truth tables for the following laws: (i) (2–7–5),   (j) (2–7–9), (k) (2–7–23),   (l) (2–7–12).

In the following, change $=$ to $\leftrightarrow$ and make a truth table of the result: (m) (2–5–6),   (n) (2–5–15),   (o) (2–5–24).

Show that the following are not laws and indicate the truth values for which they fail: (p) $(p \to q) \to (q \to p)$,   (q) $(p \to q) \to (\sim p \to \sim q)$, (r) $((p \to q) \land \sim p) \to \sim q$.

A logical formula that is a law, or a sentence in the form of such a formula, is called a *tautology*. The negation of a tautology is called a *contradiction*.

(s) Show that $p \land \sim p$ is a contradiction.   (t) Show that $p \to \sim p$ is not a tautology. Show that it is not a contradiction!   (u) Why is "I went or I did not go" a tautology?   (v) Why is "I went and I did not go" a contradiction?   (w) Show that the following is a tautology: It is snowing in Denver and it is not raining in Nashville, or it is not snowing in Denver and it is hailing in Kansas, or it is not hailing in Kansas, or it is raining in Nashville.

Whenever a statement is in the form of a tautology we know that it must be true without further consideration of the truth of its parts. Surprisingly often people try to prove a tautology by arguments about its terms, without realizing that this is not necessary. Consider, for example, the following quotation from a newspaper editorial. "There may be justification for a subsidy—defense needs, for example—but it should be clearly understood that it is never economically justifiable. So the reasons for a subsidy must be strong enough to override the drawbacks." This sounds as though the second sentence follows from the first. But the second sentence is a disguised tautology. It says that if a subsidy is a good thing its advantages must outweigh its disadvantages, i.e., subsidy is justified $\to$ reasons for it must override its drawbacks. But what do we mean by saying that anything is justified except that the reasons for it override its drawbacks? In other words, by definition, x is justified $=$ reasons for x override reasons against x. Hence the second sentence is logically equivalent to $p \to p$, with $p =$ the subsidy is justified, and is therefore true quite independently of the first sentence.

Problems

1. Make truth tables for other laws in Sections 2–5 and 2–7.

Which of 2 through 6 are laws?

2. $\sim(p \lor q) \leftrightarrow (\sim p \lor \sim q)$.
3. $[p \land (q \lor r)] \leftrightarrow [(p \land q) \lor r]$.
4. $(p \to q) \to [p \to (p \land q)]$.
5. $(p \to q) \to [s \to (p \to q)]$.
6. $(p \to q) \to [(p \land r) \to (q \land r)]$.

7. Make a truth table for (2–7–39). Show that its converse is not a law.
8. Experiment with other logical formulas.
9. Why is this a disguised contradiction? "Our speaker was born in Manhattan, the first of his family to venture north of the Mason-Dixon line."
★10. Is the following verse by John Donne a tautology?

> *I am unable, yonder beggar cries,*
> *To stand, or move, if he say true, he lies.*

11. Is an argument based on a tautology necessarily a poor argument?
12. Ogden Nash once wrote, "I regret that before people can be reformed they have to be sinners." Why is the clause beginning "before..." true?
13. Why is "Either an electron is excited to the fullest extent, or it is not excited at all" not a tautology? Why would it be a tautology if "to the fullest extent" were deleted?
14. How many columns are there in a truth table for a formula involving n variables?

Answers to Exercises

(b) $(p{:}2 = 2, q{:}3^2 = 9)$.

(e)

p	1	1	0	0
q	1	0	1	0
$p \land q$	1	0	0	0
$\sim(p \land q)$	0	1	1	1
p	1	1	0	0
q	1	0	1	0
$\sim p$	0	0	1	1
$\sim q$	0	1	0	1
$\sim p \lor \sim q$	0	1	1	1

Note how we include an additional row for each formula that is a part of the final formula, then calculate each row in order.

(k)

p	1	1	0	0
q	1	0	1	0
$q \to p$	1	1	0	1
$p \to (q \to p)$	1	1	1	1

(p) Fails when q is true and p false.   (q) Same.   (r) Same.   (w) If p = it is snowing in Denver, q = it is raining in Nashville, r = it is hailing in Kansas, then the sentence is $(p \wedge \sim q) \vee (\sim p \wedge r) \vee \sim r \vee q$, which is easily shown to be a tautology by a truth table.

Answers to Problems

4, 5, and 6 are laws.   7. The converse fails for q true, p and r false.

2–5 Logical identities.   In this section we derive a number of useful identities from a few very plausible axioms. We let $\wedge$, $\sim$, and $=$ be undefined and define $\vee$, $\underline{\vee}$, $\to$, and $\leftrightarrow$ as in Sections 2–2 and 2–3. We specify that the following are sentences: $a = b$, $\sim p$, $p \wedge q$, $p \vee q$, $p \underline{\vee} q$, $p \to q$, $p \leftrightarrow q$, and expressions obtained by substituting for their variables other variables or sentences.

(1) **Ax.** $x = x$ (*Law of identity*),

(2) **Ax.** $p \wedge q = q \wedge p$,

(3) **Ax.** $p \wedge (q \wedge r) = (p \wedge q) \wedge r$,

(4) **Ax.** $p \wedge (q \vee r) = (p \wedge q) \vee (p \wedge r)$,

(5) **Ax.** $p \wedge p = p$,

(6) **Ax.** $\sim \sim p = p$.

(a) Make the substitution (p:You can walk, q:You can swim, r:You can fly) in (1) through (6).   (b) What is the scope of each $\sim$ in (6)?   (c) Suggest names for (2), (3), and (4).

In building any mathematical theory, we can list laws in many different orders, provided, of course, that each proof uses only previously established laws. The order here has been chosen to make the proofs as short as possible and to bring out the similarity of the properties of $\vee$ and $\wedge$.

(7) $\sim(p \wedge q) = \sim p \vee \sim q$,

(8) $\sim(p \vee q) = \sim p \wedge \sim q$.

(*De Morgan's laws*)

To prove Eq. (7) informally we note that $(2\text{-}2\text{-}10)(p{:}\sim p, q{:}\sim q)$ is $\sim p \lor \sim q = \sim(\sim \sim p \land \sim \sim q)$. But by (6) $\sim \sim p = p$ and $\sim \sim q = q$. Hence $\sim p \lor \sim q = \sim(p \land q)$. In more formal style,

(9) $\sim(p \land q) = \sim(\sim \sim p \land \sim \sim q)$ $(6)(p{:}q)$

(10) $= \sim p \lor \sim q$ $(2\text{-}2\text{-}10)(p{:}\sim p, q{:}\sim q)$.

Similarly, (8) is proved by

(11) $\sim(p \lor q) = \sim[\sim(\sim p \land \sim q)]$ $(2\text{-}2\text{-}10)$

(12) $= \sim p \land \sim q$ $(6)(p{:}\sim p \land \sim q)$.

(d) Rewrite this proof informally.

Laws (7) and (8), named after the English mathematician Augustus De Morgan (1806–1871), are useful in proving other laws and in stating the negations of compound sentences. The first law states that to deny a conjunction is to affirm the disjunction of the negations of its terms. The second states that to deny a disjunction is to assert the conjunction of the negations of its terms. To apply them, one must first state a compound sentence in the form $p \land q$ or $p \lor q$. For example, John and Jim are here = (John is here) $\land$ (Jim is here). Hence the negation is (John is not here) $\lor$ (Jim is not here), i.e., (John or Jim is not here).

Simplify: (e) $\sim$(We won and we are happy).   (f) $\sim$(We won or we are happy).   (g) $\sim$(John and Jim are 17).

State the negation of the following in two ways, first using logical symbols and then using familiar English: (h) John should honor his father and mother. (i) He is neither rich nor poor.

(13) $p \lor q = q \lor p,$

(14) $p \lor (q \lor r) = (p \lor q) \lor r,$

(15) $p \lor (q \land r) = (p \lor q) \land (p \lor r),$

(16) $p \lor p = p.$

Remember Section 1–17!   (j) Interchange $\lor$ and $\land$ in (13) through (16) and state your conclusion in words.   (k) What must be the key to proving (13) through (16)?   (l) Apply $\langle \lor{:}+, \land{:}\cdot, \sim{:}-\rangle$ to (1) through (6) and (13) through (16) and state your conclusion in words.

To prove (13), we write

(17) $$p \vee q = \sim(\sim p \wedge \sim q) \qquad (2\text{-}2\text{-}10)$$

(18) $$ = \sim(\sim q \wedge \sim p) \qquad (2)(p{:}\sim p, q{:}\sim q)$$

(19) $$ = q \vee p \qquad (2\text{-}2\text{-}10)(p{:}q, q{:}p).$$

(m) Give reasons in the following proof of (16).

(20) $$p \vee p = \sim(\sim p \wedge \sim p)$$

(21) $$ = \sim(\sim p)$$

(22) $$ = p.$$

The others [(14) and (15)] are proved similarly.

From the above identities it is easy to derive many others. We list below those that are most frequently used or are required for later proofs in this book.

(23) $$p \rightarrow q = \sim p \vee q,$$

(24) $$\sim(p \rightarrow q) = p \wedge \sim q,$$

(25) $$\sim(p \leftrightarrow q) = (\sim p \wedge q) \vee (p \wedge \sim q).$$

Because of (3) and (14), parentheses are usually omitted when only $\vee$ or $\wedge$ are involved.

(26) $$\sim(p \wedge q \wedge r) = \sim p \vee \sim q \vee \sim r,$$

(27) $$\sim(p \vee q \vee r) = \sim p \wedge \sim q \wedge \sim r.$$

State the negation of: (n) If we lose, I'll eat my hat. (o) All three of our teams won today. (p) At least one of our three teams won today. (q) He will graduate if and only if he passes this test. (r) "There's neither honesty, manhood, nor good fellowship in thee." (*King Henry IV*)

(28) $$p \rightarrow q = \sim q \rightarrow \sim p.$$

Since $\sim q \rightarrow \sim p$ is called the *contrapositive* of $p \rightarrow q$, (28) asserts that an implication and its contrapositive are synonymous. To prove it, we note by (23) that $(p \rightarrow q) = (\sim p \vee q) = (q \vee \sim p) = (\sim \sim q \vee \sim p) = (\sim q \rightarrow \sim p)$, the last step being justified by $(23)(p{:}\sim q, q{:}\sim p)$.

(s) Write out the preceding proof with reasons. (t) Illustrate (23) for (p:He wins, q:I lose).

To prove an implication, we very often begin by assuming the negation of the conclusion and arguing to the negation of the hypothesis. This procedure is completely justified by (28). For example, in economics there is a theorem to the effect that when resources are allocated so as to yield maximum output, the marginal product (the additional output that would result from an additional unit of input) is the same for all enterprises. In symbols, Output is maximum → Marginal products are equal. The economist argues for this as follows. Suppose that marginal products were not equal. Then an additional unit of input in one enterprise would result in greater output than would be lost by removing this unit from another location. Since output can be increased by shifting resources, it is not maximum. What the economist has proved by this argument is that ∼ (Marginal products are equal) → ∼ (Output is maximum).

(u) If no tracks are observed in snow we usually conclude that no one has passed by. Why is this legitimate?

The following are listed for future reference:

(29) $p \leftrightarrow q = q \leftrightarrow p,$

(30) $[p \rightarrow (q \rightarrow r)] = [(p \wedge q) \rightarrow r],$

(31) $[p \rightarrow (q \rightarrow r)] = [q \rightarrow (p \rightarrow r)],$

(32) $[p \rightarrow (q \rightarrow r)] = [\sim r \rightarrow (p \rightarrow \sim q)],$

(33) $[(p \wedge q) \rightarrow p] = [q \rightarrow (p \vee \sim p)],$

(34) $[p \rightarrow (q \rightarrow [p \wedge q])] = [\sim(p \wedge q) \vee (p \wedge q)],$

(35) $[(p \wedge q) \rightarrow r] = [(p \wedge \sim r) \rightarrow \sim q],$

(36) $[(p \rightarrow q) \wedge (p \rightarrow r)] = [p \rightarrow (q \wedge r)].$

Rewrite in as simple form as possible: (v) $\sim p \rightarrow q,$ (w) $(p \rightarrow q) \vee (\sim p \rightarrow \sim q),$ (x) $\sim(p \wedge \sim q),$ (y) $\sim[p \wedge (q \vee r)],$ (z) $\sim(\sim p \vee \sim(\sim p \vee \sim(\sim a \vee \sim b))).$

PROBLEMS

Derive the identities in Problems 1 through 6.

1. $\sim p = \sim \sim \sim p.$ 2. $\sim p \wedge \sim p = \sim p.$
3. $\sim(\sim p \wedge q) = p \vee \sim q.$ 4. $\sim(p \vee \sim q) = \sim p \wedge q.$
5. (23).   6. $[p \rightarrow \sim q] = [q \rightarrow \sim p].$

In Problems 7 through 17 write the negation in good English after first writing it in logical symbols.

7. I am rich and happy.

8. John or Jim is mistaken.

9. Mathematics and physics are sciences.

10. Neither of you is right.

11. If it rains today, I'll stay home.

12. One and only one of you may go.

13. 7 and 14 are primes.

14. John, Jack, and Jim are 17.

15. New York is a big city if and only if Chicago is not a big city.

16. If he wins, I can't lose.

17. Both countries are at fault.

18. Assuming that if one cannot do mathematics he cannot be an engineer, show that if one is able to be an engineer he must be able to do mathematics.

19. Argue that if an enterprise is engaging in various activities in such a way as to maximize its profits, the additional revenue from increasing expenditure by one unit is the same in all activities.

20. Prove $a \neq 0 \rightarrow -a \neq 0$.

★21. Investigate the properties of $\underline{\vee}$, looking for laws involving it. Is it associative? Distributive over $\vee$? What is the negation of $p \underline{\vee} q$? Show that $p \underline{\vee} q = p \leftrightarrow \sim q$.

★22. Start with $\rightarrow$ and $\sim$ as undefined and define the other logical symbols in terms of them.

★23. Suppose we take p/q as our only basic term, where we take it to mean that it is false that both p and q are true, that is, $p/q = \sim(p \wedge q)$. Make a table for p/q as we did for other formulas in Section 2–2. Then define all other formulas in terms of "/". (*Note:* This is called the *stroke* and is of interest because it shows that all logical formulas can be defined in terms of just one.) Work out some laws involving the stroke.

★24. What does "p unless q" mean? Is it synonymous with $\sim q \rightarrow p$ or with $\sim q \leftrightarrow p$?

★25. Every logical identity involving only $\vee$, $\wedge$, and $\sim$ remains an identity if $\vee$ and $\wedge$ are interchanged. Verify this in particular cases. Prove that it is true for all identities derivable from our axioms.

ANSWERS TO EXERCISES

(a) There are various possibilities; for example, one rendering of (2) is: To say that you can walk and you can swim is the same as to say that you can swim and you can walk. (b) $\sim p$ and p. (c) Commutative, associative, and distributive laws. (e) We did not win and/or we are not happy. (f) We did not win and we are not happy. (g) John and/or Jim is not 17. (h) John does not have to honor both his parents, or (John need not honor his father) $\vee$ (John need not honor his mother). (i) He is either rich or poor. (j) The results are all laws of the theory. (k) The relation between $\wedge$ and $\vee$ embodied in (2–2–10), (7), and (8). (l) All but (5), (15), and (16) are laws of elementary algebra. Evidently the algebra of logic has *some* similarities to the algebra of numbers, but it *is* different.   (m) $(2\text{-}2\text{-}10)(q{:}p)$, $(5)(p{:}{\sim}p)$, (6). (n) We'll lose and I won't eat my hat.   (o) At least one of them lost.   (p) None of them won.   (q) He will graduate and not pass, or he will not graduate and will pass.   (r) Thou hast some honor, manhood, or good fellowship in thee, and possibly more than one of these virtues.   (s)

$$(37) \qquad\qquad p \to q = {\sim}p \lor q \qquad\quad (23)$$

$$(38) \qquad\qquad\qquad\quad = q \lor {\sim}p \qquad\quad (13)(p{:}{\sim}p)$$

$$(39) \qquad\qquad\qquad\quad = {\sim}\,{\sim}q \lor {\sim}p \qquad (6)(p{:}q)$$

$$(40) \qquad\qquad\qquad\quad = {\sim}q \to {\sim}p \qquad (23)(p{:}{\sim}q,\ q{:}{\sim}p).$$

(u) (28).   (v) $p \lor q$.   (w) $p \lor {\sim}p \lor q \lor {\sim}q$.   (x) $p \to q$.   (y) ${\sim}p \lor ({\sim}q \land {\sim}r)$.   (z) $p \land ({\sim}p \lor (a \land b))$.

Answers to Problems

1. $(6)(p{:}{\sim}p)$.   3. $(2\text{-}2\text{-}10)(q{:}{\sim}q)$.   5. Apply $(2\text{-}5\text{-}7)$ to $(2\text{-}3\text{-}12)$.   7. I am either not rich or not happy.   9. Not both mathematics and physics are sciences. Either one or the other or both are not sciences.   11. It rains today and I don't stay home.   13. Either 7 or 14 is not prime.   15. New York and Chicago are big cities or they are both not big cities.

2–6 Rules of proof. Everyone is familiar with the fact that what seems reasonable, even obvious, to one person may seem quite unreasonable and obscure to another. An argument that convinces some may not appear at all convincing to others. Nevertheless, in science, and particularly in mathematics, we wish to prove laws in such a way as to obtain universal agreement. How can this be done in view of the diversity of opinions and experiences of men?

To see the answer to this question, let us imagine that we wish to convince someone of the correctness of a certain statement. We might try to convince him by citing some authority in which he believes (sometimes called "proof by intimidation"), by getting him to "see" ("proof by intuition"), by appealing to his emotions ("proof by waving the red flag"), by confusing and tricking him with words and faulty logic ("elastic inference"), or by other methods familiar to everyone. These methods have two serious disadvantages: they can be used to establish false statements just as easily as to establish true ones, and they do not lead to universal agreement.

If we want a more satisfying procedure we must begin by assuming some axioms, as indicated in Section 1–16. But what if our listener does not agree that our axioms are truly laws? Is there no way out if he will not accept the axioms we propose? There *is* a way out; namely, we may ask our listener to agree merely that *if* our axioms are indeed laws, *then* our theorems are also laws. If we take this tack, we eliminate argument about the truth of both the axioms and the theorems. We ask our listeners merely to agree that the theorems really do follow from the axioms! This is precisely the procedure we follow in mathematics. The mathematician claims only that the theorems follow from the axioms, and he has no objection if someone prefers to adopt different axioms.

But suppose that our listener objects to the manner in which we derive the theorems from the axioms? Then we shall have to come to some agreement with him as to what procedures are legitimate. And if we cannot agree with him on method of proof? Then we take the same way out as before. We say to him, "Let us merely agree that if these axioms are accepted and if these methods of proof are used, then these theorems can be obtained." In constructing a formal mathematical theory we state in advance the acceptable methods of deriving theorems from axioms. Then all we claim is that the theory is derived from the axioms according to these rules. If the axioms and the rules are acceptable to anyone, the theorems should be also. If the axioms are applicable to any particular situation (i.e., are true in a particular case), the theorems may be applied to that situation.

Our first two rules of proof are the Rule of Replacement and the Rule of Substitution introduced in Section 1–11 and used frequently since. For convenience of reference we recapitulate them here.

(1)　　RULE OF REPLACEMENT: *If a term in an expression is replaced by a synonym, the resulting expression is synonymous with the original. If the original is a law, so is the result.*

To use the first part of this rule we need to cite the law that asserts the synonymity of the term and its replacement. Thus we write

$$(2) \qquad a + b = -(-a) + b \qquad \text{Rep, (1-9-33)}.$$

The reason indicates that a replacement, justified by (1–9–33), has been made in the left member to get the right member. To use the second part of the rule we must have two previously established laws, the law in which we intend to make the replacement and the law asserting that the replacement is a synonym of the term it replaces. We use the abbreviation "Rep" followed by a reference to the law in which the replacement is made and to the law asserting the synonymity. For example,

$$(3) \qquad a + b = b + (a + 0) \qquad \text{Rep, (1-9-7), (1-9-22)}.$$

The reason on the right indicates that (3) was obtained by replacement in (1–9–7) of a by $a + 0$, which is a synonym according to (1–9–22).

(4)   RULE OF SUBSTITUTION: *If a sentence is obtained by making significant substitutions for the variables in a law, the result is a law.*

When applying this rule we use the abbreviation "Sub" followed by reference to the law in which the substitution is made and an indication of the substitution. For example,

(5) $(b + c)a = a(b + c)$ Sub, $(1–9–14)(a{:}b + c, b{:}a)$.

Recall that substitutions are permitted *only for variables* and that *they must be made throughout*, whereas replacements may be made for any term and need not be made throughout.

(a) Review Section 1–11.   (b) Review previous uses of these rules.

Our next rule is based on the meaning of implication indicated in Section 2–3. As indicated in (2–3–11), when $p \to q$ and p are true, q must be true. Schematically,

(6)
$$p \to q$$
$$\frac{p}{\therefore q} \ .$$

We embody this idea in the following rule of proof.

(7)   RULE OF INFERENCE: *If an implication is a law and its hypothesis is a law, then its conclusion is a law.*

It is this rule that enables us to detach a conclusion and state it separately when we know that a hypothesis from which it follows is true. When using it we write "Inf" followed by reference to the implication and the hypothesis. For example, from

(8) $2 = \sqrt{4} \to 3 \cdot 2 = 3\sqrt{4}$ Sub, $(2–3–13)(a{:}2, b{:}\sqrt{4}, c{:}3)$,

(9) $2 = \sqrt{4}$ (assumed here),

(10) $3 \cdot 2 = 3\sqrt{4}$ Inf, (8), (9).

(c) Would we get a satisfying rule of proof by interchanging "hypothesis" and "conclusion" in (7)? Explain.

If every law were stated as a theorem, axiom, or definition, each proof would consist of a single statement justified by one of the above rules. However, many laws are not worth displaying as theorems even though they are needed to prove other more interesting laws. Accordingly, a proof usually consists of several steps, each one justified by a rule of proof.

A *complete formal proof* is a sequence of steps such that:

> I. *Each step is a sentence.*
>
> (11)　II. *Each step is an established law (axiom, definition, or previously proved theorem) or is the result of applying a rule of proof to a previous step or to an established law.*
>
> III. *The last step is the theorem to be proved.*

A *formal proof* is a complete formal proof or an abbreviation of one obtained by omitting or consolidating steps and reasons. An *informal proof* is an expression of the steps of a formal proof in paragraph style by using everyday language. In this book we use the word "proof" to apply only to formal or informal proofs. "Show" or "argue for" are used for other discussions tending to convince. We call such arguments *heuristic discussions* or *plausibility arguments*. (See Section 2–12.)

(d) Cite examples of different kinds of proofs given in this book so far.

If we relied only on the three rules of proof given above, proofs would be unreasonably long and cumbersome. It is therefore customary to cut proofs short by omitting steps, abbreviating reasons, and adopting further rules of proof. Such measures are considered perfectly legitimate provided they really do amount merely to ways of abbreviating proofs that *could* be carried out by the use of Rep, Sub, and Inf.

One very useful rule of proof, known to the reader from high-school plane geometry, applies to proving laws of the form $p \rightarrow q$. This rule permits inserting p as a step "by hypothesis." Then if q can be derived from this assumption, we consider that $p \rightarrow q$ is proved. For example, consider the theorem

(12)　　*If three sides of one triangle are equal in length respectively to three sides of another, the triangles are congruent.*

When we use letters for variables, this becomes

(12′)　$[\overline{AB} = \overline{XY}, \ \overline{BC} = \overline{YZ}, \ and \ \overline{CA} = \overline{ZX}] \rightarrow [\triangle ABC \cong \triangle XYZ].$

Here $\overline{AB}$ means the length of the segment AB. In proving the theorem,

we assume the equations in the hypothesis and then reason step by step
to the conclusion. When the conclusion is reached, the proof of (12) is
considered complete, and we write "Q.E.D.," an abbreviation for a Latin
phrase meaning "that which was to have been demonstrated."

In using this method of proof, we treat the variables in the hypothesis
as though they were constants. For example, the above theorem is supposed to hold for all values of A, B, C, X, Y, Z (that is, for all points),
but when we assume $\overline{AB} = \overline{XY}$ by hypothesis in the proof, we do not
assume that $\overline{AB} = \overline{XY}$ for all A, B, X, Y. To assume this would be to
assume that all segments are equal! What we do, rather, is the following.
We imagine A, B, C, X, Y, Z to be definite but unspecified points for which
the hypothesis holds; i.e., we treat the symbols as constants. Then we
show from this assumption that the conclusion holds for these points.
However, since we have made no special assumption about these points,
the same reasoning would enable us to reason from hypothesis to conclusion in every case. Hence the theorem is proved, since we are assured that
its conclusion is true whenever its hypothesis is true, which is all that is
required by (2-3-11).

> RULE OF HYPOTHESIS: *In order to prove an implication,*
> *its hypothesis may be introduced as a step and treated as a law,*
> (13) *except that the Rule of Substitution may not be applied to its*
> *variables. If the conclusion appears as a step, the implication*
> *is proved.*

When using this rule, we write "Hyp" after an assumed hypothesis.
When the conclusion is obtained, it is justified by whatever rule is appropriate. Then the implication is justified by writing Q.E.D. and referring
to the steps asserting the hypothesis and conclusion. Or we may omit
the statement of the implication and just write Q.E.D. when the conclusion is justified. Note that *any* sentence may be assumed by Hyp as a
step. Note, also, that a step assumed by Hyp and steps derived from it
are not necessarily laws. For this reason, substitution is not permitted
in such steps; however, Rep and Inf can be applied to them.

To illustrate the use of Hyp, we prove several laws of elementary algebra in continuation of Section 1-16.

(14) $(a = b) \rightarrow (b = a)$.

(15)   Proof of (14):

   (a) $a = a$ Sub, (2-5-1)$(x{:}a)$,

   (b) $a = b$ Hyp,

(c) $b = a$ Rep, (a), (b),

(d) $(a = b) \rightarrow (b = a)$ Q.E.D., (b), (c).

Note that (c) was obtained by replacing the first a in (a) by b, which is a synonym according to the assumption (b). We could have omitted step (d) and simply written Q.E.D. after step (c).

(e) Similarly prove one of the following.

(**16**) $(a = b) \rightarrow (a + c = b + c)$,

(**17**) $(a = b) \rightarrow (ac = bc)$,

(**18**) $(a = b) \rightarrow (-a = -b)$,

(**19**) $(a = b) \rightarrow (a - c = b - c)$,

(**20**) $(a = b) \rightarrow (1/a = 1/b)$,

(**21**) $(a = b) \rightarrow (a/c = b/c)$,

(**22**) $(a = b) \rightarrow (a^2 = b^2)$,

(**23**) $(a + c = b + c) \rightarrow (a = b)$,

(**24**) $(c \neq 0 \wedge ac = bc) \rightarrow (a = b)$,

(**25**) $(a = b \wedge c = d) \rightarrow (ac = bd \wedge a + c = b + d)$.

(f) Does (20) hold for all numbers a and b?   (g) Does (21) hold for any number c?   (h) In (15), which steps are laws?   (i) Prove that $2 = 4 \rightarrow 10 = 20$.   (j) Could we use (i) and Inf to prove that $10 = 20$?

An argument (or proof) is called *valid* when it proceeds by applying laws of logic and rules of proof, or could be justified by reference to such laws and rules. The reader should note that any law of logic may be the basis of a valid argument. He should also notice that the validity of an argument does not depend on the truth or falsity of its premises or conclusions. One cannot argue validly from true premises to false conclusions, but one can argue validly from false premises to either true or false conclusions. On the other hand, because a man's conclusions are correct it does not follow that his argument is valid.

(k) Give examples to illustrate the comments in the preceding paragraph. Show that the following two arguments are valid: (l) If the price of butter increases, demand decreases. The price has increased. Hence the demand has decreased.   (m) Free competition leads to price cutting and maximum output.   Price cutting is rare in our economy and output is usually below capacity. Hence free competition is not universal.   (n) Kant held that all proofs of the existence of God were fallacious. Does this show that he was an atheist?

An argument that is not valid is called *invalid*. A line of reasoning that is contrary to logical laws is called a *fallacy*. Many fallacies are based on misapplication of laws of logic or application of a logical formula that is not actually a law. Fallacies are often difficult to detect, but it is usually helpful to restate (or attempt to restate) a suspect argument in symbolic form. It should be kept in mind that the word "fallacy" refers to reasoning and not to the premises or conclusion taken alone.

(o) Collect and/or construct examples of invalid arguments and explain the fallacies involved.

When we construct a theory on the basis of explicitly stated axioms and rules of proof, we have not really eliminated all possibility of controversy. For one thing, there may be differences of opinion as to whether we have correctly applied the rules of proof. But this is just a question of whether we have made a mistake. Mistakes may be hard to find, but such differences of opinion can be solved by sufficiently careful examination of the theory. Unsolvable disagreement is still possible, however, on whether the axioms and rules of proof should have been adopted at all, whether the theory that results from them is a good theory, and so on. Such questions are not answered by the theory itself, but they are placed outside the theory when we agree to argue on the basis of the axioms and methods of proof. Hence we may expect universal agreement *within* the theory, but no universal agreement *about* it. To settle arguments about a theory we should have to construct a second theory about it. A theory about a theory is called a *metatheory*. Of course, there would remain areas of possible disagreement about the assumptions of any metatheory. Evidently we cannot eliminate disagreement or controversy, but we can construct a theory in such a way that disagreement is possible only about certain parts of it, namely the axioms and methods of proof. This is a great advantage because it leads to universal agreement over a considerable area, avoids arguing about matters that can be agreed upon, and identifies the really controversial issues.

Problems

1. Suppose we have proved the Pythagorean theorem: $\angle ABC$ is a right angle $\rightarrow \overline{AB}^2 + \overline{BC}^2 = \overline{AC}^2$. Now suppose that we have a particular triangle XYZ in which $\angle XYZ$ is a right angle, $\overline{XY} = 7$ and $\overline{YZ} = 5$. Give an informal proof that $\overline{XZ}^2 = 5^2 + 7^2$. Indicate the rules of proof used.

2. Prove one of (16) through (22) with a complete formal proof. Rewrite it in informal style.

3. Prove others in the list (16) through (22).

4. Prove that if the ocean is lemonade it contains citric acid, indicating your rules of proof.

5. Accepting as a fact that the ocean contains salt, prove it contains sodium, indicating rules of proof.

6. What rules of proof were used in Section 2-5? Insert reasons in the following complete proof of (2-5-7).

(a) $\sim(p \wedge q) = \sim(p \wedge q)$,
(b) $\sim\sim p = p$,
(c) $\sim\sim q = q$,
(d) $\sim(p \wedge q) = \sim(\sim\sim p \wedge \sim\sim q)$,
(e) $\sim p \vee \sim q = \sim(\sim\sim p \wedge \sim\sim q)$,
(f) $\sim(p \wedge q) = \sim p \vee \sim q$.

7. Similarly prove (2-5-8).

8. Give a complete proof of (2-5-28).

9. Explain why the following schema is valid, and state a possible rule of proof based on it.

$$(26) \qquad \begin{array}{c} p \to q \\ \sim q \\ \hline \therefore \sim p \end{array} \text{. (valid)}$$

10. Explain why the following schemata are invalid.

$$(27) \qquad \begin{array}{c} p \to q \\ \sim p \\ \hline \therefore \sim q \end{array} \text{. (fallacious!)} \qquad \begin{array}{c} p \to q \\ q \\ \hline \therefore p \end{array} \text{. (fallacious!)}$$

11. From $1.5 = 3/2$ and the law in (22), by what pattern can you conclude that $(1.5)^2 = (3/2)^2$?

12. If two lines are parallel, they do not meet. If two particular lines do meet, what conclusion can you draw and why?

13. Suppose you know that the squares of two numbers are equal. Can you draw from (22) the conclusion that the numbers are equal? Explain.

14. Suppose two numbers are not equal. Can you draw the conclusion that their squares are not equal by relying on (22)? Explain.

15. Prove (23) by using $(13)(a{:}a + c, b{:}b + c, c{:}{-}c)$.

16. Similarly prove (24).

17. Show that $(ac = bc) \to (a = b)$ is *not* a law. Detect the error in the following *fallacious* proof. By $(17)(a{:}ac, b{:}bc, c{:}1/c)$ we have $ac = bc \to ac(1/c) = bc(1/c)$ or $ac = bc \to a = b$.

18. Show that the converse of (22) is not a law.

★19. Prove

$$(28) \qquad (p = q) \to (\sim p = \sim q),$$

$$(29) \qquad (p = q) \to (p \wedge r = q \wedge r).$$

20. What conclusions can you draw from $a^2 \neq b^2$?

21. If we know that an argument is valid and that it leads to a false conclusion, what can we conclude?

22. If we know that an argument leads from correct assumptions to false conclusions, what can we conclude?

ANSWERS TO EXERCISES

(c) No; from $p \rightarrow q$ and q, p does *not* follow.   (d) In Section 1–16 we gave informal and formal, but not complete formal proofs.   (e) (16) $a + c = a + c$; $a = b$; $a + c = b + c$; $a = b \rightarrow a + c = b + c$.   (f) Not for $a = b = 0$; but (20) is a law as it stands, since a law is a sentence all of whose values are true, and (20) is not a sentence for $(a{:}0, b{:}0)$.   (g) Not for $c = 0$.   (h) (a) and (d) only.   (i) $(17)(a{:}2, b{:}4, c{:}5)$.   (j) No, since we do not know that $2 = 4$.

ANSWERS TO PROBLEMS

1. We have given as true that $\angle XYZ$ is a right angle, $\overline{XY} = 7$, $\overline{YZ} = 5$. Now $\angle XYZ$ is a right angle $\rightarrow \overline{XY}^2 + \overline{YZ}^2 = \overline{XZ}^2$ by Sub, (Pythagorean theorem)$(A{:}X, B{:}Y, C{:}Z)$. Also, XYZ is a right triangle by Hyp. Hence $\overline{XY}^2 + \overline{YZ}^2 = \overline{XZ}^2$ by Inf and the two previous steps. Hence $7^2 + 5^2 = \overline{XZ}^2$ by Rep. The equation to be proved then follows by Sub, $(14)(a{:}7^2 + 5^2, b{:}\overline{XZ}^2)$ and Inf. 2. (16) By the law of identity, $a + c = a + c$. Since $a = b$ by hypothesis, we may replace a by b in the right member to get $a + c = b + c$, which is the desired conclusion. 4. Use Hyp and others. 5. Use Inf and others. 9. (2–5–28) and Inf. 17. The proof is informal and incomplete, but it would still be in error if omitted steps were inserted. The point is that "$1/c$" is undefined for $c = 0$. Hence the substitution is not significant unless $c \neq 0$. Since $ac = bc \rightarrow a = b$ is still a sentence when $c = 0$, we must exclude this possibility by inserting $c \neq 0$ in the hypothesis, as we have done in (24).

2–7 Laws of implication. We now use our additional rules of proof to prove theorems that enable us to reason from a sentence to another that is not synonymous with it.

(1) $p \rightarrow p$ (Law of tautology).

This law is so obvious that it would seem out of the question to prove it. However, it can be proved by use of the Rule of Hypothesis.

(2)   Proof of (1):

    (a) p Hyp,

    (b) $p \rightarrow p$ Q.E.D., (a).

Note that (b) is justified, since we have assumed p and so, of course, obtained p as a step, in conformity with (2–6–13)!

(3) $(p = q) \rightarrow (p \rightarrow q).$

This is proved by assuming $p = q$ by hypothesis and then replacing p by q in the right member of (1).

(a) Does it follow from (3) that $4 \rightarrow 2^2$ since $4 = 2^2$?   (b) Prove from (3) that $\sim \sim p \rightarrow p$.   (c) Derive (4) and (5) from (1) by using (2–5–23) and (2–5–7).

(4) $p \vee \sim p$ (Law of excluded middle),

(5) $\sim (p \wedge \sim p)$ (Law of contradiction),

(6) $(p \wedge q) \rightarrow p.$

To prove (6) it is convenient to first prove $q \rightarrow (p \vee \sim p)$. Since this law is not important except for the purpose of proving (6), we do not list it as a theorem. Instead we call it a *lemma*.

(7)   LEMMA: $q \rightarrow (p \vee \sim p).$

    Proof:

    (a) q Hyp,

    (b) $p \vee \sim p$ (4),

    (c) $q \rightarrow (p \vee \sim p)$ Q.E.D., (a), (b).

(8)   Proof of (6):

    (a) $[(p \wedge q) \rightarrow p] = [q \rightarrow (p \vee \sim p)]$ (2–5–33),

    (b) $(p \wedge q) \rightarrow p$ Rep, (7), (a).

(9) $p \rightarrow (p \vee q).$

★(d) Prove (9) by showing that it is synonymous with (6)$(q{:}\sim q)$.   ★(e) Show by a truth table that (6) and (9) are not laws if we change $\rightarrow$ to $\leftrightarrow$. (f) Show that the converse of (9) is not a law.   (g) Prove that $(p \wedge q) \rightarrow q$ is a law by showing that it is synonymous with (6)$(p{:}q, q{:}p)$.

Although we cannot include the proofs of all theorems, we list those that are most interesting and not too complicated, and in an order such that each can be proved conveniently by relying on those already listed.

(10) $$p \rightarrow [q \rightarrow (p \wedge q)].$$

Theorem (10) follows from (2–5–34) and (4). It permits us to infer that $p \wedge q$ is true if we know that two propositions p and q are each true, for by (10), from p and Inf we can conclude that $q \rightarrow (p \wedge q)$. Then from q and Inf we can conclude $(p \wedge q)$.

(11) $$(p \rightarrow q) \rightarrow [(p \vee r) \rightarrow (q \vee r)].$$

Theorem (11) enables us to "add" the same proposition to both members of an implication.

(12) $\quad [(p \rightarrow q) \wedge (q \rightarrow r)] \rightarrow (p \rightarrow r) \quad$ (Law of syllogism).

Theorem (12) is the principle of logic that permits us to construct chains of reasoning. Thus, if we know $a \rightarrow b, b \rightarrow c, c \rightarrow d, \ldots, x \rightarrow y$, and $y \rightarrow z$, then we may conclude by using (12) repeatedly that $a \rightarrow c$, $a \rightarrow d, a \rightarrow x, a \rightarrow y$, and finally that $a \rightarrow z$. Arguments based on it are called "syllogistic."

★**(13)** Proof of (12):

(a) $(\sim q \rightarrow \sim p) \rightarrow [(\sim q \vee r) \rightarrow (\sim p \vee r)] \qquad$ Sub,
$$(11)(p{:}\sim q, \, q{:}\sim p),$$

(b) $(p \rightarrow q) \rightarrow [(\sim q \vee r) \rightarrow (\sim p \vee r)] \qquad$ Rep, (a), (2–5–28),

(c) $(p \rightarrow q) \rightarrow [(q \rightarrow r) \rightarrow (p \rightarrow r)] \qquad$ Rep, (b), (2–5–23),

(d) $[(p \rightarrow q) \wedge (q \rightarrow r)] \rightarrow (p \rightarrow r) \qquad$ Rep, (c), (2–5–30).

(14) $$[p \wedge (p \rightarrow q)] \rightarrow q,$$

(15) $$[\sim q \wedge (p \rightarrow q)] \rightarrow \sim p,$$

(h) How are laws (14) and (15) related to the discussion in Section 2–6?

(16) $$p \wedge (q \vee r) \rightarrow (p \wedge q) \vee r,$$

(17) $$p \vee (q \wedge r) \rightarrow (p \vee q).$$

Laws (16) and (17) illustrate how implication and identity are quite different. With $=$ in place of $\rightarrow$, they would certainly not be laws.

(18) $$p \leftrightarrow p,$$

(19) $(p \leftrightarrow q) \rightarrow (q \leftrightarrow p),$

(20) $[(p \leftrightarrow q) \wedge (q \leftrightarrow r)] \rightarrow (p \leftrightarrow r).$

(i) Which of (18) through (20) hold when $\leftrightarrow$ is changed to $\rightarrow$?   (j) to $=$?

(21) $(p = q) \rightarrow (p \leftrightarrow q).$

(k) From (21) prove that $(p \rightarrow q) \leftrightarrow (\sim q \rightarrow \sim p)$ is a law.

Note that (21) permits us to write a theorem on equivalence corresponding to each identity in logic.

(22) $[(x = y) \wedge (y = z)] \rightarrow (x = z).$

★(22)   Proof of (22):

(a) $x = y \wedge y = z$ Hyp,

(b) $x = y$ Inf, (6)$(p{:}x = y, q{:}y = z),$

(c) $y = z$ same, (2–5–2),

(d) $x = z$ Rep, (c), (b), Q.E.D.

The following laws are stated here as examples from the infinite list of theorems provable from the axioms of logic. They were selected for their own interest or because we shall use them later.

(23) $p \rightarrow (q \rightarrow p),$

(24) $\sim p \rightarrow (p \rightarrow q),$

(25) $(p \rightarrow q) \rightarrow [(p \wedge r) \rightarrow (q \wedge r)],$

(26) $[(p \rightarrow q) \wedge (r \rightarrow s)] \rightarrow [(p \wedge r) \rightarrow (q \wedge s)],$

(27) $(p \rightarrow q) \rightarrow [(p \wedge r) \rightarrow q],$

(28) $(p \rightarrow q) \rightarrow [p \rightarrow (q \vee r)].$

The first asserts that a true statement is implied by any statement. The last two say that a true implication remains true if we conjoin a statement to its hypothesis or disjoin one to its conclusion.

(29) $[(p \rightarrow r) \wedge (q \rightarrow r)] \rightarrow [(p \vee q) \rightarrow r],$

(30) $q \rightarrow [(p \wedge q) \leftrightarrow p],$

(31) $\sim q \to [(p \lor q) \leftrightarrow p]$,

(32) $q \to [(p \lor q) \leftrightarrow q]$,

(33) $\sim q \to [(p \land q) \leftrightarrow q]$,

(34) $(p \to q) \leftrightarrow [(p \land q) \leftrightarrow p]$,

(35) $(p \to q) \leftrightarrow [(p \lor q) \leftrightarrow q]$,

(36) $(p \leftrightarrow q) \to [(p \land r) \leftrightarrow (q \land r)]$,

(37) $[(p \leftrightarrow q) \land (r \leftrightarrow s)] \to [(p \lor r) \leftrightarrow (q \lor s)]$,

(38) $[(p \to r) \land (p \to s)] \to [p \to (r \land s)]$,

(39) $[[\sim p \to (q \leftrightarrow r)] \land (p \to r)] \to [r \leftrightarrow (p \lor q)]$.

Any argument based on assuming a negation is called *indirect*. The basis of such arguments is (2-5-28), but they take many forms, of which the following are typical.

(40) $(\sim q \to q) \to q$,

(41) $\{p \land [(p \land \sim q) \to q]\} \to q$,

(42) $\{p \land [(p \land \sim q) \to \sim p]\} \to q$,

(43) $[\sim q \to (p \land \sim p)] \to q$,

(44) $[(\sim q \to p) \land \sim p] \to q$,

(45) $[(p \land \sim q) \to (r \land \sim r)] \to (p \to q)$,

(46) $[(p \land \sim q) \to \sim p] = (p \to q)$,

(47) $[(p \land \sim q) \to q] = (p \to q)$.

An example of the use of (44) is the proof that two different lines parallel to the same line are parallel to each other. Let the theorem be written: $(a \parallel c \land b \parallel c \land a \neq b) \to (a \parallel b)$. Suppose $\sim(a \parallel b)$. Then a meets b, say in P. Then $(a \parallel c) \land (b \parallel c) \land (a$ and b pass through $P)$. But this is false since only one line can be drawn through a point and parallel to a given line. Letting $q = a \parallel b$, $p = (a \parallel c) \land (b \parallel c) \land (a$ and b pass through P), we have $\sim q \to p$ and $\sim p$. Hence by (44) and Inf we may conclude q, that is, $a \parallel b$. Then the Rule of Hypothesis establishes the desired implication.

(l) Use a similar argument to show that if two lines are perpendicular to the same line, then they are parallel. (m) Argue that if a man invests his money so as to maximize his return, then his rate of profit must be the same on all his investments.

Proofs by *elimination* are based on showing that the law to be proved is one term of a disjunction that is a law and each of whose other terms is false. It is based on such laws as the following.

(48) $[(p \vee q) \wedge \sim q] \to p,$

(49) $[(p \vee q \vee r) \wedge \sim q \wedge \sim r] \to p.$

(n) Argue for the plausibility of (48) and (49).   (o) If one angle of a triangle is greater than another, the side opposite the larger angle is larger. To prove this, we note that the first side must be greater than, equal to, or less than the second. The side opposite the greater angle cannot be equal to the other, since if it were the two angles would be equal. It cannot be less, for then the opposite angle would be less instead of greater. (We assume it is known that if one side of a triangle is larger than another, the angle opposite the first is greater than the angle opposite the second.) Hence it must be greater. Explain how (49) applies.   (p) What other logical ideas are involved in this argument?

It would be possible to formulate additional rules of proof based on the theorems of logic. For example, because of (6), we could say that if $p \wedge q$ is a law, then p is a law. Instead of formulating such rules, we refer directly to the theorems when they are needed.   (q) Formulate a rule of proof corresponding to (2-5-28).   (r) Do the same for (15).

Problems

1. Show that the converse of (3) is not a law.
2. Is $[(p \vee q) \leftrightarrow r] \leftrightarrow [\sim p \to (q \leftrightarrow r)]$ a law? Is it a law if the central $\leftrightarrow$ is changed to $\to$? Justify your answers.
3. Rewrite (7) informally.
★4. Explain (13) by giving a more complete proof.
★5. Rewrite (13) informally.
★6. Discover and prove some laws involving $\underline{\vee}$.
7. Prove that $(p \leftrightarrow q) \leftrightarrow (\sim p \leftrightarrow \sim q)$.
8. Formulate a rule of proof corresponding to (3).
9. Formulate rules of proof corresponding to other laws of implication.
10. Prove $(p \leftrightarrow q) \to (q \leftrightarrow p)$.
11. Assuming that there is one and only one line passing through two different points, prove that two lines intersect in at most one point.
12. A prime number is by definition a natural number greater than 1 with no factors other than itself and 1. Euclid proved that there is no greatest prime number in the following way. Suppose that there is a greatest prime. Let us call it N. Now consider all the prime numbers, 2, 3, 5, 7, ..., N. Let us construct a number M equal to the product of all of these plus 1, that is, $M = (2 \cdot 3 \cdot 5 \cdot 7 \cdot 11 \ldots N) + 1$. Now M is not divisible by any of these primes, since there is always a remainder of one after such a division. Hence M is a

prime, since if it had a factor other than itself or 1, this factor would have to be a prime or be divisible by a prime, and in either case M would be divisible by one of the primes. But obviously $M > N$, and hence N is not the largest prime! Explain the use of indirect argument in this proof.

13: Show that the converse of (21) is not a law.

ANSWERS TO EXERCISES

(a) No; *only* sentences are significant substitutes for the variables in (3). (b) $\sim \sim p = p$. Hence by (3) and $Inf \sim \sim p \to p$. (c) $(p \cdot \to p) = (\sim p \lor p) = p \lor \sim p = \sim(\sim p \land \sim \sim p) = \sim(\sim p \land p) = \sim(p \land \sim p)$. (d) $(6)(q{:}\sim q) = [(p \land \sim q) \to p] = \sim(p \land \sim q) \lor p$ [by (2-5-23)] $= \sim p \lor \sim \sim q \lor p = \sim p \lor (q \lor p) = \sim p \lor (p \lor q) = p \to (p \lor q)$. (f) When p is false and q true, it fails. (g) The substitution yields $(q \land p) \to q$, which is the same as $(p \land q) \to q$. (h) They embody the schemata (2-6-6) and (2-6-26). (i) First and third only. (j) All. (k) (2-5-28).

(l) $\sim(a \parallel b) \to [a$ meets b in some point]. Call this point P. Then through P there are two lines $\perp c$. But this is false, since only one line can be drawn from a given point perpendicular to a given line. (m) If his rates of profit are different, he can increase his return by transferring some resources to the more profitable investment, and hence his profit is not maximum. The conclusion follows by (2-5-28). (o) (49)(p:first greater than second, q:first equals second, r:first less than second). (p) Each possibility is shown to be false by showing that it implies a false statement, i.e., by applying (44)($q{:}\sim q$). (q) To prove an implication it is sufficient to prove its contrapositive. (r) To prove a statement false, it is sufficient to prove that it implies a false statement.

ANSWERS TO PROBLEMS

1. $(2 = 5) \to (2 = 2)$. 3. We may assume q by hypothesis; also, $p \lor \sim p$ is a law by (4). Hence, in accordance with the Rule of Hypothesis, the desired conclusion appears as a step in a proof in which the hypothesis is assumed, and we have proved the implication. 7. $(p \leftrightarrow q) = (p \to q) \land (q \to p) = (\sim q \to \sim p) \land (\sim p \to \sim q) = (\sim p \leftrightarrow \sim q)$. Then use (21). 12. The main argument is justified by (43)(q:There is no largest prime, p:N is the largest prime). To show M is prime we use (44)(q:M is prime, p:M is divisible by a prime).

2-8 Manipulating algebraic equations. Much scientific discourse is formulated in terms of equations. The following are some of the problems that arise.

I. *To prove that a given equation is an identity.* Numerous examples in logic and the algebra of numbers have already been given.

II. *To prove that a given equation is not an identity.* This is often most easily done by citing a *counterexample* (a single example for which it is false).

III. *To prove that a given value of the variable in an equation is a solution.* The proof is accomplished by substituting in the equation and proving that the result is an identity. Hence this problem reduces to I.

IV. *To find some solutions of an equation.* This is quite a different matter from III. Indeed, before we can try to prove that a number is a solution, we must find it! The difference between III and IV is very much like the difference between the jobs of a prosecuting attorney and a detective. A trial cannot be held without an accused! The prosecuting attorney (III) wishes to *prove* that the accused is guilty, and he is bound by very strict rules (rules of evidence, court procedure, etc.) He wishes to *prove* that the accused is a solution of the sentence "*x* committed the crime." On the other hand, the detective (IV) wishes merely to *find* the guilty party. In this search he may guess, use his intuition, listen to gossip, and do all sorts of things that would not be considered proof by any court of law. Similarly, we may use any convenient method to find solutions of an equation, but we must not imagine that this is the same thing as proving that what we have found is indeed a solution.

V. *To find all solutions of an equation,* i.e., to solve it. This involves more than III or IV.

VI. *To derive some equations from others.* This problem is more general than I, since here we do not require that the equations be identities. It arises very frequently in science when the scientist assumes that certain relations hold among values of his variables and desires to discover the consequences of this assumption.

To solve such problems we make use of manipulations of several kinds:

A. *Manipulations yielding synonymous sentences.* If we replace an expression in an equation by a synonym, the result is an equation synonymous with the original. Moreover, by (2–7–21), the original and resulting equations are logically equivalent. Because of the equivalence, any value of the variable that is a solution of one is a solution of the other. If one is an identity, so is the other. If one is not, neither is the other. If we find all solutions of one, we have found all solutions of the other.

(a) Why does an equation imply any equation found by a manipulation of Type A above?

For example, to prove that 2 is a solution of $x^2 + x - 6 = 0$, we may write

(1) $$2^2 + 2 - 6 = 0 \leftrightarrow 4 + 2 - 6 = 0 \leftrightarrow 0 = 0.$$

Here each change involves a replacement. Since the last equation is an identity, so is the first.

(b) Prove that -3 is a solution of $x^2 + x - 6 = 0$.

B. *Manipulations yielding logically equivalent sentences.* These include not only those of Type A but also other manipulations that yield sentences that are equivalent without being synonymous. For example,

$$(2) \qquad\qquad 2x - 1 = 0 \leftrightarrow 2x = 1,$$

because $2x - 1 = 0 \rightarrow 2x = 1$ by (2-6-16)(a:$2x - 1$, b:0, c:1), and $2x = 1 \rightarrow 2x - 1 = 0$ by (2-6-16)(a:$2x$, b:1, c:-1). Manipulations of this kind have the same advantages arising out of the equivalence as do those of Type A.

The most useful laws that justify manipulations of this type are the following.

$$(3) \qquad\qquad (a = b) \leftrightarrow (a + c = b + c),$$

$$(4) \qquad\qquad (c \neq 0) \rightarrow (a = b \leftrightarrow ac = bc),$$

$$(5) \qquad\qquad (ab = 0) \leftrightarrow (a = 0 \lor b = 0).$$

The first law follows from (2-6-16), (2-6-23), and (2-3-17). To prove the second, we note that $a = b \rightarrow ac = bc$. Hence by (2-7-23)($p$:$a = b \rightarrow ac = bc$, q:$c \neq 0$) and Inf, we may assert $c \neq 0 \rightarrow (a = b \rightarrow ac = bc)$. From this and (2-6-24) we have (4) by using (2-7-38) and Inf.

We can argue informally for (5) as follows. First to prove $ab = 0 \rightarrow a = 0 \lor b = 0$, we assume $ab = 0$ by hypothesis. If $a = 0$, we have the conclusion. If not, (4) justifies multiplying both members of $ab = 0$ by $1/a$ to get $b = 0$. Second, $(a = 0 \lor b = 0) \rightarrow (ab = 0)$ follows immediately from (1-9-30) and (1-9-31). Hence, by the definition of logical equivalence we have (5).

To prove (5) more rigorously we use (2-7-39)(p:$c = 0$, q:$a = b$, r:$ac = bc$) to get $[[c \neq 0 \rightarrow (a = b \leftrightarrow ac = bc)] \land (c = 0 \rightarrow ac = bc)] \rightarrow [ac = bc \leftrightarrow (c = 0 \lor a = b)]$. The hypothesis of this implication is a law, since its first term is just (4) and its second term follows immediately from (1-9-30). Hence by Inf, its conclusion is a law; that is,

$$(6) \qquad\qquad (ac = bc) \leftrightarrow (c = 0 \lor a = b).$$

This law is sometimes useful as it stands. We can get (5) from it immediately by the substitution (b:0).

We illustrate the use of these laws by the following manipulations.

$$(7) \qquad 2x^2 + 2x = 12 \leftrightarrow x^2 + x = 6 \qquad\qquad (4)$$

$$(8) \qquad\qquad \leftrightarrow x^2 + x - 6 = 0 \qquad\qquad (3)$$

(9) $\leftrightarrow (x + 3)(x - 2) = 0$ (1-12-7)

(10) $\leftrightarrow x + 3 = 0 \lor x - 2 = 0$ (5)

(11) $\leftrightarrow x = -3 \lor x = 2$ (3).

Each sentence is logically equivalent to the preceding one for the reason given. Hence we have simultaneously solved problems III, IV, V for this equation. We know that -3 and 2 are roots and all the roots.

(c) Which steps in (7) through (11) involve replacements?   (d) Which sentences are synonymous?   (e) Solve $3 - 2x = 3x + 1$ by such manipulations, giving reasons.   (f) Do the same for $x^2 - x - 6 = 0$ and for (g) $2x^2 + 8x + 8 = 0$.

An equation of the form $ax + b = 0$ (where a and b do not involve x, and $a \neq 0$) is called a *linear equation in one variable*. Equations that can be reduced to this form by replacements and manipulations like those in (3) and (4) are also called linear. The solution of such an equation is trivial.

(12) $ax + b = 0 \leftrightarrow ax = -b$

(13) $\leftrightarrow x = -b/a.$

(h) Solve, giving reasons: $x/2 + 16 - 5x = -2(x + 3)$.

An equation of the form $ax^2 + bx + c = 0$ (where $a \neq 0$, and a, b, and c are independent of x) and equations that can be reduced to this form are called *quadratic*. Manipulations similar to (7) through (11) may be used to solve any quadratic equation.

$ax^2 + bx + c = 0$

(14) $\leftrightarrow x^2 + (b/a)x + (c/a) = 0$

(15) $\leftrightarrow x^2 + (b/a)x + (b/2a)^2 - (b/2a)^2 + (c/a) = 0$

(16) $\leftrightarrow (x + b/2a)^2 - (b^2/4a^2 - c/a) = 0$

(17) $\leftrightarrow (x + b/2a)^2 - (b^2 - 4ac)/4a^2 = 0$

(18) $\leftrightarrow (x + b/2a + \sqrt{b^2 - 4ac}/2a)(x + b/2a - \sqrt{b^2 - 4ac}/2a) = 0$

(19) $\leftrightarrow x + \dfrac{b + \sqrt{b^2 - 4ac}}{2a} = 0 \lor x + \dfrac{b - \sqrt{b^2 - 4ac}}{2a} = 0$

(20) $\leftrightarrow x = \dfrac{-b - \sqrt{b^2 - 4ac}}{2a} \lor x = \dfrac{-b + \sqrt{b^2 - 4ac}}{2a}.$

(i) Give the reasons for steps (14) through (17), (19), and (20).

Step (18) is obtained by factoring (17) as a difference of two squares, that is, by using

$$[a^2 - b^2 = (a + b)(a - b)](a{:}x + b/2a, b{:}\sqrt{b^2 - 4ac}/2a).$$

However, if $b^2 - 4ac < 0$, $\sqrt{b^2 - 4ac}$ is not a real number and does not even exist in the number system we have used so far. (Why? For a proof, see Section 3–5.) Hence, for the present when $b^2 - 4ac < 0$ we say that the equation has no real roots. In Chapter 6 we extend the number system so that any number has a square root. Then (14) through (20) proves that any quadratic equation with $b^2 - 4ac \neq 0$ has two roots given by (20). When $b^2 - 4ac = 0$ there is only one root, but it is customary to say that the two roots are equal. We call $b^2 - 4ac$ the *discriminant* of the quadratic. Usually (20) is written

$$(21) \qquad ax^2 + bx + c = 0 \leftrightarrow x = \frac{-b \pm \sqrt{b^2 - 4ac}}{2a},$$

and the expression giving the two values of x is called the *quadratic formula*.

(j) Carry through (14) through (20) with $x^2 + 2x - 4 = 0$. Use the quadratic formula to find the roots of: (k) $x^2 - 2x - 4 = 0$, and (l) $x^2 + 6x + 9 = 0$. Solve: (m) $3x^2 + 5x - 1 = 0$, and (n) $3x^2 - 5x + 1 = 0$.

C. *Manipulations yielding a necessary or sufficient, but not a logically equivalent, sentence.* For example, to solve $x^2 = x$ we may divide both members by x to get $x = 1$. According to $(2{-}6{-}17)(a{:}x, b{:}1, c{:}x)$, $x = 1 \rightarrow x^2 = x$. Hence 1 is a root of $x^2 = x$. But the converse is false; that is, $x^2 = x \rightarrow x = 1$ is not a law. Indeed, its hypothesis is true and its conclusion false for $x = 0$. This manipulation yielded a sufficient but not necessary condition for the original equation, and one root was lost. Of course, (6) yields

$$(22) \qquad (x^2 = x) \leftrightarrow (x = 0 \lor x = 1).$$

Or we could solve the equation by rewriting it as $x^2 - x = 0$ and then factoring or using the quadratic formula.

As a second example, consider the equation $1/x + 3 = 1/x$. Multiplying both members by x, we have $1 + 3x = 1$, $3x = 0$, $x = 0$. But 0 is not a root of the original, since "$1/0$" is undefined. The difficulty arose because $1/x + 3 = 1/x \rightarrow 1 + 3x = 1$, but the converse is false. Hence we gained a root by this manipulation. Here we found a necessary but not sufficient condition. Again the difficulty could be avoided by sticking to manipulations of Type B. By (4) we have

$$(23) \qquad x \neq 0 \rightarrow (1/x + 3 = 1/x \leftrightarrow 1 + 3x = 1).$$

And since 0 is the only solution of the right member of the equivalence, there is no root.

(o) Show directly that $1/x + 3 = 1/x$ has no solution by showing it logically equivalent to $3 = 0$.

The above examples make clear the risks in manipulations of this type when solving equations. As long as we stick to manipulations of Type B we neither gain nor lose roots, but those of Type C may do either. However, manipulations of Type C are convenient in some situations, provided they are used carefully.

For example, to solve the equation $\sqrt{x} + 1 = \sqrt{4x}$, we may use (2–6–22) to write

$$\text{(24)} \qquad \sqrt{x} + 1 = \sqrt{4x} \rightarrow x + 2\sqrt{x} + 1 = 4x$$

$$\text{(25)} \qquad \rightarrow 2\sqrt{x} = 3x - 1$$

$$\text{(26)} \qquad \rightarrow 4x = 9x^2 - 6x + 1$$

$$\text{(27)} \qquad \rightarrow 9x^2 - 10x + 1 = 0$$

$$\text{(28)} \qquad \rightarrow (9x - 1)(x - 1) = 0$$

$$\text{(29)} \qquad \rightarrow x = 1/9 \lor x = 1.$$

Since the converse of (2–6–22) is not a law, we cannot use equivalence signs here. The implication in (29) means that if x satisfies the original equation, it must equal one of the values given by (29). But it does not follow that if x equals one of the values given by (29), then it must satisfy (24). Indeed, substitution in the original equation shows that 1 is a root but 1/9 is not. Here this procedure caused no error, since our manipulations did not lose any roots and we tested the alleged roots it yielded.

(p) Verify the statements about 1 and 1/9.   (q) Justify (24) through (29).

As an example of the usefulness of manipulations of Type C in deriving some equations from others in science, we consider the following equations taken from an article in the Autumn, 1956, issue of *Educational and Psychological Measurements*.

$$\text{(30)} \quad \text{A.L.} = \frac{(A_5 - A_4) + (A_5 - A_3) + (A_5 - A_2) + (A_5 - A_1)}{4},$$

$$\text{(31)} \quad \text{Acc} = A_1 + A_2 + A_3 + A_4 + A_5,$$

$$\text{(32)} \quad \text{A.L.} = (5A_5 - \text{Acc})/4.$$

The authors say that (30) and (31) imply (32). Now by replacements we find that (30) is equivalent to

$$(33) \qquad \text{A.L.} = (4A_5 - (A_1 + A_2 + A_3 + A_4))/4.$$

Also, (31) is equivalent to

$$(34) \qquad \text{Acc} - A_5 = A_1 + A_2 + A_3 + A_4.$$

Now we make in (33) a replacement of the right member of (34) by its left member. After slight simplification, the result is (32) as desired. We see that by assuming (30) and (31) we can derive (32). Hence [(30) $\wedge$ (31)] $\rightarrow$ (32) by the Rule of Hypothesis. But the converse certainly does not hold.

(r) Carry through all the details required to verify the above argument. Show that the converse fails by finding numerical values for which (32) holds but at least one of (30) and (31) is false.   (s) The following equations are taken from an article in the Spring, 1955, issue of *Educational and Psychological Measurements*.

$$(35) \qquad R_i = (1 - c)T_i + cn,$$

$$(36) \qquad R_i + W_i = n,$$

$$(37) \qquad T_i = R_i - \frac{c}{1 - c} W_i.$$

Derive (37) from (35) and (36).

The discussion of this section refers to equations. However, the general comments under I through VI and A through C apply more generally to any sentences. Examples will be given in later sections.

PROBLEMS

1. Prove that $(x_1^2 + y_1^2)(x_2^2 + y_2^2) = (x_1 x_2 - y_1 y_2)^2 + (x_1 y_2 + x_2 y_1)^2$ is an identity.

2. Prove that $a^2 = b \rightarrow a = \sqrt{b}$ is not a law.

3. Prove that $1/a$ is a solution of $x^2 + 1/x^2 = a^2 + 1/a^2$.

4. Find three other solutions of the equation in Problem 3.

5. Find all the solutions of $x(x^2 - 1)(x + 2)(x^2 - 3x - 4) = 0$.

6. Show that if $p = \dfrac{3P - 1}{2}$, then $P = p + (1/3)(1 - p)$.

7. Prove

$$(38) \qquad (a^2 = b^2) \leftrightarrow (a = b \vee a = -b).$$

8. Prove $(a \neq 0 \wedge b \neq 0) \rightarrow (ab \neq 0)$.

9. Prove $(a \neq 0 \wedge ab = 0) \rightarrow (b = 0)$.

10. Sometimes people try to prove an identity by showing that it implies a true statement. Thus, in place of (1) they write $2^2 + 2 - 6 = 0 \rightarrow 4 + 2 - 6 = 0 \rightarrow 0 = 0$. What is the fallacy in this argument?

11. Find the fallacy in the following: Let $2 = y$. Then $4 = 2y$ or $4 - 2y = 0$. Factoring, we find $2(2 - y) = 0$. Dividing by $2 - y$, we find $2 = 0(!)$.

12. Solve $x/2 - x/3 = 4$. 　　　13. Solve $x/3 - (1 - 2x) = 7x + 9$.

14. Solve $ax + b = cx + d$. 　　15. Solve $x^2 + x - 1 = 0$.

16. Solve $x^2 + a_1x + a_0 = 0$. 　17. Solve $5x^2 - x - 1 = 0$.

★18. Prove by substitution that the quadratic formula gives roots of the quadratic.

★19. Find (21) by the change of variable $x = y - b/2a$ in $ax^2 + bx + c = 0$, solving the result for y, and substituting back for x.

★20. Show that if r_1 and r_2 are the roots of $x^2 + a_1x + a_0 = 0$, then $r_1 + r_2 = -a_1$ and $r_1r_2 = a_0$.

21. Show $[h = (N - x_1^2)/2x_1 \wedge x_2 = x_1 + h] \rightarrow [x_2 = (N + x_1^2)/2x_1]$.

22. From the conclusion of Problem 21 and $x_3 = (N + x_2^2)/2x_2$, find a formula for x_3 in terms of x_1 and N.

23. (This and the next problem are taken from articles in *Educational and Psychological Measurements*, Summer, 1956.) Show that

$$\frac{U(n - L) - L(n - U)}{n^2} = \frac{U - L}{n}.$$

24. Solve for r_{12}:

$$1 - (r_{12}^2 + r_{13}^2 + r_{23}^2) + 2r_{12}r_{13}r_{23} = 0.$$

25. (This and Problem 26 are taken from *The Geometry of International Trade*, by J. E. Meade.) $(a + b = c) \rightarrow [a/b = 1/(c/a - 1)]$.

26. $[e' = E/Q \wedge i = I/e(Q + E)] \rightarrow [i = (1/e)(I/Q)/(1 + e') \wedge (1/e)(I/Q) = i(1 + e')]$.

27. Solve for x:

$$\frac{2x - a}{b} = \frac{bc - cx}{a} \qquad \text{for } a = 2, b = -1, c = 3.$$

28. Solve for x:

$$x - a = \frac{bc}{d} + \frac{c^2x}{de} \qquad \text{for } a = -3, b = 0, c = -2, d = -2, e = 4.$$

29. Solve for x:

$$\frac{x + a}{x - a} - \frac{x - a}{x + a} = \frac{1}{x - a} - \frac{1}{x^2 - a^2} + \frac{1}{x + a} \qquad \text{for } 6a + 7 = 0.$$

30. Solve for x:

$$\frac{2b - x - 2a}{bx} = \frac{x - 4a}{ab - b^2} - \frac{4b - 7a}{ax - bx}.$$

31. Show that if $b \neq 0$, $[(-a)(-b) = -ab] \leftrightarrow (-a = a)$.

32. Show that if $[\text{Pb}^{++}] = [\text{HSO}_4^-] + [\text{SO}_4^=]$, $[\text{Pb}^{++}][\text{SO}_4^=] = K_{sp}$, and $[\text{H}^+][\text{SO}_4^=] = K_2[\text{HSO}_4^-]$, then $[\text{Pb}^{++}]^2 = \dfrac{K_{sp}[\text{H}^+]}{K_2} + K_{sp}$. (*Note:* The brackets stand for concentrations of the chemical named within.)

33. (This and Problems 34 and 35 are taken from *Mathematical Introduction to Economics*, by G. C. Evans.) Solve for u: $2u/a - b/a - 2Au - B = 0$.

34. Solve for p: $\dfrac{b + Ba}{2 - 2Aa} = ap + b$.

35. Solve for p: $\dfrac{p - B}{2A} = ap + b$.

36. (This and the next problem are taken from *Mathematical Biology of Social Behavior*, by N. Rashevsky, pp. 44 and 139.) Solve for $x_1 + x_2$:

$$a_1 + a_2 - 2b_1(x_1 + x_2) = 2b_2x_1 + 2b_2x_2.$$

37. Find x if $1/(1 + x) - B = 0$.

38. Show that $-4(x - 1/2)(y - 1/4) + 5/2 = -4xy + x + 2y + 2$. (*Introduction to the Theory of Games*, by J. C. C. McKinsey, p. 22)

39. Show that if $\Delta y = y_1 - y_0$ and $\Delta x = x_1 - x_0$, then

$$y_0 + \frac{(x - x_0)\,\Delta y}{\Delta x} = \frac{1}{x_1 - x_0}\,[(y_1 - y_0)x + y_0x_1 - x_0y_1].$$

40. (From *An Introduction to the History of Mathematics*, by H. Eves, p. 184) Solve for x:

$$\frac{\sqrt{[(2/3)(7/4)(3x)]^2 - 52} + 8}{10} = 2.$$

41. (From *Statistical Methods*, by G. W. Snedecor) Solve for x:

$$\frac{(a - x)(d - x)}{(b + x)(c + x)} = 1.$$

42. Show that

$$P = \frac{1}{1 + a} = 1 - Q \to \frac{P}{Q} = \frac{1}{a}.$$

43. Show that

$$p = q - 1 \wedge P = Q - 1$$

$$\to \frac{(pN - PN)^2}{PN} + \frac{(qN - QN)^2}{QN} = \frac{N}{PQ}\,(p - P)^2(P + Q).$$

44. Show that

$$jd - j(A - d) = \log\frac{E}{E'} \to d = \frac{1}{2}\left[\frac{\log (E/E')}{j} + A\right].$$

(*Note:* You do *not* need to know the meaning of $\log (E/E')$ to solve this problem.)

Answers to Exercises

(a) $(p = q) \to (p \to q)$. (b) $(-3)^2 + (-3) - 6 = 0 \leftrightarrow 9 - 3 - 6 = 0 \leftrightarrow 0 = 0$. (c) (9) only. (d) (8) = (9). (e) $3 - 2x = 3x + 1 \leftrightarrow -2x = 3x - 2 \leftrightarrow -5x = -2 \leftrightarrow x = 2/5$. By adding -3 to both members, adding $-3x$ to both members, dividing both members by -5, and simplifying. See Sections 1–9 and 1–14. (f) $x^2 - x - 6 = 0 \leftrightarrow (x - 3)(x + 2) = 0 \leftrightarrow x - 3 = 0 \lor x + 2 = 0 \leftrightarrow x = 3 \lor x = -2$. (g) $x^2 + 4x + 4 = 0 \leftrightarrow (x + 2)^2 = 0 \leftrightarrow x = -2$. (h) Check solution by substitution. See Sections 1–9 and 1–14.

(i) (14):(4), (15):(1–9–20) and (1–9–22), (16):(1–12–9), (17):(1–14–2) and (1–14–3), (19):(5), (20):(3). (j) Instead of substituting $(a{:}1, b{:}2, c{:}-4)$ throughout, carry through each step. (k) $x = (2 + \sqrt{20})/2 = 1 \pm \sqrt{5}$. Note that by (1–13–22), $\sqrt{20} = \sqrt{4 \cdot 5} = \sqrt{4}\sqrt{5} = 2\sqrt{5}$. (l) Double root, -3. (m) $(-5 \pm \sqrt{37})/6$. (n) $(5 \pm \sqrt{13})/6$. (o) Add $-1/x$ to both sides. (p) By substitution. (q) (25) and (27) by (2–6–19). (s) Replace n in (35) from (36) and solve for T_i.

Answers to Problems

1. Expand each member. 2. $(a{:}-1)$. 4. $a, -a, -1/a$. 5. $0, \pm 1, -2, 4, -1$. 6. Solve for P and rearrange. 7. $a^2 = b^2 \leftrightarrow a^2 - b^2 = 0 \leftrightarrow (a + b)(a - b) = 0$. 10. (2–6–27). 11. Division by $2 - y$ is division by 0. 12. Check alleged solutions by substitution. 14. $(d - b)/(a - c)$. 15. $(-1 \pm \sqrt{5})/2$. 17. $(1 \pm \sqrt{21})/10$.

2–9 The functional notation. It is often convenient to have an abbreviation for a formula that appears several times in a discussion. For example, we might wish to talk about $x^2 - x + 1$ and its values for various values of x. To avoid having to write out the formula each time we wish to mention it, we could adopt a single letter as an abbreviation. Thus we might let $s = x^2 - x + 1$. However, if we wish to indicate what happens when substitutions are made in a formula, a single letter is unsatisfactory because it does not involve the variable explicitly. The difficulty is overcome by adopting an abbreviation in which x appears. We choose some letter, such as f, and follow it by the variable in parentheses, as in $f(x)$ (read "f of x"). Then we adopt a temporary definition, such as

(1) $$\text{Let } f(x) = x^2 - x + 1.$$

Now (1) says that $f(x) = x^2 - x + 1$ is an identity by definition. Hence we get identities from it by any significant substitution for x. We must, of course, substitute for x throughout. Thus

(2) $$f(2) = 2^2 - 2 + 1.$$

(3) $f(a) = a^2 - a + 1,$

(4) $f(y) = y^2 - y + 1,$

(5) $f(x + y) = (x + y)^2 - (x + y) + 1,$

(6) $f(x^2) = (x^2)^2 - x^2 + 1,$

(7) $f(x^3) = (x^3)^2 - x^3 + 1,$

(8) $f(a^2 + 1) = (a^2 + 1)^2 - (a^2 + 1) + 1.$

In this way we can use the notation to indicate any substitution in the original formula. Thus $f(2)$ means the value of $f(x)$ when $x = 2$; that is, $f(2) = [f(x)](x{:}2)$. Indeed, for any value of a,

(9) $f(a) = [f(x)](x{:}a).$

We read $f(x)$ as "f of x." It does *not* mean "f times x." Multiplication is not involved in any way. It means the value of a certain formula corresponding to the value of x. To find $f(a)$, where a is any significant substitute whatever for x, we simply substitute a, however complicated it may be, into the formula for which $f(x)$ has been adopted as an abbreviation. Thus if we let $f(x) = [x = 3]$, then $f(2) = [2 = 3]$, $f(3) = [3 = 3]$, $f(a^2 + 2) = [a^2 + 2 = 3]$, and so on.

(a) Let $f(x) = x - 1$. Find $f(1)$, $f(2)$, $f(-3)$, $f(b)$, $f(x - 1)$, $f(x + 2)$, $f(x + 4)$, $f(a + b + c)$.   (b) Let $N(x) = {\sim}x$. Find $N(p)$, $N({\sim}p)$, $N(N(p))$, $N(p \lor q)$, $N(p \land q)$. Comment on "$N(7)$."

How can the reader know whether an expression such as $s(x)$ means "s times x" or "s of x"? He has to tell by the context. However, he is aided by the following conventions: (1) Ordinarily letters near the center of the alphabet are used in the way explained in this section, especially f, g, h, F, G, H. (2) When multiplication is meant, and confusion might occur, we write $s \cdot x$, $(s)(x)$, or sx. Hence, in the absence of indications to the contrary, $s(x)$ stands for "s of x."

Find $f(-1)$ for: (c) $f(x) = (x - 1)^2$,   (d) $f(x) = -x$,   (e) $f(x) = x^2$, (f) $f(x) = x + y$,   (g) $f(x) = x^3$.   (h) Argue that if $n(x) = -x$, then $n[n(x)] = x$.
If $F(x) = $ the father of x, $F[F(x)] = $ the father of the father of $x = $ the paternal grandfather of x. Let $M(x) = $ the mother of x. Find: (i) $M[M(x)]$, (j) $M[F(x)]$,   (k) $F(F[F(x)])$.

Reading "f-of-x" for $f(x)$ may seem strange, but it is a natural outcome of ordinary ways of speaking. Thus, suppose we let $f(x) = x^2 = $ the square of x. Here f takes the place of "the square" and the parentheses indicate "of." Similarly, in the last exercise F stands for "the father." We shall discuss the meaning of f in $f(x)$ more fully in Chapter 5. For the present the reader may think of it as standing for the relation between the values of x and the corresponding values of the formula.

The functional notation is very widely used in mathematics, and for this reason it is worth while to develop some skill with it. It is very simple to use if we remember simply that whatever appears within the parentheses after f in $f(\ )$ is to be substituted for x in the formula for which $f(x)$ is an abbreviation. Then the result is to be interpreted as the value of the formula corresponding to this value of x. If a is a constant and the formula contains no other variables, $f(a)$ is a constant synonymous with the constant resulting from substituting a for x in the formula. For example, if $f(x) = x^2$ and $a = 2$, $f(a) = f(2) = 4$. If a is a variable or the formula involves other variables, $f(a)$ is a formula synonymous with the new formula obtained by substituting a for x. For example, if $f(x) = 2y + x$, $f(2) = 2y + 2$ and $f(x^2) = 2y + x^2$.

Let $f(x) = 3x + 2$ and $g(x) = 1 - x$. Find: (l) $f(4)$, (m) $g(4)$, (n) $f(x + 2)$, (o) $g(1 - x)$, (p) $f(f(x))$, (q) $g(g(x))$, (r) $f(g(x))$, (s) $g(f(x))$.

Often the functional notation is used to stand for an unspecified formula. When $f(x)$ is used in this way, both f and x are variables, whereas when $f(x)$ has been temporarily defined, only x is a variable. Indeed we have been using the functional notation in this way, for we intended (9) to hold no matter what formula had been abbreviated by $f(x)$.

(t) How would you interpret $p(x) \rightarrow q(x)$? What sort of formulas would be significant substitutes here for $p(x)$ and $q(x)$? (u) Suggest a substitute for $p(x)$ and $q(x)$ for which the implication is a law.

Later, when the functional notation is used to stand for any one of many formulas, we shall be able to make substitutions for the letter in front of the parentheses. For the present we treat "$f(x)$" as though it were a variable, and we consider any term to be a significant substitute for "$f(x)$" and similar expressions. But we have as yet no constants that are significant substitutes for "f." The reason we say that any term may substitute for "$f(x)$" is that we may wish to substitute a constant or an expression not involving "x." For example, suppose we let $f(x) = x - x$. Since the right member is 0, we have $f(x) = 0$ for all values of "x." This may seem strange, but it is often convenient to let "$f(x)$" stand for a constant.

(v) Let $f(x) = (x - 1)^2 - x^2 + 2x$. Find $f(3)$, $f(7)$, $f(b)$, $f(-3)$. Show that $f(x) = 1$ for all values of x.

The functional notation can easily be extended to deal with formulas in which more than one variable appears. We simply adopt some letter as before and place in parentheses after it a list of the variables in which we are interested. Thus we may let $f(x, y) = x - y$. Then we may as before make any significant substitutions throughout. For example, $f(y, x) = y - x$, $f(2, 4) = 2 - 4$, $f(0, 1) = 0 - 1$, $f(1, 0) = 1 - 0$. It is important, of course, to notice the order of the variables in the abbreviation and place the substitutes in the same order.

Let $h(x, y) = x + 2y$. Find: (w) $h(2, 3)$, (x) $h(a, b)$, (y) $h(-1, -2)$, (z) $h(0, 0)$.

Problems

In Problems 1 through 3 let $f(x) = 1 - x^2$, $g(x) = 2x$.

1. Find $f(x) + 1, f(x + 1)$.
2. Find $g(x) - 3, g(x - 3)$.
3. Find $f(x) + f(y), f(x + y)$.

4. State a conclusion from Problems 1 through 3 in terms of the notion of distributivity.

Find $f(f(x))$ for $f(x)$ defined as in 5 through 10.

5. x^2.	6. $-x$.	7. $\sim x$.
8. $1/x$.	9. $2x$.	10. $3 - x$.

Find $f(g(x))$ and $g(f(x))$ where $f(x)$ and $g(x)$ are defined as in 11 through 15.

11. $x^2, \sqrt{x}$ with $x \geq 0$.	12. $2x, x/2$.	13. $-x, -x$.
14. $2x - 1, (1 + x)/2$.	15. $1 + 1/x, 1/(x - 1)$.	

In Problems 16 through 26 let $T(x) = $ the truth value of x.

★16. What are the significant values of x?
★17. What are the values of $T(x)$?
★18. Show that $T(p \wedge q) = T(p) \cdot T(q)$.
★19. $T(\sim p) = 1 - T(p)$.
★20. $T(p \vee q) = $ the larger of $T(p)$ and $T(q)$ or either if they are equal.
★21. $T(p \wedge q) = $ the smaller of $T(p)$ and $T(q)$ or either if they are equal.
★22. $T(p \vee q) = $ the smaller of $T(p) + T(q)$ and 1.
★23. $T(p \rightarrow q) = T[T(p) \leq T(q)]$.
★24. $T(p \leftrightarrow q) = T[T(p) = T(q)]$.
★25. $T(p \vee q) = T(p) + T(q) - T(p) \cdot T(q)$.
★26. $T(p \underline{\vee} q) = |T(p) - T(q)|$.

27. If $f(x, y) = x - y$, prove $f(y, x) = -f(x, y)$.
28. Find some definitions of $f(x, y)$ so that $f(x, y) = f(y, x)$.
29. $f(x, y) = x^2 + y - xy$. Find $f(2, y), f(2, x), f(y, x), f(x, f(x, y))$.

30. Repeat Problem 29 for $f(x, y) = (y = 2x)$.

31. Repeat Problem 29 for $f(x, y) = y/x$.

In Problems 32 through 35 let $J(p, q) = p \wedge q$, $D(p, q) = p \vee q$, $N(p) = \sim p$, $C(p, q) = p \rightarrow q$.

★32. Argue that $D(p, q) = D(q, p)$.

★33. Show that $N[D(p, q)] = J[N(p), N(q)]$.

★34. Define D and C in terms of J and N.

★35. State (2–7–23) and (2–7–38) using these symbols.

36. If h = heredity, i = environment, and p = the personality, what would it signify to write $p = f(h, i)$?

37. If t = the tax rate of x, I = the income of x subject to tax, and T = the tax paid by x, what would we mean by $T = g(t, I, x)$? Would it be adequate to write $T = g(t, I)$?

38. As the reader may have observed, we are almost always omitting the quotation marks called for by the discussion in Section 1–3. He may find it interesting to decide where quotation marks could properly be inserted, especially in the present section.

Answers to Exercises

(a) $f(1) = 1 - 1$, $f(2) = 2 - 1$, $f(-3) = (-3) - 1$, $f(b) = b - 1$, $f(x - 1) = (x - 1) - 1$, $f(x + 2) = (x + 2) - 1$, $f(x + 4) = (x + 4) - 1$, $f(a + b + c) = (a + b + c) - 1$. (b) $\sim p, \sim \sim p, \sim \sim p, \sim(p \vee q), \sim(p \wedge q)$. "$N(7)$" is nonsense. (c) $(-1 - 1)^2$. (d) $-(-1)$. (e) $(-1)^2$. (f) $-1 + y$. (g) $(-1)^3$. (h) $n[n(x)] = n(-x) = -(-x) = x$. (i) The maternal grandmother of x. (j) The paternal grandmother of x. (k) The paternal great grandfather on the father's side of x.

(l) 14. (m) -3. (n) $3(x + 2) + 2$. (o) $1 - (1 - x)$. (p) $3(3x + 2) + 2$. (q) $1 - (1 - x)$. (r) $3(1 - x) + 2$. (s) $1 - (3x + 2)$. (t) Some propositional formula implies some propositional formula; sentences. (u) Let $p(x) = (x = 1)$, $q(x) = (2x = 2)$. (v) All substitutes yield 1, since $f(x) = x^2 - 2x + 1 - x^2 + 2x = 1$. (w) 8. (x) $a + 2b$. (y) -5. (z) 0.

Answers to Problems

1. $2 - x^2$, $1 - (x + 1)^2$. 2. $2x - 3$, $2(x - 3)$. 3. $1 - x^2 + 1 - y^2$, $1 - (x + y)^2$. 5. x^4. 7. x. 9. $4x$. 11. x. 13. x. 15. x. 16. Propositions. 17. 0 and 1. 27. $f(y, x) = y - x = -(x - y) = -f(x, y)$. 29. $4 + y - 2y$, $4 + x - 2x$, $y^2 + x - xy$, $x^2 + (x^2 + y - xy) - x(x^2 + y - xy)$. 31. $y/2$, $x/2$, x/y, y/x^2. 36. That h and i determine p, that is, that heredity and environment determine the personality. 37. That the tax is determined by the tax rate, the total income, and the individual. Not if we wished to indicate explicitly all the factors determining a person's tax. 38. This has been done in the paragraph preceding Exercise (v).

2-10 Quantifiers. As we have seen, a sentence may be true for all, some, or none of the significant values of its variables. If we wish to claim that a sentence is true for all values of its variables, we can simply display it as a theorem according to the convention explained in Section 1-8. However, it is convenient to have special symbols to indicate how many of the values of a sentence are true propositions. We call such symbols *quantifiers*.

We adopt the symbol $\forall x$ to mean "for all significant values of x." Thus, $\forall x\, p = $ (for all significant values of x, p). Usually we say "For all x, p" or "p is a law." If $f(x)$ is a sentence involving no variables other than x, then to claim that $\forall x\, f(x)$ is the same as to claim that $f(x)$ is a law. For example, $\forall x\, x^2 = x \cdot x$ is true because $x^2 = x \cdot x$ is a law. Similarly, $\forall p\, (p \to p)$ is true. We call $\forall$ the *universal quantifier*.

State in words: (a) $\forall x\, f(x)$,   (b) $\forall x \sim f(x)$,   (c) $\sim\forall x\, f(x)$, (d) $\sim\forall x \sim f(x)$.
True or false?   (e) $\forall x\, x - x = 0$,   (f) $\forall x\, (x+1)^2 = x^2 + 1$.

We adopt the symbol $\exists x$ to mean "for some values of x" where "some" is understood to mean "one or more." *This meaning of "some" is universal in mathematics. In mathematics, some = at least one.* We also read $\exists x$ as "there exists an x such that." For example, $(\exists x\, x^2 = 0) = $ (for some value of x, $x^2 = 0$) = (there exists an x such that $x^2 = 0$). We call $\exists$ the *existential quantifier*.

State in words: (g) $\exists x\, f(x)$,   (h) $\exists x \sim f(x)$,   (i) $\sim \exists x\, f(x)$, (j) $\sim \exists x \sim f(x)$.
True or false?   (k) $\exists x\, x = x + 1$,   (l) $\exists x\, x^2 \neq 0$.
Express, using quantifiers: (m) $\sim \sim p = p$ is a law,   (n) $p \to \sim p$ is sometimes true,   (o) $p \wedge \sim p$ is never true.

We take $\forall x\, f(x)$ as undefined and define $\exists x\, f(x)$. It is to be understood that significant substitutes for $f(x)$ in $\forall x\, f(x)$ are only sentences or formulas standing for sentences.

(1) Def. $[\exists x\, f(x)] = \sim\forall x \sim f(x).$

This definition merely reflects the idea that to claim that $f(x)$ is true for *some* x is the same as to say that it is *not false for all x!* Several examples have appeared in the exercises above.

Still further light is shed on the definition and the meaning of quantifiers by considering the special case when the quantified variable has only two significant values, say a and b. Then $\forall x\, f(x) = [f(a)$ and $f(b)$ are true] $= [f(a) \wedge f(b)]$, and $\exists x\, f(x) = [$at least one of $f(a)$ and $f(b)$ is true$] = [f(a) \vee f(b)]$. Then $\sim\forall x \sim f(x) = \sim[\sim f(a) \wedge \sim f(b)]$. But the right member, by De Morgan's law, is synonymous with $f(a) \lor f(b)$, that is, with $\exists x\, f(x)$. Hence (1) is verified in this case.

The following theorems are easily derived.

(2) $$\sim[\forall x\, f(x)] = \exists x \sim f(x),$$

(3) $$\sim[\exists x\, f(x)] = \forall x \sim f(x).$$

(p) Illustrate (1) through (3) for $f(x)$: $x^2 \geq 0$, and for   (q) $f(x)$: $x \land \sim x$. (r) Give reasons in the following proofs of (2): $\exists x \sim f(x) = \sim \forall x \sim (\sim f(x)) = \sim \forall x\, f(x)$, and (3): $\sim \exists x\, f(x) = \sim(\sim\forall x \sim f(x)) = \forall x \sim f(x)$.

We now state an axiom that embodies the essential property of the universal quantifier.

(4) **Ax.** $[\forall x\, f(x)] \rightarrow f(y).$

In words, if a sentence is true for all values of x, it is true for any chosen value. In (4) we may substitute for $f(x)$ any sentence involving x, and for y any constant that is a significant substitute for x in $f(x)$. For example, if we make the substitution $[f(x):x/x = 1,\, y:2]$, we have $[\forall x\, x/x = 1] \rightarrow 2/2 = 1$.

The axiom (4) embodies formally the Rule of Substitution, because it says that if a sentence is a law, then any sentence obtained by a significant substitution is also a law. It permits us to "drop" the universal quantifier. Our convention about displaying laws, discussed in Section 1–8, simply means that we write laws in a form ready for substitution. If we did not have such a convention, we should have quantifiers in front of all laws and be obliged to apply (4) whenever we wished to substitute.   If we did not have the convention of Section 1–8, we should write (4) as $\forall y\, [[\forall x\, f(x)] \rightarrow f(y)]$.

(5) $f(y) \rightarrow [\exists x\, f(x)].$

Theorem (5) says that if a sentence is true for a particular value of its variable, it is true for some value. With all quantifiers shown, it is $\forall y\, [f(y) \rightarrow [\exists x f(x)]]$. It is proved by the substitution $(f(x):\sim f(x))$ in (4) and then the use of $(p \rightarrow q) = (\sim q \rightarrow \sim p)$.

(s) Write out the proof just outlined.   (t) Treat (2) through (5) in the case of only two significant values of x.

A theorem of the form $\exists x\, p$ is called an *existence theorem*. By (5), to prove $\exists x\, f(x)$, it is sufficient to exhibit a significant value of x, say y, such that $f(y)$ is true. Proofs of this kind are called *constructive*. For example,

(6) $\exists x\, a + x = a.$

Proof:

(a) $a + 0 = a$ (1-9-22),

(b) $a + 0 = a \rightarrow \exists x\, a + x = a$ Sub, (5)$(f(x): a + x = a, y: 0)$,

(c) $\exists x\, a + x = a$ Inf, (b), (a).

(u) Similarly prove that $\exists x\, a \cdot x = a.$

Another way in which existence theorems are sometimes proved is to assume the contrary and derive a contradiction. Because of (1), to prove $\exists x\, p$, one may show that $\sim(\forall x \sim p)$, that is, that $\forall x \sim p$ is false. Such indirect existence proofs are called *nonconstructive*.

(v) Prove (6) using indirect proof to establish $\sim \forall x\, a + x \neq a.$

(7) $\sim f(y) \rightarrow [\sim\forall x\, f(x)],$

(8) $[\forall x\, f(x)] \rightarrow [\exists x\, f(x)].$

(w) Prove (7) and (8).

If we know a law of the form $\exists x\, f(x)$, we know that $f(x)$ for some value of x, but we may not know any particular solution. However, it would seem legitimate to adopt some name for the unknown value (or for one of the unknown values) of x, provided we do not assume anything about this value except that it is a solution of $f(x)$. For example, if we know that [$\exists x\, x$ has run the mile race in less than four minutes], we may draw some conclusions about whoever has done so without knowing who it is. In carrying on such a discussion it would be convenient to say: Let A be a person such that A has run the mile race in less than four minutes. Then we might say some things about A, such as that he must have been in good physical condition at the time. These considerations lead us to adopt the following additional rule of proof.

(9) Rule of Choice: *If a law asserts the existence of a solution of a sentence "f(x)," then "f(c)" may appear as a step in a proof, provided "c" is treated as a constant and no other assumption is made about it.*

Like the Rule of Hypothesis, this rule could be omitted at the cost of lengthening and complicating proofs considerably. It is merely a convenience, and any proof using it could be replaced by one relying only on the previous rules. We shall see later how it is used.

PROBLEMS

Prove or disprove 1 through 6.

1. $\forall x \; x/x = 1$.   2. $\exists x \; x/x = 1$.
3. $\forall x \sim (x^2 = 2x)$.   4. $\exists x \; x = 2x$.
5. $\sim \exists x \; x^2 \neq x \cdot x$.   6. $\exists x \; x^2 = x \cdot x$.

Express 7 through 12 in ordinary language.

7. $\exists x \; x = -x$.
8. $\sim \forall x \; x^2 \neq 2x$.
9. $\sim \forall x \; |x| = x$.
10. $\exists x \; |x| = 0$.
11. $\forall x \; \forall y \; x^2 = y^2 \leftrightarrow x = y \lor x = -y$.
12. $\forall x \; \exists y \; x + y = 2$.

Express 13 through 18 in symbols two ways, one using $\exists$ and the other $\forall$.

13. $x = x + 1$ is false for all x.
14. $x^2 = x$ is not always false.
15. $x^2 = x$ is sometimes true.
16. "All plant and animal tissues are made up of units known as cells. . . . All cells contain a living substance known as protoplasm." (Harvey E. Stork, *Studies in Plant Life*, 1945, p. 19)
17. $|x| = x$ is sometimes false.
18. There is an x such that for all y $y/y = x$.

19. Complete the following verbalizations of (2) and (3): (2): To say that it is false that $f(x)$ is true for all x is to say that there is some x for which $f(x)$ is ———. (3): To say that it is ——— that ——— is true for some x is to say that $f(x)$ is ——— for ——— x.
20. Write out (1) through (5), (7) and (8) for three values of x: a, b, and c.
21. What is the scope of the first $\sim$ in (1)? of $\forall x$?
22. Would (1) mean anything different if we substituted y for x in only the left member? in both members?
23. What theorem in this section expresses the idea that to prove that a sentence is not a law it is sufficient to cite a counterexample?
24. Which theorem expresses the idea that in order to prove that a sentence is sometimes true it is sufficient to cite a case in which it is true?
25. Prove that it is false that all natural sciences are based primarily on controlled experiment.
26. Prove $\sim \forall x \; (x < 1{,}000{,}000)$.
27. Prove $\forall a \; 1/a \neq 0$.

Quantifiers are helpful in clarifying the meaning of statements involving "all," "some," and "none." For example,

(10)   All squares are positive $= \forall x \; x^2 > 0 = \sim (10')$.

(10′)   Not all squares are positive $= \sim \forall x \; x^2 > 0 = \sim (10)$

$$= \exists x \sim (x^2 > 0)$$

$$= \text{Some squares are not positive.}$$

(11)   Some squares are positive $= \exists x \; x^2 > 0 = \sim (11')$.

(11′)   No squares are positive $= \sim \exists x \; x^2 > 0 = \sim (11)$

$$= \forall x \sim (x^2 > 0)$$

$$= \text{All squares are not positive.}$$

Often the last form in (11′) is used carelessly when (10′) is meant. Such usage is incorrect (see H. W. Fowler, *A Dictionary of Modern English Usage*, under) "not"). To avoid confusion it is best not to use the last form in (11′).

    28. Justify the identities (10) through (11′).

    29. Which of the sentences within the identities (10) through (11′) are true?

    30. Repeat (10) through (11′) with the sentence "All men are mortal" ($\forall x \; x$ is a man $\rightarrow x$ is mortal).

Taking the negation of a quantified statement is easy if one can express it with quantifiers as above and then make use of laws of quantification. It should be noted that because of (1) through (3) and the flexibility of the spoken language, there are many ways of stating each idea. The solution to any confusion may be found by translating into mathematical terms, re-expressing, and translating back, being sure at each translation that the meaning is unchanged. Express in normal English in two ways the negation of each of 31 through 34.

    31. All Negroes favor an end to segregation.
    32. Some Negroes favor an end to segregation.
    33. No Negroes favor an end to segregation.
    34. Not all Negroes favor an end to segregation.

Express 35 through 43 symbolically. Then express the negation in normal English in two ways.

    35. Nothing is good enough for him.
    36. Not all white people favor segregation.
    37. He is always happy.
    38. He is not always happy.
    39. He is always not happy.
    40. He is sometimes happy.
    41. He is sometimes not happy.
    42. "All that glisters is not gold."
    43. There's nothing new under the sun.

The following laws express formally the meaning of equality.

(12) $[x = y] \leftrightarrow \forall f \; [f(x) = f(y)],$

(13) $[x = y] \leftrightarrow \forall f \; [f(x) \leftrightarrow f(y)].$

44. What rule of proof corresponds to (12)?

★45. Prove that the left member of (12) implies the right.

★46. Prove the converse by letting $f(x) = x$.

★47. Why do these two results prove (12)?

★48. Prove (13).

49. Write out the laws of quantification for the case of four values of x: a, b, c, and d.

★50. Justify the following:

(14) **Ax.** $[\forall x(f(x) \wedge g(x))] = [\forall x \, f(x)] \wedge [\forall x \, g(x)],$

(15) $[\exists x(f(x) \vee g(x))] = [\exists x \, f(x)] \vee [\exists x \, g(x)],$

(16) **Ax.** $[\forall x(f(x) \rightarrow g(x))] \rightarrow [(\forall x \, f(x)) \rightarrow (\forall x \, g(x))],$

(17) $[\forall x(f(x) \rightarrow g(x))] \rightarrow \{[\exists x \, f(x)] \rightarrow [\exists x \, g(x)]\}.$

51. Show that (14) and (15) are false if $\vee$ and $\wedge$ are interchanged.

ANSWERS TO EXERCISES

(a) [For all x, $f(x)$] = [Every significant value of x is a solution of $f(x)$] = [$f(x)$ is always true] = [$f(x)$ is a law].   (b) [For all x, not-$f(x)$] = [$f(x)$ is always false] = [No value of x is a solution of $f(x)$] = [$f(x)$ is never true] = [For no x, $f(x)$].   (c) [For not all x, $f(x)$] = [Not every significant value of x is a solution of $f(x)$] = [$f(x)$ is not always true] = [$f(x)$ is not a law].   (d) [For not all x, $\sim f(x)$] = [Not every significant value of x fails to be a solution of $f(x)$] = [$f(x)$ is not always false] = [the negation of $f(x)$ is not a law] = [For some x, $f(x)$].   (e) 1.   (f) 0, since the sentence is false for $x = 1$.

(g) [For some x, $f(x)$] = [At least one significant value of x is a solution of $f(x)$] = [$f(x)$ is sometimes true] = [There exists an x such that $f(x)$] = [$\sim f(x)$ is not a law] = Exercise (d).   (h) [For some x, not-$f(x)$] = [Some significant value of x is not a solution of $f(x)$] = [$f(x)$ is sometimes false] = [There exists an x such that $\sim f(x)$] = Exercise (c).   (i) [For no x, $f(x)$] = [No significant value of x is a solution of $f(x)$] = [There exists no x such that $f(x)$] = Exercise (b).   (j) [For no x, not-$f(x)$] = [No significant value of x fails to be a solution of $f(x)$] = [There exists no x such that $f(x)$ is false] = Exercise (a).   (k) 0.   (l) 1.

(m) $\forall p \sim \sim p = p$.   (n) $\exists p \, p \rightarrow \sim p$.   (o) $\sim \exists p \, p \wedge \sim p$ or $\forall p \sim (p \wedge \sim p)$. (p) (For some x, $x^2 \geq 0$) = (It is false that for all x, $x^2 < 0$). (It is false that for all x, $x^2 \geq 0$) = (For some x, $x^2 < 0$). (There is no x such that $x^2 \geq 0$) = (For all x, $x^2 < 0$).   (r) (1) and $\sim \sim p = p$.   (s) $f(y) \rightarrow \sim[\sim f(y)] \rightarrow \sim[\forall x \sim f(x)] \rightarrow \exists x \, f(x)$ by (2–5–6), (2–7–3), (4), (2–5–28), (1). (t) (2), (3), (4), (5) become (2–5–7), (2–5–8), (2–7–6), and (2–7–9).

The results of Exercise (t) should not be surprising. The universal quantifier asserts the conjunction of all the values of the formula that follows it, since $\forall x\, f(x)$ means that every value of $f(x)$ is true.   Hence $\forall x\, f(x) = f(a) \wedge f(b) \wedge f(c) \wedge \ldots$, where the values of x are a, b, $c, \ldots$. Similarly $\exists x\, f(x) = f(a) \vee f(b) \vee f(c) \vee \ldots$.

(u) (1–9–25).   (v) $(\forall x\, a + x \neq a) \rightarrow (\forall x\, x \neq 0) \rightarrow (0 \neq 0)$; by (2–5–28), $(0 = 0) \rightarrow (\sim \forall x\, a + x \neq a)$.   (w) (7) = (4) by (2–5–28); (8) follows from (4), (5), and (2–7–12).

Answers to Problems

1. True. Zero is not a value of x here. 3. False. $2^2 = 2 \cdot 2$. 5. True, since $\forall x \sim (x^2 \neq x \cdot x)$. 7. There is a number equal to its own negative. 9. Not all numbers are equal to their absolute values. 11. If the squares of two numbers are equal, the numbers are equal or one is the negative of the other. 13. $\forall x\, x \neq x + 1; \sim \exists x\, x = x + 1$. 15. $\exists x\, x^2 = x, \sim \forall x\, x^2 \neq x$. 17. $\exists x\, |x| \neq x; \sim \forall x\, |x| = x$. 19. False; false; $f(x)$; false; all. 21. $[\forall x \sim f(x)]; \sim f(x)$.

28. Use the theorems with $(f(x){:}x^2 > 0)$. 29. $x^2 = 0$ for $x = 0$; otherwise $x^2 > 0$. 30. All men are mortal $= \sim$ (Not all men are mortal) $= \sim$ (Some men are not mortal). Some men are mortal $= \sim$ (No men are mortal). 31. Not all Negroes favor an end to segregation $=$ Some Negroes do not favor an end to segregation. 32. No Negroes favor an end to segregation $=$ All Negroes are against an end to segregation. 33. Some Negroes favor an end to segregation $=$ Not all Negroes do not favor an end to segregation. 34. All Negroes favor an end to segregation $=$ No Negroes do not favor an end to segregation. 35. $\sim \exists x\, x$ is good enough. 37. $\forall t$ He is happy at time t. 42. $\sim \forall x\, x$ glisters $\rightarrow x$ is gold. ($\neq \forall x\, x$ glisters $\rightarrow x$ is not gold). This is an example of improper usage. 44. Replacement.

★2–11 Multiple quantification.

The most useful applications of quantifiers occur when more than one variable is quantified. We have the following possibilities for two variables.

(1) $\forall y\, \forall x \ldots$ = For every y, for every $x \ldots$.

(2) $\forall x\, \forall y \ldots$ = For every x, for every $y \ldots$.

(3) $\forall x\, \exists y \ldots$ = For every x there exists a y such that $\ldots$.

(4) $\forall y\, \exists x \ldots$ = For every y there exists an x such that $\ldots$.

(5) $\exists y\, \forall x \ldots$ = There exists a y such that for every $x \ldots$.

(6) $\exists x\, \forall y \ldots$ = There exists an x, such that for every $y \ldots$.

(7) $\exists x\, \exists y \ldots$ = There exists an x such that there exists a y such that $\ldots$.

(8) $\exists y\, \exists x \ldots$ = There exists a y such that there exists an x such that $\ldots$.

Read (1) through (8) with each of the following in place of the dots, and decide on truth values: (a) $x + y = 2x$, (b) $xy = x$.

Exercises (a) and (b) suggest that (1) = (2) and (7) = (8), so that the order of quantification is immaterial when only like quantifiers are involved. But clearly no two of (3) through (6) are synonymous, so that unlike quantifiers are *not commutative*.

(9) Ax. $[\forall x \forall y \, f(x, y)] = [\forall y \forall x \, f(x, y)]$,

(10) $[\exists x \exists y \, f(x, y)] = [\exists y \exists x \, f(x, y)]$.

To avoid unnecessary parentheses, we follow the convention that the scope of a quantifier is the entire following expression or extends to the end of the smallest parentheses within which it lies. This convention is illustrated in (1) through (8) where, for example, $\forall x \forall x \ldots = \forall y[\forall x \ldots]$, in (2–10–1) where the right member means $\sim[\forall x[\sim f(x)]]$, and in (2–10–8) where the parentheses are essential. Because (1) = (2) and (7) = (8) we usually read $\forall x \forall y$ as "for all x and y" and $\exists x \exists y$ as "for some x and y." We also adopt the following abbreviations.

(11) Def. $\forall x,y \, f(x, y) = \forall x \forall y \, f(x, y)$,

(12) Def. $\exists x,y \, f(x, y) = \exists x \exists y \, f(x, y)$.

It is easy to extend the laws on negation to two quantifiers by simply applying (2–10–2) and (2–10–3) carefully.

(13) $[\sim \forall x \forall y \, f(x, y)] = [\exists x \exists y \sim f(x, y)]$,

(14) $[\sim \forall x \exists y \, f(x, y)] = [\exists x \forall y \sim f(x, y)]$,

(15) $[\sim \exists x \forall y \, f(x, y)] = [\forall x \exists y \sim f(x, y)]$,

(16) $[\sim \exists x \exists y \, f(x, y)] = [\forall x \forall y \sim f(x, y)]$.

(c) Prove (13) through (16).   (d) Negate each statement made in Exercise (a) and compare truth values.

Numerous other quantifiers could be defined, but only one is in general use. It is $\exists! x$, which is read "for one and only one value of x" or "for just one x."

(17) Def. $[\exists! x \, f(x)] = [\exists x \, f(x)] \wedge \forall x,y \, [(f(x) \wedge f(y)) \rightarrow x = y]$,

(18) $[\exists! x \, f(x)] \rightarrow [\exists x \, f(x)]$,

(19) $[\sim \exists! x \, f(x)] = \{[\forall x \sim f(x)] \vee \exists x,y \, [f(x) \wedge f(y) \wedge y \neq x]\}$.

(e) Verbalize (17) through (19) and argue for them informally.   (f) Discuss $\exists!x\, f(x)$ for the case where x has only two significant values and the case where it has only three.   (g) Prove (18) and (19).   (h) Prove $\forall y\, \exists!x\; y + x = y$. (i) Does $\exists!$ commute with $\forall$ or $\exists$?

A variable that is within the scope of a quantifier is said to be *quantified*. From the discussion in this and the previous section the reader is already aware that some care is necessary when dealing with quantified variables. Consider, for example,

(20) $\exists x\, \forall y\; x$ dislikes $y,$

which says that there exists an x such that for all y, x dislikes y; that is, there exists a person who dislikes everybody. Now if we substitute x for y we have

(21) $\exists x\, \forall x\; x$ dislikes $x,$

which says that there exists an x such that for all x, x dislikes x. If it makes sense at all, it certainly means something different from (20). On the other hand, if we substitute z for y in (20), we get

(22) $\exists x\, \forall z\; x$ dislikes $z.$

This evidently has the same meaning as (20). We see that for a quantified variable we may safely substitute another variable only if the latter does not already appear within the scope of the quantifier.

Since $\forall x\, f(x)$ means that the formula $f(x)$ is true for all values of the variable x, the substitution of a constant for x would result in nonsense. Similar remarks apply to $\exists$ and $\exists!$. We see then that we cannot substitute constants for quantified variables. Of course, if $\forall x\, f(x)$ is a law, we can substitute a constant for x in $f(x)$ standing alone, but not in the whole expression $\forall x\, f(x)$. For similar reasons it is not convenient to substitute formulas for quantified variables.

If we recall the definition of "variable" in Section 1-3, we see that in the context $\forall x\, f(x)$, x is not a variable (since constants are not to be substituted for it), even though it is a variable in the formula $f(x)$ considered alone. A symbol that is a variable in part of its context considered alone but is not a variable in the whole context is said to be a *dummy* in the whole context. For example, x is a variable in $x - x = 0$, but it is not a variable in $\forall x\; x - x = 0$, so we call it a dummy in the second context. We limit substitutions for dummies to variables not appearing elsewhere in the same context.

From the preceding discussion it is evident that a quantified variable is a dummy. In (20) both x and y are dummies. Hence (20) actually involves no variables, which fits with the fact that when we express (20) in words it is seen to be a statement and not a propositional formula. However, "$\forall y\, x$ dislikes y" involves the variable x, and it is a propositional formula.

(j) When we use the Rule of Hypothesis, any variable in the hypothesis becomes a dummy. Why?   (k) Why in using the Rule of Hypothesis is it legitimate to substitute a variable not already appearing elsewhere in the context?

PROBLEMS

Letting $S = x$ is the father of y, where x and y are men, state 1 through 18 in words and decide on truth values.

1. $\forall x\, \forall y\, S$.	2. $\forall y\, \forall x\, S$.	3. $\forall x\, \exists y\, S$.
4. $\forall y\, \exists x\, S$.	5. $\exists x\, \forall y\, S$.	6. $\exists y\, \forall x\, S$.
7. $\exists x\, \exists y\, S$.	8. $\exists y\, \exists x\, S$.	9. $\forall x\, \exists! y\, S$.
10. $\forall y\, \exists! x\, S$.	11. $\exists! x\, \forall y\, S$.	12. $\exists! y\, \forall x\, S$.
13. $\exists x\, \exists! y\, S$.	14. $\exists y\, \exists! x\, S$.	15. $\exists! x\, \exists y\, S$.
16. $\exists! y\, \exists x\, S$.	17. $\exists! x\, \exists! y\, S$.	18. $\exists! y\, \exists! x\, S$.

19. Repeat 1 through 18 for $S =$ the season x immediately follows the season y.

State 20 through 26 symbolically.

20. Every man has one and only one father.
21. Some men are brothers.
22. All men are brothers.
23. There are two numbers whose product is 16.
24. Every man has ancestors.

25. Any even number greater than 4 is the sum of 2 odd primes (Goldbach's conjecture).
26. Ideas have consequences.

27. Negate each of Problems 1 through 18.
28. Do the same for Problem 19.
29. Prove $\exists! x\, \forall y\, x + y = y + x = y$.
30. Prove $\exists! x\, \forall y\, x \cdot y = y \cdot x = y$.
31. Prove $\forall x\, \exists! y\, x + y = y + x = 0$.
32. Prove $\forall x\, x \neq 0 \rightarrow \exists! y\, xy = 1$.
33. Is $\exists! x\, \exists! y\, f(x, y) = \exists! y\, \exists! x\, f(x, y)$ a law?

ANSWERS TO EXERCISES

(a) F, F, T, T, F, F, T, T.   (b) F, F, T, T, T, T, T, T.   (c) Apply (2–10–2) and (2–10–3).   (d) Apply (13) through (16); for example, [$\sim \exists y\, \forall x\, x + y = 2x$] = [$\forall y\, \exists x\, x + y \neq 2x$] = [For every y there exists an x such that $x + y$ does not equal $2x$].   (e) The right member of (17) reads "There exists an x

such that $f(x)$, and if both x and y are solutions of $f(x)$ then x and y are identical." The right member of (17) $\rightarrow$ its first term, by (2-7-6); hence (18) holds. To get (19) we negate both members of (17). The right member of the result, by De Morgan's law, becomes $\sim \exists x\, f(x) \lor \sim \forall x,y\, [f(x) \land f(y) \rightarrow x = y]$. The first term in this conjunction is $\forall x \sim f(x)$ by (2-10-3). The second simplifies as follows: Let $p = f(x) \land f(y)$, and let $q = (x = y)$. Then we have $[\sim \forall x \forall y\, p \rightarrow q] = [\exists x\, \exists y \sim (p \rightarrow q)] = [\exists x\, \exists y\, p \land \sim q]$ by using (16) and (2-5-24).

(f) For two significant values, $[\exists!x\, f(x)] = [f(a) \lor f(b)]$.   (g) Along lines indicated in Exercise (e).   (h) $\forall y\, y + 0 = y$.   Hence $\forall y\, \exists x\, y + x = y$. $[y + x = y \land y + x' = y] \rightarrow [y + x = y + x'] \rightarrow [x = x']$.   (i) No.   (j) We excluded substitutions in a hypothesis introduced by this rule.   (k) It makes no difference what symbols we use so long as they are not otherwise involved.

Answers to Problems

1. Every man is the father of all men; 0.   2. Same as Problem 1.   3. Every man is the father of some man; 0.   4. Every man is a son of some man; 1. (This truth value and others below may be considered controversial by some, but this book is not concerned with theology.)   5. Some man is the father of all men; 0. 6. Some man is a son of all men; 0.   7. Some man is the father of some man; 1. 8. Same as Problem 7.   9. Every man has just one son; 0.   10. Every man has just one father; 1.   11. Just one man is the father of all men; 0.   12. Just one man is a son of all men; 0.   13. Some man is the father of just one son; 1.   14. There is a man who is a son of just one father; 1.   15. Just one man is the father of a son; 0.   16. Just one man is a son of some man; 0.   17. Just one man is the father of just one son; 0.   18. Same as 17.

20. $\forall x\, [x$ is a man $\rightarrow \exists!y\, y$ is the father of $x]$.   21. $\exists x\, \exists y\, x$ and y are brothers. 22. $\forall x \forall y\, x$ and y are brothers.   23. $\exists x\, \exists y\, xy = 16$.   24. $\forall x\, [x$ is a man $\rightarrow (\exists y\, y$ is an ancestor of $x)]$.   25. $\forall x\, [x$ is even $\land\ x > 4 \rightarrow \exists y\, \exists z\, x = y + z \land y$ and z are odd primes].   26. $\forall x\, [x$ is an idea $\rightarrow \exists y\, y$ is a consequence of $x]$.

2-12 Heuristic. Logic is concerned with deductions that are certain. As explained in Section 2-6, in a deductive theory all the theorems follow with certainty if the axioms, definitions, and rules of proof are accepted. Although strict logical thinking is very important in science and in daily living, it is by no means the only kind of worth-while intellectual activity. Indeed, without observation, experiment, discovery, and inspiration there would be nothing to be logical about!

Before a theorem can be proved, it must be conceived and considered plausible. Before axioms can be used, they must be formulated. Before a theory can be built about certain concepts, basic terms must be created and defined. Before a logical theory can be constructed, its subject matter must be understood to some extent. Before an answer can be justified, it must be found. Such activities are exploratory and uncertain, as are all activities of creation and discovery.

The word "heuristic" is used as an adjective to describe activities serving to discover or reveal, including arguments that are persuasive and plausible without being logically rigorous. It is used as a noun to refer to the science and art of heuristic activity.

Heuristic is obviously very important, and the high rewards that sometimes go to its most successful devotees are well deserved. Why, then, is not more attention devoted to studying and perfecting this art? The most obvious answer is that, by definition, it includes just those things that have not yet been systematized and made scientific. As soon as we discover a formula for solving a particular kind of problem, we no longer need heuristic for dealing with it. It is possible, also, that most heuristic experts prefer to practice the art rather than to talk about it!

(a) Look up "heuristic" in an unabridged dictionary. (b) Could heuristic be entirely formalized?

We do not attempt here to discuss heuristic in any detail. Instead we make some brief observations and refer the reader to the very few books on the subject. Among these, the best one for a beginning is *How To Solve It*, by G. Polya (Princeton University Press, 1945, and also in a paperback edition issued by Doubleday, 1957.) In it we read: "The first rule of discovery is to have brains and luck. The second rule of discovery is to sit tight and wait till you get a bright idea." A third rule might be: while waiting for brains, luck, and a bright idea, spend some time experimenting, some time thinking, and some time resting. If there is any rule that applies to all heuristic situations, it is: observe, experiment, think, and rest. The greatest difficulty comes in combining thought and action. Some persons act without thinking. Others are afraid to act until they know the answer. In Shakespeare's classic lines,

> *Our doubts are traitors,*
> *And make us lose the good we oft might win*
> *By fearing to attempt.* (Measure for Measure)

Polya gives a very useful outline of steps that may be helpful in problem solving. The main steps are: (1) understand the problem, (2) devise a plan, (3) carry out the plan, (4) check and reconsider the results. Although the bare outline may not seem to be very helpful, Polya gives many practical suggestions under each step. The methods are applicable to all kinds of heuristic activity, both to problems where a simple answer is sought and problems consisting of discovering a proof. Perhaps a fifth step might be added: (5) clean up the problem. This suggestion would involve reformulating the problem and its solution in a logical and clear way after eliminating all the experiments and confusion of the heuristic process. Step (5) has usually been performed on the mathematics that students see in books, which is the reason there is often no clue to how the results were discovered.

It is possible to formulate heuristic rules that serve as guides in creative work. They differ from rules of proof or theorems of mathematics since they suggest possible lines of activity and do not claim to be laws. For example,

Heuristic Rule: An identity may sometimes be proved by writing an identity whose left member is the left member of the identity to be proved and then modifying the right member by replacements until it is the same as the right member of the identity to be proved.

(c) Give an example of our use of the procedure described in the preceding rule.   (d) Suggest an obvious modification of the rule.   (e) Formulate a heuristic rule to cover the cases in which we work with both sides of an identity. (f) Formulate other heuristic rules for doing proofs. Keep a list and add to it. (g) Read Polya's *How to Solve It.*

Among the very few books on heuristic, the student will find the following most helpful:

How to Study—How to Solve, by H. M. Dadourian, Addison-Wesley, 1951. Emphasizes problems in elementary mathematics.

The Psychology of Invention in the Mathematical Field, by J. Hadamard, Princeton University Press, 1949. A fascinating little book by a great mathematician.

Also interesting are:

Mathematics and Plausible Reasoning (two volumes), by G. Polya, Princeton University Press, 1954.

The Creative Process, by B. Ghiselin, University of California Press, 1952.

Problem-Solving Processes of College Students, by B. S. Bloom and L. J. Broder, University of Chicago Press, 1950.

Productive Thinking, by M. Wertheimer, Harper & Brothers, 1945.

PROBLEMS

Problems 1 through 12 are to be solved in the light of Polya's discussion of heuristic and his outline. The answer should include an analysis of how the problem was solved as well as a justification of the solution. Of course, all problems in this book are exercises in heuristic.

1. How much must a merchant mark up an article costing him $12 in order that he may later mark it down 25% from its sale price and still make a profit of 20% on it?

2. A man agrees to build a garage to be owned jointly by himself and his neighbor, each to share equally in the cost. The man does all the labor himself,

working 200 hours, and the neighbor contributes $500 worth of material. If the man's labor is worth $1.50 an hour, how should settlement be made?

3. According to a recent news item, a mail-order house installed a $500,000 electronic computer with which ten girls could do the work formerly done by 150. Is this a money-saving move?

4. Which would you rather have, four annual salary increases of $100 each or two salary increases of $200 each at two-year intervals?

5. Did Greek authors of the pre-Christian era use footnotes?

6. A lady asks how long she should make a valance to go all around a circular table of diameter 3 feet 6½ inches. What do you tell her?

7. Fill in the blank in the following quotation from the *Wall Street Journal*. "Alden's, Inc., voted a quarterly of 30 cents on the increased common, payable July 1, to stock of record June 14. A 50% common stock dividend was paid May 7. Before that 37 1/2 cents was paid quarterly on the old common. In announcing the increase, Robert W. Jackson, president, said the 'previous common quarterly dividend of 37 1/2 cents is equivalent to 25 cents based on outstanding shares reflecting a 50% distribution made May 7, 1954. The current rate of 30 cents, therefore, represents a ____% increase in the common stock dividend.'"

8. A reed growing in the middle of a pond 8 feet in diameter just touches the edge when it is pulled over. How deep is the pool if one foot of the reed is above the water? (An old Chinese problem.)

9. Find a formula for the long diagonal of a rectangular box of sides a, b, and c.

10. Show that in chess it is impossible to reach a position in which all pawns of one color are in the same file.

★11. In chess, a "won position" for White (Black) is one from which White (Black) can force a win no matter how Black (White) plays. A "draw position" is one in which either player can force at least a draw. It has been proved that every position in chess is a won position for Black, a won position for White, or else a draw. There is no known method of deciding which of these possibilities is the actual one in any case, however. If we modify the rules to permit a player not to make a move when it is his turn, it follows that the initial position is either a draw or a won position for White. Prove this last statement.

12. A chess board can easily be completely covered by dominoes of a size that cover two adjacent squares. Show that this cannot be done if two diagonally opposite corner squares are removed from the board.

Let S, C, T, s, c, t be undefined terms satisfying the following axioms and subject to the laws of algebra of real numbers.

(1) $$S^2 + C^2 = 1,$$

(2) $$T = S/C,$$

(3) $$t = C/S,$$

(4) $$s = 1/S,$$

(5) $$c = 1/C.$$

Prove 13 through 20.

13. $T^2 + 1 = c^2$.

14. $1 + t^2 = s^2$.

15. $CT = S$.

16. $Ss = Cc = Tt = 1$.

17. $(1 - S)(c + T) = C$.

18. $c/(T + t) = S$.

19. $T^2 = S^2T^2 + S^2$.

20. $(c^2 + s^2)/(sc) = T + t$.

21. An auditorium with 800 seats was not full one evening. Exactly \$340 was taken in. Adults paid 50¢, children 20¢. Prove that more than three times as many adults as children were present.

If a sentence is true in a particular case we say that it is *verified* in that case. Verification is important. In empirical sciences it is the way in which scientific hypotheses are checked, but it should not be confused with proof. The physicist checks his hypotheses by attempting to verify them experimentally. If he finds a counterexample, he has to reject an alleged law. If he does not, he accepts the law tentatively. Usually he tries to prove the alleged law by mathematical methods from other laws thought to be correct. In short, he tries to prove his hypotheses by constructing a deductive theory, but this process is quite different from the process of checking by experiment and observation.

22. Write a sentence that is true for 100,000,000 values of its variable but is not a law.

23. Write one that is true for an infinite number of values of its variable, and also false for an infinite number!

24. Cite an example of an accepted scientific law and explain why it could not be verified for every case.

25. Could verification by numerous cases ever prove a law?

According to Webster, evidence is "that which furnishes, or tends to furnish proof, . . ." If $p \rightarrow q$, we call p *conclusive evidence* of q, but evidence for q includes also any proposition that verifies, makes plausible, or otherwise supports belief in q. For example, F. H. Elwell in *Elementary Accounting* (p. 83) writes: "The fact that a trial balance 'proves' does not necessarily mean that the ledger accounts accurately record all transactions of the period. . . . In general, however, the fact that the bookkeeper obtains a trial balance is usually accepted as evidence that the ledger accounts are correct."

26. Letting p = ledger accounts are correct, and q = books balance, which is true, $p \rightarrow q$ or $q \rightarrow p$?

27. In what sense is q evidence for p?

Since evidence includes so much, it is impossible to indicate all types. Certainly, however, when $p \rightarrow q$, q is used as evidence for p, and this is justified on the ground that the truth of q at least establishes conclusively that p *may* be true.

28. Justify the last remark by showing the implications of q being false in such a situation.

29. If $p \rightarrow q$, is $\sim q$ conclusive evidence for $\sim p$?

In legal practice evidence includes only such matter as conforms to the established rules of evidence. *Direct evidence* is evidence that "immediately establishes the fact to be proved by it," while *indirect evidence* is that which "establishes immediately collateral facts from which the main fact may be inferred."

30. If q is the fact to be proved, p expresses the direct evidence, and r the indirect evidence, indicate symbolically the relations between p, q, and r.

31. X is accused of stabbing Y. A testifies that he saw X stabbing Y. Direct or indirect? Conclusive?

32. Z testifies that he saw X running from the scene with a bloody knife, that Y gasped out X's name, and that Z saw no one else in the area. Direct or indirect? Convincing? Conclusive?

The last question indicates the frequency, and indeed the necessity, of making decisions on the basis of evidence that is, strictly speaking, not conclusive. The law requires merely certainty "beyond a reasonable doubt," which is much less certain than strict deductive proof. For example, in Problem 32 there may be doubts about Z's honesty, his memory, his accuracy of observation, the meaning of Y's last words, and the possibility of someone else having been present and committed the crime without Z's knowing it. Nevertheless, the evidence is convincing because we feel it very unlikely that these things be accompanied by Z's testimony. We ignore here the aspects of evidence that involve probability judgments, and consider merely examples in which (the evidence) $\rightarrow$ (the fact) and those in which it does not. Evidence of the latter type is usually called *circumstantial evidence.*

33. Would you call Z's testimony circumstantial?

Let $a = (X$ committed the crime$)$, and $b = (X$ was present at the time of the crime$)$. An *alibi* is a claim on the part of X to have been elsewhere, i.e., a claim that $\sim b$.

34. Which is correct, $a \rightarrow b$ or $b \rightarrow a$?

35. Why is a valid alibi conclusive proof of innocence?

Let $G =$ the defendant is guilty, $I = \sim G$, $C =$ evidence for G such that $G \rightarrow C$, but $\sim(C \rightarrow G)$, and $E =$ the evidence such that $E \rightarrow G$.

36. Write various relations between G, I, and E.

37. What is the basis in logic for the requirement that the prosecution must prove G ("assumed innocent until proven guilty") rather than the alternative of requiring the defense to prove I ("guilt by accusation")? In this connection consider the possibility of both truth and falsity for E and C, and also the possibility that all evidence is inconclusive.

CHAPTER 3

ELEMENTARY THEORY OF SETS

3-1 The set concept. When we think about several things, we may consider them individually or we may view them as a single whole, class, collection, or aggregate. Mathematicians call such a whole a *set*, and they refer to the individual objects of which it is composed as *members* that *belong* to the set.

Words that designate sets are called collective nouns by grammarians. In ordinary discourse, collective nouns are used sometimes to refer to a set and sometimes to its members. For example, we may say "That bunch of grapes is heavy" or "That bunch of grapes have thick skins." The first sentence refers to the set, the second to the individual members. In mathematics, the name of a set refers only to the set as a whole.

An example of a set is the United States Senate. Its members are senators. Other examples of sets are the Los Angeles Dodgers, the planets, the living ex-Presidents, the even numbers, the possible opening moves in a game of chess, and the possible outcomes of an experiment.

(a) Describe the members of each of the sets named in the last sentence.
(b) Among words used as variables to stand for sets are "collection," "flock," "herd," "bunch," and "crowd." Indicate the kinds of sets described by these words, and give other examples of collective common nouns.

The set concept plays a very important role in mathematics, and indeed in all clear thinking. It is one of the basic concepts in terms of which other mathematical ideas are explained. The purposes of this chapter are (1) to familiarize the student with the concepts and notations for dealing with sets, (2) to show the relations between sets and sentences, and (3) to apply set theory to further development of elementary algebra and to linear analytic geometry.

Answers to Exercises

(a) Baseball players; planets; Hoover, Truman; ... −6, −4, −2, 0, 2, 4, 6, ... ; any one of 20 different moves; say, different readings on a dial.

3-2 Designating sets. Since we frequently wish to say that an object is a member of some set, we adopt the special notation $a \in C$ to mean that a belongs to C. This is an informal definition. The symbol $a \in C$ will be introduced as a basic term of the theory. We write $a \notin C$ to mean that a is not a member of C.

(a) Define $a \notin C$ formally in terms of $a \in C$ and logical symbolism.

Suppose we wish to designate the set whose members are just the numbers 1, 2, and 3. We wish to distinguish between the set considered as a whole and the individual members. Hence it would be unsatisfactory to write "the set 1, 2, and 3." We could, of course, always write "the set whose members are 1, 2, 3," but it is convenient to have an abbreviation. We adopt the device of writing a list of the members within braces. For example, $\{1, 2, 3\}$ = the set whose only members are 1, 2, and 3. Note that the braces are read "the set whose members are . . . ."

Describe in words: (b) $\{1, 2\}$,   (c) $\{2\}$,   (d) $\{a, b, c\}$,   (e) $\{0, 1\}$, (f) $\{0\}$.

We call the above way of designating sets the *roster notation*. We refer to the list of members that appear within the braces as the *roster* of the set. Note that the roster is an expression made up of names of the members; it is not a name of the set.

Use the roster notation to designate the following sets.   (g) The positive integers less than 5.   (h) The solutions of $2x = 5$.   (i) The three major religious faiths in the U. S.   (j) The digits greater than 2 and less than 7.

The roster notation is very convenient for sets with only a few members. However, there are sets with many members (e.g., the population of the world) and even with an infinite number of members (e.g., the points on a line). For such sets the roster notation is awkward or impossible. To think of some things as a set we do not need to know or list every member. All we need to know is some sentence that tells us which things are members and which are not. That is, we need to know of some sentence like "A thing is a member of the set if and only if the thing . . . ," for example, "A person is a citizen of the U.S. if and only if the person is born in the U.S. or is naturalized." This sentence completely determines the set called "the citizens of the U.S." even though we would hardly care to list all the members. Replacing "a thing" by "x," a rule for designating a set takes the form "x is a member of the set if and only if x . . . ," where the dots indicate the rest of some sentence of which x is the subject. In our example, we have "x is a citizen if and only if x is born in the U.S. or x is naturalized." Letting $f(x)$ = (x is born in the U.S. or is naturalized), we may say that (the citizens of the U.S.) = (the set of x's such that $f(x)$).

Complete the following: (k) The student body of this college = the set of x's such that x _____.   (l) The living ex-Presidents = the set of x's such that x _____ and x _____.

We now adopt an abbreviation for the formula "the set of x's such that $f(x)$", namely, $\{x \mid f(x)\}$. In this notation, the braces are read "the set" and the vertical bar is read "such that." For example, $\{x \,|\, 2x = 1\} = \{\frac{1}{2}\}$, since the only x such that $2x = 1$ is $\frac{1}{2}$. (It should be noted that $\{\frac{1}{2}\} \neq \frac{1}{2}$; the first being the set whose one member is $\frac{1}{2}$, the second being $\frac{1}{2}$ itself.) When a set is designated in this way, we call the sentence that appears after the vertical bar *a defining sentence* of the set. We say *a* defining sentence because the same set may have many defining sentences. Thus $\{x \,|\, 2x = 1\} = \{x \,|\, 1/x = 2\} = \{x \,|\, 3x - 1 = x\} = \ldots.$

Designating a set by a defining sentence is an entirely general method that applies to any set. If we know the roster of the set we can always use the designation "the set of all x such that x is one of the items listed on the roster." Thus $\{1, 2, 3\} = \{x \,|\, x = 1 \lor x = 2 \lor x = 3\}$. If we do not know the roster, we must know some characteristic that determines the members of the set (unless we do not know what set we are talking about), and we need merely state a sentence embodying the requirement that members possess this characteristic.

Designate the following by words or by the roster notation:
(m) $\{x \,|\, x + 4 = 5\}$.   (n) $\{x \,|\, x = 3 \lor x = 4\}$.   (o) $\{x \,|\, x$ has no living parents, x is a child, and x is living$\}$.   (p) $\{x \,|\, x$ has been convicted of a crime$\}$.

The set of solutions of a sentence $f(x)$, that is, the set $\{x \,|\, f(x)\}$, is called the *truth set* of $f(x)$. For example, the truth set of $x^2 = 4$ is $\{2, -2\}$.

The set of all values of a term is called the *range* of the term. For example, the range of x in $1/x$ is all numbers other than zero. If the range of x is the real numbers, the range of x^2 is the non-negative real numbers. As indicated in Section 1–3, the range of a variable or other term may be specified explicitly or indicated by the context. The range of a variable in a formula containing just one variable is called the *domain* of the formula. For example, the set of values of x in "the father of x" is called the range of x and the domain of "the father of x."

(q) If the domain of a sentence is the same as the truth set of the sentence, what can we conclude?   (r) What is the range of x in $1/(1 - x)$?   (s) in "Mr. x"?

If we look at the equation $\{1, 2\} = \{x \,|\, x = 1 \lor x = 2\}$, we note that there are no variables on the left. Indeed, the left member is a name of a quite definite set and is therefore a constant. Hence the right member must also be a constant. How do we explain the presence of what is apparently the variable "x"? Certainly "x" is a variable in the sentence $x = 1 \lor x = 2$, but in the context of the entire right member it is not a variable. When we say "the x's such that $x = 1 \lor x = 2$" we mean "the solutions of the sentence $x = 1 \lor x = 2$." To substitute a constant for "x" would make no sense. Accordingly, "x" is a dummy as defined in Section 2–11. We may substitute for it any variable not already appearing in the context without changing the meaning of the expression.

Thus $\{x|2x = 1\} = \{y|2y = 1\}$. But we may not substitute for it a constant or some variable already appearing, for example, $\{x|2x = a\} = \{a/2\} = \{y|2y = a\} \neq \{a|2a = a\} = \{0\}$.

Show that the following are laws.   (t) $\{x\,|\,ax + b = 0\} = \{-b/a\}$.
(u) $\{x\,|\,(x - a)(x - b) = 0\} = \{a, b\}$.   (v) $\{x\,|\,x^2 = a^2\} = \{a, -a\}$.
(w) $\{a\,|\,a^2 = a^2\} = $ all numbers.   (x) $\{a\,|\,a^2 = x^2\} = \{x, -x\}$.

We take "$a \in S$" and "$\{x|f(x)\}$" as basic undefined formulas. An expression of the form "$a \in S$" is a sentence if the substitute for "S" is a term whose values are sets, provided that certain limitations are placed on the substitutes for "a." It is necessary to restrict the substitutes for "a" to avoid nonsense and paradoxes, but the subject is too difficult to present here.*

Common sense will suffice for our purposes. In "$\{x|f(x)\}$" any sentence is a significant substitute for "$f(x)$," and x is a dummy, as indicated above.

Complete the following: (y) The values of $a \in S$ are _____ if S is a _____.   (z) The values of $\{x|f(x)\}$ are _____ if $f(x)$ is a _____.

PROBLEMS

Identify the sets in Problems 1 through 9.

1. $\{x\,|\,x = 2 \lor x = 3 \lor x = 4 \lor x = 5 \lor x = 6 \lor x = 7 \lor x = 8 \lor x = 9 \lor x = 0 \lor x = 1\}$.
2. $\{x\,|\,x^2 = 0\}$.
3. $\{x\,|\,x^2 + 2x - 1 = 0\}$.
4. $J = \{x\,|\,x = 0 \lor (\exists y\, y \in J^+ \land (x = y \lor x = -y))\}$, where $J^+ = $ the positive integers.
5. $E = \{x\,|\,x \in J \land \exists y\, y \in J \land x = 2y\}$, where $J = $ the integers.
6. $D = \{x\,|\,x \in J \land x \notin E\}$.
7. $\{x\,|\,x \in J \land \exists y\, y \in J \land x = 2y + 1\}$.
8. $Ra = \{x\,|\,\exists m, n\, m \in J \land n \in J \land x = m/n\}$.
9. $P = \{x\,|\,x \in J \land x > 1 \land \forall y, z\, [(y \in J^+ \land z \in J^+ \land x = yz) \to (y = 1 \lor z = 1)]\}$.

10. Does $\{x\,|\,x + 2 = 4\} = \left\{x\,\Big|\,\dfrac{x^2 - 4}{x - 2} = 4\right\}$?

★11. It is said that a set can be defined either connotatively or denotatively. Which of these terms applies to the definition by roster and which to definition by a defining sentence?

12. Which is correct: "$2 \in \{1, 3\}$" is false, or "$2 \in \{1, 3\}$" is not a sentence?

13. Argue that the domain of $f(x)$ is $\{x\,|\,f(x) = f(x)\}$ and the range of $f(x)$ is $\{y\,|\,\exists x\, y = f(x)\}$.

---

* For example, if we admit "$S \in S$" to be a sentence, we can prove any sentence, e.g., "$(p \land \sim p)$," to be a law. See "On the Theory of Types," by W. V. Quine in *The Journal of Symbolic Logic*, Vol. 3, No. 4, December, 1938.

In Problems 14 through 20 use the defining-sentence notation to name the indicated set.

14. The set whose numbers are 1 and 2.
15. The positive integers between $\frac{1}{2}$ and $\frac{3}{2}$.
16. The positive integers that are greater than 1 and also less than 2.
17. The States of the U. S. touching an ocean.
18. The integers that are multiples of 3.
19. The integers that leave remainders of 1 when divided by 7.
20. Integers that are perfect squares.

Answers to Exercises

(a) $a \not\in c = \sim(a \in c)$.   (b) The set whose members are just 1 and 2.
(c) The set whose one member is 2.   (d) The set whose members are just a, b, and c.   (e) The set whose members are just 1 and 0.   (f) The set whose one member is 0.   (g) $\{1, 2, 3, 4\}$.   (h) $\{2.5\}$; note that we speak of "the solutions" even though the set turns out to have only one member.   (i) $\{$Catholicism, Protestantism, Judaism$\}$.   (j) $\{3, 4, 5, 6\}$.   (k) is registered here.   (l) is living; was but is no longer President.   (m) $\{1\}$.   (n) $\{3, 4\}$.   (o) All orphans.   (p) All felons.   (q) The sentence is a law.   (r) All numbers except 1.   (s) Set of all men.   (t) $-b/a$ is unique solution.   (u) through (x) See Section 2–8.   (y) sentences; set.   (z) sets; sentence.

Answers to Problems

1. The digits.   3. $\{-1 + \sqrt{2}, -1 - \sqrt{2}\}$.   5. The even integers.   7. The odd integers.   9. The prime numbers.   13. Since $f(x) = f(x)$ for just the values of x that have meaning in the context.   15. $\{x \mid x \in J^+ \wedge 1/2 < x < 3/2\}$.   17. $\{x \mid x$ is a State $\wedge x$ touches an ocean$\}$.   19. $\{x \mid x \in J \wedge \exists y \, y \in J \wedge x = 7y + 1\}$.

3–3 Special sets. In any discussion, we talk about some set of objects. The set of all things under discussion in any context is called the *universe of discourse* or *universal set*, and we shall designate it by U. For example, in elementary algebra the universe of discourse is usually the real numbers. From the members of U we can construct sets, which we call *subsets* of U. Then $a \in S$ if a is a member of S and S is a subset of U. Also $\{x \mid f(x)\}$ is a subset of U if the values of x in $f(x)$ are members of U, that is, if the domain of $f(x)$ is a subset of U.

Suppose, now, that the domain of $f(x)$ is a subset of U but that $f(x)$ is false for all values of x. We still consider $\{x \mid f(x)\}$ to be a set and a subset of U even though it has no members. We call a set with no members *empty* or *vacuous*. We define $\emptyset$ as follows.

(1) Def. $\emptyset = \{x \mid x \neq x\}$.

Clearly $\emptyset$ is vacuous; that is, we have the law

(2) $$\sim(x \in \emptyset).$$

(a) Why?

We call $\emptyset$ *the null set*. The word "the" is used because any vacuous set is equal to $\emptyset$. [See (8) below.]

To prove (2) or any other theorem about sets, we need some axioms containing our undefined terms. The first axiom below embodies the essential idea in our notion of $\{x \mid f(x)\}$, namely that it is just the set of objects satisfying $f(x)$.

(3) **Ax.** $$[a \in \{x \mid f(x)\}] = f(a).$$

(b) In (3) what are the variables, and what terms, if any, are dummies?

Our next axiom indicates that we regard a set as just the collection of its members. To say that sets A and B are identical is to say that they have the same members.

(4) **Ax.** $$(A = B) = \forall x\,(x \in A \leftrightarrow x \in B).$$

Now we can prove

(5) $$A = \{x \mid x \in A\},$$

(6) $$[\{x \mid f(x)\} = \{x \mid g(x)\}] = [\forall x(f(x) \leftrightarrow g(x))].$$

This theorem tells us that two logically equivalent sentences define the same set, and conversely. Evidently every set A has many defining sentences, among which, by (5), $x \in A$ is one.

The following axiom indicates the nature of the universal set.

(7) **Ax.** $$U = \{x \mid x = x\}.$$

(c) Could the second $=$ in (6) be changed to $\leftrightarrow$?   (d) Give an example of three sentences defining the same set.   (e) Give a sentence that defines the set of all statements.

We can now prove that any vacuous set is identical with the null set $\emptyset$; that is,

(8) $$[\forall x\; x \notin A] \leftrightarrow [A = \emptyset].$$

Indeed, by (6), $A = \{x \mid x \in A\} = \{x \mid x \neq x\} = \emptyset$ if and only if $\forall x[x \neq x \leftrightarrow x \in A]$. But by hypothesis, the right member of this equivalence is always false, and we know that $x \neq x$ is always false. Hence $x \neq x$ and $x \in A$ are logically equivalent for all x, as required.

(f) Show that $\exists! A \; \forall x \; x \notin A$. (g) We have seen that the null set is unique. Is this true of the universe of discourse U; that is, does (7) determine a unique U? (h) What is $\{x \mid 1/x = 0\}$?

Now we introduce the roster notation by definition.

(9) **Def.** $\qquad\qquad \{a\} = \{x \mid x = a\}$,

(10) **Ax.** $\qquad\qquad a \neq \{a\}$,

(11) $\qquad\qquad\qquad a \in \{a\}$,

(12) $\qquad\qquad\qquad \exists! x \; x \in \{a\}$,

(13) $\qquad\qquad\quad (\{a\} = \{b\}) \leftrightarrow (a = b)$.

We call a set with just one number a *singleton*. Note how the "oneness" of a singleton is expressed by (12) without using numbers.

(i) Why did we include (10) as a law? (j) What is the variable in (12) and what is the dummy? (k) Write each of (9) through (13) with quantifiers as you would if we did not have the convention that we omit quantifiers that stand before a law.

(14) **Def.** $\quad \{a, b\} = \{x \mid x = a \lor x = b\}$,

(15) **Def.** $\quad \{a, b, c\} = \{x \mid x \in \{a, b\} \lor x = c\}$.

When $a \neq b$, we call $\{a, b\}$ a *pair*. Similarly, when $a \neq b$, $b \neq c$, and $a \neq c$, we call $\{a, b, c\}$ a *triple*. When designating sets by their rosters, it is customary to name each member only once in the roster. Hence, ordinarily $\{a, b\}$ is a pair and $\{a, b, c\}$ is a triple.

(l) Define $\{a, b, c, d\}$ and $\{a, b, c, d, e\}$. (m) Distinguish between a, $\{a\}$, "a," $\{$"a"$\}$, and "$\{a\}$."

(16) $\qquad\qquad \{a, a\} = \{a\}$,

(17) $\qquad\qquad \{a, b, c\} = \{x \mid x = a \lor x = b \lor x = c\}$,

(18) $\qquad\qquad \{a, b\} = \{b, a\}$,

(19) $\quad [\{a, b\} = \{c, d\}] \leftrightarrow [(a = c \land b = d) \lor (a = d \land b = c)]$,

(20) $\quad [\{a\} = \{b, c\}] \leftrightarrow (a = b = c)$.

★*Special legislation.* Constitutions often prohibit special legislation, that is, laws that apply to a single community, individual, or group. One way of getting around this is to pass a law applying to a certain set of communities, but to word the defining sentence of the set so that it actually defines a singleton! Professor Ralph Fjelstad, after a study of the Minnesota legislature, reported that in the 1953 session there were hundreds of "special bills couched in general terms." For example, "Any village having a population of more than 1,300 but less than 1,500 persons according to the last applicable state or federal census and having an assessed valuation of real and personal property in excess of $1,000,000 may ...." It "so happened" that only one village, Aurora, met these qualifications. The formula for a law of this kind is "Any x such that $f(x)$ may ...," where one takes care that $\exists!x \, f(x)$ and $f(a)$, where a is the individual community, person, or group for which one wishes to pass special legislation "couched in general terms."

(n) Express the defining sentence in the previous example, using x as the variable.   (o) Using the expression in Exercise (n), write symbolically that there is only one such village.   (p) Use the symbols of this section to express the fact that this one solution is the village of Aurora.   (q) Suggest other kinds of sentences that might define a singleton whose one member is a town. (r) Give, if you can, examples of similar methods of evading prohibition of legislation directed against special religious, racial, national, or political groups.

PROBLEMS

In Problems 1 through 6 identify the sets.

1. $\{x \,|\, x^2 = 16\}$.   2. $\{x \,|\, 2x + 1 = 0\}$.
3. $\{x \,|\, 1/x = x\}$.   4. $\{x \,|\, (x - 2)(x + 1)(x + 1000) = 0\}$.
5. $\{x \,|\, 1/x = 1/x\}$·   6. $\{x \,|\, x = \sqrt{9}\}$.

7. Does $\emptyset = 0$?
8. Show that $\{x \,|\, x > x\} = \{x \,|\, x \cdot 0 = 2\}$.
9. Prove (2).
10. Prove (5) and (6).
11. Prove some of (9) through (13) and (16) through (20).
12. Show that $[A \neq B] = \exists x \, (x \in A \land x \notin B) \lor (x \in B \land x \notin A)$.
13. The following laws indicate the relation between quantifiers and set membership. Justify them.

(21) $[\forall x \, f(x)] = [\{x \,|\, f(x)\} = \{x \,|\, f(x) = f(x)\}]$,

(22) $[\exists x \, f(x)] = [\{x \,|\, f(x)\} \neq \emptyset]$.

14. Prove the following:

(23) $$[x = y] \leftrightarrow \forall S \, (x \in S \leftrightarrow y \in S),$$

(24) $$[x = y] \leftrightarrow [\{S \,|\, x \in S\} = \{S \,|\, y \in S\}].$$

15. Verbalize (23) and (24).
16. Prove (25).

(25) $$x \in U.$$

17. Define $A \neq B$ (for sets) without using $=$ in the defining term.

Axiom (10) formalizes our distinction between a set and its members. Failure to make this distinction results in confusion. The fallacy of taking a statement about the members of a set to be about the whole set is called the *fallacy of composition*. The opposite fallacy of mistaking a statement about a set to be a statement about its members is called the *fallacy of division*.

★18. Let A = the barrel of apples is ripe. Does A = the barrel $\in$ ripe things?
★19. Let B = the barrel of apples is heavy. Does B = the barrel $\in$ heavy things?
★20. What is the fallacy in the following? The even numbers, E, are divisible by 2; hence, E is divisible by 2, and so E is an even number.
★21. Explain the fallacy in the following: The Russians claim to have liquidated the kulaks, but this is obviously false since some of them escaped.
★22. What two sets of Americans were eliminated by the Civil War? Were their members eliminated?
★23. A corporation may be sued for debt. Mr. Jones is a member of the corporation, hence he may be sued. Valid?
★24. In the eyes of the law a corporation is a person. A person may be sued, hence a corporation may be sued. Valid?

Answers to Exercises

(a) $\forall x \, x = x$, hence $\forall x \sim (x \neq x)$ and every x fails to satisfy the defining sentence.   (b) "a" and "f" are variables, "x" is a dummy.   (c) Yes, since $(p = q) \rightarrow (p \leftrightarrow q)$.   (d) $x - 3 = 0$, $(x - 3)^2 = 0$, $2x = 6$.   (e) $x \leftrightarrow x$.
(f) Recall (2–11–17). $\exists A \, \forall x \, x \notin A$ by (2). Suppose $\forall x \, x \notin A$ and $\forall x \, x \notin B$. Then by (8) $A = \emptyset$ and $B = \emptyset$. Hence the second term in the right member of (2–11–17) with $(a{:}A, y{:}B)$ is satisfied.   (g) No. U depends on the context.   (h) Null set.   (i) To be sure that "$a \in \{a\}$" is a sentence.   (j) "a" is a variable, "x" a dummy.   (k) For example, (9) becomes $\forall a \, \{a\} = \{x \,|\, x = a\}$.
(l) $\{a, b, c, d\} = \{x \,|\, x \in \{a, b, c\} \lor x = d\}$.   (m) The object a, the singleton whose one member is a, the letter "a," the singleton whose member is the first letter of the alphabet, and a name of the singleton whose one member is a.
(n) x is a village having a population . . . and having . . . \$1,000,000.
(o) $\exists! x \, x$ is a . . . .   (p) $\{\text{Aurora}\} = \{x \,|\, x \text{ is a village} \ldots\}$.   (q) There are many, but the trick is to find one that does not appear to name a single village!
(r) Literacy tests, country-of-origin tests, etc., in certain circumstances.

Answers to Problems

1. $\{4, -4\}$. 3. $\{1, -1\}$. 5. All nonzero numbers. 7. No! 0 is a number. It is the number of members in $\emptyset$, but it is not $\emptyset$ itself! 14. To prove that the right member of (23) implies the left member, assume the right member by Hyp, use (2–10–4) with $(y:\{x\})$ to get $x \in \{x\} \leftrightarrow y \in \{x\}$. Since $x \in \{x\}$, $y \in \{x\}$, from which $y = x$. 17. See Problem 12.

3–4 Subsets. We call the set A a *subset* of the set B if every member of A is a member of B. We use the abbreviation $A \subseteq B$. An alternative verbalization is that *"B includes A."*

(1) Def. $A \subseteq B = \forall x\,[x \in A \rightarrow x \in B]$.

(a) Read definition (1).   (b) Show that:

(2) $[\{x\,|\,f(x)\} \subseteq \{x\,|\,g(x)\}] = \forall x\,[f(x) \rightarrow g(x)]$.

We may illustrate conveniently by using small sets. Let $U = \{1, 2, 3, 4, 5\}$, $A = \{1, 2, 3\}$, $B = \{2, 3\}$, $C = \{1, 4, 5\}$. Then $A \subseteq U$, $B \subseteq U$, $B \subseteq A$, but $\sim(A \subseteq B)$ and $\sim(A \subseteq C)$.

An alternative illustration arises from letting $U =$ the set of points in the interior of the square in Fig. 3–1 and letting A and B equal the sets of points within the labeled circles. In the figure, $A \subseteq B$, $A \subseteq U$, and $B \subseteq U$.

FIGURE 3–1

True or false? (Give reasons.)   (c) $\{1\} \subseteq \{1, 2\}$;   (d) $\{4, 5\} \subseteq \{1, 2, 3, 4, 5\}$; (e) $\{2\} \subseteq \{3, 4\}$;   (f) $\{2\} \in \{2, 3\}$;   (g) $\{2\} \subseteq \{2, 3\}$.

It is important to distinguish carefully between membership and inclusion; as suggested in Exercises (f) and (g), they are quite different. It is possible to construct examples in which a member of a set is also one of its subsets. Thus $\{2\} \in \{\{2\}, 2, 3\}$ and $\{2\} \subseteq \{\{2\}, 2, 3\}$. The first is true because "$\{2\}$" appears in the roster. The second is true because "2" appears in the roster, and hence $\{2\}$ is a set made up of members of the given set. It should be kept in mind that $\in$ is used to indicate a relation between a member and a set, whereas $\subseteq$ indicates a relation between two sets. Of course, a set may be a member of another set, as indicated in the example.

Given any set, we may form various subsets. For example, $\{1, 2, 3\}$ has the subsets $\emptyset$, $\{1\}$, $\{2\}$, $\{3\}$, $\{1, 2\}$, $\{1, 3\}$, $\{2, 3\}$, $\{1, 2, 3\}$. The set of all subsets of this set is then $\{\emptyset, \{1\}, \{2\}, \{3\}, \{1, 2\}, \{1, 3\}, \{2, 3\}, \{1, 2, 3\}\}$. Note that the null set and the set itself are members of this set. We call the set of all subsets of A the *power set* of A. Let A be a women's club. Then the power set of A is the set of all possible committees, including the committee of the whole (the club itself) and the committee to which no one belongs. We designate the power set by $\mathcal{P}(A)$.

Find the power set of: (h) $\{1, 2\}$,   (i) $\{2, 3\}$,   (j) $\{1\}$,   (k) $\emptyset$, (l) $\{1, 2, 3, 4\}$.

The examples given suggest that the number of members in the power set of A is $2^{\mathfrak{N}(A)}$, where $\mathfrak{N}(A)$ is the number of members in A. Such is indeed the case. This is the origin of the name "power set."

Linear graphs. Let U be the set of all points on an axis. Because of the one-to-one correspondence between the points of U and the real numbers, we may think of U also as the set of all real numbers. Let $f(x)$ be a sentence whose domain (range of x) is a subset of U. Then the truth set of $f(x)$ is a set of points on the axis, a subset of U. We call this set of points the *linear graph* of $f(x)$. Thus the graph of $x^2 = 1$ is the set of points (real numbers) $\{1, -1\}$. The graph of a sentence is then just a geometrical picture of its truth set. Here the graph of $x^2 = 1$ is a geometrical picture of $\{x \mid x^2 = 1\}$. The graph may be a finite set or it may contain an infinite number of points. For example, the graph of $x < 2$ contains all points to the left of 2, as suggested in Fig. 3-2. We often use the defining sentence as a name for its graph.

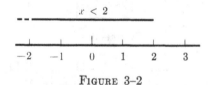

FIGURE 3-2

Sketch the linear graph of each of the following sentences: (m) $x^2 = 9$, (n) $x > -1$,   (o) $x < 3 \lor x > 4$,   (p) $x < 4 \land x > 3$,   (q) $(x - 1)(x + 5) > 0$,   (r) $x^3 \leq 0 \land x^3 \geq 8$.   (s) Describe $\mathcal{P}(U)$ where U = the axis of real numbers.

Intervals of real numbers. We say that x lies *between* a and b, where $a < b$, if $a < x \land x < b$. It is natural to adopt the following abbreviation.

(3) **Def.** $(a < x < b) = (a < x \land x < b)$.

We call $\{x \,|\, a < x < b\}$ the *open interval* from a to b and designate it by $(a\_\_b)$. The word "open" refers to the fact that it does not contain its endpoints.

(4) Def. $(a\_\_b) = \{x \,|\, a < x < b\}$.

It is convenient to think of the set of all points to the left of a given point (all numbers less than a given one) as an open interval stretching "to infinity" toward the left. We designate it by $(-\infty\_\_b)$, but the student should keep in mind that ∞ is not a numeral. Similarly, $(a\_\_\infty)$ is the set of all points (numbers) to the right of a.

(5) Def. $(-\infty\_\_b) = \{x \,|\, x < b\}$,

(6) Def. $(a\_\_\infty) = \{x \,|\, x > a\}$.

Describe in set notation, in words, and sketch: (t) $(2\_\_4)$,   (u) $(-1\_\_1)$, (v) $(a - \epsilon\_\_a + \epsilon)$.

When the endpoints belong to an interval, it is called *closed* and is designated by $(a\_\_b)$.

(7) Def. $(a \leq x \leq b) = (a \leq x \wedge x \leq b)$,

(8) Def. $(a\_\_b) = \{x \,|\, a \leq x \leq b\}$.

When only one of the endpoints belongs, the interval is said to be closed on the right or on the left. Thus $(a\_\_b)$ is *closed on the left* and *open on the right*, $(a\_\_b)$ is *open on the left* and *closed on the right*. An infinite interval cannot be closed, since it does not have a first or last point in the direction in which it "stretches to infinity."

Give definitions of: (w) $a < x \leq b$,   (x) $a \leq x < b$,   (y) $a > c > b$, (z) $(a\_\_b)$,   (aa) $(a\_\_b)$,   (bb) $(a\_\_\infty)$,   (cc) $(-\infty\_\_b)$. Sketch and describe in words: (dd) $(3\_\_4)$,   (ee) $(-1\_\_1)$,   (ff) $(0\_\_\infty)$,   (gg) $(-\infty\_\_0)$, (hh) $(-100\_\_100)$,   (ii) $(a - e\_\_a + e)$.

PROBLEMS

1. Find $\mathcal{P}\{0\}$.   2. Find $\mathcal{P}\{1, 2, 3, 4, 5\}$.

Graph Problems 3 through 10.

3. $|x| \leq 1$.   4. $x^2 + 2x - 3 = 0$.
5. $x^2 + 4x + 4 = 0$.   6. $3x - 1 = 0$.
7. $3x - 1 \leq 0$.   8. $x \notin (-1\_\_1)$.
9. $x^2 < 0$.   10. $x \in (-\infty\_\_\infty)$.

In Problems 11 through 19 write inequalities synonymous to each expression.

11. $x \in (2\_\_3)$.   12. $x \in (2\_\_3)$.   13. $x \in (2\_\_3)$.

14. $x \in (\overline{-1}\_\_\infty)$.   15. $x \in (-\infty\_\_3)$.   16. $x \in (\overline{-1}\_\_5)$.

17. $x \notin (2\_\_8)$.   18. $x \notin (-\infty\_\_\overline{-1})$.   19. $x \notin (-1\_\_\infty)$.

Tell whether each of Problems 20 through 23 is true or false.

20. $(2\_\_3) \subseteq (2\_\_3)$.   21. $(\overline{-1}\_\_0) \subseteq (-1\_\_1)$.

22. $(3\_\_5) \subseteq (4\_\_5)$.   23. $(\underline{1\_\_5}) \subseteq (2\_\_6)$.

24. Identify $(3\_\_3)$.

25. Comment on $(3\_\_2)$, $(2\_\_2)$, and $(2\_\_ -\infty)$.

26. Is $2 \subseteq (2\_\_3)$?

★27. Name several subsets of the U. S. Senate.

ANSWERS TO EXERCISES

(a) "A is included in B" means that any member of A is also a member of B. (b) (1) and (3–3–3).   (c) T.   (d) T.   (e) F.   (f) F.   (g) T. (h) $\{\emptyset, \{1\}, \{2\}, \{1, 2\}\}$.   (i) $\{\emptyset, \{2\}, \{3\}, \{2, 3\}\}$.   (j) $\{\emptyset, \{1\}\}$. (k) $\{\emptyset\}$.   (l) $\{\emptyset, \{1\}, \{2\}, \{3\}, \{4\}, \{1, 2\}, \{1, 3\}, \{1, 4\}, \{2, 3\}, \{2, 4\},$ $\{3, 4\}, \{1, 2, 3\}, \{1, 2, 4\}, \{2, 3, 4\}, \{1, 3, 4\}, \{1, 2, 3, 4\}\}$.   (m) Two points, 3 and -3.   (n) All points to right of -1.   (o) All points to left of 3 and all points to the right of 4.   (p) All points between 3 and 4.   (q) All points to the left of -5 and all points to the right of 1.   (r) Null set.   (s) The set of all linear graphs.   (t) $\{x \,|\, 2 < x < 4\}$; points between 2 and 4.   (w) $a < x \wedge x \leq b$.   (z) $\{x \,|\, a < x \leq b\}$.   (bb) $\{x \,|\, x \geq a\}$.   (ii) The closed interval from $a - e$ to $a + e$.

ANSWERS TO PROBLEMS

1. $\{\emptyset, \{0\}\}$.   3. $(\overline{-1}\_\_1)$.   5. $\{-2\}$.   7. $(-\infty\_\_1/3)$.   9. $\emptyset$.   11. $2 < x < 3$. 13. $2 < x \leq 3$.   15. $x < 3$.   17. $x < 2 \vee x > 8$.   19. $x \leq -1$.   21. F. 23. F.   25. $\emptyset$.

3–5 Ordering of the real numbers. In Sections 1–7 and 3–4 we dealt with inequalities in terms of their geometric interpretation without definitions or proof. In this section we use simple set theoretic ideas to give a more logical theory.

Let Re be the set of all real numbers. We assume that there is a subset of Re, which we call Re^+, with the following properties.

(1) **Ax.** $(a \in Re^+ \wedge b \in Re^+) \rightarrow (a + b) \in Re^+$,

(2) **Ax.** $(a \in Re^+ \wedge b \in Re^+) \rightarrow ab \in Re^+$,

(3) **Ax.** $a \in Re \rightarrow [(a \in Re^+) \vee (a = 0) \vee (-a \in Re^+)]$,

(4) **Def.** $a < b = (b - a) \in Re^+$,

(5) Def. $a > b = b < a$,

(6) $\quad a > 0 = a \in Re^+$.

(7) $\quad a < 0 = -a \in Re^+$.

(a) Prove (7).

From (6), which follows immediately from (4)($a{:}0$, $b{:}a$), we see that Re^+ is just the set of positive real numbers. We may read "$a > 0$" ($a \in Re^+$) as "a is positive" and "$a < 0$" ($-a \in Re^+$) as "a is negative." With this intuitive idea of Re^+, (1) through (3) are most plausible. However, the proofs that follow do not depend on any interpretation.

(b) Justify (8) through (13).

(8) $\qquad (a > 0 \wedge b > 0) \rightarrow (a + b > 0)$,

(9) $\qquad (a > 0 \wedge b > 0) \rightarrow (ab > 0)$,

(10) $\qquad (a < 0) \vee (a = 0) \vee (a > 0)$,

(11) $\qquad (a > 0) = (-a < 0)$,

(12) $\qquad (a < 0) = (-a > 0)$,

(13) $\qquad (a < b) = (b - a > 0)$.

We can now prove the following frequently used laws.

(14) $\qquad [(a < b) \wedge (b < c)] \rightarrow (a < c)$,

(15) $\qquad (a < b) \leftrightarrow (a + c < b + c)$,

(16) $\qquad c > 0 \rightarrow (a < b \leftrightarrow ac < bc)$,

(16′) $\qquad c < 0 \rightarrow (a < b \leftrightarrow ac > bc)$.

The proof of (14) rests directly on (4). By hypothesis we have $a < b$ and $b < c$, that is, $b - a \in Re^+$ and $c - b \in Re^+$. From (1) we have, then, $(b - a) + (c - b) \in Re^+$ or $c - a \in Re^+$. But this last is just our conclusion by (4).

★(c) Write out the preceding proof, with steps and reasons, in formal style.
(d) Prove (15). (*Suggestion:* Write hypothesis and conclusion in terms of Re^+.)

To prove (16) and (16′), we begin with $a < b$, that is, $b - a \in Re^+$, by hypothesis. Now, if in addition $c > 0$, that is, $c \in Re^+$, we have $(b - a)c \in Re^+$, $bc - ac \in Re^+$, or $ac < bc$, as desired in (16). But if $c < 0$, $-c > 0$ or $-c \in Re^+$ by (12), and we have $(b - a)(-c) \in Re^+$, $ac - bc \in Re^+$, or $ac > bc$, as required in (16′).

An understanding of (16') is essential for work with inequalities. Because of it, inequalities cannot be treated as though they were equations!

(e) Illustrate (16) and (16') for $(a{:}2, b{:}4, c{:}2)$, $(a{:}2, b{:}4, c{:}-2)$, $(a{:}-2, b{:}4, c{:}-1)$. (f) Restate (8) through (16') in geometric terms by reading $a < b$ as "a lies to the left of b."

Laws (15) through (16') can be used to solve simple inequalities. For example,

(17) $\qquad\qquad 3x + 4 < 5x + 1 \leftrightarrow -2x < -3 \qquad$ (15)

(18) $\qquad\qquad\qquad\qquad\qquad \leftrightarrow x > 3/2 \qquad$ (16').

(g) Show that $mx + b < 0 \leftrightarrow x < -b/m$ is not a law, and write a correct law giving the solution of $mx + b < 0$. (h) Solve $1 - 5x > 2x + 4$ as in (17) and (18). (i) What is the geometric interpretation of $a \le b$? of $a \ge b$?

(19) **Def.** $\qquad (a \le b) = (a < b \vee a = b)$,

(20) **Def.** $\qquad (a \ge b) = (a > b \vee a = b)$.

We can now easily prove that the square of any real number is nonnegative; that is,

(21) $\qquad\qquad\qquad x \in Re \rightarrow x^2 \ge 0$.

(j) According to (10), there are only three possible cases: $x < 0$, $x = 0$, $x > 0$. If $x < 0$, $-x > 0$, and $(-x)(-x) > 0$ by (12) and (9). But $(-x)(-x) = x^2$, so $x^2 > 0$ in this case. Dispose of the other two cases.

Suppose we wish to solve the inequality $x^2 - 4 < 0$. We have $(x + 2)(x - 2) < 0$. If $x + 2 > 0$, we may divide by it to get $x - 2 < 0$. Hence if both $x + 2 > 0$ and $x - 2 < 0$, that is, if $x > -2$ and $x < 2$, x is a solution. Evidently any point in the open interval $(-2\_\_2)$ is a solution. But suppose $x + 2 < 0$. Then division yields $x - 2 > 0$. Hence if both $x + 2 < 0$ and $x - 2 > 0$ are satisfied, x is a solution. But this means $x < -2 \wedge x > 2$, which is impossible. Another way of looking at the inequality is to observe that for $(x + 2)(x - 2)$ to be negative, one factor must be positive and the other negative, which yields the same result.

(k) Solve $x^2 - 1 \le 0$ in this way. (l) Solve $x^2 - 4 > 0$.

The following theorems are helpful in solving quadratic inequalities.

(22) $\qquad ab > 0 \leftrightarrow [(a > 0 \wedge b > 0) \vee (a < 0 \wedge b < 0)]$,

(23) $\qquad ab < 0 \leftrightarrow [(a > 0 \wedge b < 0) \vee (a < 0 \wedge b > 0)]$.

Use (22) and (23) to solve: (m) $2x^2 - 1 < 0$, (n) $x^2 \geq 17$, (o) $x^2 + 3x + 2 < 0$, (p) $x^2 + 3x + 1 > 0$, (q) $(x - 3)(x + 4) < 0$.

Problems

1. Show from (10) that if $a \neq 0$, then $a > 0$ or $a < 0$.

2. Consolidate in a single axiom the entire assumption ending with (1) through (3), including all quantifiers required.

3. Show that $a + 1 > a$ is a law by proving that it is logically equivalent to $1 > 0$.

4. Show that if $c > 0$, $a + c > a$.

5. Show that $a - 1 < a$.

6. Interpret the following verbally, argue for their plausibility, and give examples.

(24) $\sim(a < a)$,

(25) $a < b \rightarrow \sim(b < a)$,

(26) $a \leq a$,

(27) $(a \leq b \wedge b \leq a) \rightarrow (a = b)$,

(28) $(a \leq b \wedge b \leq c) \rightarrow (a \leq c)$,

(29) $(a < b) \leftrightarrow (-a > -b)$,

(30) $(a < b) \leftrightarrow (a - c < b - c)$,

(31) $(a < 0) \leftrightarrow (1/a < 0)$,

(32) $(a > 0) \leftrightarrow (1/a > 0)$,

(33) $(a < b) \vee (a = b) \vee (a > b)$,

(34) $(ab > 0) \leftrightarrow (a/b > 0)$,

(35) $(ab > 0) \rightarrow (a < b \leftrightarrow 1/a > 1/b)$,

(36) $(ab < 0) \rightarrow (a < b \leftrightarrow 1/a < 1/b)$,

(37) $a^2 + b^2 \geq 2ab$,

(38) $(a \geq 0 \wedge b \geq 0) \rightarrow \left[\dfrac{a + b}{2} \geq \sqrt{ab}\right]$,

(39) $(a > 1) \rightarrow (a^2 > a)$,

(40) $(0 < a < 1) \rightarrow (a^2 < a)$,

(41) $(0 < a < b \wedge 0 < c < d) \rightarrow (ac < bd)$,

(42) $(a < b \wedge c < d) \rightarrow (a + c < b + d)$,

(43) $(0 \leq a \wedge 0 \leq b) \rightarrow [a \leq b \leftrightarrow a^2 \leq b^2]$.

★7. Any sentence involving $<, >, \leq$, or $\geq$ can be rewritten by using the definitions (5), (19), and (20). For example, (29) yields $b > a \leftrightarrow -b < -a$. Also, laws concerning $\leq$ and $\geq$ can be constructed by combining laws of equality and inequality. For example, (29), (2-6-11), and (2-7-37) yield $a \leq b \leftrightarrow -a \geq -b$. Use these and other methods to find further theorems about inequalities.

★8. The *arithmetic mean* of a and b is $(a + b)/2$, and the *geometric mean* is $\sqrt{ab}$. State (38) verbally. Prove (37) by showing that it is equivalent to $(a - b)^2 \geq 0$. Prove (38) by a substitution in (37).

In Problems 9 through 12 identify the set.

9. $\{x \mid x^2 - 2 < 0\}$.
10. $\{x \mid x^2 - 2x + 4 > 0\}$.
11. $\{x \mid (x - 7)(x + 4) \leq 0\}$.
12. $\{x \mid x(x + 1)(x - 2) \geq 0\}$.

13. Show that (21) is equivalent to the statement that no negative number has a real square root.

Problems 14 through 17 refer to the discriminant of a quadratic equation and the quadratic formula discussed in Section 2–8.

14. Without solving, determine whether the following have real roots: $x^2 - x + 1 = 0$, $2x^2 + 5 + 1 = 0$, $x^2 - 10x + 25 = 0$.
15. For what values of k are the roots of $x^2 + 2x + k = 0$ real?
16. For what values of k are the roots of $2kx^2 + x - 1 = 0$ not real?
17 For what values of k are the roots of $x^2 + kx + 2 = 0$ real?

★18. Show that if $A^2 > K$, $B^2 > K$, $A > 0$, and $B > 0$, then $AB > K$.
★19. Show that $0 < a < b \wedge x > 0 \rightarrow a/b < (a + x)/(b + x) < 1$.

★20. Suppose that $Y = C + S$, when Y = national income, C = consumption, and S = savings. Show that with national income constant, a decrease in savings means an increase in consumption.
★21. Prove $0 < v < 1 \rightarrow 1 - v < 1 - v^2$.

★22. Prove $\left(\dfrac{a}{a + b} > \dfrac{c}{c + d}\right) \leftrightarrow \left(\dfrac{a}{a + c} > \dfrac{b}{b + d}\right)$.

★23. Prove $\left(-1 \leq \dfrac{r_{12} - k_1}{k_2} \leq 1\right) \rightarrow (k_1 - k_2 \leq r_{12} \leq k_1 + k_2)$.

24. Under what conditions is $4.36\sigma > 2.65\sigma$?
★25. Prove that $(ax + by)^2 \leq (a^2 + b^2)(x^2 + y^2)$.
26. Solve for x: $ax - 2 + (2 - a) < x + a + (2 - a)$.
27. Solve $\dfrac{3}{x} + 4 \leq \dfrac{2}{x}$.

28. "The square, divided by 10, of any number larger than 0 and smaller than 10 ... is a positive quantity smaller than the number itself." ("Predicting Supreme Court Decisions Mathematically," by Fred Kort, in the *American Political Science Review*, Vol. LI, No. 1, March, 1957.) Prove this.

ANSWERS TO EXERCISES

(a) $(a < 0) = (0 - a) \in Re^+ = -a \in Re^+$.   (b) From (6) and the axioms.   (d) $(a < b) \leftrightarrow (b - a) \in Re^+ \leftrightarrow (c - c) + (b - a) \in Re^+ \leftrightarrow b + c - (a + c) \in Re^+ \leftrightarrow a + c < b + c$.   (f) For example, (11): If a is to the right of the origin, $-a$ is to the left.   (g) False for any $m < 0$; $m > 0 \to (mx + b < 0 \leftrightarrow x < -b/m) \wedge m < 0 \to (mx + b < 0 \leftrightarrow x > -b/m)$.
(h) $x < -3/7$.   (i) $a \leq b$: the point a coincides with or is to the left of b.
(j) $x = 0 \to x^2 = 0$; $x > 0 \to x^2 > 0$.   (k) $(x + 1)(x - 1) \leq 0 \leftrightarrow$ $[(x + 1) \leq 0 \wedge (x - 1) \geq 0] \vee [(x + 1) \geq 0 \wedge (x - 1) \leq 0] \leftrightarrow$ $[x \leq -1 \wedge x \geq 1] \vee (x \geq -1 \wedge x \leq 1] \leftrightarrow -1 \leq x \leq 1$; and the graph of $x^2 - 1 \leq 0$ is the closed interval $(-1 \quad 1)$.   (l) $x < -2 \vee x > 2$.
(m) $-\dfrac{1}{\sqrt{2}} < x < \dfrac{1}{\sqrt{2}}$.   (n) $x \leq -\sqrt{17} \vee x \geq \sqrt{17}$.   (o) $-2 < x < -1$.
(p) $x < (-3 - \sqrt{5})/2 \vee x > (-3 + \sqrt{5})/2$.   (q) $-4 < x < 3$.

ANSWERS TO PROBLEMS

1. From (2-2-11), (2-7-48), and other laws of logic.   3. (15)$(a{:}0, b{:}1, c{:}a)$.
5. (15)$(a{:}{-}1, b{:}0, c{:}a)$.   9. $(-\sqrt{2} \quad \sqrt{2})$.   11. $(-4 \quad 7)$.   13. $\forall x [x \in Re \to x^2 \geq 0] = \sim \exists x \sim [x \in Re \to x^2 \geq 0] = \sim \exists x [x \in Re \wedge x^2 < 0]$.
15. $\{k \mid 2^2 - 4k \geq 0\} = \{k \mid k - 1 \leq 0\} = (-\infty \quad 1)$.   17. $k \leq -\sqrt{8} \vee k \geq \sqrt{8}$.

3–6 Operations on sets. We consider a universe of discourse U together with all its subsets. If A is the subset, the *complement of A*, A' (read "A prime"), is the set of all elements of U that do not belong to A. Formally,

(1) Def. $A' = \{x \mid \sim(x \in A)\}$.

For example, if $U = \{0, 2, 4, 6, 8, 10\}$, $\{0, 4, 8\}' = \{2, 6, 10\}$.

(a) With the same U, $\{2, 4, 6\}' = ?$ (b) Justify (2) and (3).

(2) $x \in A' = \sim(x \in A)$,

(3) $\{x \mid f(x)\}' = \{x \mid \sim f(x)\}$.

If A and B are subsets, their *intersection*, $A \cap B$, is the set of elements of U that belong to both of them.

(4) Def. $A \cap B = \{x \mid x \in A \wedge x \in B\}$.

For example, $\{0, 2, 6, 8\} \cap \{2, 8, 10\} = \{2, 8\}$.

(c) $\{a, b, c, d\} \cap \{a, c, e, f\} = ?$ (d) Justify (5) and (6).

(5) $x \in (A \cap B) = x \in A \land x \in B,$

(6) $\{x\,|\,f(x)\} \cap \{x\,|\,g(x)\} = \{x\,|\,f(x) \land g(x)\}.$

If A and B are subsets, their *union*, $A \cup B$, is the set of elements that belong to either or both of them.

(7) Def. $A \cup B = \{x\,|\,x \in A \lor x \in B\}.$

For example, $\{0, 2, 6, 8\} \cup \{2, 8, 10\} = \{0, 2, 6, 8, 10\}.$

(e) $\{a, b, c, d\} \cup \{a, c, e, f\} = ?$ (f) Justify (8) and (9).

(8) $x \in (A \cup B) = x \in A \lor x \in B,$

(9) $\{x\,|\,f(x)\} \cup \{x\,|\,g(x)\} = \{x\,|\,f(x) \lor g(x)\}.$

(g) The symbols $\cap$ and $\cup$ are sometimes read "cap" and "cup." Suggest a reason for the choice of these symbols.

The intersection, union, and complement may be conveniently visualized geometrically if we think of the universe of discourse as a region in a plane and the other sets as regions within it. In Fig. 3–3 we show the intersection as the region common to A and B, the union as the region consisting of those points in either or both A and B, and the complement as the region outside A. Note that A' is not everything not in A, but everything in U that is not in A. Diagrams like Fig. 3–3 are called *Venn diagrams*. They are useful aids to the imagination; however, like the truth tables in Chapter 2, they are not to be confused with proofs.

$$A \cap B \qquad\qquad A \cup B \qquad\qquad A'$$

FIGURE 3–3

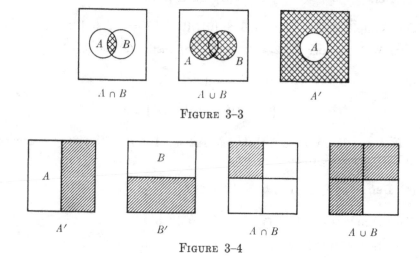

$$A' \qquad\qquad B' \qquad\qquad A \cap B \qquad\qquad A \cup B$$

FIGURE 3–4

A special kind of Venn diagram, due to Lewis Carroll, is illustrated in Fig. 3–4. It is very convenient for visualizing the subsets formed from a universe by two sets.

(h) Copy the diagram for $A \cap B$ in Fig. 3–4 and label the three unshaded squares properly, using the following labels: $A' \cap B'$, $A' \cap B$, $A \cap B'$. On a similar figure, shade (i) $A' \cup B'$, (j) $A' \cup B$, (k) $A \cup B'$, (l) $(A \cap B)'$. (m) Repeat Exercises (i) through (l) using a Venn diagram like that of Fig. 3–3.

Let U = the human race, A = the residents of New York City, and B = the citizens of the U. S. A. Express the following symbolically and draw Venn diagrams: (n) the American citizens who live in N. Y.,   (o) aliens, (p) people who live outside N. Y.,   (q) citizens who do not live in N. Y., (r) the set of all people,   (s) residents of N. Y. together with all aliens.

With symbols defined as in the preceding paragraph, express in words: (t) $A' \cap B$,   (u) $A \cap B$,   (v) $A \cup B$,   (w) $A' \cap B'$,   (x) $A' \cup B'$.

Complements, unions, and intersections of intervals are very convenient for specifying the solutions of inequalities. For example, we found in Section 3–5 that $\{x \,|\, x^2 - 4 < 0\} = (-2\_\_2)$. Then

(10) $\{x \,|\, x^2 - 4 \geq 0\} = (-2\_\_2)' = \{x \,|\, x \leq -2 \ \vee \ x \geq 2\}$

(11) $= \{x \,|\, x \leq -2\} \cup \{x \,|\, x \geq 2\}$

(12) $= (-\infty\_\_\_-2) \cup (2\_\_\infty).$

By combining intervals we can get a very large variety of sets of points on the axis of real numbers. Among these sets are those consisting of single points, since $\{x\} = (a\_\_x) \cap (x\_\_b)$ for any $a < b$. Thus we may think of a set consisting of a single point as a "degenerate" interval. We might even write $\{x\} = (x\_\_x)$.

Find a simple expression for and sketch the following sets: (y) $(2\_\_\infty)'$, (z) $(3\_\_5) \cap (4\_\_6)$,   (z') $(3\_\_5) \cup (4\_\_6)$.

Problems

An insurance company with policyholders U is studying characteristics of certain kinds of policyholders. Let A = adult policyholders, B = male policyholders, C = married policyholders. Describe in words:

1. A'.
2. B'.
3. C'.
4. $A \cap B$.
5. $A \cup B$.
6. $B \cap C'$.
7. $B' \cap C'$.
8. $B' \cup C'$.
9. $A \cap (B \cap C)$.
10. $(A \cap B) \cap C$.
11. U'.
12. $A' \cap (B \cap C)$.
13. $A' \cup (B \cap C)$.
14. $(B' \cap C) \cap A$.
15. $(A \cup B)'$.
16. $B \cup B'$.

Using three circles, draw Venn diagrams of the following:

17. $A \cap (B \cap C)$.   18. $A \cup (B \cup C)$.   19. $(A \cap B')'$.
20. $(A \cup B) \cap C$.   21. $B \cup (C \cap A')$.   22. $A' \cap (B' \cap C')$.

★23. Suggest ways in which diagrams of the type of Fig. 3–4 could be adapted to representing three or more sets.

Sketch:

24. $(2 - \epsilon\_2 + \epsilon)$ where $\epsilon > 0$.   25. $(\underline{3\quad 5}) \cup (\underline{5\quad 6})$.
26. $(\underline{3\quad 4}) \cap (\underline{4\quad 5})$.   27. $(-\underline{1}\quad 3) \cup (\underline{0\quad 2})$.
28. $(-\underline{1}\quad 3) \cup (\underline{2\quad 5})$.   29. $(-\underline{1}\quad 3) \cap (\underline{2\quad 5})$.
30. $(-\underline{1}\quad 3) \cap (\underline{0\quad 2})$.   31. $\{1\} \cup (\underline{3\quad 4})$.

Show that

32. $(-\infty\_b)' = (\underline{b}\quad \infty)$.   33. $(\underline{b}\quad \infty)' = (-\infty\_b)$.
34. $(\underline{a\_b})' = (-\infty\_a) \cup (\underline{b}\_\infty)$.   35. $(\underline{a\quad b})' = (-\infty\_a) \cup (\underline{b}\_\infty)$.
36. $(\underline{a\quad b})' = (-\infty\_a) \cup (\underline{b}\_\infty)$.   37. $(\underline{a\quad b})' = (-\infty\_a) \cup (\underline{b}\_\infty)$.
38. $(\underline{a\quad a}) = \{a\}$.   39. $(\underline{a\_a}) = \emptyset$.
40. $a > b \rightarrow [(\underline{a\quad b}) = \emptyset]$.
41. $(\underline{a\quad b}) \cap (\underline{c\quad d}) = (\underline{c\quad b})$ is not a law.

For Problems 42 through 45, let $A \underline{\cup} B$ be the elements in A or B, but not both.

★42. Give a formal definition.
★43. Draw a Venn diagram and a Lewis Carroll type.
★44. Show that $A \underline{\cup} B = (A \cup B) \cap (A \cap B)'$.
★45. Under what conditions is $A \underline{\cup} B = A \cup B$?

The *relative complement* $B - A$ of A with respect to B is defined as the set of elements of B that are not in A.

★46. Give a formal definition.
★47. Show that the complement of a set is its relative complement with respect to the universe of discourse.

Express 48 through 55 in terms of intervals.

48. $\{x \mid (x + 1)(x - 3) \geq 0\}$.   49. $\{x \mid (x + 1)(x - 3) \leq 0\}$.
50. $\{x \mid x^2 - 3 < 0\}$.   51. $\{x \mid x^2 - 3 > 0\}$.
52. $\{x \mid x^2 + 5x - 6 \geq 0\}$.   53. $\{x \mid x^2 + x + 2 > 0\}$.
54. $\{x \mid x^2 + 2x + 5 \leq 0\}$.   55. $\{x \mid x^2 + 2x + 1 < 0\}$.

ANSWERS TO EXERICSES

(a) $\{0, 8, 10\}$.   (b) $(3\text{-}3\text{-}3)$.   (c) $\{a, c\}$.   (d) $(3\text{-}3\text{-}3)$.   (e) $\{a, b, c, d, e, f\}$.
(f) $(3\text{-}3\text{-}3)$.   (g) Similarity to $\wedge$ and $\vee$ in terms of which they are defined.
(h) See Fig. 3–5.   (i) Fig. 3–6.   (j) Fig. 3–7.   (k) Fig. 3–8.   (l) See Fig. 3–6.
(n) $A \cap B$.   (o) B'.   (p) A'.   (q) $A' \cap B$.   (r) U.   (s) $A \cup B'$.   (t) Citizens who do not live in N. Y.   (u) Citizens who live in N. Y.   (v) Those who are citizens or who live in N. Y., i.e., citizens and New Yorkers.   (w) Aliens

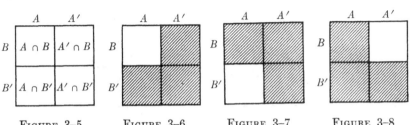

	A	A'
B	$A \cap B$	$A' \cap B$
B'	$A \cap B'$	$A' \cap B'$

FIGURE 3–5   FIGURE 3–6   FIGURE 3–7   FIGURE 3–8

who live outside N. Y.   (x) Those who live outside N. Y. or who are aliens. (y) $(-\infty\_2)$.   (z) (4_5).   (z') (3_6).

ANSWERS TO PROBLEMS

1. Minor policyholders.   3. Unmarried policyholders.   5. All adults and all males.   7. Unmarried females.   9. Married adult males.   11. Empty set.   13. Minors and all married males.   15. Those who are neither adults nor males, i.e., the minor females.   25. (3_6).   27. (-1_3).   28. (-1_5). 31. A point and an interval.   32 through 41. Use definitions from Section 3–4. 42. $\{x \mid x \in A \underline{\vee} x \in B\}$.   45. $A \cap B = \emptyset$.   46. $B - A = A' \cap B$. 49. (-1_3).   51. $(-\infty\_-\sqrt{3}) \cup (\sqrt{3}\_\infty)$.   53. $(-\infty\_\infty)$.   55. $\emptyset$.

★3–7 Algebra of sets. Since operations on sets are defined in terms of the corresponding logical operations, we might expect that the sets would satisfy laws similar to those of logic.

(1)	$(A \cup B)' = A' \cap B'$	
(2)	$(A \cap B)' = A' \cup B'$	(*De Morgan's laws*),
(3)	$A \cup B = B \cup A$	
(4)	$A \cap B = B \cap A$	(*Commutative laws*),
(5)	$A \cup (B \cup C) = (A \cup B) \cup C$	
(6)	$A \cap (B \cap C) = (A \cap B) \cap C$	(*Associative laws*),
(7)	$A \cup (B \cap C) = (A \cup B) \cap (A \cup C)$	
(8)	$A \cap (B \cup C) = (A \cap B) \cup (A \cap C)$	(*Distributive laws*),
(9)	$A \cup A = A$	
(10)	$A \cap A = A$	(*Idempotent laws*),
(11)	$(A')' = A$	(*Involution law*).

Diagrams are helpful in appreciating these and other laws. By making separate diagrams for each side, we can see that different procedures are

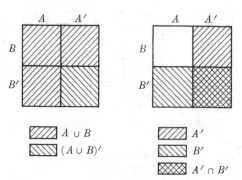

FIGURE 3–9

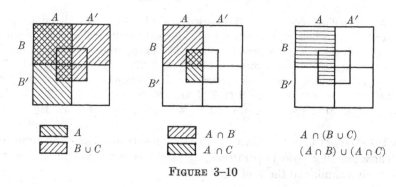

FIGURE 3–10

involved and that the final result is the same. Different kinds of shading should be used for each set that is of importance, and the sets should be clearly labeled. Figure 3–9 shows Lewis Carroll diagrams of the two sides of (1). In Fig. 3–10 we sketch the sides of (8), letting the central square represent C.

(a) Let $U = $ people, $A = $ students, $B = $ males, $C = $ females, so that $A \cap (B \cap C) = $ students who are both male and female, and $A \cup (B \cup C) = $ those who are either students or men or women. State each law above in words in terms of these sets.

The proofs of the above laws are easy. One simply applies the definitions and then the identities of Section 2–5. For example,

(12)   Proof of (3):

   (α) $A \cup B = \{x \,|\, x \in A \lor x \in B\}$ (3–6–7)

   (β) $= \{x \,|\, x \in B \lor x \in A\}$
                $(p \lor q = q \lor p)(p{:}x \in A, q{:}x \in B)$

   (γ) $= B \cup A$ (3–6–7)$(A{:}B, B{:}A)$.

(b) What rules of proof were used in step (β)?   (c) in step (γ)?

One can also argue informally for these identities by showing, as called for by (3–3–4), that every element of one set belongs to the second, and conversely. To argue for (3) in this way we suppose that an element belongs to $A \cup B$. Then it must belong to A or to B or both. But then it must belong to B or to A or to both, and hence to $B \cup A$. Similarly, beginning with an element belonging to $B \cup A$, we argue that it must belong to $A \cup B$. Of course this is merely an informal paraphrase of (12) with the logical identities remaining implicit.

(d) Argue informally for (1) and prove it.

(13) Def. $A \cup B \cup C = A \cup (B \cup C)$,

(14) Def. $A \cap B \cap C = A \cap (B \cap C)$,

(15) $\qquad A \cup \emptyset = A$,

(16) $\qquad A \cap \emptyset = \emptyset$,

(17) $\qquad A \cup U = U$,

(18) $\qquad A \cap U = A$,

(19) $\qquad A \cup A' = U$,

(20) $\qquad A \cap A' = \emptyset$,

(21) $\qquad U' = \emptyset$,

(22) $\qquad \emptyset' = U$,

(23) $\qquad (A = B') \leftrightarrow (A' = B)$,

(24) $\qquad A = (A \cap B) \cup (A \cap B')$,

(25) $\qquad A \cup B = (A \cap B') \cup (A' \cap B) \cup (A \cap B)$,

(26) $\qquad U = (A \cap B) \cup (A' \cap B) \cup (A \cap B') \cup (A' \cap B')$,

(27) $\quad A \cup (A \cap B) = A$,

(28) $\quad A \cap (A \cup B) = A$.

(e) Argue for the above from diagrams and informally from the meanings of the terms.   (f) Prove (15).   (g) Prove (23).

Simplify: (h) $B \cap (A \cup A')$,   (i) $A \cup (C \cup C')$,   (j) $(S \cap T) \cap (T \cap T')$, (k) $(C \cap C') \cap (C \cap C')$,   (l) $(B \cup C) \cup (B' \cap C')$,   (m) $(A \cup B \cup C) \cup (A' \cap B' \cap C')$,   (n) $U \cap \emptyset'$,   (o) $U' \cup \emptyset$,   (p) $(B \cap C) \cup (B \cap C')$. (q) $(S \cap T) \cup (S \cap T') \cup (S' \cap T)$,   (r) $A' \cup (B \cap C)'$.

Using a diagram like Fig. 3–10, sketch each of the following: (s) $A' \cap (B \cup C)$, (t) $A \cap B \cap C$, (u) $A' \cap B' \cap C'$, (v) $A' \cup B' \cup C'$, (w) $A \cap B' \cap C$, (x) $A \cup (B' \cap C')$, (y) the set in Exercise (r).

PROBLEMS

In the introduction to a recent dictionary there appears a diagram similar to the one in Fig. 3–11, in which X = formal literary English, Y = colloquial English, and Z = illiterate English.

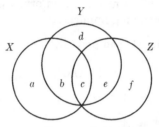

FIGURE 3–11

1. Name a, b, c, d, e, f, $a \cup b$, $b \cup c$, $a \cap X$, and $b \cup d \cup e$ in terms of X, Y, and Z and verbally.

2. What assumption is implicit in the diagram? Express it in symbols and words.

3. For A, B, and C defined as in Problem 4, find the sets of Exercises (s) through (y).

4. Illustrate the laws of this section for $U = \{1, 2, 3, 4, 5\}$, $A = \{1, 2, 3\}$, $B = \{2, 3, 4\}$, $C = \{2, 4, 5\}$.

5. Find the complement of $(A' \cup B') \cap (A' \cup C')$.

6. Prove some of the laws of this section.

7. Derive (26) by applying the distributive law to the right member.

★8. Discover and prove some laws involving the relative complement.

★9. Do the same for $A \underline{\cup} B$.

★10. Interpret the following definitions:

(29) **Def.** $\bigcup S = \{x \mid \exists y \ y \in S \land x \in y\}$,

(30) **Def.** $\bigcap S = \{x \mid \forall y \ y \in S \rightarrow x \in y\}$.

★11. Show that $\bigcup \{A, B\} = A \cup B$, and $\bigcap \{A, B\} = A \cap B$.

★12. Let K = the set of points in a plane within and on the boundary of a square. Let S consist of all circular regions that include K. Argue that $\bigcap S = K$.

★13. Let C = the set of all points in a plane within and on the circumference of a circle. Let T be the set of all polygons inscribed in the circle. Argue that $\bigcup T = C$.

14. Aristotle expressed the fundamental laws of logic as follows: (1) All A is A. (2) Nothing is both A and not A. (3) Everything is either A or not A. Express these laws in set terminology.

15. Prove (28).

Answers to Exercises

(a) (1) Those who are not either students or males = those who are non-students and nonmales. (2) Those who are not male students = those who are either not students or not males. (3) Those who are either students or males = those who are either males or students. (4) Student males = male students. (5) Those who are either students, or males or females = those who are students or males, or females. An alternative: Students together with males and females = students and males together with females. (6) Students who are males and females = student males who are females. (7) Students together with the male females = those who are both student males and student females. (8) Students who are males or females. (9) Students and students = students. (10) Students who are students = students. (11) Those who are not non-students = students. *Note:* Either "and" or "or" may be involved in a correct translation of $\cup$. Thus $A \cup B$ = the A's and the B's = those who are in A or in B = those who are in A and those who are in B. But $A \cap B$ = the A's in B = the B's in A = those who are in A and in B. It is best not to rely on rules or memory, but to understand the meaning and translate into the best English in the particular context.

(b) Rule of Substitution in $p \vee q = q \vee p$ to get $x \in A \vee x \in B = x \in B \vee x \in A$, then Rule of Replacement to insert the second for the first. (c) Rule of Substitution in (3–6–7), then Rule of Replacement, to replace the right member of (β) by (γ). (d) If an element belongs to $(A \cup B)'$, it does not belong to $A \cup B$. Hence it cannot belong to A or to B. Hence it does not belong to A and it does not belong to B, that is, it belongs to $A' \cap B'$. The converse is shown similarly. The proof goes as follows: $(A \cup B)' = \{x \mid x \in (A \cup B)'\} = \{x \mid \sim(x \in A \cup B)\} = \{x \mid \sim(x \in A \vee x \in B)\} = \{x \mid \sim(x \in A) \wedge \sim(x \in B)\} = \{x \mid x \in A' \wedge x \in B'\} = \{x \mid x \in (A' \cap B')\} = A' \cap B'$. An alternative way of working the proof is to write $x \in (A \cup B)' = \sim(x \in (A \cup B)) = \ldots = x \in (A' \cap B')$, and then use (3–3–4).

(e) For example, (15) says that the objects that belong to a set or to the null set are just those that belong to the set. (f) $A \cup \emptyset = \{x \mid x \in A \vee x \in \emptyset\}$. By (3–3–2), $\sim(x \in \emptyset)$, and by (2–7–31) $\sim(x \in \emptyset) \to [(x \in A \vee x \in \emptyset) \leftrightarrow (x \in A)]$. Hence by the Rule of Inference $(x \in A \vee x \in \emptyset) \leftrightarrow (x \in A)$. Hence $\{x \mid x \in A \vee x \in \emptyset\} = \{x \mid x \in A\}$ by (3–3–6), and $\{x \mid x \in A\} = A$. (g) By hypothesis $A = B'$. Hence $A' = (B')'$. But $(B')' = B$ by (11)(A:B). Hence $A = B' \to A' = B$. The converse is proved similarly. (h) B. (i) U. (j) $\emptyset$. (k) $\emptyset$. (l) U. (m) U. (n) U. (o) $\emptyset$. (p) B. (q) $S \cup T$. (r) $(A \cap B \cap C)'$.

Answers to Problems

1. $a = X \cap Y' \cap Z'$, $b = X \cap Y \cap Z'$, $c = X \cap Z$, $d = X' \cap Y \cap Z'$ $e = X' \cap Y \cap Z$, $f = X' \cap Y' \cap Z$, $a \cup b = X \cap Z'$, $b \cup c = X \cap Y$, $a \cap X = a$, $b \cup d \cup e = Y \cap (X \cap Z)'$. 2. $X \cap Y' \cap Z = \emptyset$. There is no formal literary illiterate English that is not colloquial; i.e., all formal literary English that is illiterate is also colloquial. 3. (s) $\{4, 5\}$. (t) $\{2\}$. (u) $\emptyset$. (v) $\{2\}'$. (w) $\emptyset$.

(x) $\{1, 2, 3\}$.   (y) $\{1, 3, 4, 5\}$.   5. $A \cap (B \cup C)$.   7. Right member =
$[A \cap (B \cup B')] \cup [A' \cap (B \cup B')] = (A \cap U) \cup (A' \cap U) = A \cup A' = U$.
15. $A \cap (A \cup B) = (A \cup \emptyset) \cap (A \cup B) = A \cup (\emptyset \cap B) = A \cup \emptyset = A$.

★**3–8 Relations between sets.** In Section 3–4 we defined and illustrated
briefly the relation of inclusion. With the aid of set algebra we can analyze inclusion further and consider some other possible relations between
sets.

From the definition (3–4–1) we have immediately

(1) $A \subseteq A$,

(2) $[A \subseteq B \wedge B \subseteq C] \to A \subseteq C$,

(3) $\emptyset \subseteq A$,

(4) $A \subseteq U$.

From (3–3–4),

(5) $[A \subseteq B \wedge B \subseteq A] \to A = B$.

Additional laws are

(6) $(A \cap B) \subseteq A$,

(7) $A \subseteq (A \cup B)$,

(8) $A \subseteq B \leftrightarrow B' \subseteq A'$,

(9) $A \subseteq B \to (A \cap C) \subseteq (B \cap C)$,

(10) $A \subseteq B \to (A \cup C) \subseteq (B \cup C)$,

(11) $A \subseteq B \leftrightarrow [A \cap B = A]$,

(12) $A \subseteq B \leftrightarrow [A \cup B = B]$,

(13) $A \subseteq (B \cap C) \leftrightarrow [A \subseteq B \wedge A \subseteq C]$,

(14) $(A \cup B) \subseteq C \leftrightarrow [A \subseteq C \wedge B \subseteq C]$,

(15) $[A \subseteq B \wedge \sim (A \subseteq C)] \to \sim (B \subseteq C)$,

(16) $A \subseteq B \leftrightarrow [A \cap B' = \emptyset]$.

These theorems are intuitively evident if they are expressed verbally
or illustrated by Venn diagrams. They are also easy to prove by simply
using the definitions and applying theorems from Chapter 2. For example, to prove (6), we assume by hypothesis that $x \in A \cap B$. Then
$x \in A \wedge x \in B$. But by the law $[p \wedge q] \to p$ and the Rule of Inference

it follows that $x \in A$. Hence $\forall x \, (x \in A \cap B \to x \in A)$, which is equivalent to (6) by the definition of inclusion. Or we may argue informally that if an element lies in both A and B, it must lie in A, and hence every member of $A \cap B$ is a member of A.

(a) Argue informally for the above laws in terms of Venn diagrams and the meaning of terms.   (b) Prove (7).

To prove (15) we note that it is equivalent to (2) by the identity

$$[(p \wedge q) \to r] = [(p \wedge {\sim}r) \to {\sim}q]$$

with

$$(p{:}A \subseteq B, \; q{:}B \subseteq C, \; r{:}A \subseteq C).$$

(c) Prove the logical identity (2–5–35).   (d) Make the substitution $(A{:}C, B{:}A, C{:}B)$ in (2) and then prove (17).

(17) $$[A \subseteq B \wedge {\sim}(C \subseteq B)] \to {\sim}(C \subseteq A).$$

To show the plausibility of (16) we note that A is a subset of B if and only if every member of A is a member of B, that is, if and only if no members of A are in B', that is, $A \cap B' = \emptyset$.

To prove it, we write

$$\begin{aligned}
A \subseteq B &= [\forall x \, (x \in A \to x \in B)] \\
&= [{\sim}\exists x \, {\sim}(x \in A \to x \in B)] \\
&= [{\sim}\exists x \, (x \in A \wedge x \in B')] \\
&= [{\sim}\exists x \, x \in (A \cap B')] \leftrightarrow [A \cap B' = \emptyset].
\end{aligned}$$

(e) Justify each step in the preceding argument.

If two sets have no members in common we say that they are *disjoint*. This relation may be symbolized by $A \mathbin{//} B$, the notation being suggested by the fact that two straight lines in a plane are disjoint if and only if they are parallel.

(18) **Def.** $A \mathbin{//} B = [A \cap B = \emptyset].$

(f) Draw a Venn diagram showing two disjoint sets.   (g) Show

(19) $$A \mathbin{//} B \to B \mathbin{//} A.$$

(h) Prove (20).

(20) $$A \mathbin{//} B \leftrightarrow A \subseteq B'.$$

(i) Prove that $A//B \leftrightarrow B \subseteq A'$.

(21) $[A // B \wedge A // C] \rightarrow A // (B \cup C)$.

(j) Show that $[A//B \wedge B//C] \rightarrow A//C$ is not a law.

We define two further relations, leaving their properties to be worked out by the reader.

(22) **Def.** $A \subset B = [A \subseteq B \wedge A \neq B]$.

When $A \subset B$ we say that A is a *proper subset* of B.

(k) Give an example for which $A \subset B \wedge A \subseteq B$.

(23) **Def.** $A \between B = \sim[A \subseteq B \vee B \subseteq A \vee A // B]$.

When $A \between B$ we say that A *overlaps* B.

(l) Draw a Venn diagram illustrating $A \between B$ and several examples in which $\sim(A \between B)$.   (m) Illustrate the definitions and theorems of this section using $U = \{1, 2, 3, 4, 5\}$ and its subsets.

Problems

1. Complete the reasons in the following proofs and supply informal arguments for the laws.

(24) Proof of (2):

    (a) $x \in A \rightarrow x \in A$ (?),

    (b) $A \subseteq A$ (3-4-1)($B{:}A$).

(25) Proof of (3):

    (a) $A \subseteq B \wedge B \subseteq C$ Hyp,

    (b) $(x \in A \rightarrow x \in B) \wedge (x \in B \rightarrow x \in C)$ (1), (2-10-14),

    (c) $x \in A \rightarrow x \in C$ (?),

    (d) $A \subseteq C$ (?),

    (e) (3)   Q.E.D., (a), (d).

2. Prove (8) by using (2-5-28).

3. Prove (9) by using (2-7-25).

4. Prove (10) by using (2-7-11).

5. Prove (11) by noting that $A \cap B \subseteq A$ by (6), and proving that $A \subseteq A \cap B$ from the hypothesis $A \subseteq B$ and (9)($C{:}A$). Why is this sufficient?

6. Prove (12) from (7) and (10).

7. Prove that $A \subseteq B \leftrightarrow A//B'$.

8. Prove (26).

(26)　　　　　　　　$A \subseteq B \leftrightarrow [A \subset B \lor A = B]$.

9. Discover, argue for, and prove other laws that involve proper inclusion.

10. Prove (27).

(27)　　　　　　　　$[A//B \land C \subseteq A] \to C//B$.

11. Prove that the sets in the right member of (3–7–25) are disjoint two by two.

12. Do the same for (3–7–26).

13. Prove (28).

(28)　　　　　　　　$A \between B \to B \between A$.

14. Do Problem 9 for overlapping.

15. Show that $\exists A \ \exists B \ A//B \land A \subseteq B$!

16. Show that $A = B \lor A \subset B \lor B \subset A \lor A//B \lor A \between B$ is not a law. Show that it is a law if the hypothesis $A \neq \emptyset \land B \neq \emptyset$ is added.

★17. Prove $[\{x \mid a \leq x \leq b\} = \{x \mid c \leq x \leq d\}] \to (a = c \land b = d)$.

★18. Prove $\{x \mid a \leq x < b\} \neq \{x \mid a \leq x \leq b\}$.

★19. Prove $\{x \mid 1 \leq x \leq 2\} \subseteq \{x \mid 0.5 \leq x \leq 7\}$.

Under what conditions does

★20. $(a\_\_b) \between (c\_\_d)$?

★21. $(a\_\_b)//(c\_\_d)$?

★22. $(a\_\_b) \subset (c\_\_d)$?

★23. Prove $A \subseteq \emptyset \to A = \emptyset$.

★24. Consider the following sets of figures in plane geometry: T = triangles, O = obtuse triangles, R = right triangles, E = equilateral triangles, S = scalene triangles, I = isosceles triangles, A = acute triangles. What relation holds between each pair? (Pair each one in order with every following one.)

★25. Prove $\exists x \ \exists y \ x \in y \land x \subset y$.

★26. Prove $[A \cup X = U \land A \cap X = \emptyset] \to [X = A']$.

Answers to Exercises

(a) For example, (7) is plausible, since any member of A is certainly in A or in B. Again, (8) is true because if every element of A is in B, any element not in B cannot be in A.　　(b) $x \in A \to x \in A \lor x \in B \to x \in (A \cup B)$ by $(p \to p \lor q)(p{:}x \in A, q{:}x \in B)$.　　(c) $[(p \land q) \to r] = [\sim(p \land q) \lor r] = \sim p \lor \sim q \lor r$; $[(p \land \sim r) \to \sim q] = [\sim(p \land \sim r) \lor \sim q) = \sim p \lor \sim \sim r \lor \sim q = \sim p \lor \sim q \lor r$.　　(d) It yields $C \subseteq A \land A \subseteq B \to C \subseteq B$; then by (2–5–35) $(p{\cdot}A \subseteq B, q{:}A \subseteq B, r{:}C \subseteq B)$ we have (17).

(e) (3–4–1); (2–10–3) $(f(x){:}\sim(x \in A \to x \in B))$; (2–5–24); (3–6–4); (3–3–8).

FIGURE 3-12

$A \between B$

FIGURE 3-13

(f) See Fig. **3-12**.   (g) Since $[A \cap B = \emptyset] = [B \cap A = \emptyset]$.   (h) (16), (18).
(i) (19), (20).   (j) $(A:\{1, 2\}, B:\{3, 4\}, C:\{1, 2\})$.   (k) $(A:\{1, 2\}, B:\{1, 2, 3\})$.
(l) See Fig. 3-13.

ANSWERS TO PROBLEMS

1. (24a) $p \to p$ (25c) $[(p \to q) \land (q \to r)] \to (p \to r)$.   (25d)   (3-4-1).
Note that quantifiers are omitted.   3. $(p:x \in A, q:x \in B, r:x \in C)$.   5. (5).
7. $A \subseteq B \leftrightarrow A \cap B' = \emptyset \leftrightarrow A//B'$ by (16) and (18).   9. Some examples:
$\sim(A \subset A)$, $A \subset B \to \sim(B \subset A)$, $A \subset B \land B \subset C \to A \subset C$.   11. Any
product of two of these sets has $A \cap A'$ or $B \cap B'$ as a factor.   13. (23).
15. $(A:\emptyset. \ B:\emptyset)$.   21. $b \leq c \lor a \geq d$.

3-9 Descriptions. It is often convenient to refer to something as the
object that satisfies a certain condition. For example, "the set that has
no members" is a description of the empty set. Such descriptions take
the form "the x such that . . . ," where the dots stand for some sentence
involving x that is satisfied by one and only one value of x. We introduce
the symbol $\ni$ to stand for "such that" and adopt the formula $x \ni f(x)$ to
stand for the unique solution of $f(x)$. For example, $[x \ni 2x = 1] = 1/2$.
We read $x \ni f(x)$ as "the x such that $f(x)$." Expressions of this form have
meaning if and only if there is one and only one solution of the condition
$f(x)$, that is, if and only if $\exists! x \, f(x)$. The essential property of this undefined
formula is given by the following axiom.

(1) Ax. $[a = x \ni f(x)] = [f(a) \land \forall y \, f(y) \to y = a]$.

(a) Read (1) in words.   (b) Why is "x" in "$x \ni f(x)$" a dummy?   (c) Justify
$a = (x \ni x = a)$.   (d) Comment on "$x \ni x^2 = 1$."   Justify the following:
(e) $1 = (x \ni \forall z \, xz = z)$,   (f) $0 = (x \ni \forall y \, y + x = y)$,   (g) $a/b = (x \ni xb = a)$,
(h) $a - b = (x \ni x + b = a)$,   (i) $\{x | f(x)\} = S \ni \forall x \, [x \in S \leftrightarrow f(x)]$.

We illustrate the utility of the description notation by using it to
formulate several definitions.

Given a set of real numbers S, we say that x is the *maximum* of S if x
is in S and is larger than any other member of S. Similarly we call x the
minimum of S if it is in S and is smaller than any other member of S.

Geometrically the maximum of S is the point in S farthest to the right, and the minimum is the point farthest to the left.

(2) Def. $\max S = x \ni x \in S \land \forall y \ [y \in S \to x \geq y]$,

(3) Def. $\min S = x \ni x \in S \land \forall y \ [y \in S \to x \leq y]$.

Find the following: (j) $\max \{1, 2, 3\}$;   (k) $\max \{5\}$;   (l) $\min \{4, -3, 8\}$;   (m) $\max \{-3, 3\}$;   (n) $\min \{x \mid x^2 \leq 5\}$.

In Section 1–4 the absolute value of x, symbolized by $|x|$, was introduced intuitively as the length of the vector x. In (1–13–17) we defined it by considering separately the cases $x \geq 0$ and $x < 0$. We now can give a simpler definition.

(4) Def. $|x| = \max \{x, -x\}$.

(o) Use (4) to find $|0|$, $|-3|$, $|10|$.

We can now easily prove the following, which are equivalent to (1–13–17).

(5) $x \geq 0 \leftrightarrow |x| = x$,

(6) $x \leq 0 \leftrightarrow |x| = -x$.

The reader should check that the definition is consistent with the intuitive concept. Recalling from Section 1–4 that $b - a$ is interpreted as the vector from a to b or as the directed distance from a to b, we see that $|b - a|$ is the length of this vector. Accordingly we describe $|b - a|$ as the *undirected distance between a and b*. In particular, $|x|$, which is equal to $|x - 0|$, is the undirected distance between the origin and the point x. These geometric interpretations are very helpful in dealing with absolute values.

The following laws are evident from the geometric interpretation. They can, of course, be proved from (4).

(7) $|a| \geq 0$,

(8) $(|a| = 0) \leftrightarrow (a = 0)$,

(9) $|a| = |-a|$,

(10) $(|a| = |b|) \leftrightarrow (a = b \lor a = -b)$,

(11) $|a + b| \leq |a| + |b|$,

(12) $|a| - |b| \leq |a + b|$,

(13) $$|ab| = |a| \cdot |b|,$$

(14) $$\max \{a, b\} = (1/2)[a + b + |b - a|],$$

(15) $$\min \{a, b\} = (1/2)[a + b - |b - a|].$$

The most useful law for solving equations involving absolute values is (10). Thus $(|x - 1| = 3) \leftrightarrow (x - 1 = 3 \vee x - 1 = -3) \leftrightarrow (x = 4 \vee x = -2)$. A still easier way to solve this equation is to note that it requires that x lie at an undirected distance of 3 from 1. Hence it must lie at 4 or $-2.(!)$ We sketch this in Fig. 3–14.

Solve each of the following and sketch as in Fig. 3–14 [note that $x + 2 = x - (-2)$]: (p) $|x - 3| = 1$, (q) $|4 - x| = 3$, (r) $|x + 2| = 1$.

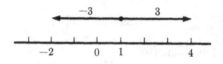

FIGURE 3–14

Solving inequalities involving absolute values is most easily done from the geometric interpretation. Thus to solve $|x - 3| \leq 2$, we note that it requires x to lie within an undirected distance of 2 from 3. Hence it lies between 1 and 5 inclusive, i.e., in the closed interval $(1 \quad 5)$, which is the solution. Similarly, the solution of $|x - 3| \geq 2$ is $(-\infty \quad 1) \cup (5 \quad \infty)$. Also, since

$$|x + 3| \leq 1 \leftrightarrow |x - (-3)| \leq 1 \leftrightarrow -4 \leq x \leq -2,$$

then

$$\{x \mid |x + 3| \leq 1\} = (\underline{-4 \quad -2}).$$

Solve and sketch the solution for: (s) $|x| \leq 3$, (t) $|x| \geq 3$, (u) $|x - 5| \leq 3$, (v) $|x - 8| > 2$.

The following laws generalize the results of the previous exercises.

(16) $$\{x \mid |x| < b\} = (-b\_b),$$

(17) $$\{x \mid |x| > b\} = (-\infty\_\_-b) \cup (b\_\infty),$$

(18) $$\{x \mid |x - a| < b\} = (a - b\_a + b),$$

(19) $$\{x \mid |x - a| > b\} = (-\infty\_a - b) \cup (a + b\_\infty).$$

Use absolute values to write brief defining sentences for: (w) $(-\epsilon\_\epsilon)$, (x) $(3 - \delta\_3 + \delta)$, (y) $(\underline{2 \quad 4})'$, (z) $(a - \delta\_a + \delta)'$.

PROBLEMS

1. In the light of Exercise (g), explain briefly why "$a/0$" is meaningless.

2. Comment on "Whenever they disagree, I choose the version which, in my judgment, is the more credible and at the same time the more interesting of the two."

3. Show that $[a = x \ni f(x)] = [\{a\} = \{x|f(x)\}]$.

4. Comment on "max $\{x \,|\, x < 1\}$."

5. Let $T(x) =$ the truth value of x. Show that $T(p \lor q) = \max\{T(p), T(q)\}$ and $T(p \land q) = \min\{T(p), T(q)\}$.

6. What is the range of significance of "S" in "max S"?

7. Cite several sets S such that max S or min S does not exist.

Solve and sketch:

8. $|2x| = 4$.

9. $|2x + 3| = 7$.

10. $|x| = -1$.

11. $|x| = 0$.

12. $|x + 4| < 1$.

13. $|x + 4| \geq 5$.

14. $|x - 3| > -2$.

15. $|2x - 1| < 2$.

16. $|x - 5| < 0.01$.

17. $|x - 2| < 0.02$.

18. $|x + 0.01| < 5$.

19. $|1 - 3x| = |x|$.

20. Find c and δ such that $(a\_\_b) = \{x \,|\, |x - c| < \delta\}$.

21. Use absolute values to write brief defining sentences for $(3\_\_8)$, $(-1\_\_5)$, and $(2\_\_4)$.

22. Often results of scientific measurements are given in the form $x = a \pm e$, meaning that x probably differs from a by at most e. Express this in each of the forms $\alpha \leq x \leq \beta$, $|x - \gamma| \leq \delta$, and $x \in (r\_\_s)$.

23. $x \ni \forall y \sim (y \in x) = ?$

24. $x \ni \forall y \, y \in x = ?$

25. Use a description to define $\sqrt{x}$. (See 1–13–16.)

★26. Prove (5) by proving that $x \geq 0 \leftrightarrow x \geq -x$.

★27. Prove (6) similarly.

★28. Prove (7) and (8) by brief arguments.

★29. Prove (9) by showing that max $\{a, -a\} = \max\{-a, -(-a)\}$.

★30. Prove (10) by noting its equivalence to max $\{a, -a\} = \max\{b, -b\} \leftrightarrow a = b \lor a = -b$.

★31. Writing (11) in the form $|a - (-b)| \leq |a| + |b|$, interpret it geometrically, sketch, and argue for its plausibility.

★32. Prove (11) by considering the cases $a > 0, b > 0$; $a < 0, b < 0$; $a > 0, b < 0$; and $a < 0, b > 0$.

★33. Treat (12) according to Problems 31 and 32.

★34. Symbolize by a description "the greatest good for the greatest number." Does it exist?

ANSWERS TO EXERCISES

(a) "a is the x such that f of x" means that f of a is true, and if f of y is true, then y equals a; that is, a is the one and only solution of $f(x)$.   (b) $x \ni f(x)$ stands for the solution for the variable x of the sentence $f(x)$. Without a vari-

able present, this would make no sense. Hence x cannot be replaced by a constant.
(c) By (1), since $(a = a) \wedge \forall y (y = a \rightarrow y = a)$.   (d) Nonsense, since the
sentence has two solutions.   (e) $\forall z\, 1 \cdot z = z \wedge \forall y\, [\forall z\, yz = z \rightarrow y = 1]$.
(f) To avoid confusion due to y appearing in two quantifiers, we substitute $(y\!:\!z)$
here to get $0 = (x \ni \forall z\, z + x = z)$. Then $\forall z\, z + 0 = z \wedge \forall y\, [\forall z\, z + y = z \rightarrow y = 0]$.   (g) $(a/b)b = a \wedge \forall y\, yb = a \rightarrow y = a/b$.   Note that this
holds only if $b \neq 0$.   (i) $[x \in \{x\,|\,f(x)\} \rightarrow f(x)] \wedge \forall y\, [(\forall x\, x \in y \leftrightarrow f(x)) \rightarrow y = \{x\,|\,f(x)\}]$.
  (j) 3.   (k) 5.   (l) -3.   (m) 3.   (n) $-\sqrt{5}$.   (o) 0, 3, 10.   (p) $\{2, 4\}$.
(q) $\{1, 7\}$.   (r) $\{-1, -3\}$.   (s) $(-3\underline{\quad}3)$.   (t) $(-\infty\underline{\quad}-3) \cup (3\underline{\quad}\infty)$.
(u) $(2\underline{\quad}8)$.   (v) $(-\infty\underline{\quad}6) \cup (10\underline{\quad}\infty)$.   (w) $|x| < \epsilon$.   (x) $|x - 3| < \delta$.
(y) $|x - 3| > 1$.   (z) $|x - a| \geq \delta$.

Answers to Problems

1. $x \cdot 0 = a$ has no solution or more than one according to $a \neq 0$ or $a = 0$,
as explained in Section 1-14.   3. To say that $\{x\,|\,f(x)\}$ is the singleton $\{a\}$ is
to say that a is the unique solution of $f(x)$.   5. Refer to truth tables and consider cases.   7. $(2\underline{\quad}5)$ has neither max nor min.   9. $\{2, -5\}$.   11. $\{0\}$.
13. $(-9\underline{\quad}1)'$.   15. $(-0.5\underline{\quad}1.5)$.   17. $(1.98\underline{\quad}2.02)$.   19. $\{1/4, 1/2\}$.
  21. $|x - 5.5| < 2.5$, $|x - 2| \leq 3$, $|x - 3| < 1$.   23. $\emptyset$.   25. $y \ni y^2 = x \wedge y \geq 0$.   27. $x \leq 0 \rightarrow -x \geq 0 \rightarrow -x \geq x \rightarrow \max\{x, -x\} = -x$.
  29. Since $\{-a - (-a)\} = \{-a, a\} = \{a, -a\}$!   31. The distance between
$-b$ and a is not greater than the sum of the distances between the origin and
a and b.

★3-10 Sets and sentences.

When we say that a number is prime, we
are asserting that it belongs to the set of prime numbers or that it has the
property of being an integer greater than 1 whose only factors are itself
and 1. In this way every statement asserting set membership can be
reformulated as a claim that the object has the property that is peculiar
to members of the set. This correspondence between sets and properties
is embodied in (3-3-3). Because of it we can formulate any sentence in
terms of set membership, or, conversely, we can formulate any sentence
about set membership in terms not explicitly involving sets.

(a) Consider the statement "3 is odd." Complete the following equivalent
sentences: "$\underline{\qquad} \in \underline{\qquad}$"; "3 is an integer $\underline{\qquad}$."   (b) Consider
the statement "He is my father." Complete the following equivalent sentences:
"$\underline{\qquad} =$ my father"; "$\underline{\qquad} \in \underline{\qquad}$"; "$\underline{\qquad} x \ni x \underline{\qquad}$."
(c) Consider "There is an even prime number." Complete the following
equivalent sentences: "$\{x\,|\,x$ is prime and x is even$\} \neq \underline{\qquad}$"; $\exists x \underline{\qquad}$."

Exercises (a), (b), and (c) suggest the variety of ways of expressing
ideas in terms of or without reference to sets. To systematize the relations
between sets and sentences, let $P = \{x\,|\,p(x)\}$, $Q = \{x\,|\,q(x)\}$. Below
we give a table of statements in terms of the sets and the corresponding
synonymous statements involving the defining sentences.

(1) $a \in P$ $p(a)$,

(2) $P = Q$ $\forall x \, p(x) \leftrightarrow q(x)$,

(3) $P \subseteq Q$ $\forall x \, p(x) \rightarrow q(x)$,

(4) $P \,//\, Q$ $\forall x \sim (p(x) \wedge q(x))$,

(5) $a \in P'$ $\sim p(a)$,

(6) $a \in P \cap Q$ $p(a) \wedge q(a)$,

(7) $a \in P \cup Q$ $p(a) \vee q(a)$,

(8) $P = \emptyset$ $\forall x \sim p(x)$,   or $\sim \exists x \, p(x)$,

(9) $P = U$ $\forall x \, p(x)$.

Translate into set terminology: (d) He is neither happy nor wealthy. (H = happy people; W = wealthy people.)   (e) He is healthy, wealthy, and wise. (S = wise people.)   (f) He is a happy man. (M = men.)   (g) If $\angle ABC$ is a right angle, then $\overline{AB}^2 + \overline{BC}^2 = \overline{AC}^2$, and conversely. ($R$ = right triangle; P = triangles such that the square of the length of one side is the sum of the squares of the lengths of the other two sides.)   (h) A number cannot be both even and odd.   (i) If x is a man then x is mortal.   (j) No one lives forever.   (k) If a triangle has two equal angles then it has two equal sides.

There are always many ways of stating an idea verbally. As the previous exercises illustrate, the essential thing is to understand the meaning and then restate in precise terms. Sometimes the set language is simpler. This is evidently the case in (2), (3), (4), (8), and (9). There the set language enables us to dispense with explicit use of quantifiers. For example, it is easier to say that all men are mortal (men $\subseteq$ mortals) than to say that if x is a man then x is mortal. On the other hand, the set terminology is sometimes less convenient. This appears to be the case in (1), (5), (6), and (7). For example, it is easier to say that John loves Mary than to say that John belongs to the set of people that love Mary. In many cases, either terminology is used conveniently. For example, we say that 3 is prime or that 3 is a prime number.

(l) Review Section 2–10, especially (10) and (11) and Problems 28 through 43. State the following in terms of sets.   (m) All triangles are polygons. (n) All equiangular triangles are equilateral, and conversely.   (o) No oblique triangles are right triangles.   (p) Some isosceles triangles are equilateral. (q) Not all odd numbers are prime.   (r) Every integer has one as a factor.

In classical (presymbolic) logic, much reasoning was carried on in terms of "all," "some," and "none," a kind of primitive reasoning in terms of sets, which was called "syllogistic reasoning." The rules for such reasoning are complicated, and all such problems can be handled more easily in modern symbolic terms. The easiest method is to express all statements in terms of set inclusion, then use the laws of this chapter.

The following table indicates the appropriate translations:

(10)   All P are Q, $P \subseteq Q$.

(11)   Not all P are Q, *or*

              Some P are not Q, $\sim(P \subseteq Q)$.

(12)   No P are Q, *or*

              All P are not Q, $P \subseteq Q'$.

(13)   Some P are Q, $\sim(P \subseteq Q')$.

In the typical syllogistic problem, we are given two premises of the forms (10) and (11) and asked to derive a third if this is possible. For example, given that all collies are dogs ($C \subseteq D$) and that all dogs are mortal ($D \subseteq M$), we have immediately that all collies are mortal ($C \subseteq M$) by (3–8–2). All syllogistic reasoning can be reduced to the application of (3–8–2) or its variants (3–8–15) and (3–8–17). For example, given that all primes greater than 2 are odd ($P \subseteq D$) and that some numbers greater than two are not odd ($\sim(N \subseteq D)$), it follows that some numbers greater than 2 are not prime ($\sim(N \subseteq P)$) by (3–8–17).

To test a syllogistic argument for validity, simply express its premises and conclusion in set terms and see whether it follows from the laws of set inclusion. For example, given that all primes greater than 2 are odd and that some numbers greater than 2 are not prime, does it follow that some numbers greater than 2 are not odd? In symbols,

$$[(P \subseteq D) \wedge \sim (N \subseteq P)] \rightarrow \sim(N \subseteq D)?$$

The reader can easily check that this is not in the form (3–8–15) or (3–8–17). Recalling that a valid argument must be in the form of a law, we can easily show that the above is invalid by the counterexample ($P:\{1, 2\}$, $D:\{1, 2, 3, 4\}$, $N:\{3, 4\}$).

Of course we could handle syllogistic problems directly in terms of laws of logic, using quantifiers, or we could express the inclusion relations in various ways by using (3–8–16) or (3–8–20). In any case, by symbolizing such problems we avoid both the complexity and ambiguity of the traditional logic.

(s) "If the first premise is the proposition that all human beings are motivated by self-interest in their actions, the conclusion that all rulers tend to serve their interests can readily be obtained by means of a syllogism." ("The Issue of a Science of Politics in Utilitarian Thought," by F. Kort, *American Political Science Review*, December, 1952) Do this.

What conclusion can be drawn from: (t) all A is B and no B is C, (u) all A is B and some A is C, (v) no A is B and some C is B?

(w) Some laws are complicated; no confusing laws are satisfactory; every complicated law is confusing. Draw all conclusions you can. (x) No A is B, no B is C. Can you draw a conclusion?

PROBLEMS

Express Problems through 12 in terms of sets and using quantifiers.

1. All dogs are animals.
2. No dogs are able to talk.
3. No one can be both a man and a woman.
4. Not all men are good men.
5. Some women are blond and blue-eyed, some are blond and not blue-eyed, and some are blue-eyed and not blond.
6. Some men are both rich and happy.
7. If a triangle has two equal sides, it has two equal angles.
8. If a triangle has no equal sides, it has no equal angles.
9. Not all primes are odd.
10. No perfect squares are primes. (Is it true?)
11. Some even numbers are prime.
12. Some A is not B.

With P, Q, R defined as for (1) through (9), continue the table of synonyms by inserting missing entries in 13 through 18.

13. $P \subset Q$, ? 14. $P \between Q$, ?
15. $P \neq \emptyset$, ? 16. $P \neq U$, ?
17. ? $p(a) \lor q(a)$. 18. ? $p(a) \land q(a)$.

★19. When two sentences cannot both be true, they are said to be *contradictory*. Suppose $p(x)$ and $q(x)$ are contradictory. Express this in terms of quantifiers and in terms of sets.

20. How is the discussion of this section illustrated in Section 3–5?

In Problems 21 through 26, draw any conclusion you can or decide on the validity of the reasoning.

21. All A is B, some A is C.
22. No A is B, all C is A. ∴ No C is B.
23. All A is B, no C is A. ∴ No C is B.
24. No A is B, no B is C.

25. "Of the prisoners who were put on their trial at the last Assizes, all, against whom the verdict 'guilty' was returned, were sentenced to imprisonment; some, who were sentenced to imprisonment, were also sentenced to hard labor. Hence, some, against whom the verdict 'guilty' was returned, were sentenced to hard labor." [This and Problem 26 are among the many highly realistic and practical logical problems composed by Lewis Carroll and published in his *Symbolic Logic* in 1897.]

26. "No kitten that loves fish is unteachable; no kitten without a tail will play with a gorilla; kittens with whiskers always love fish; no teachable kitten has green eyes; no kittens have tails unless they have whiskers." Draw all possible conclusions.

27. Go through this chapter, formulating the laws of logic corresponding to each law of set theory.

28. Do the reverse, translating laws of Chapter 2 into set terminology.

29. In this section we have violated our agreement about the meaning of "equals." Where?

30. Read "The Maneuvers in Set Thinking," by W. L. Duren, in *The Mathematics Teacher*, May, 1958, for an interesting review of the ideas of this chapter and a preview of some ways in which they will be used later in the book.

ANSWERS TO EXERCISES

(a) 3, odd numbers; that is not divisible by 2.   (b) He; He, {my father}; He, is my father.   (c) $\emptyset$; x is prime and x is even.   (d) He $\in (H' \cap W')$ (e) He $\in (H \cap W \cap S)$.   (f) He $\in (H \cap M)$.   (g) $R = P$.   (h) Evens $\cap$ odds $= \emptyset$.   (i) Men $\subseteq$ mortals.   (j) $F = \emptyset$ where $F = \{x \mid x$ lives forever$\}$. (k) $A \subseteq B$ where $A =$ triangles with two equal angles, $B =$ triangles with two equal sides.

(m) Tr $\subseteq$ Po. (n) Eq $=$ El. (o) Ob $\subseteq$ R'. (p) $\sim$(Is $\subseteq$ Eq'). (q) $\sim$(Od $\subseteq$ P). (r) $J \subseteq F$, where $J =$ integers, $F =$ numbers having one as a factor. (s) [Kings $\subseteq$ humans $\wedge$ humans $\subseteq$ beings motivated by self-interest] $\rightarrow$ [kings $\subseteq$ BMSI].   (t) No A is C, by $[A \subseteq B \wedge B \subseteq C'] \rightarrow (A \subseteq C')$. (u) Some B is C, by $[A \subseteq B \wedge \sim(A \subseteq C')] \rightarrow \sim(B \subseteq C')$ by (3-8-15).   (v) Not all C is A, by $[A \subseteq B' \wedge \sim(C \subseteq B')] \rightarrow \sim(C \subseteq A)$ by (3-8-17).   (w) Some laws are not satisfactory. We have $\sim$(Comp $\subseteq$ L'), (Conf $\subseteq$ Sat'), (Comp $\subseteq$ Conf). From the last two by (3-8-2), (Comp $\subseteq$ Sat'). From this and the first by (3-8-15), $\sim$(Sat' $\subseteq$ L') or $\sim$(L $\subseteq$ Sat).   (x) No.

ANSWERS TO PROBLEMS

1. $D \subseteq A$, $\forall x\, x \in D \rightarrow x \in A$.   3. $M \cap W = \emptyset$, $\sim\exists x\, x \in M \wedge x \in W$. 5. $W \cap$ Blo $\cap$ Blu $\neq \emptyset \wedge W \cap$ Blo $\cap$ Blu' $= \emptyset \wedge W \cap$ Blo' $\cap$ Blu $= \emptyset$. $[\exists x\, x \in W \wedge x \in$ Blo $\wedge x \in$ Blu$] \wedge [\exists x\, x \in W \wedge x \in$ Blo $\wedge x \in$ Blu'$] \wedge$ $[\exists x\, x \in W \wedge x \in$ Blo' $\wedge x \in$ Blu$]$. 7. Es $\subseteq$ Ea. $\forall x$ [x has two equal sides $\rightarrow x$ has two equal angles]. 9. $\sim(P \subseteq D)$. 11. $\sim(E \subseteq P')$. 12. $\sim(A \subseteq B)$.

13. $\forall x\, [p(x) \rightarrow q(x)] \wedge \sim \forall x\, [q(x) \rightarrow p(x)]$.   15. $\exists x\, p(x)$.   17. $a \in P \cup Q$. 19. (4). 21. Some B is C. 22. Valid. 23. Invalid. 24. No conclusion. 25. Invalid! No conclusion [see *Symbolic Logic* by Lewis Carroll (C. L. Dodgson) p. 63].   26. No kitten with green eyes will play with a gorilla.   29. In Problems 7 and 8, and in Exercise k.

CHAPTER 4

PLANE ANALYTIC GEOMETRY

4–1 Ordered pairs. Only rarely are we concerned with a single object in isolation; usually we are interested in objects related to other objects. In the simplest case we have two objects paired. For example, with each finite set, we may pair the number of its members; with each item in a store, we may pair its price; with each time, we may pair the velocity of a moving body; and so on. For definiteness, let us think of the items in a store and their prices. We number the item and give price in cents, so that item 1 has price 5, item 2 has price 103, and so on. We could make a table showing this information. Each pair, consisting of an item and its price, makes up a set of two members, for example $\{1, 5\}$. However, suppose that the number and price of an item were the same, for example, that item 6 has price 6. Then we should have for this item $\{6, 6\}$ or $\{6\}$, which is no longer a pair. Or suppose that item 10 has price 15 and item 15 has price 10. Then we wish to distinguish these two pairs, but $\{10, 15\} = \{15, 10\}$. Accordingly, we see that the simple concept of a pair of numbers is not adequate in this situation. We have to be able to distinguish which member of the set is the item and which the price.

A pair in which we distinguish one of the members as the first and the other (which need not be different) as the second is called an *ordered pair*. The ordered pair whose first member is x and whose second member is y is symbolized by (x, y) and is read "x, y" or "the ordered pair x, y." The essential property of an ordered pair is given by the following axiom.

(1) Ax. $\qquad [(a, b) = (c, d)] = [a = c \wedge b = d].$

(a) Show that $[\{a, b\} = \{c, d\}] = [a = c \wedge b = d]$ is not a law. (b) Prove (2). ★(c) We could define (x, y) by $(x, y) = \{\{x, y\}, \{x\}\}$ and then prove (1). Do this.

(2) $\qquad [(a, b) = (b, a)] = [a = b].$

In Section 1–2 we associate single numbers with points on an axis in such a way that to each real number there corresponds one and only one point, and to a point one and only one real number. In Section 1–4 and later sections we make use of this one-to-one correspondence to interpret numbers and operations on them in terms of points and vectors. For example, we interpret $b - a$ as the directed distance from a to b and its

173

absolute value as the undirected distance between a and b. Also, we think of a sentence as determining a graph, namely, the set of points corresponding to the truth set of the sentence. For example, we view $\{x \mid a \leq x \leq b\}$ as the closed interval between a and b. (See Section 3–4.)

(d) Review previous sections, noting the way in which we used the correspondence between numbers and points on an axis to interpret numbers, operations on numbers, equations, and inequalities.   (e) Is the correspondence between sentences and sets one-to-one; i.e., is there just one defining sentence for each set of points and just one set of points for each defining sentence?

The significant thing about the correspondence between single numbers and points or vectors on an axis is that it permits us to deal with geometric problems by algebra and to visualize algebraic ideas geometrically. The idea can easily be generalized by establishing a correspondence between ordered pairs of numbers and points in a plane. The interpretations of operations generalize naturally. As before, sentences have geometric interpretations, and geometric figures (sets of points) can be dealt with through their defining sentences. The idea of establishing a correspondence between real numbers and points, developed by Descartes and others in the early seventeenth century, is the key idea of analytic geometry. It brought about the reconciliation of algebra and geometry, which had been separated since Greek times, and led directly to the invention of the calculus, which has played such a significant role in modern science.

The purposes of this chapter are (1) to show how the key idea of analytic geometry applies to ordered pairs of real numbers and points in the plane, (2) to introduce the student to the algebra of plane vectors, (3) to familiarize the student with the graphs of certain frequently occurring equations and inequalities, and (4) to utilize analytic geometry to develop a skill in the algebra of real numbers, logic, and set theory. Additional topics in analytic geometry are included also in later chapters.

PROBLEMS

Solve the following sentences:

1. $(x, 3) = (4, 3)$.
2. $(x, 1) = (1, x)$.
3. $x^2 + y^2 = 0$.
4. $(x, y) = (2x, 3y)$.

ANSWERS TO EXERCISES

(a) See (3–3–19).   (b) Immediate from (1). Note the use of $p \wedge p = p$. (e) No. There is just one set of points corresponding to each sentence, but any two logically equivalent sentences define the same set.

ANSWERS TO PROBLEMS

1. $x = 4$. 3. $x = y = 0$.

4–2 Plane coordinates. There are many ways of associating points in a plane with numbers. The one most commonly used is based on setting up two axes with common origin and with positive directions to the right and up. As indicated in Fig. 4–1, it is customary to call the horizontal axis the *x-axis* and the vertical axis the *y-axis*. The axes divide the plane into four regions called *quadrants*, numbered as indicated in the figure.

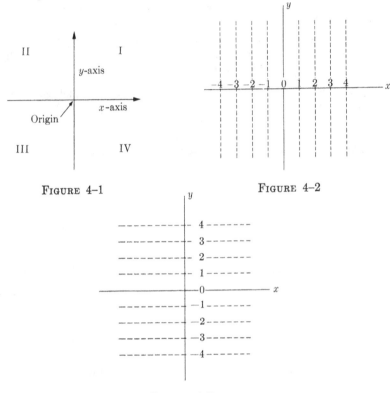

FIGURE 4–1   FIGURE 4–2

FIGURE 4–3

Now we imagine a line drawn parallel to the *y*-axis (perpendicular to the *x*-axis) through each point on the *x*-axis (Fig. 4–2). Similarly, we imagine lines parallel to the *x*-axis (perpendicular to the *y*-axis) through the points on the *y*-axis (Fig. 4–3). Each point in the plane lies on one and only one of the vertical lines and one and only one of the horizontal lines.

(a) Why?

The vertical line on which a point lies indicates the directed distance of the point from the *y*-axis. The directed distance is precisely the linear coordinate of the point where the vertical line crosses the *x*-axis, i.e., the number corresponding to the point on the *x*-axis. We call this number the *x-coordinate* of the point. Similarly, the horizontal line indicates the directed distance from the *x*-axis. This directed distance is the linear coordinate of the point where the line crosses the *y*-axis. We call this number the *y-coordinate* of the point. Evidently, to any point there corresponds a unique ordered pair of real numbers (x, y), where x and y are respectively the *x*- and *y*-coordinates.

Conversely, to each ordered pair (x, y) there corresponds a unique point lying at the intersection of the vertical line at a directed distance x from the *y*-axis and the horizontal line at a directed distance y from the *x*-axis.

In Fig. 4–4 we sketch the point corresponding to $(5, -2)$. The directed distances are suggested by vectors.

For each of the following draw and label a pair of axes; plot, circle and label points; and show directed distances, as in Fig. 4–4:   (b) $(5, 2)$,   (c) $(-2, 5)$, (d) $(-5, 2)$,   (e) $(2, 5)$,   (f) $(-5, -2)$.

(g) through (n) Label the points in Fig. 4–5 so as to indicate approximately the ordered pair of real numbers corresponding to each one.

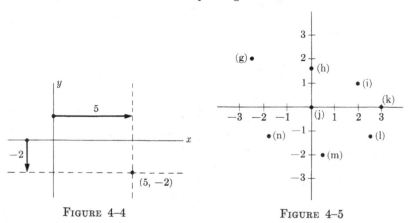

FIGURE 4–4   FIGURE 4–5

We can now say precisely what we mean by a solution of a sentence involving two variables. We call an ordered pair (a, b) a *solution of a sentence* $f(x, y)$ if and only if $f(a, b)$. We then say that (a, b) *satisfies* the sentence $f(x, y)$. The set of all solutions of a sentence $f(x, y)$ is then a set of ordered pairs, which is the truth set of the sentence. We designate the set of all solutions of $f(x, y)$ by $\{(x, y) \mid f(x, y)\}$, which is read "the set of ordered pairs (x, y) such that $f(x, y)$." This notation may be formally defined as follows.

(1) **Def.** $\{(x, y) \mid f(x, y)\} = \{z \mid \exists x \, \exists y \, f(x, y) \wedge [z = (x, y)]\}.$

Its essential property is

(2) $[(a, b) \in \{(x, y) \mid f(x, y)\}] \leftrightarrow f(a, b).$

Just as we used a numeral or a single numerical variable to represent both a number and its corresponding point, so now we use an expression of the form (x, y) to stand for any ordered pair of numbers or for the corresponding point. Similarly, just as we used $\{x \mid f(x)\}$ to represent both a set of numbers and a set of points, so now we think of $\{(x, y) \mid f(x, y)\}$ as both a set of ordered pairs and as the set of corresponding points. We call the set of ordered pairs (points) defined by $f(x, y)$ the *plane graph* of $f(x, y)$.

The plane graph of a sentence is simply the set of points whose coordinates satisfy the sentence. *A point lies on the graph if and only if its coordinates satisfy the sentence.* This criterion can be used directly to find the graphs of simple sentences. For example, the sentence $x = 3 \wedge y = 4$ is satisfied only by the pair $(3, 4)$. Hence the graph consists of a set of points with only one member, namely, the point $(3, 4)$. Symbolically, $\{(x, y) \mid x = 3 \wedge y = 4\} = \{(3, 4)\}$. Now $\{(x, y) \mid x = 3\}$ is the set of all pairs with first component 3. The corresponding points make up the line parallel to the y-axis and three units to the right of it, as sketched in Fig. 4–6. In the same figure we have graphed $\{(x, y) \mid y = 4\}$. But $\{(x, y) \mid x = 3 \wedge y = 4\} = \{(x, y) \mid x = 3\} \cap \{(x, y) \mid y = 4\} = \{(3, 4)\}$, since the two graphs intersect in the single point already found. Note that we use the defining sentence of a graph as a name for it.

Sketch the following: (o) $x = -1$, $y = 3$, $x = -1 \wedge y = 3$; (p) $x = -1 \vee y = 3$; (q) $y = x$.

The reader has observed that when one variable is missing from a defining sentence, this means simply that no limitations are placed on the values of that variable. For example, the plane graph of $x < 3$ consists of all those points whose x-coordinate is less than 3, i.e., all those points to the left of the line $x = 3$, as sketched in Fig. 4–7.

Sketch and describe in words: (r) $y \geq 1$, (s) $x > 0$, (t) $x^2 + y^2 = 0$.

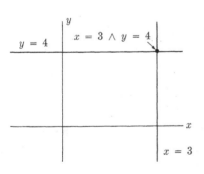

FIGURE 4–6

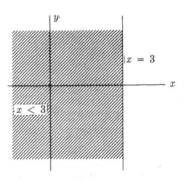

FIGURE 4–7

It is often difficult to discover the nature of the graph when the range of the variables in a sentence is the entire set of real numbers R_e. Even in the simple cases we have considered above, the graph has usually been of infinite extent, so that we could sketch only a part of it. However, if the variables are limited to a range consisting of a finite number of values, the graph of any sentence may be determined by testing each point to see whether it satisfies the sentence. Usually many points can be eliminated as obviously not solutions. Suppose, for example, that we limit the range of the variables to $\{0, 1, 2, 3, 4\}$. Then

$$\{(x, y) \mid x = 3\} = \{(3, 0), (3, 1), (3, 2), (3, 3), (3, 4)\}.$$

The graph is sketched in Fig. 4–8.

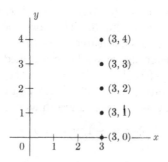

FIGURE 4–8

Let the range of variables be $\{0, 1, 2, 3, 4\}$. Graph: (u) $x = 2 \lor y = 3$, (v) $x > 2$,   (w) $x < 2 \land y > 1$,   (x) $y \leq x$,   (y) $2 \leq x \leq 3 \land 2 \leq y \leq 3$,   (z) $2 \leq x \leq 3 \lor 2 \leq y \leq 3$.

PROBLEMS

In which quadrant does (a, b) lie if

1. $a < 0 \land b > 0$.   2. $-a < 0 \land b > 0$.
3. $a < 0 \land b < 0$.   4. $a > 0 \land b < 0$.

In which quadrant does $(-a, b)$ lie if

5. $a > 0 \land b > 0$.   6. $a < 0 \land b < 0$.
7. $a > 0 \land -b > 0$.   8. $-a > 0 \land -b > 0$.

9. Complete the following: The members of an ordered pair are called the _____ of the corresponding point.

In Problems 10 through 21 the range of the variables is $\{0, 1, 2, 3, 4\}$. Determine the set of ordered pairs defined by each sentence, and graph.

10. $x \geq 3$.

11. $y < 2$.

12. $0 < y < 3$.

13. $x < 2 \land y < 2$.

14. $x = -1 \lor y = 2$.

15. $x = 0 \lor y = 0$.

16. $y = 2 \land 0 \leq x \leq 2$.

17. $y = x^2$.

18. $y = 2x$.

19. $x + y = 4$.

20. $y = (1/2)x$.

21. $y = \sqrt{x}$.

In Problems 22 through 31 the range of the variables is the real numbers. Sketch the graph of each sentence, using whatever method you wish.

22. $x = 0$.

23. $y = 0$.

24. $xy = 0$.

25. $x = 1 \lor y \leq 0$.

26. $x = 1 \land y \leq 0$.

27. $y = 2 \land -1 \leq x \leq 1$.

28. $xy > 0$.

29. $x + y = 4$.

30. $|x| = 2$.

31. $x + y = 4 \land x \geq 0 \land y \geq 0$.

★32. Show without reference to geometry: $\{(x, y) \mid x = 3\} \; // \; \{(x, y) \mid x = 4\}$.

Graph

★33. $\{(x, y) \mid xy > 0\}'$.

★34. $\{(x, y) \mid xy > 0\} \cap \{(x, y) \mid |x| \leq 1\}$.

★35. $\{(x, y) \mid xy > 0\} \cup \{(x, y) \mid |x| \leq 1\}$.

ANSWERS TO EXERCISES

(a) Through a given point not on a line there is one and only one line parallel to a given line. This is Euclid's famous parallel axiom.   (b) through (f) Fig. 4–9.   (g) $(-2.5, 2)$.   (h) $(0, 1.6)$.   (i) $(2, 1)$.   (j) $(0, 0)$.   (k) $(3, 0)$.   (l) $(2.5, -1.2)$.   (m) $(0.5, -2)$.   (n) $(-1.8, -1.2)$.   (o) First is line parallel to y-axis one unit to left; second is the parallel to x-axis three units above; third is point of intersection of these.   (p) Figure consisting of the two lines graphed in Exercise (o).   (q) Line bisecting first and third quadrants.   (r) All

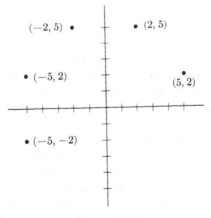

FIGURE 4–9

of plane on and above line $y = 1$.   (s) All of plane to right of y-axis.   (t) The origin.   (u) 9 points.   (v) 10 points.   (w) 6 points.   (x) 15 points forming triangle.   (y) 4 points forming a square.   (z) 16 points forming a cross.

ANSWERS TO PROBLEMS

1. II. 3. III. 5. II. 7. III. 9. coordinates. 13. $\{(0, 0), (0, 1), (1, 0), (1, 1)\}$. 15. 9 points on the axes. 17. $\{(0, 0), (1, 1), (2, 4)\}$. 19. $\{(0, 4), (1, 3), (2, 2), (3, 1), (4, 0)\}$. 21. $\{(0, 0), (1, 1), (4, 2)\}$. 23. x-axis. 25. Fig. 4–10. 26. Fig. 4–11. 27. Fig. 4–12. 29. Straight line through $(0, 4)$ and $(4, 0)$. 31. Portion of 29 in quadrant I.

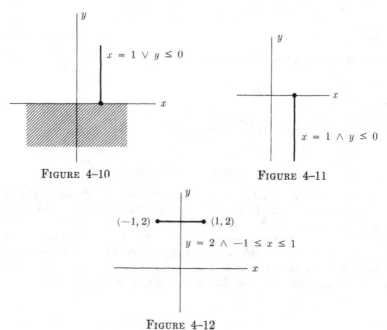

FIGURE 4–10   FIGURE 4–11

FIGURE 4–12

4–3 Plane vectors.   In Section 1–4 we established a correspondence between single real numbers and vectors on an axis, which we used to give vector interpretations of alegbraic operations. We now generalize this by establishing a correspondence between ordered pairs of real numbers and vectors in a coordinate plane.

An ordered pair of real numbers (a, b) determines a vector from the origin to the point (a, b). As in Section 1–4, we think of the vector as moving about without changing its length or direction, which is suggested in Fig. 4–13. Wherever it is located, the vector corresponding to (a, b) is determined by the property that if one starts at the initial point and goes a horizontal directed distance a and then a vertical directed dis-

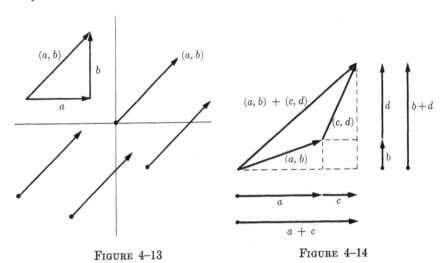

FIGURE 4–13   FIGURE 4–14

tance b, one arrives at the terminal point. The axes are required only to determine the horizontal and vertical directions and the scale of measurement. Conversely, if we have any vector, we can draw lines through its initial and terminal points parallel to the axes and so determine the corresponding ordered pair. Hence there is a one-to-one correspondence between ordered pairs and vectors. Because of this we use a name of an ordered pair as a name of the corresponding vector, and we think of (a, b) as an ordered pair, a point, or a vector, as convenient. We call a and b the *components* of the vector (a, b). They are indicated graphically in Fig. 4–13.

Sketch the following vectors, showing their components as in Fig. 4–13: (a) $(1, 3)$; (b) $(-3, 7)$; (c) $(4, 0)$; (d) $(0, -1)$; (e) $(4, -1)$; (f) $(-3, -3)$.

Our procedure for adding vectors in the plane is the same as on a line. To find $(a, b) + (c, d)$ we place the initial point of (c, d) on the terminal point of (a, b) and then find the vector from the initial point of (a, b) to the terminal point of (c, d), as in Fig. 4–14. Note from the figure that the components of the sum are sums of corresponding components.

Sketch the following sums as in Fig. 4–14:   (g) $(3, 5) + (4, 2)$; (h) $(3, 5) + (-1, -2)$;   (i) $(4, -1) + (-2, -3)$.

The above discussion suggests the following definition of the *sum of two ordered pairs of real numbers.*

(1) **Def.** $(a, b) + (c, d) = (a + c, b + d)$.

By $(b{:}0, c{:}0, d{:}b)$ we have immediately

(2) $(a, b) = (a, 0) + (0, b)$.

Since $(a, 0)$ is a horizontal vector, and $(0, b)$ is a vertical one, (2) says that any vector is the sum of a vertical and a horizontal vector. We call $(a, 0)$ and $(0, b)$ the *component vectors* of (a, b). They are simply the vectors labeled a and b in Fig. 4–13.

We see immediately from (1) that

(3) $$(a, b) + (0, 0) = (a, b),$$

which corresponds to (1–16–7). Hence it appears that $(0, 0)$ plays the same role in addition of vectors that 0 did in the addition of real numbers. We call $(0, 0)$ the *null vector*.

(j) Show that $(0, 0)$ is the only vector with the property (3); that is, that $[(a, b) + (x, y) = (a, b)] \rightarrow (x, y) = (0, 0)$.

We saw in Section 1–4 that the vector $-a$ has the same length as a but is opposite in direction. Figure 4–15 suggests that to reverse the direction of a vector is to negate its components. Accordingly, we define *the negative of a vector.*

(4) Def. $$-(a, b) = (-a, -b).$$

We have immediately that $(a, b) + (-(a, b)) = (a, b) + (-a, -b) = (0, 0)$, which corresponds to (1–16–9). Hence the negative plays the same role here as for real numbers.

We define *the difference of two vectors* by analogy with (1–9–18), which is equivalent to (1–14–19).

(5) Def. $$(a, b) - (c, d) = (a, b) + (-(c, d)).$$

This gives immediately

(6) $$(a, b) - (c, d) = (a - c, b - d).$$

It is not hard to see that $(a, b) - (c, d)$ is the vector from the point (c, d) to the point (a, b), that is, the directed distance from (c, d) to (a, b).

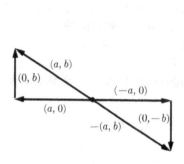

FIGURE 4–15

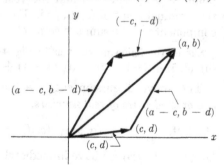

FIGURE 4–16

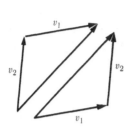

FIGURE 4-17

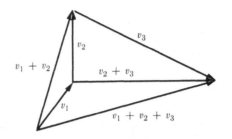

FIGURE 4-18

In Fig. 4-16 we see that the vector $(a - c, b - d)$ obtained by adding the vectors (a, b) and $(-c, -d)$ is the same as the vector from the point (c, d) to the point (a, b). Hence, just as for single real numbers, we interpret a difference as a directed distance, $(a, b) - (c, d)$ being the directed distance (the vector) from the point (c, d) to the point (a, b).

Sketch the following points and the directed distance from the first to the second in each case: (k) $(2, -5)$, $(5, 1)$;   (l) $(5, 1)$, $(2, -5)$;   (m) $(3, 5)$, $(6, 10)$.

Do vectors satisfy the commutative and associative laws of addition? We leave it to the reader to prove that they do. Figures 4-17 and 4-18 suggest their plausibility.

(n) Prove that $(a, b) + (c, d) = (c, d) + (a, b)$.   (o) Prove that $(a, b) + ((c, d) + (e, f)) = ((a, b) + (c, d)) + (e, f)$.

The above discussion shows that with our definitions, ordered pairs of real numbers (plane vectors) follow the axioms of algebra relating to addition, subtraction, and negation. Could we define multiplication of ordered pairs so that all the other axioms are satisfied also? The answer is shown to be yes in Section 6-13, but in this section we shall merely define the product of an ordered pair of reals by a real number.

In Section 1-4 we interpreted the product ab as the vector found when the vector b was laid out a times. Figure 4-19 shows a vector added to

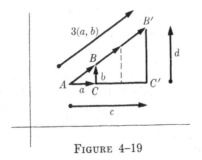

FIGURE 4-19

itself 3 times. Because of the similarity of the triangles ABC and $AB'C'$, we have $\overline{AB'}/\overline{AB} = c/a = d/b = 3$. Here the components of the product are obtained by multiplying the original components by 3. This suggests the definition.

(7) **Def.** $k(a, b) = (ka, kb)$.

(p) Sketch $(3, 4)$, and $2(3, 4)$, and $(3, 4) + (3, 4)$.   (q) Show that the vectors (a, b) and $k(a, b)$ always lie in the same or parallel lines.

Although we have not defined the product of two vectors, we have defined addition, subtraction, and multiplication by a real number. Since addition and subtraction were shown to satisfy the same axioms as the real numbers, we can use laws relating to real numbers that were derived from these axioms. We can also use any laws we can derive from the definition (7) and the multiplicative properties of real numbers.

(r) Prove that

(8) $k[(a, b) + (c, d)] = k(a, b) + k(c, d)$.

PROBLEMS

Simplify the expressions in Problems 1 through 6 and draw diagrams indicating the operations and result.

1. $(2, -4) + (3, 1)$.   2. $(4, -1) - (2, -7)$.
3. $2(5, -3)$.   4. $(-5)(1, -1)$.
5. $(1/2)(4, -6)$.   6. $(-1, 2) + (3, -5) + (-2, -2)$.

7. Prove that

(9) $(a, b) = a(1, 0) + b(0, 1)$.

(This result is described by saying that every vector is a linear combination of horizontal and vertical unit vectors.)

8. Let $(1, 0) = e_1$ and $(0, 1) = e_2$. Express the following in the form $ae_1 + be_2$: $(a, -1)$, $(4, 3)$, $(-5, -1.7)$.

9. Show that

(10) $1 \cdot (a, b) = (a, b)$,

(11) $(-1)(a, b) = -(a, b)$.

10. To what laws for real numbers do (10) and (11) correspond?
11. How would you define division of a vector by a real number?
12. From the definition in Problem 11 show that $(1-9-29)(a:(a, b))$ is a law.
13. Similarly show that $(1-9-30)$ through $(1-9-32)$ hold for vectors.
14. Prove that $(1-9-33)$ through $(1-9-37)$ hold if we let the range of the first variable be the real numbers and let the range of the second be the set of all ordered pairs of real numbers.

15. Show that (2–8–3) and (2–8–4)(0:(0, 0)) hold for plane vectors.

In Problems 16 through 21, solve for the unknown ordered pair.

16. $2(x, y) = (3, 5)$.
17. $-3(x, y) = 2(x, y) + (1, -3)$.
18. $x(x, y) - (2, 4) = (3, -1) - (x, y)$.
19. $(x, y) - (5, 1) = 3(4, 3) - (x, y)$.
20. $z - (-3, -2) = (4, -10)$.
21. $5z = z + (2, -1)$.

ANSWERS TO EXERCISES

(a) through (f) See Fig. 4–20.　(g) (7, 7).　(h) (2, 3).　(i) (2, −4).
(j) $(a, b) + (x, y) = (a, b) \rightarrow (a + x, b + y) = (a, b) \rightarrow a + x = a \wedge b + y = b$.
(k) (3, 6).　(l) (−3, −6).　(m) (3, 5).　(n) $(a, b) + (c, d) = (a + c, b + d) = (c + a, d + b) = (c, d) + (a, b)$.　(o) Similarly.

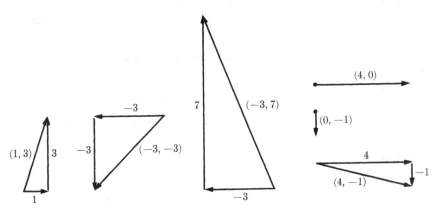

FIGURE 4–20

(p) $2(3, 4) = (3, 4) + (3, 4)$.　(q) The triangles formed by the vectors and their component vectors are similar since they are right triangles whose legs are proportional. Since their legs are respectively parallel (or in the same line), so are their hypotenuses.

(r) Left member $= k(a + c, b + d) = (ka + kc, kb + kd) = (ka, kb) + (kc, kd) =$ right member.

ANSWERS TO PROBLEMS

1. (5, −3).　3. (10, −6).　5. (2, −3).　7. Simplify right member.　9. From definition (7).　11. $(a, b)/k = (1/k)(a, b)$.　13. Immediate from definition.　15. The proofs would be unchanged.　17. (−1/5, 3/5).　19. (8.5, 5).　21. (1/2, −1/4). Note that there is no objection to using a single variable to stand for an ordered pair!

4–4 Distance in plane geometry. On an axis the distance between two points a and b is the length of the vector, $b - a$, from one to the other. Similarly, in the plane the distance between two points (a, b) and (c, d) is the length of a vector joining them. Such a vector is the difference of the vectors (a, b) and (c, d), that is, $(c, d) - (a, b)$ or $(c - a, d - b)$. But how shall we define the length of a vector? Figure 4–13 and the Pythagorean theorem suggest the answer: since the length of a vector is the length of the hypotenuse of a right triangle whose other two sides have lengths equal to the absolute values of the components, the length of a vector is the principal square root of the sum of the squares of its components. Representing the length of the vector (a, b) by $|(a, b)|$, we have in Fig. 4–13, $|(a, b)|^2 = |a|^2 + |b|^2 = a^2 + b^2$. This suggests

(1) Def. $|(a, b)| = \sqrt{a^2 + b^2}.$

(a) Why did we say that the lengths of the sides in Fig. 4–13 are the absolute values of a and b?   (b) Why do these absolute value symbols not appear in (1)? (c) Find the length of the vector $(-3, 5)$.

There is no reason why we should not use variables whose ranges are sets of ordered pairs, points, or vectors. We shall adopt the convention of using boldface type for this purpose. Thus **u** is a variable for which we may substitute other boldface variables, formulas whose values are vectors, or expressions of the form (a, b). Then $|\mathbf{u}|$ is the length of the vector **u**. We let

(2) Def. $\mathbf{0} = (0, 0).$

Then

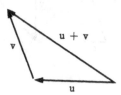

(3) $|\mathbf{u}| \geq 0,$

(4) $(|\mathbf{u}| = 0) \leftrightarrow (\mathbf{u} = \mathbf{0}),$

(5) $|k\mathbf{u}| = |k| \cdot |\mathbf{u}|,$ Figure 4–21

(6) $|\mathbf{u} + \mathbf{v}| \leq |\mathbf{u}| + |\mathbf{v}|.$

The geometric interpretations are as follows: (3) The length of a vector is non-negative. (4) The length of a vector is zero if and only if the vector is the null vector. (5) The length of a multiple of a vector is the absolute value of the multiple times the length. (6) If the vectors **u** and **v** do not lie in the same straight line, this becomes the famous "triangle inequality" of Euclid: the length of any side of a triangle is less than the sum of the lengths of the other two sides. (See Fig. 4–21.)

These laws are "evident" geometrically, but they can be proved algebraically without any reference to geometry, and then the results interpreted as proofs of the corresponding geometric laws! The proofs of the first three are easy.

Prove: (d) (3),   (e) (4),   (f) (5).

To prove (6) we let $\mathbf{u} = (a, b)$ and $\mathbf{v} = (c, d)$. Then

$$\mathbf{u} + \mathbf{v} = (a + c, b + d),$$

and (6) becomes

(6′) $\sqrt{(a + c)^2 + (b + d)^2} \leq \sqrt{a^2 + b^2} + \sqrt{c^2 + d^2}.$

But (6′) is equivalent to

(7) $(a + c)^2 + (b + d)^2 \leq a^2 + b^2 + c^2 + d^2 + 2\sqrt{(a^2 + b^2)(c^2 + d^2)}$

by applying (3–5–43), which is legitimate since the left member of (6′) is non-negative, and using (1–12–9) and (1–13–22). Continuing, (7) is equivalent to

(8) $a^2 + 2ac + c^2 + b^2 + 2bd + d^2$
$$\leq a^2 + b^2 + c^2 + d^2 + 2\sqrt{(a^2 + b^2)(c^2 + d^2)}$$

Now subtracting common terms from both sides and dividing by 2, we have the equivalent inequality,

(9) $ac + bd \leq \sqrt{(a^2 + b^2)(c^2 + d^2)}.$

Now if $ac + bd < 0$, (9) is certainly a law, and the theorem is proved. If $ac + bd \geq 0$, then we can apply (3–5–43) again to get the equivalent inequality,

(10) $a^2 c^2 + 2abcd + b^2 d^2 \leq a^2 c^2 + a^2 d^2 + b^2 c^2 + b^2 d^2.$

Simplifying and subtracting $2abcd$, we find

(11) $0 \leq a^2 d^2 - 2abcd + b^2 c^2,$

(12) $0 \leq (ad - bc)^2.$

The last sentence is certainly a law. Since we have proved that (6′) is logically equivalent to it, the proof is complete. (Compare Section 2–8.)

(g) Indicate the substitutions in (3–5–43) to get the logical equivalence of (6′) and (7).   (h) How are (1–12–9) and (1–13–22) used?   (i) Indicate the

laws and substitutions used to get from (7) to (8), (j) from (8) to (9).
(k) Justify in detail (9) ↔ (10). (l) Why is (11) ↔ (12)? (m) Prove (13).

(13) $$|-\mathbf{u}| = |\mathbf{u}|$$

We can now easily derive a formula for the distance between two points
(a, b) and (c, d). This distance is the length of either of the vectors joining the two points, and these vectors are $(c, d) - (a, b)$ and its negative
$(a, b) - (c, d)$. By (13) it makes no difference which we take. Hence the
distance is given by the length of $(c - a, d - b)$, that is, by

(14) $$|(c, d) - (a, b)| = \sqrt{(c - a)^2 + (d - b)^2}.$$

Thus we get the distance between two points by subtracting their coordinates, squaring, adding, and taking the principal square root.

Find the directed distance from the first to the second point and the distance
between them: (n) $(3, -5)$, $(7, 8)$;   (o) $(\sqrt{2}, -3)$, $(1 + \sqrt{2}, \sqrt{5} - 3)$.

In plane geometry a *locus* is described as a set of points satisfying some
condition. In analytic geometry, we often have the problem of finding a
sentence defining a certain locus, i.e., such that the locus is the graph of
the sentence. For example, a circle is defined as a locus of points equidistant from a fixed point. Suppose the center is at (h, k) and the radius is r.
Then a point (x, y) lies on the circle if and only if the distance between
(x, y) and (h, k) is r, that is, if and only if $|(x - h, y - k)| = r$, or,
since both members are non-negative $|(x - h, y - k)|^2 = r^2$. Hence a
defining equation of the circle with center (h, k) and radius r is

(15) $$(x - h)^2 + (y - k)^2 = r^2.$$

A point lies on the circle if and only if its coordinates satisfy equation (15).

Write the equations of a circle with given centers and radii, and sketch them
in the coordinate plane: (p) $(3, 4)$, 1;   (q) $(0, 0)$, 1;   (r) $(4, -3)$, 2.

A standard problem of plane geometry is to bisect a segment. Let us
see how we can do this with vectors. Figure 4–22 shows the segment
joining (x_1, y_1) and (x_2, y_2) with the midpoint $(\bar{x}, \bar{y})$, and also shows the
vectors from the origin to the points, where $\mathbf{u}_1 = (x_1, y_1)$, $\mathbf{u}_2 = (x_2, y_2)$,
and $\bar{\mathbf{u}} = (\bar{x}, \bar{y})$. Note that we think of the same symbols as representing
points or vectors, as convenient. Now $\bar{\mathbf{u}}$ is the midpoint if and only if

(16) $$\mathbf{u}_2 - \mathbf{u}_1 = 2(\bar{\mathbf{u}} - \mathbf{u}_1),$$

which follows from our discussion of the meaning of multiplication of a

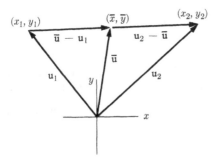

FIGURE 4-22

vector by a real number [see Fig. 4-19 and (4-3-7)]. We can solve (16) for $\bar{u}$ by the usual rules of algebra. We find

(17) $$\bar{u} = (1/2)(u_1 + u_2).$$

Note that the midpoint is simply the average of the endpoints. To get formulas for the coordinates, we substitute the coordinate forms for the variables in (17) to get

(17') $$(\bar{x}, \bar{y}) = (1/2)[(x_1, y_1) + (x_2, y_2)]$$

(17'') $$= [(1/2)(x_1 + x_2), (1/2)(y_1 + y_2)].$$

From this we have, as expected,

(18) $$\bar{x} = (x_1 + x_2)/2, \quad \bar{y} = (y_1 + y_2)/2.$$

Find the midpoints of: (s) $(3, -4)$, $(2, 16)$;   (t) $(-1, -1)$, $(-17, -3)$. (u) Carry through the algebra to get (17) from (16).   (v) Derive (17) from the observation that we must have $u_2 - \bar{u} = \bar{u} - u_1$.

PROBLEMS

In 1 through 4 find the distance from the origin to the point.

1. $(3, -4)$.   2. $(1, 0)$.   3. $(0, -4)$.   4. $(-1, -7)$.

5. Under what conditions will the equality in (6) hold?

In 6 through 9 sketch the two points, the directed distance (vector) from the first to the second, and its component vectors. Find the distance between them.

6. $(4, -10)$, $(0, 1)$.   7. $(350, 0)$, $(352, 4)$.
8. $(1.6, -3.7)$, $(2, -3.3)$.   9. $(\sqrt{2}, \sqrt{5})$, $(\sqrt{2}, -\sqrt{5})$.

Sketch the circles in 10 through 13.

10. $(x - 1)^2 + y^2 = 16$.   11. $(x + 2)^2 + (y - 4)^2 = 1$.
12. $(x - 2)^2 + y^2 = 4$.   13. $x^2 + (y - 3)^2 = 9$.

Sketch the graphs in 14 through 17.

14. The union of the truth sets of Problems 10 and 11.

15. The complement of the graph of Problem 12.

16. The intersection of the graphs of Problems 12 and 13.

17. The intersection of the graph of Problem 10 with $\{(x, y) \mid x \leq 0\}$.

18. Show that $(x - h)^2 + (y - k)^2 \leq r^2$ defines the region interior to and on the circle.

Sketch the graphs of the sentences in 19 through 22.

19. $x^2 + y^2 \leq 1 \wedge x > 0$.   20. $\sim(x^2 + y^2 < 1)$.

21. $x^2 + y^2 = 9 \vee x^2 + (y - 1)^2 \leq 16$.

★22. $x^2 + y^2 \neq 9 \wedge y < x$.

23. Show that if two points lie on the x-axis, then (14) gives the distance between them as the absolute value of the difference of their coordinates. Generalize.

★24. In what sense can it be said that the length of a vector behaves "the same" as the absolute value of a real number?

★25. If a point u lies $\frac{2}{3}$ of the way from one point u_1 to another u_2, then $u - u_1 = (\frac{2}{3})(u_2 - u_1)$. Use this idea to find the point $\frac{2}{3}$ of the way from $(3, 5)$ to $(8, 2)$. Sketch.

★26. Find the point $\frac{3}{4}$ of the way from $(1, -4)$ to $(2, -8)$. Sketch.

★27. Derive a formula for the coordinates of a point P lying on the line determined by the points P_1 and P_2 in such a position that the directed distance from P_1 to P is r times the directed distance from P to P_2.

★28. What is r for the bisector? for Problem 25?

★29. Where is P if $r < 1$? $r > 1$? $r < 0$? $-1 < r < 0$? $r < -1$? Illustrate with examples.

★30. Let capital letters be variables standing for points and $\overline{PQ}$ = the distance between P and Q. Prove that

(19) $\overline{PQ} \geq 0$,

(20) $(\overline{PQ} = 0) \leftrightarrow (P = Q)$,

(21) $\overline{PQ} = \overline{QP}$,

(22) $\overline{PQ} \leq \overline{PR} + \overline{RQ}$.

ANSWERS TO EXERCISES

(a) Lengths are never negative, but a and b could be.   (b) $(-a)^2 = a^2$. (c) $\sqrt{34}$.   (d) Principal square root is never negative.   (e) Squares are nonnegative, so sum of squares is zero if and only if each is zero.   (f) Put (ka, kb) for (a, b) in (1).   (g) The left and right members of (6') are substituted respectively for a and b.   (h) To evaluate square of right member of (6'). (i) (1-12-9).   (j) (3-5-15).   (k) (3-5-43), (1-13-19).   (l) (1-12-10). (m) $(1)(a:-a, b:-b)$.

(n) $\sqrt{185}$.   (o) $\sqrt{6}$.   (p) $(x - 3)^2 + (y - 4)^2 = 1$.   (q) $x^2 + y^2 = 1$.
(r) $(x - 4)^2 + (y + 3)^2 = 4$.   (s) $(2.5, 6)$.   (t) $(-9, -2)$.   (u) $\mathbf{u}_2 -$
$\mathbf{u}_1 = 2\overrightarrow{\mathbf{u}} - 2\mathbf{u}_1; \mathbf{u}_2 + \mathbf{u}_1 = 2\overrightarrow{\mathbf{u}}$.   (v) Add $\overrightarrow{\mathbf{u}} + \mathbf{u}_1$ to both members.

ANSWERS TO PROBLEMS

1. 5. 3. 4. 5. When the vectors have the same direction. 7. $2\sqrt{5}$. 9. $2\sqrt{5}$.
11. Center at $(-2, 4)$, radius 1. 13. Center at $(0, 3)$, radius 3. 15. All the
plane but the circle. 17. The portion of the circle (an arc) lying to the left
of and on the y-axis. 19. The part of the interior and circumference of the
circle to the right of the y-axis. 21. Both loci, a circle and a circle together
with its interior. 23. Here $d = b = 0$ and we use (1–13–20). 25. $(19/3, 3)$.
27. $(\mathbf{u}_1 + r\mathbf{u}_2)/(1 + r)$. 29. Left of midpoint; right of midpoint; outside segment; to left of segment; to right of segment.

4–5 The straight line. To study the straight line by analytic methods,
we must find defining sentences for straight lines. For this purpose we
need to specify what we mean by a straight line. We recall that if $\mathbf{u}$ is a
vector, then $k\mathbf{u}$ is a vector in the same line with $\mathbf{u}$ and of length $|k| \, |\mathbf{u}|$,
in the same direction as $\mathbf{u}$ if $k > 0$ and in the opposite direction if $k < 0$.
Figure 4–19 sketched the case $k = 3$. Figure 4–23 shows cases $k = -3$
and $k = 0.5$. If we accept the theorems of plane geometry, we can prove
that a point (x, y) lies on the straight line through (x_1, y_1) and (x_2, y_2)
if and only if the vector $(x, y) - (x_1, y_1)$ is some multiple of the vector
$(x_2, y_2) - (x_1, y_1)$, as sketched in Fig. 4–24. Instead, we *define* the

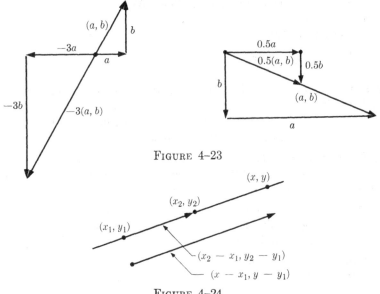

FIGURE 4–23

FIGURE 4–24

straight line through two points to be the set of all points obtained in this way; that is,

The straight line through (x_1, y_1) and (x_2, y_2) is by definition

(1) $\{(x, y) \mid \exists k \, (x - x_1, y - y_1) = k(x_2 - x_1, y_2 - y_1)\}.$

We will show that straight lines so defined do have the properties we expect.

First we work with the defining sentence
$$\exists k \, (x - x_1, y - y_1) = k(x_2 - x_1, y_2 - y_1).$$

(2) $\leftrightarrow \exists k \, [x - x_1 = k(x_2 - x_1)] \wedge [y - y_1 = k(y_2 - y_1)],$

(3) $\leftrightarrow \exists k \, \dfrac{x - x_1}{x_2 - x_1} = \dfrac{y - y_1}{y_2 - y_1} = k,$

(4) $\leftrightarrow y - y_1 = \dfrac{y_2 - y_1}{x_2 - x_1} (x - x_1)$ *(two-point form).*

The equivalent sentence (2) results simply from using the definition of multiplication and (4-1-1). The next step results from dividing both members of the first term in the conjunction by $x_2 - x_1$ and the second by $y_2 - y_1$. To see the equivalence of (4), we note that if (3) is true then the two fractions are equal, from which (4) follows. Conversely, if (4) is true the two fractions are equal, and so are equal to *some* k! The argument leading to (3) fails if $x_2 = x_1$ or $y_2 = y_1$, that is, if the two points are in the same vertical or horizontal line. We ignore these cases for the moment.

The equations in (2),

(5) $x = x_1 + k(x_2 - x_1)$, $y = y_1 + k(y_2 - y_1)$ *(parametric form),*

define the straight line in the sense that we get the points on the line by assigning values to k. The variable k is called a *parameter* to distinguish it from the variables x and y whose values are coordinates of points on the line. The equations in (5) are called *parametric equations* of the line.

(a) Sketch the points $(1, 3)$ and $(2, 7)$ and write the parametric equations of the line through them; find the points corresponding to $k = 0, 1, 3, -1$ and graph them.   (b) Do the same for $(0, 0)$ and $(-1, -5)$.

Equation (4) is called the *two-point form* of the equation of the straight line. It is a single defining equation which is a necessary and sufficient condition for a point to lie on the line through (x_1, y_1) and (x_2, y_2).

(c) Write the two-point form for the line in Exercise (a). Find points for which $x = 1, 2, 4, 0$ and plot them.   (d) Do the same for the line in Exercise (b).

Suppose that the two points are in the same vertical line. Then $x_1 = x_2$ and (2) becomes $\exists k\ x - x_1 = 0 \land y - y_1 = k(y_2 - y_1)$. Now for any y, y_1, and y_2 there exists a k satisfying the second equation (remember, since we assume that the points are different, we cannot have $y_2 = y_1$ as well as $x_2 = x_1$), so that the sentence is equivalent to $x = x_1$ (see 2–7–30). Hence, as we expect, the line is parallel to the y-axis and has the equation $x = x_1$. Similarly, if $y_1 = y_2$, the equation of the line is $y = y_1$ (compare Section 4–2).

We can now find parametric equations or a single equation defining any line. But suppose we are given an equation. How can we tell whether it defines a straight line? For example, does $3x + 2y = 5$, or $y = x^2$, or $y = |x|$? The answer to this question is given by (6).

(6)   *Any straight line has a defining equation of the form $Ax + By + C = 0$ where $\sim(A = B = 0)$, and any equation of this form is a defining equation of some straight line.*

The first part of the theorem is immediate, since (4) is equivalent to $(y_2 - y_1)x + (x_1 - x_2)y + (x_2 - x_1)y_1 - (y_2 - y_1)x_1 = 0$, which is of the desired form, and so are both $x - x_1 = 0$ and $y - y_1 = 0$.

To prove the second part, we suppose first that $B = 0$. Then $A \neq 0$, and the equation becomes $x = -C/A$, which we know is the equation of a straight line. Now, if $B \neq 0$, the equation is equivalent to $y = mx + b$, where $m = -A/B$, $b = -C/B$. We find by substitution that $(0, b)$ and $(-b/m, 0)$ satisfy this equation and hence lie on the graph of $Ax + By + C = 0$. Now, by (4)$(x_1{:}0,\ y_1{:}b,\ x_2{:}{-}b/m,\ y_2{:}0)$, the straight line through $(0, b)$ and $(-b/m, 0)$ is $y - b = [-b/(-b/m)](x - 0)$, which is logically equivalent to $y - b = mx$ or $y = mx + b$. Since our original equation is equivalent to an equation that defines a straight line, it is a defining equation of a straight line, as we wished to prove.

An equation of the form $Ax + By + C = 0$ is called a *linear equation*, for obvious reasons. With theorem (6) we can easily graph such an equation. We simply locate two points by substituting for one variable and solving for the other, then draw the corresponding line. The easiest points to find are those with one zero coordinate. For example, to graph $2x - 5y = 15$ we substitute $(x{:}0)$ to get $-5y = 15$ or $y = -3$. Hence $(0, -3)$ is one point. From $(y{:}0)$ we find $(7.5, 0)$ as a second point. The line is sketched in Fig. 4–25.

Find three points on the graphs of the following and sketch the line for:
(e) $4x - y = 2$, and   (f) $-3x + 2y - 7 = 0$.
(g) Choose two points on the graph of Exercise (e) and use them to write parametric equations of the line.

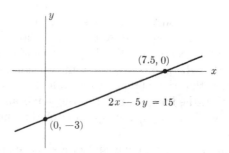

FIGURE 4–25

Slope. The proof of (6) showed that any nonvertical line has an equation of the form $y = mx + b$. Consider two points on the graph of this equation, say (x_1, y_1) and (x_2, y_2). Then

(7) $\quad y_1 = mx_1 + b \land y_2 = mx_2 + b$

(8) $\quad\quad\quad \rightarrow y_1 - mx_1 = b \land y_2 - mx_2 = b$

(9) $\quad\quad\quad \rightarrow y_1 - mx_1 = y_2 - mx_2$

(10) $\quad\quad\quad \rightarrow m(x_2 - x_1) = y_2 - y_1$

(11) $\quad\quad\quad \rightarrow m = (y_2 - y_1)/(x_2 - x_1).$

(h) Justify steps (8) through (11).

The above argument shows that if we choose *any* two points on the graph of $y = mx + b$, the ratio of the difference of the y-coordinates to the difference of the corresponding x-coordinates is the same, namely m. This is what we would expect, since this ratio is precisely the ratio of the components of the vector joining the two points. As suggested in Fig. 4–26, all non-null vectors lying in the same straight line have components in the same ratio. (Note that when the direction of the vector is reversed, *both* components change sign.)

It follows that associated with every nonvertical line there is a real number m given by (11). We call this number the *slope* of the line. If

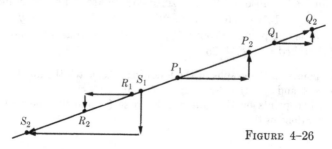

FIGURE 4–26

we know the slope of a line, we know its orientation though not its exact position.

Consider two points whose x-coordinates differ by 1, that is, $x_2 - x_1 = 1$. Then from (11), $y_2 - y_1 = m$. This means that if we move along the line with slope m so that x increases by 1, then y will change by m. For example, if $m = 2$, y changes by 2 when x changes by 1. If $m = -3$, y changes by -3 when x changes by 1. For this reason the slope is sometimes called the *change in y per unit change in x*. If $m > 0$, the line slopes upward from left to right; if $m = 0$, it is horizontal; if $m < 0$, it slopes downward from left to right. These cases are sketched in Fig. 4–27.

Find the slopes of the lines through: (i) $(2, 5)$, $(4, 6)$;   (j) $(-3, 2)$, $(4, -5)$;   (k) $(2a, 1)$, $(3a, b)$.

(l) What is the slope of the graph of $y - 3x = 0$?   (m) of $2x + 3y - 6 = 0$? (*Suggestion:* Put in form $y = mx + b$.)   (n) Show that the slope of a line is equal to the ratio of the second to the first components of any vector lying in it.

An intersection of a locus with an axis is called an *intercept*, an *x-intercept* or a *y-intercept*. Since $(0, b)$ satisfies $y = mx + b$, we have the following result.

(12)   *The graph of* $y = mx + b$ *is the straight line with y-intercept* $(0, b)$ *and slope m.*

Theorem (12) can be used to graph quickly any equation of this form. We simply locate $(0, b)$, then find a second point from the fact that for every unit motion to the right we must move m units vertically. For example, to graph $y = -2x + 6$, we plot $(0, 6)$, then go 1 unit to the right and -2 units vertically to get the point $(1, 4)$. The line is sketched in Fig. 4–28. The equation in (12) is called the *slope-intercept form*.

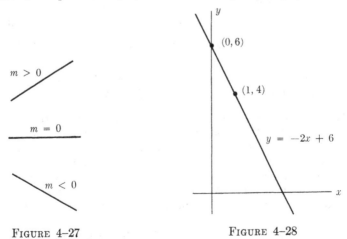

FIGURE 4–27   FIGURE 4–28

Find slopes and intercepts, then sketch: (o) $y = -5x$, and   (p) $14x +$ $2y - 7 = 0$.   (q) Show that

(13) $y - y_1 = m(x - x_1)$ (*point-slope form*)

is an equation of the line with slope m passing through (x_1, y_1). Find the equation of the lines with slope and point as indicated for: (r) 2, $(5, -1)$;   (s) 0, $(0, 0)$;   (t) 0, $(5, 4)$;   (u) -7, $(-1, -3)$.
    (v) Sketch lines with slopes 0, 10, -10, 1/2, 1/10, $-1/10$.   (w) What kind of line has no slope?

PROBLEMS

Graph 1 through 6 by finding the intercepts. In each case find several other points on the graph and verify that they appear to lie on the straight line joining the intercepts. Calculate the slope.

1. $3x - 4y = 12$.   2. $3x + 4y = 12$.
3. $x + y = 10$.   4. $x - 6y = 18$.
5. $2x - 10 = 5y$.   6. $y = 3x + 1$.

7. Suppose that an individual is going to spend \$100 on two commodities whose prices are \$25 and \$10 per unit. Letting x and y be the number of units of each that he buys, we have $25x + 10y = 100$. Graph the points representing his possible purchases, assuming that he can buy only whole units.
    8. Graph the possibilities of Problem 7 on the assumption that the individual can buy any real number of units.

Show that each of 9 through 11 is the equation of a straight line by showing that it is logically equivalent to an equation of the form $Ax + By + C$ with $\sim(A = B = 0)$, and graph.

    9. $y = x$.   10. $x = 41$.   11. $y = -7$.

Graph 12 through 14 by using the intercept and slope.

    12. $y = 10x + 1$.   13. $y = -x$.   14. $y = (-3/10)x + 5$.

Find the slopes and defining equations of the lines through the points in 15 through 18.

    15. $(-3, -4)$, $(2, -1)$.   16. $(\sqrt{2}, 5)$, $(5, \sqrt{2})$.
    17. (a, b), (b, a).   18. (a, b), $(-a, b)$.

19. Graph the locus $\{(x, y) \mid \exists t \, x = 2 + 3t \wedge y = 3 + t\}$.
    20. Prove that the parametric equations $x = x_0 + At$ and $y = y_0 + Bt$, where t is the parameter, define a straight line through (x_0, y_0) with slope B/A.
    21. Graph $\{(x, y) \mid \exists t \, x = 1 + 3t \wedge y = -1 + 2t\}$.
    22. Graph the parametric equations $x = -3 - 5t$, $y = 1 + 4t$.
    23. When does a line have an x-intercept but no y-intercept? a y-intercept but no x-intercept? an infinite number of intercepts? x-intercept and y-intercept coincident?
    24. Show that when $A \neq 0$ and $B \neq 0$ the intercepts of $Ax + By + C$ $= 0$ are $(0, -C/B)$ and $(-C/A, 0)$.

★25. Show that $x/a + y/b = 1$ is an equation of a straight line with intercepts $(0, b)$ and $(a, 0)$. (This is called the *intercept form*.)

★26. Show that the intercepts of $y = mx + b$ are $(0, b)$ and $(-b/m, 0)$. What assumption do we make here?

27. Explain why finding the x-intercept of the graph of $y = mx + b$ is the same as solving the equation $mx + b = 0$.

28. Solve $7x - 10 = 0$ graphically by considering $y = 7x - 10$.

Linear inequalities

29. Show that the graph of $y < mx + b$ is the set of all points below the graph of $y = mx + b$.

In 30 through 33, graph the sentence.

30. $y < 2x - 1$.   31. $y > -x + 3$.

32. $x + 2y < 1$.   33. $3x - y \geq 5$.

34. Graph the possibilities in Problem 7 if the individual can spend any amount less than or equal to $100.

In 35 through 40 let A, B, and C be the graphs of the sentences in Problems 30, 31, and 32, respectively. Graph.

35. A'.   36. B'.

37. $A \cap C$.   38. $A \cup C$.

39. $A \cap B' \cap C'$.   40. $A' \cup B \cup C$.

★Segments

To write a sentence defining a straight line segment, i.e., a portion of a straight line between two points, we need merely write the sentence for the entire straight line joining the points and, in conjunction with it, a sentence requiring one of the variables to lie between the values at the two endpoints. For example, $y = 2x + 1 \land -2 \leq x \leq 2$ defines the segment joining $(-2, -3)$ and $(2, 5)$ as sketched in Fig. 4–29.

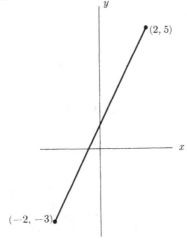

FIGURE 4–29

Graph the sentences in 41 through 44.

★41. $y = x + 1 \land 0 \leq x \leq 3$.

★42. $2x - 3y = 9 \land 1 \leq x \leq 5$.

★43. $\exists t\, 0 \leq t \leq 2 \land x = 1 + 3t \land y = 2 - t$.

★44. $y = 5x + 3 \land 1 \leq y \leq 2$.

★45. Show that $\{(x, y) \mid \exists t\, 0 \leq t \leq 1 \land (x, y) = (x_1, y_1) + t(x_2 - x_1, y_2 - y_1)\}$ is the closed segment joining (x_1, y_1) and (x_2, y_2).

★46. Write parametric equations of the segments joining the points in Problems 15 through 18.

★47. By means of tests students are given scores in linguistic aptitude (x) and mathematical aptitude (y), where scores may vary from 0 to 800 continuously. If we consider general intellectual ability to be measured by the sum of these scores, sketch and describe the locus of points corresponding to a total score of 1100. Do the same if we consider $2x + y$ as a measure of ability. (*The Language of Social Research*, by P. F. Lazarsfeld and M. Rosenberg, p. 47)

Graph

★48. $|x| + |y| = 1$.

★49. $|x| + |y| < 1$.

★50. $|x + y| < 1$.

ANSWERS TO EXERCISES

(a) $x = 1 + k, y = 3 + 4k$.   (b) $x = -k, y = -5k$.   (c) $y - 3 = 4(x - 1)$.
(d) $y = 5x$.   (e) $(0, -2), (1/2, 0)$.   (f) $(0, 3.5), (-7/3, 0)$.   (g) The parametric equations are not unique.   (h) (8) subtracting mx_1 and mx_2; (9) $a = b \land b = c \to a = c$; (10) adding $mx_2 - y_1$ and factoring; (11) dividing by $x_2 - x_1$.   (i) 1/2.   (j) -1.   (k) $(b - 1)/a$.   (l) 3.   (m) $-2/3$.   (n) (11).
(o) $-5, (0, 0)$.   (p) $-7, (0, 3.5), (0.5, 0)$.   (q) It is equivalent to $y = mx + (y_1 - mx_1)$, and (x_1, y_1) satisfies it.   (r) $y + 1 = 2(x - 5)$.   (s) $y = 0$.
(t) $y = 4$.   (u) $y + 3 = -7(x + 1)$.   (v) See Fig. 4–30.   (w) A vertical line. Note that having no slope is quite different from having a slope of zero.

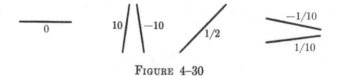

FIGURE 4–30

ANSWERS TO PROBLEMS

1. $3/4, (0, -3), (4, 0)$.   3. $-1, (0, 10), (10, 0)$.   5. $2/5, (0, -2), (5, 0)$.
7. $\{(0, 10), (2, 5), (4, 0)\}$.   8. Segment joining $(0, 10)$ and $(4, 0)$, since $x \geq 0$, $y \geq 0$.   9. $1 \cdot x + (-1)y + 0 = 0$.   11. $0 \cdot x + 1 \cdot y + 7 = 0$.   13. Through origin.   15. $3/5$.   17. -1.   19. Straight line with slope 1/3 through $(2, 3)$.
21. Straight line through $(1, -1)$ with slope 2/3.   23. Parallel to y-axis; parallel

to x-axis; coincident with an axis; through origin. 24. (6); and substitute coordinates to show it passes through the points. 27. x-intercept is found by $(y: 0)$. 29. If $y = mx + b$, (x, y) is on the line. If y is less, (x, y) lies below. 31. Above the line. 33. Below and on the line. 35. Above and on the line. 37. A wedge widening to the right and down. 39. Interior of a triangular region and part of its boundary. 41. [(0, 1) (3, 4)]. 43. [(1, 2) (7, 0)]. 45. The endpoints are given by $t = 0$ and $t = 1$.

4-6 Simultaneous linear equations. A great many problems in science lead to finding the intersection of two straight lines. Suppose, for example, that p is the price and q the amount of a commodity. The amount that consumers are willing to buy (the demand) may be related to the price by an equation such as $q = -0.5p + 10$. (See Fig. 4-31.) The amount that sellers will be willing to offer for sale (the supply) may be given by an equation such as $q = p - 5$. The actual price and amount sold (both offered and demanded) must satisfy both equations, i.e., must satisfy $q = -0.5p + 10 \wedge q = p - 5$.

The coordinates of the intersection may be found exactly by manipulating the two equations together. Equating the two expressions for q gives $p - 5 = -0.5p + 10$, from which $p = 10$. Substituting this value of p in either equation and solving for q, we find $q = 5$. Hence the solution is $(10, 5)$

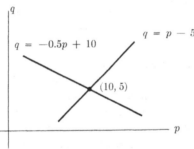

FIGURE 4-31

(a) Solve the problem if supply is given by $q = 2p - 7$ and demand by $q = 11 - p$. (b) Solve for the couple (x, y) that satisfies both $2x - y = 4$ and $x + 2y + 3 = 0$. (c) Find and graph $\{(x, y) \mid x + y = 25 \wedge y = 3x\}$.

When we wish to find ordered pairs that satisfy both of two equations, we call the equations *simultaneous*. The same term is used when two or more equations, inequalities, or other sentences are involved. To solve several simultaneous sentences is to find the set of solutions that satisfies their conjunction, i.e., all of them at the same time. It amounts to finding the intersection of their truth sets, i.e., the set of points common to the sets defined by the individual sentences.

Since the graphs of linear equations are straight lines, the solution of a pair of simultaneous linear equations is a single point (ordered pair) if the corresponding lines meet in one point. If the corresponding lines are parallel, there is, of course, no solution. It may also happen that the two linear equations define the same line. In that case any point on this line is a solution.

We dispose of this last case first. Suppose the two equations are $Ax + By + C = 0$ and $A'x + B'y + C' = 0$. Assuming first that none of the coefficients is zero, the intercepts are $(0, -C/B)$, $(-C/A, 0)$ and $(0, -C'/B')$, $(-C'/A', 0)$, respectively. Clearly the lines are identical if and only if $C/A = C'/A'$ and $C/B = C'/B'$, or $A/A' = B/B' = C/C'$. We describe this by saying that the coefficients are proportional. If any of the coefficients are zero, some of these expressions may be undefined. However, the condition may be restated as follows: there exists a k such that $A = kA'$, $B = kB'$, and $C = kC'$. We leave it to the reader to show that this condition is also necessary and sufficient in the cases where some of the coefficients are zero. To summarize,

(1) $[\{(x, y) \mid Ax + By + C = 0\} = \{(x, y) \mid A'x + B'y + C' = 0\}]$

$\leftrightarrow \exists k\, A = kA' \wedge B = kB' \wedge C = kC'$.

(d) Write several equations equivalent to $3x - 5y + 1 = 0$.   (e) Find the intercepts of $2x - 3y + 1 = 0$ and of $4x - 6y + 2 = 0$.   (f) Show that $C/A = C'/A' \wedge C/B = C'/B' \leftrightarrow A/A' = B/B' = C/C' \leftrightarrow \exists k\, A = kA' \wedge B = kB' \wedge C = kC'$.

When a single solution exists, simultaneous linear equations can always be solved by manipulations such as those described in Section 2–8, except that now we must carry along two equations instead of just one. We may operate on either equation alone by replacing any term by a synonym (see manipulation Type A in Section 2–8) or by multiplication or addition as indicated by (2–8–4) and (2–8–3). In addition to these manipulations, we may add corresponding members according to the law

(2) $(a = b \wedge c = d) \leftrightarrow (a + c = b + d \wedge c = d)$.

(g) Illustrate (2) and prove it.   (h) Show that (2) would be false if $c = d$ were omitted from the right member.

We could have solved the equations in the first paragraph of this section by applying (2) with $(a{:}q,\, b{:}{-}0.5p + 10,\, c{:}{-}q,\, d{:}{-}(p - 5))$, after first multiplying both members of $q = p - 5$ by -1. This amounts to subtracting the corresponding members of one equation from the other. Indeed, $(2)(c{:}{-}c,\, d{:}{-}d)$ yields

(3) $(a = b \wedge c = d) \leftrightarrow (a - c = b - d \wedge c = d)$.

The standard method for solving linear equations is to apply (3) in order to eliminate one of the variables, after multiplying by appropriate constants so that the coefficients of one of the variables are the same in the equations. For example, to solve $3x - 5y = 15 \wedge 2x + y = 5$ we multiply both members of the first equation by 2 and of the second by 3 to get $6x - 10y = 30 \wedge 6x + 3y = 15$. Then using (3), we have $-13y = 15 \wedge 2x + y = 5$. From this we find $y = -15/13 \wedge 2x + y = 5$. Now we may find x by substituting $(y:-15/13)$ in $2x + y = 5$, or by eliminating y from the original equations. This could be done, for example, by multiplying the second equation by 5, then adding to the first by (2). This gives $3x - 5y = 15 \wedge 10x + 5y = 25$, $13x = 40 \wedge 2x + y = 5$, and $x = 40/13 \wedge 2x + y = 5$.

(i) Check by substitution that $(40/13, -15/13)$ satisfies both the original equations.

We have arranged these manipulations below in a systematic manner and also modified the details slightly.

(4)
$$3x - 5y = 15$$
$$2x + y = 5$$
$$\begin{pmatrix} 3 & -5 & 15 \\ 2 & 1 & 5 \end{pmatrix}$$

(5)
$$6x - 10y = 30$$
$$6x + 3y = 15$$
$$\begin{pmatrix} 6 & -10 & 30 \\ 6 & 3 & 15 \end{pmatrix}$$

(6)
$$-13y = 15$$
$$2x + y = 5$$
$$\begin{pmatrix} 0 & -13 & 15 \\ 2 & 1 & 5 \end{pmatrix}$$

(7)
$$-13y = 15$$
$$26x + 13y = 65$$
$$\begin{pmatrix} 0 & -13 & 15 \\ 26 & 13 & 65 \end{pmatrix}$$

(8)
$$-13y = 15$$
$$26x = 80$$
$$\begin{pmatrix} 0 & -13 & 15 \\ 26 & 0 & 80 \end{pmatrix}$$

(9)
$$y = -15/13$$
$$x = 40/13$$
$$\begin{pmatrix} 0 & 1 & -15/13 \\ 1 & 0 & 40/13 \end{pmatrix}$$

(j) Explain each transition from (4) to (9).

Each pair of equations is logically equivalent to the preceding one in (4) through (9), so that the last pair gives the solution. However, the reader must have noticed that the x, y, $=$, and $+$ play only a minor role.

It is the coefficients that are manipulated. Indeed, the variables and equals sign serve only to indicate the proper location of the coefficients. This suggests the idea of writing down only the coefficients and manipulating them without having to rewrite all the other symbols. We have done this to the right of equations (4) through (9). Each rectangular array of numbers is an abbreviation for the corresponding pair of equations. Rectangular arrays of objects are called *matrices*. There are six matrices above. In each one, the *row* indicates the equation, and the *column* indicates whether the number is a coefficient of x, y, or is the right-hand member of an equation.

Multiplying all members of a row corresponds to multiplying both members of an equation. Adding one row to another, term by term, corresponds to adding the members of one equation to those of the other. Hence we may solve the equations by manipulating the matrix until we arrive, as in (9), at a matrix of the form

(10) $$\begin{pmatrix} 0 & 1 & r_2 \\ 1 & 0 & r_1 \end{pmatrix} \quad \text{or} \quad \begin{pmatrix} 1 & 0 & r_1 \\ 0 & 1 & r_2 \end{pmatrix}$$

Then (r_1, r_2) is the solution. The operations permitted are

    I. *Multiplying all elements of a row by the same number.*

(11)   II. *Adding to all elements of a row a fixed multiple of the corresponding element in another row.*

    III. *Interchanging rows.*

Solve the following sentences along the lines of (4) through (9), using both methods, and check by graphing and substitution: (k) $2x + 3y = 6 \wedge x + y = 3$, (l) $3x - 4y = 10 \wedge -2x + 5y = 11$.

The matrix method applies only to linear equations. Others must be attacked by the use of (2) and other laws. For example, suppose we wish to find the intersection of the graphs of $x + 2y = 2$ and $x^2 + y^2 = 9$, as sketched in Fig. 4–32. Solving the first one for x we get $x = 2(1 - y)$. Replacing x in the second equation by $2(1 - y)$, we find $4(1 - y)^2 + y^2 = 9$. Simplification yields $5y^2 - 8y - 5 = 0$, from which we have $y = (4 \pm \sqrt{41})/5$. The corresponding values of x may be found by substituting in the linear equation. We find in this way that the two unlabeled points are $((2 + 2\sqrt{41})/5, (4 - \sqrt{41})/5)$, $((2 - 2\sqrt{41})/5, (4 + \sqrt{41})/5)$.

(m) Express the coordinates as single decimals and verify that the corresponding points are approximately as suggested by the graph. (n) Sketch and solve $2x + y = 2 \wedge x^2 + y^2 = 10$.

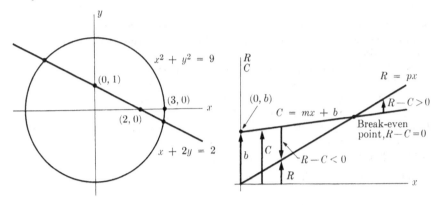

FIGURE 4-32   FIGURE 4-33

PROBLEMS

Solve 1 through 4 by the matrix method.

1. $5x + 2y = 6,$
   $x - 3y = 1.$

2. $5x - 4y = 7,$
   $2x - 3y = 10.$

3. $1.5x + 3.6y = 10,$
   $2x - 1.2y = 15.$

4. $x - y = 5,$
   $-17x + 3y = 1.$

If the sales of a firm are x and the price per unit is p, then the revenue from sales R is given by $R = px$. If the overhead is b, and the variable cost per unit is m, then the cost of producing and selling x units is given by $C = mx + b$. For definiteness let $R = 0.5x$ and $C = 0.25x + 10$. Measuring R and C along the y-axis, we may graph the two equations as in Fig. 4-33. At any particular point along the x-axis, the distance up to the revenue curve gives the revenue, the distance up to the cost curve gives the cost, and the vertical vector from the cost curve to the revenue curve $(R - C)$ gives the profit. At the point where the two curves meet, the profit is zero, and the firm "breaks even." To the left, it loses money. To the right it makes a profit. Such charts, called *break-even charts*, are widely used in business management.

★5. Make a break-even chart for $R = 1.5x$, $C = 50 + x$, and find the break-even point by algebra.

★6. Do the same for $R = 1.5x$, $C = 25 + 2x$.

★7. Do the same for $R = 1.5x$, $C = 25 + 1.2x$.

★8. Two businesses selling the same product at price p have cost functions given by $C_1 = 200 + 0.5x$ and $C_2 = 100 + x$. Which is in the better position?

★9. Find a formula for the coordinates of the break-even point in terms of p, m, and b for $R = px$, $C = mx + b$.

★10. Show that the break-even point is at the sales where total cost per unit, C/x, equals the price.

★11. Show analytically that two lines cannot have more than one point in common.

★12. Three lines are said to be *concurrent* if they have a point in common. Find a necessary and sufficient condition for this in terms of the coefficients of their equations.

★13. Show that the medians of a triangle are concurrent.

★14. Graph $2x - y + 4 = 0$, $5x + 3y + 15 = 0$, and $k_1(2x - y + 4) + k_2(5x + 3y + 15) = 0$ for $(k_1, k_2) = (1, 1)$, $(1, -1)$, and $(2, 3)$.

★15. Show that for any k_1 and k_2, the graph of $k_1(ax + by + c) + k_2(a'x + b'y + c') = 0$ is a straight line passing through the intersection of $ax + by + c = 0$ and $a'x + b'y + c' = 0$.

★16. What happens if the two lines are parallel?

Find the line through the intersection of $2x - y = 7$ and $51x + 33y = 2$ and

★17. having a slope of 1,

★18. passing through the origin,

★19. passing through $(-5, 7)$.

★20. Prove (12).

(12) $$ac \neq 0 \rightarrow [(a = b \land c = d) \leftrightarrow (ac = db \land c = d)].$$

★21. Show that $[k^2 = k^2u^2 + x_1^2 \land 2z/(1 - u^2) = 2k] \rightarrow [z^2 = x_1^2(1 - u^2)]$.

★22. Let $h_1 = -C + \theta + \phi$, $h_2 = -C - \theta + \phi$. Find θ in terms of h_1 and h_2 only. (Astronomy.)

★23. Prove the identity $[(x - y)/2]^2 = [(x + y)/2)^2 - xy$, and show how it can be used to solve $xy = 1 \land x + y = a$. (This method appears in a mathematical treatise of about 300 B.C.)

★24. Given $a(\alpha - \alpha_0) = h_1\lambda$, $a(\beta - \beta_0) = h_2\lambda$, $a(\gamma - \gamma_0) = h_3\lambda$, $\alpha^2 + \beta^2 + \gamma^2 = 1$, and $\alpha_0^2 + \beta_0^2 + \gamma_0^2 = 1$, find a formula not involving α, β, γ that gives λ.

Answers to Exercises

(a) $(6, 5)$.   (b) $(1, -2)$.   (c) $(6.25, 18.75)$.   (d) $6x - 10y + 2 = 0$, $5y = 3x + 1$, etc.   (e) $(-1/2, 0)$, $(0, 1/3)$ for both.   (f) Multiply both members of first equation by A/C' and of second by B/C'; then these ratios are equal if and only if they are equal to some k for which the three equations hold. (g) $(2\text{-}7\text{-}36)$ and $(2\text{-}8\text{-}3)$.   (h) $(a\text{:}2, b\text{:}3, c\text{:}1, d\text{:}0)$.   (i) Review the comments about $(2\text{-}8\text{-}1)$.   (j) Multiply both members of first equation by 2, second by 3; subtract second from first, and then divide second by 3; multiply second by 13; subtract first from second; divide first by -13, second by 26.   (k) Check by substitution. From now on answers will not be given where this can be done. (n) Two solutions.

Answers to Problems

8. The second for output less than 200, the first beyond that. 11. Show that conditions (1) are satisfied. 13. Find midpoints of sides and use two-point form to get equations of medians. Solve two of them and show that the third is satisfied by the result. Or use the result of Problem 12.

4-7 Parallel and perpendicular lines. To investigate the possibility of parallelism we attempt to solve $A_1x + B_1y = C_1 \land A_2x + B_2y = C_2$. To save writing we use the matrix method.

(1)
$$\begin{pmatrix} A_1 & B_1 & C_1 \\ A_2 & B_2 & C_2 \end{pmatrix}$$

(2)
$$\begin{pmatrix} A_1B_2 & B_1B_2 & C_1B_2 \\ B_1A_2 & B_1B_2 & B_1C_2 \end{pmatrix}$$

(3)
$$\begin{pmatrix} (A_1B_2 - B_1A_2) & 0 & (C_1B_2 - B_1C_2) \\ A_2 & B_2 & C_2 \end{pmatrix}$$

(a) Explain how (2) and (3) were obtained.

The first line of (3) means $(A_1B_2 - B_1A_2)x = (C_1B_2 - B_1C_2)$. Clearly this gives a unique solution for x *unless* $A_1B_2 - B_1A_2 = 0$. Hence,

(4) $A_1x + B_1y = C_1 \land A_2x + B_2y = C_2$ *has a unique solution if and only if* $A_1B_2 - B_1A_2 \neq 0$.

What if $A_1B_2 - B_1A_2 = 0$? Then there is no value of x satisfying the first equation of (3) and hence no solution if $C_1B_2 - B_1C_2 \neq 0$. Hence,

(5) $A_1x + B_1y = C_1 \land A_2x + B_2y = C_2$ *has no solution and the corresponding lines are parallel if and only if* $A_1B_2 - B_1A_2 = 0 \neq C_1B_2 - B_1C_2$.

On the other hand, if $A_1B_2 - B_1A_2 = 0 = C_1B_2 - B_1C_2$, then any value of x is a solution of the first equation in (3). Hence any pair that satisfies the second equation satisfies both, and there is only one line. We have found (4-6-1) again.

(b) Show that $A_1B_2 - B_1A_2 = 0 = C_1B_2 - B_1C_2$ is equivalent to the condition given in (4-6-1).

The result (5) can easily be given an interpretation in terms of slope. If the lines are nonvertical, the equations can be put in the form

$$y = (-A_1/B_1)x + (C_1/B_1) \quad \text{and} \quad y = (-A_2/B_2)x + (C_2/B_2).$$

Now $[-A_1/B_1 = -A_2/B_2] \leftrightarrow [A_1B_2 - B_1A_2 = 0],$

and $[C_1/B_1 \neq C_2/B_2] \leftrightarrow [C_1 B_2 - B_1 C_2 \neq 0.]$

Hence,

(6)   *Two nonvertical lines are parallel if and only if their slopes are equal and their y-intercepts different.*

Note that if, in addition, the y-intercepts are equal, there is only one line.

(c) Justify the asserted equivalences.

When there is no solution, the equations are called *inconsistent*. When there is one solution, they are called *consistent* and *independent*. When there is an infinity of solutions (coincident lines), the equations are called *dependent*.

Characterize each of the following pairs of equations precisely, using the above terminology; graph; attempt to solve; and check the theorems:
(d) $2x - 3y = 4 \wedge 2x - 3y = 2$,   (e) $2x - 3y = 4 \wedge x - 1.5y = 3$,
(f) $2x - 3y = 4 \wedge x - 1.5y = 2$,   (g) $2x - 3y = 4 \wedge 2x + 3y = 4$.

We can express the condition for parallelism in terms of vectors. We recall from Exercise (n) in Section 4–5 that for any vector (r, s) in a line of slope m, $s/r = m$. Hence from (6), nonvertical vectors (x, y) and (x', y') lie in parallel lines or in the same line if and only if $y/x = y'/x'$. Assuming $y' \neq 0$, this is equivalent to $y/y' = x/x'$, or to $\exists k\, y = ky' \wedge x = kx'$. But this last condition says simply that $(x, y) = k(x', y')$, that is, that one vector is a multiple of the other. In case $y' = 0$, we have immediately $y = 0$, $(x, y) = (x, 0)$, $(x', y') = (x', 0)$, and again one vector is a multiple of the other, since $(x, 0) = (x/x')(x', 0)$. Since any vertical vector is a multiple of any other, we have in all cases:

(7)   *If $\mathbf{u}$ and $\mathbf{v}$ are non-null vectors, $\mathbf{u}$ and $\mathbf{v}$ lie in the same or parallel lines if and only if $\exists k\, \mathbf{u} = k\mathbf{v}$.*

(h) Prove that any vertical vector is a multiple of any other.   (i) Why "non-null" in (7)?

We apply (7) to prove analytically the well-known theorem of plane geometry that the segment joining the midpoints of two sides of a triangle is parallel to and of length one-half the third side. Figure 4–34 shows the triangle determined by three vectors, $\mathbf{u}_1, \mathbf{u}_2, \mathbf{u}_3$. We have $0.5\mathbf{u}_3 + \mathbf{u}_4 = \mathbf{u}_1 + 0.5\mathbf{u}_2$. Solving for $\mathbf{u}_4$ and using $\mathbf{u}_3 = \mathbf{u}_1 + \mathbf{u}_2$, we have $\mathbf{u}_4 = \mathbf{u}_1 + 0.5\mathbf{u}_2 - 0.5(\mathbf{u}_1 + \mathbf{u}_2) = 0.5\mathbf{u}_1$. Since $\mathbf{u}_4$ is a multiple of $\mathbf{u}_1$, it is parallel to it. Moreover, by (4–4–5), the length of $\mathbf{u}_4$ is one-half that of $\mathbf{u}_1$, and the theorem is proved!

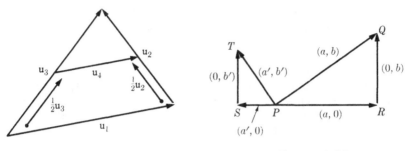

FIGURE 4–34   FIGURE 4–35

Under what conditions are two vectors perpendicular? Figure 4–35 shows $(a, b) \perp (a', b')$, together with their component vectors. The sum of the three angles at P is two right angles. Since $\angle TPQ$ is a right angle, it follows that $\angle SPT$ and $\angle RPQ$ are complementary. But since $\angle PST$ is a right angle, $\angle SPT$ and $\angle STP$ are complementary. It follows that $\angle RPQ$ and $\angle STP$ are congruent. But now we have shown that the two right triangles PST and QRP have congruent angles. Hence they are similar and their corresponding sides are proportional; that is, $\overline{QR}/\overline{PR} = \overline{PS}/\overline{TS}$. Now, in Fig. 4–35, $\overline{QR} = b$, $\overline{PR} = a$, $\overline{TS} = b'$, $\overline{PS} = -a'$, since $a' < 0$ and we wish the lengths. Hence we have $b/a = -a'/b' = -1/(b'/a')$, or $(b/a)(b'/a') = -1$; that is, the product of the slopes is -1. We describe this by saying that the slopes are *negative reciprocals*. Does our argument depend on the particular position of the vectors in Fig. 4–35? Yes, it does; but the same result holds regardless of the positions of the vectors, provided that neither of them is vertical.

(j) What happens if one of the vectors is vertical? (k) Justify the algebraic manipulations above. (l) Show by rotating two perpendicular vectors that we always have either one or three of their components negative, and hence that the proportionality of sides of the right triangles always leads to the above result.

We have justified the following theorem.

(8)   *Two nonvertical lines are perpendicular if and only if their slopes are negative reciprocals.*

We can restate the above results so as to avoid the limitation to non-vertical lines. The condition $b/a = -a'/b'$ is equivalent to $aa' + bb' = 0$. This condition is also necessary and sufficient when one of the vectors is vertical, for if one vector is vertical they are perpendicular if and only if they are of the form $(a, 0)$ and $(0, b')$, in which case clearly $aa' + bb' = a0 + 0b = 0$. Hence we have:

(9)   *A necessary and sufficient condition that two non-null vectors*
        (a, b) and (a', b') are perpendicular is that $aa' + bb' = 0$.

Which of the following pairs of vectors are perpendicular? Check by sketching.   (m) $(2, 3)$, $(3, -2)$.   (n) $(1, 2)$, $(-3, -1)$.   (o) $(2, -6)$, $(3, 1)$.

(p) Why did we exclude null vectors in (9)?   (q) Show analytically that no vector can be perpendicular to itself.   (r) Show that $\mathbf{u} \perp \mathbf{v} \to \mathbf{v} \perp \mathbf{u}$. (s) Show that $\mathbf{u} \perp \mathbf{v} \to -\mathbf{u} \perp \mathbf{v}$.   (t) Prove analytically that the diagonals of a square are perpendicular.

PROBLEMS

In Problems 1 and 2 sketch all lines on the same plane.

1. $y = 3x - b$ for $b = 1, 3, -2, 0$.
2. $y = mx + 3$ for $m = 5, 2, 1, 1/3, -2, 0$.
3. Find the line parallel to $5x - 2y = 7$ and passing through $(0, -1)$.
4. Same as in Problem 3 but passing through $(-5, 3)$.
5. Same as in Problem 3 but perpendicular to the original line.

Show 6 through 13 analytically.

6. If the opposite sides of a quadrilateral have equal length, it is a parallelogram.
7. The converse of the previous theorem.
8. The diagonals of a parallelogram bisect each other.
9. The locus of points equidistant from two points is the perpendicular bisector of the segment joining them.
10. The altitudes of a triangle are concurrent.
11. The perpendicular bisectors of the sides of a triangle are concurrent.
12. If a chord of a circle is perpendicular to a radius, it is bisected by it.
13. The diagonals of a rhombus bisect each other at right angles.

14. Find the equation of the line through (x_0, y_0) perpendicular to the line $Ax + By + C = 0$.
15. Find the point where the perpendicular in Problem 14 meets the line $Ax + By + C = 0$.
16. Use the result in Problem 15 to show that the length d of the perpendicular from the point (x_0, y_0) to $Ax + By + C = 0$ (i.e., the distance between the point and line) is given by

(10) $$d = \frac{|Ax_0 + By_0 + C|}{\sqrt{A^2 + B^2}}.$$

17. Sketch $3x - y - 16 = 0$ and the points $(5, 1)$, $(1, -13)$, $(0, 0)$. Find the distance of each point from the line.
18. Find the distance from the origin of each line in Problems 1 through 6 in Section 4-5.
19. Show that $(1, 5)$ and $(-1, 1)$ are equidistant from the line $y = 4x + 3$.

★20. Write a defining sentence for the set of points within a distance of 2 of the line $y = x - 4$. Sketch the locus.

★21. What is the distance between the lines $y = 3x + 2$ and $y = 3x - 4$?

ANSWERS TO EXERCISES

(a) Multiplying first row by B_2 and second by B_1; subtracting second row from first. (b) Divide first equation by A_2B_2, the second by C_2B_2. (c) See Exercise (b), and use $[p \leftrightarrow q] \leftrightarrow [{\sim}p \leftrightarrow {\sim}q]$. (d) Inconsistent. (e) Inconsistent. (f) Dependent. (g) Consistent and independent. (h) $(0, b) = (b/b')(0, b')$. (i) $\forall x,y\ (0, 0) = 0(x, y)$. (j) It has no slope. (k) See Section 1–9. (l) Draw a series of figures, or rotate a figure consisting of just the vectors, observing how the components change. (m) Yes. (n) No. (o) Yes. (p) The condition is always satisfied if one of the vectors is null, but a null vector has no direction. (q) For (9) would require $a^2 + b^2 = 0$, which is true only for the null vector. (r) From (9). (s) From (9). (t) Let the sides be $(a, 0)$ and $(0, a)$, so that one diagonal is $(a, 0) + (0, a) = (a, a)$ and the other is $(a, 0) - (0, a) = (a, -a)$. Then the condition (9) becomes $a^2 - a^2 = 0$.

ANSWERS TO PROBLEMS

1. Parallel lines all of slope 3. 2. Lines with different slopes through the point $(0, 3)$. 3. $5x - 2y = 2$. 5. $2x + 5y + 5 = 0$. 7. For simplicity let one side be parallel to x-axis. Use vectors. 9. Find equation by equating distances from fixed points $(a, 0)$ and $(b, 0)$ to (x, y) on the locus. Simplify to show that the resulting line is vertical and passes through $((a + b)/2, 0)$.

4–8 The circle. We have seen already that the graph of

$$(1) \qquad\qquad (x - h)^2 + (y - k)^2 = r^2$$

is the circle with center (h, k) and radius r.

(a) Review (4–4–15) and related portions of Section 4–4.

If we expand (1) and rearrange, we find

$$(2) \qquad x^2 + y^2 - 2hx - 2ky + h^2 + k^2 - r^2 = 0.$$

This is in the form

$$(3) \qquad x^2 + y^2 + Dx + Ey + F = 0.$$

Is any equation of form (3) a defining equation of a circle? To answer, let us see whether we can determine h, k, and r in (2) in terms of D, E, and F in (3) so that the two equations are equivalent. We wish to have $D = -2h$, $E = -2k$, and $h^2 + k^2 - r^2 = F$. Solving the first two for h and k, and substituting in the third, we arrive at

$$(4) \qquad h = -D/2, \qquad k = -E/2, \qquad r^2 = D^2/4 + E^2/4 - F.$$

Hence for any D, E, F we can find h, k, and r so that (3), (2), and (1) are logically equivalent. Hence,

(5)   *The graph of* $x^2 + y^2 + Dx + Ey + F = 0$ *is the circle with center* $(-D/2, -E/2)$ *and radius* r *given by* $r^2 = D^2/4 + E^2/4 - F$, *provided* $4F < D^2 + E^2$.

We illustrate by finding the center and radius of the graph of $x^2 + y^2 + 4x - 2y - 1 = 0$. Here $D = 4$, $E = -2$, $F = -1$. Then $h = -2$, $k = 1$, $r^2 = 4 + 1 + 1 = 6$. Hence the center is at $(-2, 1)$ and the radius is $\sqrt{6}$.

Find the center and radius of the graphs of: (b) $x^2 + y^2 + 2x - 4y - 1 = 0$, and   (c) $x^2 + y^2 - 3x + \sqrt{2}y + 2 = 0$.
(d) Why "provided $4F < D^2 + E^2$" in (5)?   (e) What is the graph of $x^2 + y^2 + 2x - 4y + 5 = 0$?   (f) of $x^2 + y^2 + 2x - 4y + 7 = 0$?

Replacing h, k, and r^2 in (1) by the expressions given in (4), we have

(6) $(x + D/2)^2 + (y + E/2)^2 - D^2/4 - E^2/4 + F = 0$.

Expanding, we find

(7) $x^2 + Dx + D^2/4 + y^2 + Ey + E^2/4 - D^2/4 - E^2/4 + F = 0$.

We see that (7) is synonymous with (3), as expected. Indeed, (7) may be obtained by adding and subtracting $D^2/4$ and adding and subtracting $E^2/4$ in (3). This observation suggests a method of putting an equation of the form (3) into the form (1) without using (4). Beginning with (3), we group the terms in x and y together.

(8) $x^2 + Dx + \quad y^2 + Ey \quad + F = 0$.

Then we add a term to make a perfect square out of the terms in x, and compensate by subtracting the same term.

(9) $x^2 + Dx + (D/2)^2 + y^2 + Ey - D^2/4 + F = 0$.

We add the square of half the coefficient of x, since

$$(x + D/2)^2 = x^2 + 2(D/2)x + D^2/4 = x^2 + Dx + D^2/4.$$

We do the same for the terms in y to arrive at (7). From (7) to (6) is a matter of factoring the perfect squares. And (6) is in the form (1). We illustrate.

(10) $x^2 + y^2 + 4x - 2y - 1 = 0,$

(11) $(x^2 + 4x + \quad) + (y^2 - 2y + \quad) - 1 = 0,$

(12) $(x^2 + 4x + 4) + (y^2 - 2y + 1) - 4 - 1 - 1 = 0,$

(13) $(x + 2)^2 + (y - 1)^2 = 6.$

The process of adding a term to an expression of the second degree to make it into a perfect square is called *completing the square*. In (10) through (13) we completed the square on x and on y.

(g) Find the center and radius of the circles in Exercises (b) and (c) by completing the square. (h) Where did we use completing the square in a previous chapter?

We can now prove geometric theorems and solve geometric problems about circles by analytic methods. For example, we can easily prove that a straight line meets a circle in two, one, or no points. For suppose the circle is defined by (3) and the line by $y = mx + b$. Solving the two simultaneously by replacing y by $mx + b$ in (3), we find

(14) $(1 + m^2)x^2 + (2mb + D + Em)x + b^2 + Eb + F = 0.$

Since this is a quadratic equation, it has two real roots, one real root, or none, according as $(2mb + D + Em)^2 - 4(1 + m^2)(b^2 + Eb + F)$ is positive, zero, or negative.

(i) Explain the last statement. (j) What do we call the line if there is only one root of (14)? (k) Have we covered all possible lines in our proof? (l) Find the intersections of $x^2 + y^2 = 9$ with $y = 2x$, and (m) with $y = 2x + 1$.

Regions bounded by circles play a role in many applications. If we put $<$ in place of $=$ in (1), we get an inequality defining the interior of the circular region bounded by the graph of (1).

(n) Why? (o) What if we put $>$ in place of $=$? (p) $\leq$? (q) $\geq$? (r) Graph $x^2 + y^2 < 9$.

To illustrate the application of these ideas in economic theory, imagine two factories producing the same commodity at the same cost but located in different communities. The cost of distributing the product depends on the distance of the consumers from the plants. For simplicity, we place the axes so that the plants lie at the origin and at the point $(a, 0)$, as indicated in Fig. 4–36. For maximum efficiency we wish to have each plant supply those customers for whom its transportation costs are smaller. Suppose that the plant at the origin (plant P_1) spends c_1 per

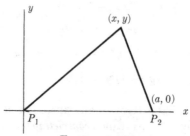

FIGURE 4-36

unit per mile and that the other plant (plant P_2) spends c_2 per unit per mile to deliver the product. For simplicity, assume that delivery may be made by travel in a straight line, so that for a customer at (x, y) the cost of the delivery of a unit is given by $C_1 = c_1\sqrt{x^2 + y^2}$ for the first plant, and $C_2 = c_2\sqrt{(x - a)^2 + y^2}$ for the second plant. Accordingly, the first plant can supply the commodity more cheaply if and only if $C_1 < C_2$, and the second if and only if $C_1 > C_2$. The points where the cost is the same are those satisfying $C_1 = C_2$. Now, $C_1 = C_2$ is

(15) $$c_1\sqrt{x^2 + y^2} = c_2\sqrt{(x - a)^2 + y^2}.$$

Since c_1 and c_2 are non-negative, this is equivalent to

(16) $$c_1^2(x^2 + y^2) = c_2^2(x^2 - 2ax + a^2 + y^2),$$

which simplifies to

(17) $$(c_1^2 - c_2^2)x^2 + (c_1^2 - c_2^2)y^2 + 2ac_2^2x - a^2c_2^2 = 0.$$

Dividing both members by the coefficient of x^2 and letting

$$s = c_2^2/(c_1^2 - c_2^2),$$

we have

(18) $$x^2 + y^2 + 2asx - a^2s = 0,$$

which is, by (5), the equation of the circle with center at $(-as, 0)$ and radius $a\sqrt{s^2 + s}$. Hence the boundary is a circle. One plant should serve customers within the circle and the other those outside the circle. The location of the circle depends on the values of a, c_1, and c_2.

Assume, as in Fig. 4-36, that $a > 0$.   ★(s) Suppose that the first firm has the lower transportation costs so that $c_1 < c_2$. Show that $s < 0$, $-as > 0$, $-as > a$, and $a\sqrt{s^2 + s} < -as$, so that the center of the circle is to the right of P_2 and the second plant serves just its interior, while the first serves all points outside.   ★(t) Sketch the previous situation for $a = 10$, $c_1 = 1$, $c_2 = 2$. ★(u) Draw similar conclusions for the case where $c_1 > c_2$, and sketch this situation for $a = 10$, $c_1 = 2$, $c_2 = 1$.   ★(v) What happens when $c_1 = c_2$?

PROBLEMS

In Problems 1 through 10, complete the square and sketch the locus.

1. $x^2 + y^2 - 2x + 3y - 1 = 0$.
2. $x^2 + y^2 + 5x - 6y - 10 = 0$.
3. $x^2 + y^2 - 5x = 0$.
4. $x^2 + y^2 + 2y = 1$.
5. $2x^2 + 2y^2 + 5x - 13y = 3$.
6. $x^2 + y^2 - 2x - 3.2y = 0$.
7. $3x - x^2 + 4y - y^2 = 1$.
8. $10 - x - y - 2x^2 - 2y^2 = 0$.
9. $x^2 + y^2 - x + 25 = 0$.
10. $x^2 + y^2 + 2x - 4y = -5$.

★11. Find a formula for the solution of $x^2 + Ax + B = 0$ by completing the square.

In Problems 12 through 21, sketch the locus.

12. $x^2 + y^2 \leq 16$.
13. $x^2 + y^2 > 16$.
14. $(x - 3)^2 + (y - 2)^2 < 4$.
15. $x^2 + y^2 + 2x > -1$.
16. $x^2 + y^2 \leq 0 \wedge -2 \leq x \leq 2$.
17. $x^2 + y^2 \leq 25 \wedge y < x < 1$.
18. $4 < x^2 + y^2 < 9$.
19. $x^2 + y^2 \leq 0$.
20. $x^2 + y^2 < 0$.
21. $x^2 + y^2 > 25 \wedge y > 2$.

★22. Show analytically that a tangent to a circle is perpendicular to the radius at the point of tangency.

★23. State and prove the converse of Problem 22.

★24. Show analytically that two circles meet in at most two points.

★25. Show analytically that the segment joining the centers of two circles is the perpendicular bisector of their common chord.

★26. Show that the length of a tangent line to the circle $G(x, y) = x^2 + y^2 + Dx + Ey + F = 0$ from the point (x_0, y_0) is $\sqrt{G(x_0, y_0)}$.

★27. Prove that for any value of m the lines $y = mx \pm a\sqrt{1 + m^2}$ are tangent to the circle $x^2 + y^2 = a^2$.

★28. Prove that $x_0x + y_0y = a^2$ is tangent to the circle $x^2 + y^2 = a^2$ at the point (x_0, y_0).

★29. Prove that a line through the center of a circle perpendicular to a chord bisects the chord.

★30. Prove the converse of Problem 29.

ANSWERS TO EXERCISES

(b) $(-1, 2)$, $\sqrt{6}$.   (c) $(1.5, -\sqrt{2}/2)$, $\sqrt{3}/2$.   (d) If this is not satisfied, we have no points on the graph, or only one, according as the right member of the equation for r^2 in (4) is negative or zero.   (e) The single point $(-1, 2)$. (f) Empty.   (h) (2-8-14) through (2-8-16).   (i) This is the discriminant; see Section 2-8.   (j) Tangent.   (k) No, not vertical lines; but the statement clearly holds for them, since letting $x = c$ in (2) we have a quadratic for y. (l) $(3/\sqrt{5}, 6/\sqrt{5})$, $(-3/\sqrt{5}, -6/\sqrt{5})$.   (m) Substitute to check.   (n) Points that satisfy it are at a distance from the center of less than the radius.   (o) Exterior.   (p) Interior and boundary.   (q) Exterior and boundary.   (r) Interior of circle with center at origin and radius 3.

ANSWERS TO PROBLEMS

1. $(1, -1.5)$, $\sqrt{4.25}$.   3. $(2.5, 0)$, 2.5.   5. $(-5/4, 13/4)$, $\sqrt{218}/4$.   7. $(1.5, 2)$, $\sqrt{21}/2$.   9. Empty.   11. Check by substitution and compare with quadratic formula $(a{:}1, b{:}A, c{:}B)$.   13. Exterior of circle with radius 4.   23. A line perpendicular to a radius at a point on a circle is tangent. Write equation and show that it meets the circle in only the point.   25. Extend the idea of Problem 15 in 4-6 to show that the equation of the common chord is obtained by subtracting the left members of the equations.   27. Solve with equation of the circle and show that there is just one root.

4-9 Directed arcs and angles. Imagine a *unit circle* (circle of unit radius) with center at the origin. Consider an axis tangent to the unit circle at the point $(1, 0)$ with the positive direction upward, as sketched in Fig. 4-37. Let this axis be "wrapped around the circle" without stretching or contracting, so that lengths are preserved. The idea is not as simple as it seems, but we ignore the mathematical difficulties here. We take for granted the theorem of high-school plane geometry that tells us that the length of the circumference of a circle of radius r is $2\pi r$, where π is a real number (approximately 3.14), the same for all circles. Hence the length of the circumference of the unit circle is 2π. The interval $(0 \quad 2\pi)$ on the axis wraps around the circle only once, so that the point 2π goes into the point $(1, 0)$.

(a) Explain the points indicated in Fig. 4-37.   (b) Sketch a unit circle and indicate the points into which the following points on the axis are carried by the above maneuver: $\pi/4$, $3\pi/4$, 4, $-\pi$, $-\pi/4$, 3π.   (c) How many different points on the axis go into each point on the circle?

The circumference of the unit circle is now a kind of map of the real numbers and of the axis. To every real number (point on the axis) there corresponds just one point on the circumference. To every point on the circumference there corresponds an infinite number of real numbers (points on the axis). Suppose that the real number θ corresponds to a

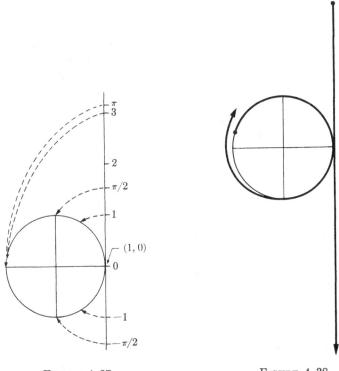

FIGURE 4–37   FIGURE 4–38

certain point on the circumference. The other numbers that correspond
are all those obtained from the formula $\theta + 2k\pi$, where k is any integer
(positive or negative), since an interval of length 2π wraps around the
circle just once. Corresponding to each segment on the axis we have an
arc of the circle. But again, each arc corresponds to an infinite number of
different segments on the axis. These correspondences are not one-to-one.

Corresponding to each vector on the axis we have a directed arc on the
circumference. We consider a *directed arc* to be determined by its length
and direction, regardless of its position on the circle, just as we consider
a vector on the axis to be determined solely by its length and direction.
Then there is a one-to-one correspondence between the vectors and
directed arcs. Indeed, the arc corresponding to a vector is obtained by
wrapping the vector around a portion of the circumference. We suggest
this in Fig. 4–38, where we have chosen a vector of length greater than
2π in order to illustrate the convention of leaving the wrapping incomplete
to suggest the length of the directed arc.

From now on we shall consider only directed arcs and shall refer to them
simply as *arcs*. When an arc is placed so that its initial point is on the

x-axis, it is said to be in *standard position*. Arcs in standard position correspond to vectors with initial point at the origin.

Because of the one-to-one correspondence between arcs and vectors on an axis, and the one-to-one correspondence between these vectors and real numbers, we have also a one-to-one correspondence between real numbers and arcs. Indeed, the arc corresponding to a real number may be found by starting at any point on the unit circle (the point $(1, 0)$ if we want the arc in standard position) and going a directed distance on the circumference equal to the real number. If θ is the real number, we go a distance $|\theta|$, in the counterclockwise direction if $\theta > 0$ and in the clockwise direction if $\theta < 0$. Because of this one-to-one correspondence we often use the same symbol to stand for a real number and the corresponding arc.

For each of the following sketch a unit circle and a directed arc corresponding to the number: (d) π,   (e) 2π,   (f) -1,   (g) $-5\pi/2$.

We can interpret algebraic operations on real numbers by arcs on a unit circle just as we did by vectors on a line. Indeed, the picture on the circle may be obtained by wrapping the corresponding linear picture around the circle. In particular, any arc equals the difference of two arcs in standard position, the arc to its terminal point less the arc to its initial point, as illustrated in Fig. 4–39, where $\theta = A - B$.

(h) Illustrate the identity $a + (-a) = 0$, using arcs on a unit circle.
(i) Sketch the following arcs: 1, 4.14, -5.28, -3π, -2.14, 2.14.

The angle concept in plane geometry is usually rather vague. We visualize a figure such as that drawn in Fig. 4–40. If we define a *ray* as a point and all points on one side of it on a straight line, we might define

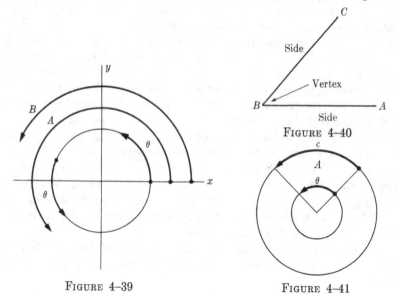

FIGURE 4–40

FIGURE 4–39   FIGURE 4–41

an angle as a figure consisting of two rays with a common endpoint. The rays are called the *sides* of the angle, and the common point is called the *vertex*. But the reader recalls that the angle concept involves more than this, namely, the idea of how much we would have to rotate one side in order to make it coincide with the other. Hence Fig. 4–40 is ambiguous, since it does not indicate in which direction we are to rotate. To get a precise definition, we define directed angles in terms of directed arcs.

A *directed angle* is a figure consisting of two rays (called *sides*) with a common endpoint (called the *vertex*), a unit circle with the vertex as center, and on this circle a directed arc from one side to the other. The side on which the initial point of the arc lies is called the *initial side*. The other side is called the *terminal side*. The angle is said to be in *standard position* when its arc is in standard position. We shall usually refer to directed angles simply as angles.

An angle is determined uniquely as soon as its arc is given, regardless of the position in which the figure is placed. Conversely, an angle determines a unique arc; there is evidently a one-to-one correspondence between arcs and angles. Indeed, the angle is precisely the arc with rays drawn from the center through its initial and terminal points. The real number that corresponds to the arc is considered to be the measure of the arc and of the angle. Because of the one-to-one correspondence we often use names of real numbers to stand for the corresponding angles. Two directed angles are called *congruent* if the undirected angles obtained by omitting the directed arcs are congruent.

Sketch the following angles in standard position: (j) 3, (k) $5\pi/3$, (l) $-\pi/3$.

Among the many applications of these ideas, the most obvious are those to arc lengths, circular areas, and circular motion. Let c be an arc on a circle and let θ be the corresponding arc on the concentric unit circle, as suggested in Fig. 4–41. We say that the arc is *intercepted* by the angle θ. A theorem of plane geometry tells us that the lengths of the arcs are proportional to the lengths of the radii; that is, (length of arc)$/|\theta| = r/1$.

(1)　　　$C = r|\theta|$, *where C is the length of an arc intercepted by the angle θ on a circle of radius r.*

(m) A boy is flying a model plane on the end of a string of length 60 ft. If the string turns through an angle of $\pi/6$, how far does the plane move?　(n) The plane moved through a distance of 125 ft. Through what angle has the string moved?

From plane geometry we know that the area of the interior of a circle is given by πr^2, where r is the radius, and that the areas of circular sectors (portions of the interior between two radii) are proportional to their central angles (arcs). Hence, if a sector has an angle of θ, we have (area of sector)$/\pi r^2 = |\theta|/2\pi$, or

(2) $A = (1/2) |\theta| r^2$, where A is the area of a sector of central angle θ and radius r.

(o) What area has been swept out by the string in Exercise (m)?   (p) in Exercise (n)?

(q) A flywheel is turning at 300 revolutions per minute. If it has a radius of 6 ft, how many feet per minute is a point on the circumference traveling?

The above system of measuring angles is called the *radian* system, and an angle is said to be given in radians when we give the real number corresponding to its arc on the unit circle. The system is the most convenient one in mathematics because of (1), but other systems are in use. They may all be defined by taking some other circle in place of the unit circle on which to use directed arcs as measures of the intercepting angles.

If we take a circle of unit circumference, an arc of 1 is intercepted by an angle of 2π radians. If we use arcs on this circle to measure angles, our unit is *revolutions*. For example, an angle of 0.25 revolutions is the same as one of $\pi/2$ radians.

If we take a circle of circumference 360, the angles are said to be measured in *degrees*. Parts of a degree may be given in the decimal notation or in minutes (1/60 of a degree) and seconds (1/60 of a minute).

If we take a circle of circumference 6400, angles are said to be measured in *mils*, a system widely used in the armed forces.

(r) What is the radius of the circle for the revolutions system?   (s) for the degree system?   (t) Since 1 revolution is 2π radians, x revolutions is $2\pi x$ radians, and $1/2\pi$ revolutions is 1 radian. Complete the following table that gives equivalents in the different systems. (*Note:* When no units are given, radian measure is understood.)

(3)

Radians	Revolutions	Degrees	Mils
2π	1 rev	360°	6400 mils
1	?	$(180/\pi)°$	$(3200/\pi)$ mils
$\pi/180$	1/360 rev	1°	?
?	?	?	1

(u) How many degrees in the angle in Exercise (m)?   (v) in Exercise (j)? (w) Express the angles in an equilateral triangle in each of the systems of measuring angles.

PROBLEMS

In each of Problems 1 through 9 sketch a unit circle with the angle corresponding to the given real number.

1. $\pi/5$.   2. $2\pi/3$.   3. -1.1π.

4. 5.   5. 7π.   6. $\pi/6$.

7. $-5\pi/6$.   8. $-3\pi/4$.   9. $2\pi/9$.

10. Express each angle in Problems 1 through 9 in degrees.

11. The sun is about 93,000,000 miles away from the earth. Suppose two spots on the sun are separated by an angle of 0.1° when observed from the earth. Approximately how far apart are they?

12. A vector of length 5 rotates around its initial point through an angle of 35°. Through how long an arc does its terminal point move?

★13. Write a defining sentence of the ray making an angle of 45° with the x-axis.

★14. Write a defining sentence of the ray from the origin passing through (a, b).

★15. Do the same for the ray with endpoint (x_0, y_0) and through the point (x_1, y_1).

★16. Write a defining equation of the angle of Problem 1 in standard position.

★17. The area of a sector of a circle of radius 31 is 1000. What is the central angle in radians? Sketch.

18. What is a necessary and sufficient condition that two angles are congruent?

Answers to Exercises

(a) Since the whole circumference has length 2π, half of it has length π, one-quarter has length $\pi/2$.   (b) See Fig. 4–42.   (c) An infinite number.   (d) through (g) See Fig. 4–43.   (h) See Fig. 4–44.   (j) Just less than a straight angle.   (k) 5/6 of a revolution.   (l) 1/6 of a revolution in clockwise direction.   (m) About 31.4 ft.   (n) Just over 2.   (o) 300π ft^2, (p) 3750 ft^2.   (q) 3600π ft/min.   (r) $1/2\pi$.   (s) $360/2\pi$.   (t) $1/2\pi$ rev, 160/9 mils, $\pi/3200$, 1/6400 rev, $(9/160)°$.   (u) 30°.   (v) $(540/\pi)°$.   (w) 60°, 1/6 rev, $\pi/3$, 1066.6 mils.

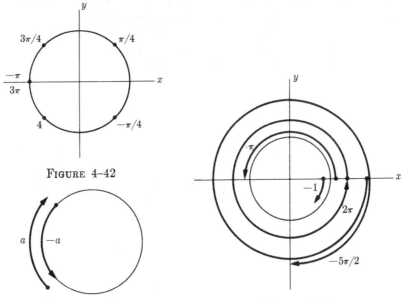

FIGURE 4–42

FIGURE 4–44

FIGURE 4–43

ANSWERS TO PROBLEMS

1. 1/10 rev. 3. Just over 1/2 rev in negative direction. 5. 3.5 rev. 7. −150°.
9. 40°.   11. $(93{,}000{,}000)(0.1)(\pi/180) \doteq 162{,}000$.   13. $y = x \wedge x > 0$.
15. $\exists t\, t \geq 0 \wedge (x, y) = (x_0, y_0) + t(x_1 - x_0, y_1 - y_0)$. 17. About 2.

4–10 Polar coordinates. The rectangular coordinate system used so far is based on a grid consisting of vertical and horizontal lines. A point is located by giving the vertical line (x-coordinate) and horizontal line (y-coordinate) on which it lies. Many other grid systems are possible. Polar coordinates are based on a grid consisting of circles with center at the origin and rays issuing from the origin. The circle on which a point lies determines its distance from the origin, as suggested in Fig. 4–45. The ray on which a point lies tells its direction from the origin, as suggested in Fig. 4–46, where direction is indicated by an arc on the unit circle. It is customary to indicate the distance from the origin by r, often called the *polar distance*, and the direction by θ (theta), often called the *direction angle*. The x-axis serves as the basic reference line on which to indicate the distances and is often called the *polar axis*. The unit circle serves as the basic reference curve on which to measure angles (arcs).

To be more precise, suppose we are given any two real numbers r and θ. We lay out the vector r from the origin along the polar axis, then rotate it through an angle (arc) θ. Its terminal point determines the corresponding point. From the previous section it is obvious that many different ordered pairs (r, θ) determine the same point. Figure 4–47 shows the point $(3, \pi/3)$ and two alternative pairs of coordinates. To determine the polar coordinates of a given point, we find its distance from the origin, then determine any angle from the positive side of the polar axis to the vector from the origin to the point. Other coordinates may be found by

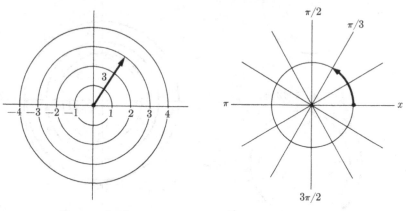

FIGURE 4–45   FIGURE 4–46

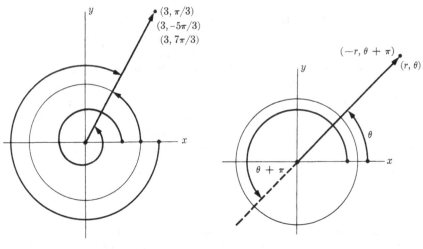

FIGURE 4-47   FIGURE 4-48

adding any integral multiple of 2π to the angle or by taking the negative of r and adding π to the angle, as suggested in Fig. 4-48.

Plot the following pairs considered as polar coordinates: (a) $(1, 1)$, (b) $(-1, 1)$,   (c) $(1, \pi/2)$,   (d) $(-1, 3\pi/2)$,   (e) $(0, 0)$,   (f) $(0, 5)$, (g) $(3, -2\pi/3)$.

If we are given an ordered pair of real numbers, say $(2, 3)$, there is no way to tell, except from the context, whether we mean the point with rectangular coordinates 2 and 3 or the point with polar coordinates 2 and 3. To avoid confusion we write $(a, b)_p$ for the point with polar coordinates a and b. With this notation we have the following laws.

(1) $(r, \theta)_p = (-r, \theta + \pi)_p,$

(2) $(r, \theta)_p = (r, \theta + 2k\pi)_p$ (k-integral),

(3) $(0, \theta)_p = (0, 0)_p.$

(h) Justify (1), (2), and (3). Why must k be integral in (2)?

Any sentence in r and θ now determines a set of points corresponding to the ordered pairs (r, θ) that satisfy the sentence. As for rectangular coordinates, we call this locus the graph of the sentence. Certain loci have very simple defining sentences.

Describe and graph: (i) $r = 2$,   (j) $r = -2$,   (k) $\theta = \pi/2$,   (l) $\theta = -1$, (m) $\theta = 3.14$.

In working with sentences involving polar coordinates, we must keep in mind that *the direction angle θ is always measured in radians.* Thus the locus defined by $\theta = 1$ is the line making an angle of 1 radian with the polar axis (cutting off an arc of length 1 on the unit circle) and not a line making an angle of 1 degree or 1 mil.

Polar coordinates are particularly convenient for defining spirals. For example, the locus of $r = \theta$ is a double spiral each of whose points has a distance from the origin equal to its direction angle, as sketched in Fig. 4-49. Note that the second spiral corresponds to negative values of r and θ.

Graph: (n) $r = 2\theta$,   (o) $r = |\theta|$,   (p) $r = \theta^2$.

PROBLEMS

1. Suggest a way in which we could limit the ranges of r and θ to subsets of the real numbers so that there would be a one-to-one correspondence between points and ordered pairs of polar coordinates.

In Problems 2 through 11 make a table of values of r and θ and sketch the curve.

2. $r^2 = 1$.   3. $r^2 + 2r - 1 = 0$.
4. $r > 0$.   5. $3 < r < 4$.
6. $\theta = -r \wedge 0 \leq \theta \leq 2\pi$.   7. $r = 3\theta$.
8. $r = \theta/2$.   9. $r = 1/\theta$.
10. $r < \theta \wedge 0 \leq \theta \leq 4$.   11. $r < \theta$.

★12. Show that when $r > 0$, $(r, \theta)_p = (r, \phi)_p \leftrightarrow \exists k\ k \in J \wedge \theta = \phi + 2k\pi$. ($J = $ the integers.)

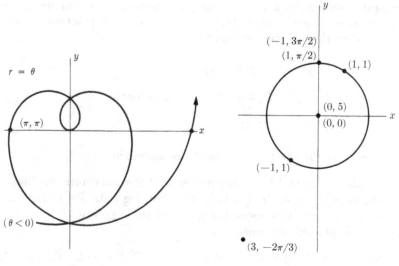

FIGURE 4–49   FIGURE 4–50

ANSWERS TO EXERCISES

(a) through (g) See Fig. 4–50.   (h) Adding π to the angle is equivalent to reversing the direction of r. Any point with $r = 0$ must be at the origin. In (2) k must be integral since we arrive at the same point only by complete rotations, i.e., integral multiples of 2π.   (i) Circle of radius 2 with center at origin. (j) Same.   (k) y-axis.   (l) Line in quadrants II and IV through origin making angle of about 57° with positive x-axis.   (m) Approximately the x-axis. (n) Spiral like that of Fig. 4–49 except that it "expands" more rapidly. (o) Same as Fig. 4–49 for $\theta \geq 0$, for $\theta < 0$, curve is mirror image in x-axis of curve for $\theta > 0$.   (p) Similar to Exercise (o) but "expands" at increasing rate.

ANSWERS TO PROBLEMS

1. Limit r to non-negative values and θ to values in $(0 \quad 2\pi)$, with the condition that only $\theta = 0$ is to be paired with $r = 0$. 3. Two circles, $r = -1 + \sqrt{2}$ and $r = -1 - \sqrt{2}$. 5. Region between two circles. 7. Similar to Fig. 4–49. 9. Approaches the origin as θ gets large, and approaches the x-axis as θ approaches 0. 11. Entire plane except origin.

4–11 The sine and cosine. Since every point in the plane determines a unique vector from the origin to itself, we can designate a vector by polar coordinates of this point as well as by the rectangular coordinates. If we limit ourselves to non-negative polar distances, the polar distance is simply the length of the vector. A direction angle is an angle from the positive x-axis to the vector. In this sense a vector is determined precisely by its length and direction, even though many congruent angles may be used to indicate the direction. Ordinarily we choose for direction angle the smallest number in the interval $(0 \quad 2\pi)$.

Figure 4–51 shows a vector with components x and y, length r, and direction angle θ. We know from (4–4–1) that

$$(1) \qquad\qquad r = \sqrt{x^2 + y^2}.$$

Hence we can easily find the length of a vector when we know its components. But how can we find the direction angle? And if we know r and

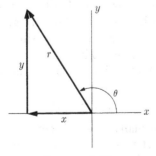

FIGURE 4–51

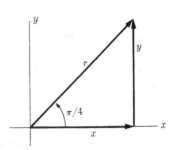

FIGURE 4–52

θ, how can we find x and y? Problems of this kind arise in many applications. In certain simple cases the questions can easily be answered. For example, if $\theta = \pi/2$, then $x = 0$ and $y = r$. If $x = 0$, then

$$\theta = \pi/2 \ \vee \ \theta = 3\pi/2$$

(or θ is some angle congruent to one of these) and $y = r \ \vee \ y = -r$ correspondingly. If we know that $x = 0 \ \wedge \ y = -1$, we have immediately that $r = 1 \ \wedge \ \theta = 3\pi/2$.

Find the missing coordinates: (a) $\theta = \pi$, $r = 2$;   (b) $x = 2$, $y = 0$, (c) $x = -1, y = 0$.

Such simple cases are not limited to those where the terminal point lies on one of the axes. For example, suppose $\theta = \pi/4$, as illustrated in Fig. 4–52. Then clearly the triangle is isosceles, $x = y$, and $x^2 + y^2 = r^2$ becomes $2x^2 = r^2$, which implies $x = r/\sqrt{2} = y$.

Find the missing coordinates: (d) $\theta = 3\pi/4$, $r = 1$;   (e) $x = y = 2$; (f) $x = 1, y = -1$.

The reader can soon convince himself that except in such special cases (of which we shall consider others later), there seems to be no easy method of finding x and y from r and θ or of finding θ when x and y are given. Nevertheless, it is clear that x and y determine θ uniquely (in the sense that they determine a set of congruent angles) and that θ and r determine x and y uniquely.

We begin by considering the special case $r = 1$. Then we have a unit vector which we imagine placed initially along the positive x-axis. If we rotate it in the counterclockwise direction, its terminal point covers an arc θ. To each value of θ there corresponds a unique x and y. We call the value of y corresponding to a given value of θ *the sine of θ* (abbreviated sin θ), and the value of x corresponding to a given value of θ *the cosine of θ* (abbreviated cos θ). The history of these names is of interest. Figure 4–53 shows the angle 2θ with the chord equal to $2y$. At least two thousand years ago, the problem was posed in terms of finding $2y$ when 2θ was known. Because of the resemblance of the figure to a bow, the chord came to be called a *sinus*, a Latin translation of a Sanskrit word for bowstring. The term *cosine* was suggested by the fact that the cosine of θ is the sine of the angle $\pi/2 - \theta$, called the complementary angle. This is suggested in Fig. 4–54. (For formal definitions of sine and cosine, see Problem 34 at the end of the chapter.)

(g) Explain Fig. 4–54.   (h) Argue that sin $\theta = \cos(\pi/2 - \theta)$.

Tables of sine and cosine. Because of their many uses, it is convenient to have tables of the sine and cosine. Table II in the Appendix gives sin θ and cos θ approximately for certain values of θ between 0 and $\pi/2$ (1.57

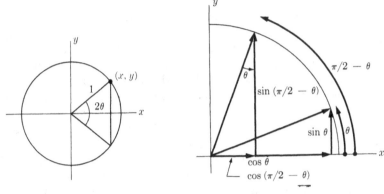

FIGURE 4–53   FIGURE 4–54

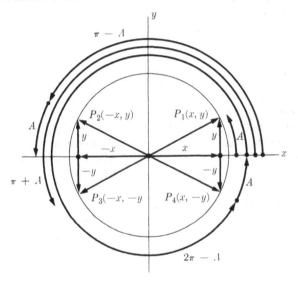

FIGURE 4–55

approximately).   The corresponding angles in degrees are given also. Although the table could be constructed by careful measurement of a large figure, actually the entries were found by consulting a much more accurate table that was constructed by methods beyond the scope of this book.

From the table we can easily find the sine and cosine of *any* angle. To see how this can be done, study Fig. 4–55. There are four points, P_1, P_2, P_3, P_4, on the unit circle determining four vectors making equal angles A with the x-axis. Thus the vector P_2 makes an angle of A with the negative x-axis and its direction angle is $\pi - A$. Similarly, the direction angles of P_3 and P_4 are $\pi + A$ and $2\pi - A$. Hence the four right triangles determined by the vectors and their components are congruent. Therefore the sines and cosines of these four angles have the same absolute values. We see that the sine and cosine of *any* angle have the same absolute values as those of *some* angle in the first quadrant, this angle being the smallest non-negative angle between the corresponding vector and the x-axis. Then we determine the sign by noting the quadrant in which the angle lies and seeing whether its sine and cosine are positive or negative.

For example, to find the sine and cosine of 2.54 (145.6°) we note that $2.54 = 3.14 - 0.60$, and hence that the angle lies in the second quadrant, similarly to point P_2 in Fig. 4–55. Hence its sine and cosine are the same in absolute value as those of 0.60, that is, approximately 0.56 and 0.83 from Table II. But we can see from the figure that the sine and cosine of an angle in the second quadrant are, respectively, positive and negative. Hence $\sin 2.54 \doteq 0.56$ and $\cos 2.54 \doteq -0.83$.

Find sine and cosine of: (i) 2.64,   (j) 3.94.

Consider the reverse problem: given the sine and cosine, to find the angle. To do this we sketch the angle by locating the unit vector with x- and y-components respectively equal to the given cosine and sine. Then we find the angle between this vector and the x-axis by looking in Table II under the absolute values of the given sine and cosine. The desired angle is then found by reference to Fig. 4–55. For example, suppose $\cos \theta = 0.70$ and $\sin \theta = -0.72$. The vector is sketched in Fig. 4–56. From Table II, $A \doteq 0.80$ (45.8°). Hence $\theta = 2\pi - 0.80 = 6.28 - 0.80 = 5.48$. Note that we need both sine and cosine. If just sine or cosine were given, there would be two angles in the interval $(0 \quad 2\pi)$. Of course, the angle found, plus any integral multiple of 2π, is a congruent angle with the same sine and cosine.

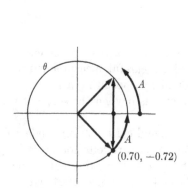

FIGURE 4–56

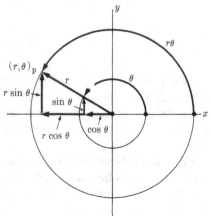

FIGURE 4–57

Find θ whose sine and cosine are: (k) -0.91, 0.42,   (l) -0.20, -0.98. Find two positive angles less than 2π whose (m) sine is 0.50, (n) cosine is -0.57,   (o) sine is -0.91.

Having dealt with the problem for unit vectors, we find it easy to handle others, for any vector is a multiple of a unit vector. If we are given a vector $(r, \theta)_p$, its components are evidently $r \cos \theta$ and $r \sin \theta$, as sketched in Fig. 4-57. This enables us to find immediately the components of any vector from its length and direction angle. On the other hand, it is evident that given (x, y) we can find the length from (1) and a direction angle by finding an angle θ such that $\cos \theta = x/r$ and $\sin \theta = y/r$. We may summarize by

$$(2) \quad [(x, y) = (r, \theta)_p] \leftrightarrow [x^2 + y^2 = r^2 \wedge \cos \theta = x/r \wedge \sin \theta = y/r].$$

These relations may be used to go from (x, y) to $(r, \theta)_p$, or from the latter to the former.

Find the components of the vector (p) of length 3 and direction angle $60°$, (q) of length 10 and direction angle 4.54.   (r) Find a direction angle of $(2, -3.5)$.

Another way of formulating the problem of this section is to ask for polar coordinates of a point whose rectangular coordinates are given, or vice versa. Hence this problem is solved also by (2).

(s) Find polar coordinates of $(-1, -1)$.   (t) Find the rectangular coordinates of $(2, -3\pi/4)_p$.

Interpolation. Table II may be used with angles, sines, or cosines that do not appear in it. Suppose, for example, that we wish the sine of 0.54. We have $\sin 0.50 = 0.48$ and $\sin 0.60 = 0.56$. Since 0.54 is $4/10$ of the way from 0.50 to 0.60, we estimate that $\sin 0.54$ is $4/10$ of the way from 0.48 to 0.56. The total distance being 0.08, we find $(4/10)(0.08) \doteq 0.03$, and estimate that $\sin 0.54 \doteq 0.51$. This process is called *linear interpolation*. It is based on assuming that the changes in θ and $\sin \theta$ or $\cos \theta$ are proportional. The assumption is not true, but the process gives sufficiently accurate results if the differences are not very large.

Estimate: (u) $\cos 0.54$,   (v) $\sin 1.32$,   (w) $\sin 5.30$.   (x) Find $\sin 0.54$ by using $\sin 0.52$ and $\sin 0.60$.

Linear interpolation can be used in reverse to estimate an angle from its sine or cosine. For example, suppose $\sin \theta = 0.74$. We have $\sin 0.80 = 0.72$ and $\sin 0.87 = 0.77$. Since 0.74 is $2/5$ of the way from 0.72 to 0.77, we estimate θ to be $2/5$ of the way from 0.80 to 0.87, that is, $\theta \doteq 0.80 + (2/5)(0.07) \doteq 0.80 + 0.03 = 0.83$.

(y) Find approximately the angle in the first quadrant whose sine is 0.88.

The reader may find it convenient to arrange in a table the numbers involved in an interpolation. We do this for the examples above.

$$\theta \qquad \sin \theta$$

(3)
$$10 \left|\ \begin{matrix} 4 \end{matrix}\ \right| \begin{matrix} 0.50 \\ \downarrow\ 0.54 \\ 0.60 \end{matrix} \qquad \begin{matrix} 0.48 \\ ?\ \ \downarrow h \\ 0.56 \end{matrix} \right|\ 8$$

$$\sin \theta \qquad \theta$$

(4)
$$5 \left|\ \begin{matrix} 2 \end{matrix}\ \right| \begin{matrix} 0.72 \\ \downarrow\ 0.74 \\ 0.77 \end{matrix} \qquad \begin{matrix} 0.80 \\ ?\ \ \downarrow h \\ 0.87 \end{matrix} \right|\ 7$$

In each case h is the unknown distance to be estimated. In (3) we have $h/8 = 4/10$ or $h = (2/5)8$. In (4) we have $h/7 = 2/5$ or $h = (2/5)7$. Note that the differences have been multiplied by 100 to avoid decimals. Common sense dictates placement of the decimal point in the result.

(z) Tabulate as in (3) and (4) your interpolations in Exercises (u) through (y).

PROBLEMS

1. A plane flies 20° north of east for 90 min at 200 mi/hr. How far east and north of its origin is it?
2. A plane is 100 mi west and 250 mi south of its starting point. What is its distance and direction from the starting point?
3. Sketch four angles between 0 and 2π whose sine in absolute value is 0.5.
4. Sketch four angles between 0 and 2π whose cosine in absolute value is 0.75.

In Problems 5 through 13 find the rectangular coordinates of the point and sketch.

5. $(1, 60°)_p$.
6. $(3, 2)_p$.
7. $(-1, 45°)_p$.
8. $(2.5, 2\pi/3)_p$.
9. $(5, 310°)_p$.
10. $(4, -3\pi/4)_p$.
11. $(100, 5.1)_p$.
12. $(255, 345°)_p$.
13. $(24, -93°)_p$.

In Problems 14 through 16 find the components of the vector with length and direction angle as given and sketch.

14. $15, 5\pi/4$.
15. $500, 215°$.
16. $1.2, 3.7$.

In Problems 17 through 20 find the length and direction angle of the vector and sketch.

17. $(-5, 3.4)$.
18. $(100, 3)$.
19. $(-2, -2)$.
20. $(1000, 2000)$.

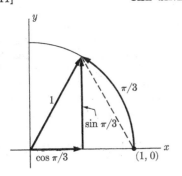

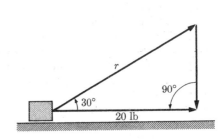

FIGURE 4-58   FIGURE 4-59

21. Figure 4-58 shows an angle of $\pi/3$ (60°). By connecting the terminal point of the arc with (1, 0), we form an equilateral triangle whose altitude has length sin $\pi/3$. From the figure, find cos $\pi/3$ and sin $\pi/3$. Compare Table II.

22. Similarly, find the sine and cosine of $\pi/6$, $2\pi/3$, $5\pi/6$, $7\pi/6$, $4\pi/3$, $5\pi/3$, and $11\pi/6$. Make drawings. Check your results against Table II.

23. Make a table showing the values of sine and cosine for 0 and all multiples of $\pi/6$ and $\pi/4$, giving exact expressions for each value.

It is an empirically established law of physics that if forces be represented by vectors (the length representing the magnitude of the force, the direction representing the direction of the force), two forces applied at a point have the same effect as a single force (called the *resultant*) corresponding to the vector sum. In brief, forces add like vectors. This law is all that is needed to solve Problems 24 through 29.

★24. Suppose a force of 60 lb is exerted in a direction 50° above the horizontal. What two forces in the horizontal and vertical directions would have the same effect?

★25. A block is dragged along a plane by a rope making an angle of 30° with the plane. What force on the rope is required to yield a force of 20 lb in the direction of motion? (See Fig. 4–59.)

★26. Two forces, one of 20 lb in the direction 140°, the other of 50 lb in the direction −80°, act on a body situated at the origin. Find the magnitude and direction of the resultant. [*Suggestion:* Resolve each force into its components, add according to (4–3–1), then find the length and direction of the sum.]

★27. Forces are said to balance or be in equilibrium when their resultant is the null vector. What force must be added to those in Problem 26 to create equilibrium?

★28. Find the resultant of three forces: (10, 300°); (−25, 200°); (5, 90°).

★29. A force eastward of 20 lb, a force northward of 50 lb, together with a third force, just balance a force of 100 lb in the southwest direction. What is the third force?

★30. From a plane 10 mi above the earth the angle between the line of sight to the horizon and the vertical is 85.6°. From this calculate the radius of the earth.

★31. In one of Leonardo da Vinci's notebooks appears the following: "If you would ascertain the exact distance of the breadth of a river proceed as follows: plant a staff upon the river bank at your side and let it project as far from the ground as your eye is from the ground; then withdraw yourself as far as the span of your arms and look at the other bank of the river, holding a thread from the top of the staff to your eye, or if you prefer it a rod, and observe where the line of sight to the opposite bank meets the staff." Draw a sketch and write a formula for calculating the width.

★32. Figure 4–60 shows a body exerting a force of 50 lb hanging from cables attached to a horizontal rigid support at A and B. From the dimensions, find the forces exerted on the support at A and B.

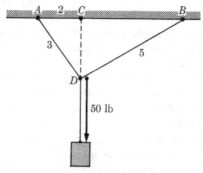

FIGURE 4–60

★33. A soldier walks 1 mi 30° east of north, 2 mi 50° south of east, 1 mi 40° west of north, and 5 mi 45° south of west. Where is he?

★34. We could have defined sin θ formally as follows:

(5) $$\sin \theta = [y \ni \exists x\ (x, y) = (1, \theta)_p].$$

Formulate a similar definition for cos θ. Justify and explain these definitions.

★35. Show that the equation $r \cos \theta = 1$ is the equation of a straight line by transforming it to rectangular coordinates.

★36. What is the polar graph of $r = 1/\sin \theta$?

★37. In Table II, $\sin \theta = \theta$ for small θ. How does this happen?

ANSWERS TO EXERCISES

(a) $(-2, 0)$.　(b) $(2, 0)_p$.　(c) $(1, \pi)_p$.　(d) $(-1/\sqrt{2}, 1/\sqrt{2})$.　(e) $(2\sqrt{2}, \pi/4)_p$.　(f) $(\sqrt{2}, 7\pi/4)_p$.　(g) The two triangles formed by the vectors with angles θ and $\pi/2 - \theta$ are congruent since they have hypotenuses of equal length and an equal angle. Hence the sine of $\pi/2 - \theta$ is the same as the cosine of θ. (h) Similarly from Fig. 4–54.　(i) $2.64 = 3.14 - 0.50$. Hence $|\sin 2.64| = \sin 0.50 = 0.48$ from Table II. But the sine is positive in quadrant II. Hence $\sin 2.64 = 0.48$. Similarly, $|\cos 2.64| = 0.88$, but the cosine is negative in quadrant II, and hence $\cos 2.64 = -0.88$.　(j) $3.94 = 3.14 + 0.80$; -0.72, -0.70.　(k) $\sin 1.13 = 0.91$, $\cos 1.13 = 0.42$. Since sine is negative and cosine positive, angle is in fourth quadrant. Hence $\theta = 2\pi - 1.13 = 6.28 -$

1.13 = 5.15.   (l) 3.34.   (m) 0.52, 2.62.   (n) 2.18, 4.10.   (o) 4.27, 5.15.
(p) (1.5, 2.6).   (q) (−1.7, −9.85).   (r) −1.08.   (s) $(\sqrt{2}, 5\pi/4)_p$.
(t) $(−\sqrt{2}, −\sqrt{2})$.   (u) 0.86.   Note that the cosine is decreasing so that
we go 4/10 of the distance from 0.88 to 0.83, that is, of a directed distance
−0.05.   (v) 0.965. Ordinarily we keep no more decimal places in interpolation
than are in the table, but here the differences are very small.   (w) sin 5.30 =
−sin 0.98 $\doteq$ −0.83.   (x) 2/8 of the way from 0.50 to 0.56, or 0.515, which
might be rounded off to either 0.51, as found before, or to 0.52.   (y) 1.09.
(z) Note that the differences are negative in some cases, since they are directed
distances. For proportionality, the arrows must all go the same way on the
table.

ANSWERS TO PROBLEMS

1. 282 mi east, 102 mi north.   3. 30°, 150°, 210°, 330°.   5. (0.50, 0.87).
7. (−0.71, −0.71).   9. (3.20, −3.85).   11. (38, −92).   13. (−1.2, −23.98).
15. (−410, −285).   17. $(6.1, 2.53)_p$.   19. $(2\sqrt{2}, 5\pi/4)_p$.   23. See (6).

(6)

θ	$\sin \theta$	$\cos \theta$	θ	$\sin \theta$	$\cos \theta$
0	0	1	π	0	−1
$\pi/6$	1/2	$\sqrt{3}/2$	$7\pi/6$	−1/2	$−\sqrt{3}/2$
$\pi/4$	$1/\sqrt{2}$	$1/\sqrt{2}$	$5\pi/4$	$−1/\sqrt{2}$	$−1/\sqrt{2}$
$\pi/3$	$\sqrt{3}/2$	1/2	$4\pi/3$	$−\sqrt{3}/2$	−1/2
$\pi/2$	1	0	$3\pi/2$	−1	0
$2\pi/3$	$\sqrt{3}/2$	−1/2	$5\pi/3$	$−\sqrt{3}/2$	1/2
$3\pi/4$	$1/\sqrt{2}$	$−1/\sqrt{2}$	$7\pi/4$	$−1/\sqrt{2}$	$1/\sqrt{2}$
$5\pi/6$	1/2	$−\sqrt{3}/2$	$11\pi/6$	−1/2	$\sqrt{3}/2$

4–12 Trigonometric identities. From the definitions of sin θ and cos θ
as coordinates of a point on a unit circle, it follows immediately that
$(\sin \theta)^2 + (\cos \theta)^2 = 1$. Writing $\sin^2\theta =_d (\sin \theta)^2$ and $\cos^2\theta =_d (\cos \theta)^2$,
we have the identity

(1) $$\sin^2\theta + \cos^2\theta = 1.$$

(a) Verify (1) for selected values of θ, including some from Table II.

The sine and cosine equal the ratios y/r and x/r, where $r = |(x, y)|$.
It is natural to give names to other ratios of x, y, and r.

(2) **Def.** $\tan \theta = \sin \theta/\cos \theta$,

(3) **Def.** $\cot \theta = \cos \theta/\sin \theta$,

(4) **Def.** $\sec \theta = 1/\cos \theta$,

(5) **Def.** $\csc \theta = 1/\sin \theta$.

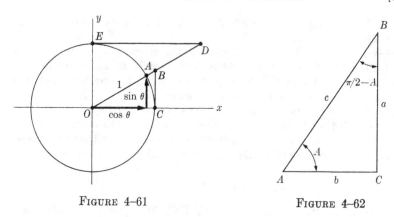

FIGURE 4-61   FIGURE 4-62

The symbols on the left of definitions (2) through (5) are abbreviations for *tangent* θ, *cotangent* θ, *secant* θ, and *cosecant* θ, respectively. The names have a geometrical significance suggested by Fig. 4–61. There, by similar triangles, $\overline{CB}/\overline{OC} = \sin\theta/\cos\theta = \tan\theta$. But $\overline{OC} = 1$. Hence $\tan\theta$ is just the length of the tangent from C to the terminal side of the angle θ.

(b) Show that $\cot\theta = \overline{ED}$, $\sec\theta = \overline{OB}$ (the length of the secant), and $\csc\theta = \overline{OD}$.   (c) For what values of θ is $\tan\theta$ undefined?   (d) $\cot\theta$? (e) $\sec\theta$?   (f) $\csc\theta$?   (g) In Table II verify that the values in the tangent and cotangent columns conform to (2) and (3).   (h) Argue for the identity $\cot\theta = 1/\tan\theta$.   (i) Show that the slope of a straight line is the tangent of any direction angle of any vector lying in the line.

With definitions (2) through (5) we have a name for the ratio of any two sides of a right triangle. Figure 4–62 shows a right triangle with sides and angles as indicated. If we imagine this triangle placed with angle A at the origin and the side AC along the x-axis, we see that

(6)

$$\sin A = a/c = \text{opposite side/hypotenuse,}$$

$$\cos A = b/c = \text{adjacent side/hypotenuse,}$$

$$\tan A = a/b = \text{opposite side/adjacent side,}$$

$$\cot A = b/a = \text{adjacent side/opposite side,}$$

$$\sec A = c/b = \text{hypotenuse/adjacent side,}$$

$$\csc A = c/a = \text{hypotenuse/opposite side.}$$

These relations make the tables of these ratios very useful in finding sides or angles of right triangles. For example, suppose we wish to find the distance across a lake, as pictured in Fig. 4–63. We sight from C to a point B on the opposite shore. Then we measure 100 ft at right angles to

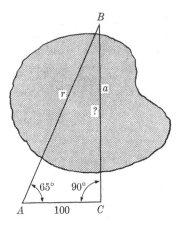

FIGURE 4–63

this line of sight to the point A. We find by sighting to B again (and using a protractor or some more accurate surveying instrument) that the angle at A is 65°. Then we have $a/100 = \tan 65°$ or $a = 100 \tan 65°$. From Table II $\tan 65° \doteq 2.1$. Hence $a \doteq 210$ ft to the nearest 10 ft.

Finding the remaining sides and angles of a triangle when some of them are given is called *solving* the triangle. It was for solution of such problems that the ratios of sides were originally given names and tabulated. The name "trigonometry" for this branch of mathematics comes from Latin words meaning "triangle measurement." However, trigonometry expanded far beyond the original purpose. The relations in (6) have meaning only for angles less than $\pi/2$, whereas we have defined $\sin \theta$ for all real numbers. Many of the most important applications of trigonometry have nothing to do with triangles.

(j) Solve the lake problem if the base line AC is 50 yd long and the angle A is 34°.

Numerous identities connect the six trigonometric ratios. Some are suggested by Figs. 4–61 and 4–62. For example,

(7)
$$\tan^2\theta + 1 = \sec^2\theta$$

(8)
$$1 + \cot^2\theta = \csc^2\theta$$

are suggested by considering the right triangles OCB and OED in Fig. 4–61.

(k) Prove (7) by dividing both members of (1) by $\cos^2\theta$. (l) Prove (8).

Further identities can be constructed without limit by manipulating the definitions and the above identities. If the reader did Problems 13 through 20 in Section 2–12, he has already proved a number of these, for (2–12–1) through (2–12–5) are just (1) through (5), where S, C, T, t, c, and s are abbreviations for sin θ, cos θ, tan θ, cot θ, sec θ, and csc θ. These identities are often useful for simplifying trigonometric expressions.

Simplify: (m) $\sin^2\theta/(1 - \cos \theta)$,   (n) (sin x sec x)/(tan x cot x).

The following identities, which are not so evident as (7) and (8), play a very important role in trigonometry.

(9) $\sin (\theta + \phi) = \sin \theta \cos \phi + \cos \theta \sin \phi,$

(10) $\cos (\theta + \phi) = \cos \theta \cos \phi - \sin \theta \sin \phi.$

(o) Show by a counterexample that sin $(\theta + \phi) = \sin \theta + \sin \phi$ is not a law.

In order to prove (9) and (10), we find it convenient to prove first

(11) $\cos (\theta - \phi) = \cos \theta \cos \phi + \sin \theta \sin \phi.$

For this purpose we visualize the angles θ and ϕ in standard position in Fig. 4–64. We have chosen ϕ in the second quadrant and θ in the third, but everything in the proof remains the same regardless of the size of the angles involved. By definition, the coordinates of P, the terminal point of the unit vector with direction angle ϕ, are as indicated. Similarly, the point Q has coordinates (cos θ, sin θ). The directed arc $\overset{\frown}{PQ}$ is $\theta - \phi$. Now we rotate the angle $\theta - \phi$ (consisting of the vectors OP and OQ, and the arc PQ) around the origin until OP lies along the x-axis. Then P goes into the point $(1, 0)$ and Q goes into Q'. Since Q' is the terminal point of a vec-

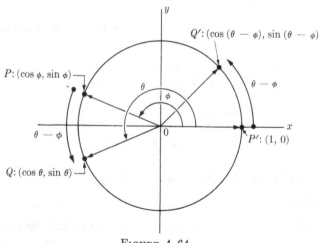

FIGURE 4–64

tor with direction angle $\theta - \phi$, its coordinates are, by definition, the cosine and sine of $\theta - \phi$. Now we assume that this rotation has not changed the straight line distance between the point P and Q, i.e., that $\overline{P'Q'} = \overline{PQ}$. But

(12) $\overline{PQ}^2 = (\cos \theta - \cos \phi)^2 + (\sin \theta - \sin \phi)^2,$

(13) $\overline{P'Q'}^2 = (\cos (\theta - \phi) - 1)^2 + (\sin (\theta - \phi) - 0)^2.$

Expanding the right members, simplifying by using identities, and setting the two results equal, we find (11).

(p) Do this algebra!

Now the derivation of (9), (10), and many other identities, is easy if we make use of our knowledge of the sine and cosine of special angles such as 0 and $\pi/2$. First $(11)(\theta{:}\pi/2)$ is

(14) $\cos (\pi/2 - \phi) = \cos \pi/2 \cos \phi + \sin \pi/2 \sin \phi.$

But $\cos \pi/2 = 0$ and $\sin \pi/2 = 1$. Hence we have proved

(15) $\cos (\pi/2 - \phi) = \sin \phi.$

(q) From (15) by $(\phi{:}\pi/2 - \phi)$, derive (16).

(16) $\sin (\pi/2 - \phi) = \cos \phi.$

Now $(15)(\phi{:}\theta + \phi)$ yields (9) if we note that $\pi/2 - (\theta + \phi) = (\pi/2 - \theta) - \phi$ and use (11).

(r) Do this!

Similar devices enable us to prove

(17) $\cos (-\phi) = \cos \phi,$

(18) $\sin (-\phi) = -\sin \phi.$

With these results we can derive (10) by writing

$$\cos (\theta + \phi) = \cos (\theta - (-\phi))$$

and using already proven identities.

(s) From (9) derive a formula for $\sin (\theta - \phi)$ by noting that $\theta - \phi = \theta + (-\phi)$.

The entire theory of trigonometry may be derived from the identities proved above. A number of important results are included in the exercises. These laws are important for their role in mathematics, and their proofs provide good practice in heuristic (Section 2-12) and in methods of deriving identities (Section 2-8).

PROBLEMS

1. What two angles in $(0\ \ 2\pi)$ have tangent 1?
2. Draw a diagram and show that $\tan \pi/4 = 1$.
3. A vector has x-component -2 and direction angle $125°$. What is its length and y-component?

In Problems 4 through 7 find the missing elements from among x, y, r, and θ of each vector.

4. $y = -3, \theta = -39°$.   5. $x = 2, r = 3$.
6. $r = 4, y = -6$.   7. $x = -10, \theta = 237°$.

8. Velocities may be represented by vectors whose lengths give the speeds of motion and whose directions give the directions of motion. A projectile is moving at an angle of $40°$ with the horizontal. Its height above the ground is decreasing at the rate of 50 ft/sec. What is its speed? What is the horizontal component of its velocity?

9. Prove that $[P \sin \theta = mv^2/R \wedge P \cos \theta = mg] \rightarrow [\tan \theta = v^2/Rg]$. (Sears and Zemansky, *University Physics*, p. 105)

10. Why would it be sufficient to replace Table II by one covering only angles in the range $(0\ \ \pi/4)$?

11. Argue for the plausibility of (17) and (18) by drawing a figure showing angles ϕ and $-\phi$ in standard position.

12. Prove (19), (20), and (21). (*Note:* These properties are described by saying that the sine and cosine are *periodic* of *period* 2π and that the tangent is periodic of period π.)

(19) $\sin (\theta + 2\pi) = \sin \theta,$

(20) $\cos (\theta + 2\pi) = \cos \theta,$

(21) $\tan (\theta + \pi)\ \ = \tan \theta.$

13. The following are called the double-angle formulas. Derive them from (9) and (10).

(22) $\sin 2\theta = 2 \sin \theta \cos \theta,$

(23) $\cos 2\theta = \cos^2\theta - \sin^2\theta = 1 - 2\sin^2\theta = 2\cos^2\theta - 1.$

14. Use (23) to derive

(24) $\sin \theta = \pm\sqrt{(1 - \cos 2\theta)/2},$

(25) $$\cos \theta = \pm\sqrt{(1 + \cos 2\theta)/2}.$$

15. Using (25), find an exact expression for sin 15°.

16. How does one choose which sign to use in (24) and (25)?

17. Verify (24) and (25) for selected values in Table II.

18. By using the definition of tan θ and previous identities, derive the following.

(26) $\tan (\theta - \phi) = (\tan \theta - \tan \phi)/(1 + \tan \theta \tan \phi)$,

(27) $\tan \phi/2 = \pm\sqrt{(1 - \cos \phi)/(1 + \cos \phi)} = (1 - \cos \phi)/\sin \phi$.

19. Find formulas for tan $(\theta + \phi)$ and tan 2ϕ.

20. Prove

(28) $$\sin \theta + \sin \phi = 2 \sin \left(\frac{\theta + \phi}{2}\right) \cos \left(\frac{\theta - \phi}{2}\right),$$

(29) $$\sin \theta - \sin \phi = 2 \cos \left(\frac{\theta + \phi}{2}\right) \sin \left(\frac{\theta - \phi}{2}\right),$$

(30) $$\cos \theta + \cos \phi = 2 \cos \left(\frac{\theta + \phi}{2}\right) \cos \left(\frac{\theta - \phi}{2}\right),$$

(31) $$\cos \theta - \cos \phi = -2 \sin \left(\frac{\theta + \phi}{2}\right) \sin \left(\frac{\theta - \phi}{2}\right).$$

★21. Show that $\forall a,b \; \exists A,\phi \; a \sin \theta + b \cos \theta = A \sin (\theta + \phi)$.

22. Show that $\tan (x/2) = \csc x - \cot x$.

23. A plane is flying on a course 80° east of north at an air speed of 200 mi/hr. The wind is blowing from the south at a speed of 45 mi/hr. What are the actual ground speed and direction of flight?

24. Prove that $\sec \theta = \pm\sqrt{1 + \cot^2\theta}/\cot \theta$.

Prove the following identities.

25. $\cos A/(1 + \sin A) = (1 - \sin A)/\cos A$,

26. $(\cot x + \cot y)/(\tan x + \tan y) = \cot x \cot y$,

27. $\tan \theta/2 = \sin \theta/(1 + \cos \theta)$,

28. $\tan(-\theta) = -\tan \theta$.

Answers to Exercises

(b) $\overline{ED}/\overline{OE} = \cos \theta/\sin \theta$; but $\overline{OE} = 1$.   (c) Odd multiples of $\pi/2$.
(d) Multiples of π.   (e) Same as tan θ.   (f) Same as cot θ.   (h) $a/b = 1/(b/a)$.   (i) Imagine the origin at the initial point of the vector (x, y); then $\tan \theta = y/x =$ the slope of the line.   (j) 34 yd.   (l) Divide both members of (1) by $\sin^2 \theta$.   (m) Multiply numerator and denominator by $1 + \cos \theta$; final result is $1 + \cos \theta$.   (n) tan x.   (o) $\theta = \phi = \pi/2$.

Answers to Problems

1. $45°$, $225°$. 3. $y = 2.8$, $r = 3.5$. 5. $y = 2.2$, $\theta = 48°$ or $y = -2.2$, $\theta = -48°$. 7. $y = -16$, $r = 19$. 13. $(\phi{:}\theta)$. 15. $0.5\sqrt{2} - \sqrt{3}$. 21. $a \sin \theta + b \cos \theta = A[(a/A) \sin \theta + (b/A) \cos \theta]$, where $A = \sqrt{a^2 + b^2}$; then let $\phi = $ a direction angle of the vector (a, b) so that $\cos \theta = a/A$, $\sin \theta = b/A$. 26. Multiply both members by the denominator of the left member.

★**4-13 Angles and the dot product.** We can now derive a formula giving the angles formed by any two lines. We first define the *inclination* of a line as the smallest non-negative angle through which the x-axis must be rotated to be parallel with the line. As indicated in Fig. 4–65, the inclination is a real number in the interval $(0 \quad \pi)$. It follows from the definition and Exercise (i) in Section 4–12 that the tangent of the inclination is the slope of the line.

(a) Explain and justify the last statement with the aid of a diagram.

We define the *directed angle from a line L_1 to a line L_2* as $\phi_2 - \phi_1$, where ϕ_1 and ϕ_2 are the inclinations of the lines. It is an angle through which L_1 may be rotated to coincide with L_2. Now we have $\tan (\phi_2 - \phi_1) = (\tan \phi_2 - \tan \phi_1)/(1 + \tan \phi_1 \tan \phi_2)$ from (4–12–26). But $\tan \phi_1 = m_1$, the slope of L_1, and $\tan \phi_2 = m_2$, the slope of L_2. Hence

(1)
$$\tan (\phi_2 - \phi_1) = \frac{m_2 - m_1}{1 + m_1 m_2}.$$

Suppose for example, that $m_1 = 1$ and $m_2 = 2$. Then $\tan (\phi_2 - \phi_1) = 1/3 \doteq 0.33$ and $\phi_2 - \phi_1 \doteq 18°$. Hence the angle from the first to the second line is about $18°$. That is, if we rotate the first line through $18°$, it will coincide with the second. Of course we could also rotate it through $-162°$, and $\tan (-162°) = \tan 18°$, so that both angles are given by (1).

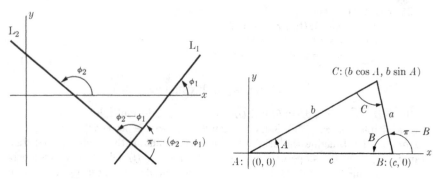

FIGURE 4–65   FIGURE 4–66

(b) Sketch the lines $y = x + 1$ and $y = 2x - 3$, showing the angles just found.   (c) What is the angle from the line $y = -x + 4$ to the line $y = -3x + 6$?   (d) Derive (4–7–6) from (1).   (e) Derive (4–7–8) from (1).

Let ABC be a triangle with angles A, B, C and opposite sides a, b, and c. We place the triangle with the angle A at the origin and the side AB along the positive x-axis. Then, as indicated in Fig. 4–66, the direction angles of the vectors AB, BC, and AC are, respectively, 0, $\pi - B$, and A. Since the lengths of the vectors are c, a, and b, respectively, we have from (4–11–2), $AB = (c, 0)$, $BC = [a \cos (\pi - B), a \sin (\pi - B)] = (-a \cos B, a \sin B)$, and $AC = (b \cos A, b \sin A)$. Since $AB + BC = AC$, we find

(2) $(c - a \cos B, a \sin B) = (b \cos A, b \sin A).$

(f) Prove that $\cos (\pi - B) = -\cos B$ and $\sin (\pi - B) = \sin B$.

From (2) we have immediately $a \sin B = b \sin A$ or $(\sin A)/a = (\sin B)/b$. Since we could have labeled the sides and angles differently, this relation must hold for any pair of sides and corresponding angles of a triangle. We have derived the *law of sines*.

(3) $$\frac{\sin A}{a} = \frac{\sin B}{b} = \frac{\sin C}{c} \qquad \text{(Law of sines).}$$

Again from (2) we have $c - a \cos B = b \cos A$ or $c = a \cos B + b \cos A$. But since we could have labeled the figure differently, we may write $b = c \cos A + a \cos C$ and $a = b \cos C + c \cos B$. Multiplying both members of these equations by c, b, and a, respectively, we have $c^2 = ac \cos B + bc \cos A$, $b^2 = bc \cos A + ab \cos C$,   and $a^2 = ab \cos C + ac \cos B$. Subtracting the last two from the first, we find $c^2 - b^2 - a^2 = -2ab \cos C$. We have derived the *law of cosines*.

(4) $$c^2 = a^2 + b^2 - 2ab \cos C \qquad \text{(Law of cosines).}$$

(g) Show that (4) reduces to the Pythagorean theorem when the triangle is a right triangle.   (h) In a certain triangle, two sides and the included angle are, respectively, 2, 3, and 25°. Find the third side from (4). Then find the other two angles approximately by using (3).   (i) In a certain triangle, $A = 35°$, $B = 40°$, $b = 10$. Find C, a, and c by using (3).

As illustrated in the exercises, the laws of sines and cosines are useful for solving triangles. The interested reader will find details for the best techniques for using them and other formulas for that purpose in books on trigonometry and surveying. The laws also appear occasionally in scientific literature, and the law of cosines may be used to derive another formula for finding the angle between two lines. Indeed, let $\mathbf{u}_1 = (x_1, y_1)$, $\mathbf{u}_2 = (x_2, y_2)$, and $\mathbf{u}_2 - \mathbf{u}_1 = (x_2 - x_1, y_2 - y_1)$ form a triangle, as in Fig. 4–67. By the law of cosines

$$|\mathbf{u}_2 - \mathbf{u}_1|^2 = |\mathbf{u}_1|^2 + |\mathbf{u}_2|^2 - 2|\mathbf{u}_1|\,|\mathbf{u}_2|\cos\phi,$$

where ϕ is the angle between the vectors $\mathbf{u}_1$ and $\mathbf{u}_2$. Using (4–4–1), we find

$$(x_2 - x_1)^2 + (y_2 - y_1)^2$$
$$= x_1^2 + y_1^2 + x_2^2 + y_2^2 - 2\sqrt{x_1^2 + y_1^2}\,\sqrt{x_2^2 + y_2^2}\,\cos\phi.$$

Expanding and simplifying, we find

$$(5) \qquad \cos\phi = \frac{x_1 x_2 + y_1 y_2}{\sqrt{x_1^2 + y_1^2}\,\sqrt{x_2^2 + y_2^2}}.$$

(j) Carry through the algebra!   (k) Derive (4–7–8) from (5).   (l) Derive (4–7–6) and (4–7–7) from (5).

Formula (5) can be simplified by the definition of a kind of multiplication of vectors. We define the *dot product* $\mathbf{u} \cdot \mathbf{v}$ of two vectors $\mathbf{u}$ and $\mathbf{v}$ to be the sum of the products of the corresponding components; that is,

(6) **Def.** $(x_1, y_1) \cdot (x_2, y_2) = x_1 x_2 + y_1 y_2.$

(m) Prove (7).

$$(7) \qquad\qquad |\mathbf{u}| = \sqrt{\mathbf{u} \cdot \mathbf{u}}.$$

Then (5) may be written

$$(5') \qquad\qquad \cos\phi = \frac{\mathbf{u}_1 \cdot \mathbf{u}_2}{|\mathbf{u}_1|\,|\mathbf{u}_2|}.$$

In words, the cosine of the angle between two vectors is their dot product divided by the product of their lengths.

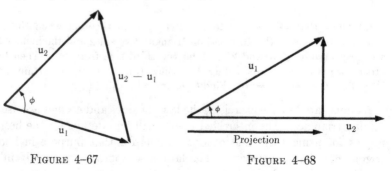

FIGURE 4–67   FIGURE 4–68

(n) Show that

(8) $(\mathbf{u}_1 \perp \mathbf{u}_2) \leftrightarrow (\mathbf{u}_1 \cdot \mathbf{u}_2 = 0)$.

The last result is one that makes the dot product very useful. Another useful property results from writing (5′) in the form

(9) $\mathbf{u}_1 \cdot \mathbf{u}_2 = |\mathbf{u}_1|\,|\mathbf{u}_2|\cos \phi$,

from which we see that the dot product of two vectors is the product of their lengths and the cosine of the angle between them.

A vector may always be expressed as the sum of two perpendicular vectors, one of which is parallel to any given vector. As in Fig. 4–68, we can simply drop a perpendicular to the second vector from the terminal point of the first and so determine the two perpendicular vectors, which are then called *components*. It is said that the given vector $\mathbf{u}_1$ has been *resolved* into the two components. The vector parallel to the second vector is said to be the *projection* of $\mathbf{u}_1$ on $\mathbf{u}_2$.

Now, as suggested in Fig. 4–68, $|\mathbf{u}_1| \cos \phi$ is in absolute value the length of the projection of $\mathbf{u}_1$ on $\mathbf{u}_2$, being positive or negative according as ϕ is less than or greater than 90°. Hence if we multiply a unit vector in the direction of $\mathbf{u}_2$ by $|\mathbf{u}_1| \cos \phi$, we get the vector projection of $\mathbf{u}_1$ on $\mathbf{u}_2$. Such a unit vector is $\mathbf{u}_2/|\mathbf{u}_2|$. Hence the projection of $\mathbf{u}_1$ on $\mathbf{u}_2$, that is, the component of $\mathbf{u}_1$ in the $\mathbf{u}_2$ direction, is $[|\mathbf{u}_1| \cos \phi][\mathbf{u}_2/|\mathbf{u}_2|]$, or by (9), $(\mathbf{u}_1 \cdot \mathbf{u}_2/\mathbf{u}_2 \cdot \mathbf{u}_2)\mathbf{u}_2$.

(10)   *The projection of* $\mathbf{u}_1$ *on* $\mathbf{u}_2$ *is* $\dfrac{\mathbf{u}_1 \cdot \mathbf{u}_2}{\mathbf{u}_2 \cdot \mathbf{u}_2} \mathbf{u}_2$.

This result enables us to find the component of a vector in any direction without using trigonometry. We need know only a vector in the desired direction. To illustrate the formula in a trivial case, the component of (x, y) in the direction of $(1, 0)$, the x-axis, is

$$((1 \cdot x + 0 \cdot y)/1)(1, 0) = (x, 0),$$

as expected.

(o) When the projection of $\mathbf{u}_1$ on $\mathbf{u}_2$ has been found from (10), how can the component perpendicular to it be found?   (p) Find the projection of $(1, 2)$ on $(3, 1)$, and sketch.   (q) The force of gravity on a body resting on an inclined plane is 15 lb. If the slope of the inclined plane is $\frac{1}{3}$, find the force parallel to it. Make a sketch.   (r) Resolve the vector $(3, 4)$ into components parallel and perpendicular to the vector $(1, 1)$. Sketch.

Problems

In Problems 1 through 4 find the angle from the line with the first slope to that with the second.

1. $1; 5.$ 2. $1; -2.$ 3. $3; 2.$ 4. $-2; -4.$

5. Find the line through the origin bisecting the angle made by the positive x-axis and the vector $(1, 3)$.
6. What can you say about the inclination of a line with negative slope?

In Problems 7 through 10 find the slope of the line with the given inclination.

7. $135°.$ 8. $3.$ 9. $\pi/2.$ 10. $1.$

In Problems 11 through 14 find the angle from the first to the second line:

11. $3x - 4y + 2 = 0; 4x = y.$
12. $2x + 4y = 3; x = -1.$
13. $y = -x + 2; x + y = 5.$
14. $9x + 10y = 1; 10x + 11y = 2.$

15. A gun changes its aim by a small angle and the point of impact shifts by 100 yd. If the range of the gun is 5000 yd, what was the angle of shift?
16. Show that the vector (A, B) is perpendicular to the line $Ax + By + C = 0$.
★17. Find a formula for the tangent of the angle between $Ax + By + C = 0$ and $A'x + B'y + C' = 0$.
★18. Write a necessary and sufficient condition that the lines in Problem 17 be perpendicular.
★19. Prove the following laws.

(11) $$\mathbf{u} \cdot \mathbf{v} = \mathbf{v} \cdot \mathbf{u},$$

(12) $$\mathbf{u} \cdot (\mathbf{v} + \mathbf{w}) = \mathbf{u} \cdot \mathbf{v} + \mathbf{u} \cdot \mathbf{w},$$

(13) $$\mathbf{u} \cdot (k\mathbf{v}) = k(\mathbf{u} \cdot \mathbf{v}).$$

★20. Laws (11), (12), and (13) are similar to certain laws for addition and multiplication of real numbers. Does the dot product of vectors satisfy other laws analogous to those for real numbers? For example, does (1–16–1) hold for the dot product of vectors? Does (1–16–5)? Does (2–8–5) hold for the dot product of vectors if the two zeros in the right member are displaced by **0**?
★21. Derive (4–7–10) by noting that d is the length of the projection on a vector perpendicular to the line of any vector joining a point on the line to (x_0, y_0).
★22. The work done by a force **F** of constant magnitude and direction moving a directed distance **s** in the direction of **F** is defined as $|\mathbf{F}| \, |\mathbf{s}|$. Suppose a force **F** is exerted on a body and causes it to move a directed distance **s** at an angle θ with **F**. Show that the work done is $\mathbf{F} \cdot \mathbf{s}$.
★23. Find the interior angles of the triangle whose vertices are $(1, 0)$, $(3, -4)$, $(-5, 7)$. Check by sketching.

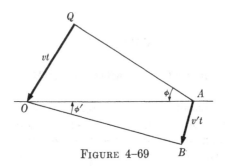

FIGURE 4–69

★24. Show that $(|\mathbf{u}|\mathbf{v} + |\mathbf{v}|\mathbf{u})/(|\mathbf{u}| + |\mathbf{v}|)$ bisects the angle between $\mathbf{u}$ and $\mathbf{v}$.
★25. Figure 4–69 is adapted from Sears and Zemansky, *University Physics*, p. 731. $\angle AQO$ and $\angle ABO$ are right angles. From $v/v' = n'/n$ show that $n \sin \phi = n' \sin \phi'$ (Snell's Law).

ANSWERS TO EXERCISES

(a) Any vector in the line has direction angles equal to the inclination θ or to $\theta + \pi$; but, (4–12–21). (b) Note that both angles are from first to second line. (c) $-26.4°$. (d) $m_1 - m_2 = 0 \leftrightarrow \tan(\phi_2 - \phi_1) = 0 \leftrightarrow \phi_2 = \phi_1 + k\pi$. (e) $1 + m_1 m_2 = 0 \leftrightarrow \cot(\phi_2 - \phi_1) = 0 \leftrightarrow \phi_2 - \phi_1 = \pi/2 + k\pi$. (f) Use (4–12–9), (4–12–10). (g) $C = \pi/2 \rightarrow \cos C = 0$. (h) 2.1, 24°, 131°; draw sketch to see which values to choose from among those satisfying (3). (i) $C = 105°$, $a = 8.9$, $c = 15$. (k) $\phi = \pi/2 + k\pi \leftrightarrow \cos \phi = 0 \leftrightarrow x_1 x_2 + y_1 y_2 = 0 \leftrightarrow \left(\dfrac{y_1}{x_1}\right)\left(\dfrac{y_2}{x_2}\right) = -1$.
(l) $\phi = 0 + k\pi \leftrightarrow |\cos \phi| = 1 \leftrightarrow (x_1 x_2 + y_1 y_2)^2 = (x_1^2 + y_1^2)(x_2^2 + y_2^2) \leftrightarrow x_1^2 y_2^2 - 2 x_1 y_2 x_2 y_1 + x_2^2 y_1^2 = 0 \leftrightarrow (x_1 y_2 - x_2 y_1) = 0 \leftrightarrow y_2/x_2 = y_1/x_1$.
(m) $(x_1 : x, y_1 : y, x_2 : x, y_2 : y)$ in (6). (n) See Exercise (k). (o) By subtracting projection from original. (p) (1.5, 0.5). (q) $\mathbf{u} = (-4.5, -1.5)$, $|\mathbf{u}| = 1.5\sqrt{10} \doteq 4.8$. (r) (3.5, 3.5), (−0.5, 0.5).

ANSWERS TO PROBLEMS

1. 34°. 3. −8°. 5. $y = 0.7x$ approximately. 7. −1. 9. Undefined. 11. 39°. 13. 0. 15. $\theta \doteq 20$ mils. 17. $(AB' - A'B)/(AA' + BB')$.

CHAPTER 5

RELATIONS AND FUNCTIONS

5–1 What is a relation? In the sentence "I have good relations with my relations," the word "relations" is used in two quite different senses. In its first appearance it refers to the relationship that I have with certain people. In the second it refers to these people. In mathematics, "relation" is used only in the first sense, i.e., in the sense of relationship.

A relationship, or as we shall say from now on, a relation, is something that holds or does not hold between two things. For example, if x and y are real numbers, the relation of greater than, symbolized by $>$, either holds between x and y (in which case we write $x > y$) or it does not hold. This suggests viewing a relation as a property of pairs. But we note that the order of the two things is important. For example, if $x > y$, then $\sim(y > x)$. This suggests that we might view a relation between two things as a property of ordered pairs.

As we pointed out in Section 3–10, there is a set corresponding to every sentence. Hence, if we define a relation by a sentence involving ordered pairs, there corresponds a set of ordered pairs. We think of a relation as a set of ordered pairs, namely the set of pairs whose first component stands in the given relation to its second component. For example, we view $>$ as the set of couples (x, y) for which $x > y$.

The purposes of this chapter are (1) to show how relations may be understood as sets of ordered pairs, (2) to introduce the concept of a function as a special kind of relation, and (3) to apply these ideas to algebra, analytic geometry, and simple scientific problems. When we say that the relation $>$ holds between x and y, we mean simply that $x > y$, or $(x, y) \in >$, where $> = \{(x, y) \mid x > y\}$.

5–2 The cartesian product. The *cartesian product*, $A \times B$, of two sets A and B is defined as the set of all ordered pairs whose first components are members of A and whose second components are members of B.

(1) Def. $A \times B = \{(x, y) \mid x \in A \land y \in B\}$.

For example, if $A = \{a, b, c\}$ and $B = \{d, e\}$, then $A \times B = \{(a, d), (a, e), (b, d), (b, e), (c, d), (c, e)\}$. If $A = \{-1, 0, 1\}$, $A \times A = \{(-1, -1), (-1, 0), (-1, 1), (0, -1), (0, 0), (0, 1), (1, -1), (1, 0), (1, 1)\}$.

Find $A \times B$ for: (a) $A = \{1, 2, 3\}$, $B = \{4, 5\}$; (b) $A = B = \{0, 1, 2\}$. (c) How is the limitation that A and B are sets "built in" to the definition?

(d) Argue for $(x, y) \in A \times B \leftrightarrow x \in A \wedge y \in B$.

The cartesian product is a useful concept when we wish to talk about ordered pairs whose components are chosen from certain sets. For example, in plane analytic geometry we identified points with pairs of real numbers. Hence each point is identified with a member of $Re \times Re$, where Re = the real numbers, and we may view the plane as a geometric picture of $Re \times Re$. The cartesian product is named after Descartes, one of the inventors of analytic geometry, and a plane with axes is often called a *cartesian plane*.

The fundamental idea of plane analytic geometry is that there is one and only one point corresponding to each member of $Re \times Re$ and one and only one member of $Re \times Re$ corresponding to each point. (See Section 4–1.) A similar conception can be applied to pairs that belong to cartesian products other than $Re \times Re$. Suppose, for example, that we are interested only in the digits G, $\{0, 1, 2, \ldots, 9\}$. Then $G \times G$ consists of the 100 ordered pairs (points) whose components (coordinates) are digits. We are not even limited to numbers. For example, we may wish to visualize the pairs chosen from 5 persons, a, b, c, d, and e. Letting $S = \{a, b, c, d, e\}$, $S \times S$ is a set of 25 ordered pairs. We may assign a point on the x-axis and on the y-axis to each member of S. Then each ordered pair is represented by a point in the lattice of 25 points. Suppose now we wish to graph the sentence "x dominates y," knowing that each person dominates those and only those whose names precede his on the roster. The graph would appear as in Fig. 5–1 if we choose the points on the axes as indicated.

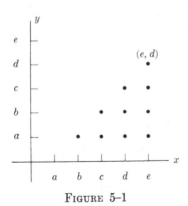

FIGURE 5–1

Abbreviating "dominates" by "D," graph x D y in the following cases. (e) Each person dominates those and only those whose names follow his on the roster. (f) The only dominations are e D a, c D b, and e D d. (g) Only a dominates and he dominates all others. (h) Only a is dominated and he is dominated by all others.

Consider the sentence $x = y$. If the range of the variables is Re, its graph is the straight line bisecting the angles between the axes in the first and third quadrants. If the range of the variables is G, the graph is the set of ten points $\{(0, 0), (1, 1), \ldots, (9, 9)\}$. If the range of the variables is the set of five people considered above, then the graph is the set of ordered pairs $\{(a, a), (b, b), (c, c), (d, d), (e, e)\}$. Evidently the set defined by a sentence depends on the ranges of the variables. If no indication is given as to the ranges, the set $\{(x, y) \mid y = x\}$ is the set of all ordered pairs of things about which we are talking and such that the two components are the same.

To have a convenient way of indicating the ranges of the variables, we define the *truth set of a sentence* $s(x, y)$ *in* $A \times B$ as the set of all solutions of $s(x, y)$ that are members of $A \times B$. This makes the range of x a subset of A and the range of y a subset of B.

(2) *Truth set* $s(x, y)$ *in* $A \times B = \{(x, y) \mid s(x, y)\} \cap (A \times B)$.

Then we can characterize *the plane cartesian graph* of a sentence as its graph in $Re \times Re$. More generally, the *graph* of $s(x, y)$ *in* $A \times B$ is the geometric picture of its truth set in $A \times B$.

In practice, when we wish to graph a sentence in the cartesian plane, $Re \times Re$, we find a few points of its graph in some subset of $Re \times Re$, most often $J \times Re$, where $J =$ the integers, and then link these points

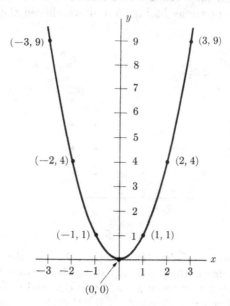

FIGURE 5-2

by a smooth curve or in some other way that appears plausible. For example, to graph $y = x^2$ in the cartesian plane we may locate all its points in $\{-3, -2, -1, 0, 1, 2, 3\} \times Re$, as in Fig. 5–2, and then draw the indicated smooth curve.

Graph the following in $\{-3, -2, -1, 0, 1, 2, 3\} \times Re$ and then link the points with a smooth curve or shade a region to suggest the graph in $Re \times Re$:
(i) $y = x^3$,　(j) $y = \sqrt{x}$ (in this one include a few more integral values of x),
(k) $x > y$,　(l) $y = 1/x$,　(m) $y < 1/x$,　(n) $y = x^2 + 1$.

PROBLEMS

In Problems 1 through 4 sketch each cartesian product in the plane.

1. $\{0, 1\} \times \{0, 1\}$.　　　　　　2. $\{-1, 0, 1\} \times \{-1, 0, 1\}$.
3. $\{0, 1\} \times \{0, 1, 2\}$.　　　　　4. $\{0, 1, 2\} \times \{0, 1\}$.

5. Show that $A \times B = B \times A$ is not a law.
6. Under what conditions does $A \times B = B \times A$?
★7. Prove that

(3) $$A \times (B \cap C) = (A \times B) \cap (A \times C).$$

★8. State and prove as many other laws as you can about the cartesian product.
★9. Argue that the number of members of $A \times B$ is the product of the number of members of A and the number of members of B.

ANSWERS TO EXERCISES

(a) $\{(1, 4), (1, 5), (2, 4), (2, 5), (3, 4), (3, 5)\}$.　(b) $\{(0, 0), (0, 1), (0, 2),$ $(1, 0), (1, 1), (1, 2), (2, 0), (2, 1), (2, 2)\}$.　(c) The right member makes no sense unless A and B are sets.　(d) (3-3-3).　(e) through (h) See Fig. 5–3. (i) $\{(-3, -27), (-2, -8), (-1, -1), (0, 0), (1, 1), (2, 8), (3, 27)\}$.　(j) See Fig. 5–4. Note that only one value of y corresponds to each value of x, and that there are no values corresponding to negative values of x.　(k) All points below $y = x$ on lines $x = -3$, $x = -2$, etc.　(l) See Fig. 5–5.　(m) See Fig. 5–6.　(n) Fig. 5–2 with graph shifted up one unit.

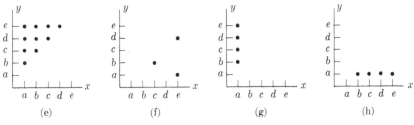

FIGURE 5–3

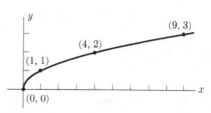

FIGURE 5-4

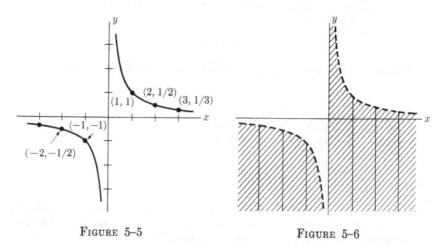

FIGURE 5-5   FIGURE 5-6

ANSWERS TO PROBLEMS

1. Four points. 3. Six points. 5. Choose $A \neq B$.
7. $A \times (B \cap C) = \{(x, y) \mid x \in A \land y \in B \cap C\} = \{(x, y) \mid x \in A \land y \in B \land y \in C\} = \{(x, y) \mid (x \in A \land y \in B) \land (x \in A \land y \in C)\} = \{(x, y) \mid x \in A \land y \in B\} \cap \{(x, y) \mid x \in A \land y \in C\} = (A \times B) \cap (A \times C)$.

5-3 Binary relations. As indicated in Section 5-1, we call any set of ordered pairs a relation. More specifically, we call any subset of $A \times B$ *a relation in* $A \times B$. Relations of this kind are often called *binary* to distinguish them from relations involving three or more objects, but we usually omit this adjective. Any set of points in the plane is a geometric picture of a relation in $Re \times Re$. We call a relation in $A \times A$ a *relation on* A. Since any sentence $s(x, y)$ defines a subset of $A \times B$, where A and B are the ranges of the variables in the sentence, any sentence in two variables defines a binary relation.

Consider, for example, the sentence $x < y$. This defines the relation $\{(x, y) \mid x < y\}$ on the real numbers. It seems natural to use $<$ as a name of this relation, i.e., to write that $< = \{(x, y) \mid x < y\}$. Consistent with this we adopt the following definition.

(1) Def. $\rho = \{(x, y) \mid x \rho y\}.$

This is really a definition *form*, since we may substitute for ρ (the Greek *rho*, pronounced like "roe") any constant such that $x \rho y$ is a sentence. Of course other substitutions would yield nonsense on the right, and hence (1) does not mean that anything is equal to a set of pairs.

   (a) List those symbols introduced so far that are significant substitutes for ρ in (1).   (b) Describe the relation named by each symbol in Exercise (a). (c) Suggest additional significant substitutes for ρ in $x \rho y$.

   According to (1) and (4–2–2),

(2) $(a, b) \in \rho \leftrightarrow a \rho b.$

We say that a is in the relation ρ to b $(a \rho b)$ if and only if the pair (a, b) belongs to the relation ρ. For example, $[(3, 2) \in >] \leftrightarrow [3 > 2]$.

   Use $\in$ to write a sentence equivalent to each of the following: (d) $3 < 4$, (e) $2 \geq 2$,   (f) $p \rightarrow q$,   (g) $x // y$.

   Even if a sentence is not in the form $x \rho y$, it defines a binary relation to which we can assign a name. For example, consider the sentence "y is a multiple of x." Using the abbreviation $x \mid y$, we give the precise definition

(3) Def. $x \mid y = [x \in J \wedge y \in J \wedge \exists z\, z \in J \wedge y = zx].$

Then $x \mid y \leftrightarrow (x, y) \in \mid$. We may read $x \mid y$ also as "x divides y," "x is a factor of y," "y is divisible by x," or "x is a divisor of y."

   The *graph of a relation* ρ in $A \times B$ is just the graph of the sentence $x \rho y$ in $A \times B$. For example, the graph of $>$ in $Re \times Re$ is the set of all points satisfying $x > y$. On the other hand, the graph of $>$ in $G \times G$ is the set of points sketched in Fig. 5–7.

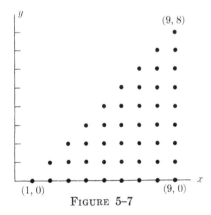

FIGURE 5–7

Graph the following relations in $G \times G$:   (h) $\leq$,   (i) $<$,   (j) $\geq$,   (k) $=$, (l) $\mid$ (divides).

For a given value of x, the sentence $x \rho y$ may have one, several, or no solutions for y. For example, let Br be a relation on people defined by

(4) x Br $y =_\text{d} y$ is a brother of x.

Then x Br y has several, one, or no solutions for y, depending on whether x has several, one, or no brothers. We call any value of y such that $x \rho y$ an *image of x under ρ*.

(5)   *Images of x under ρ $=_\text{d} \{y \mid x \rho y\}$.*

We see that for a given relation a value of x determines a set of images (which may be the null set).

Describe the images of 4 under each of the following relations: (m) $=$, (n) $>$,   (o) the relation defined by $y^2 = x$,   (p) ρ defined by Fig. 5–8.

The set of all objects that have at least one image under a relation ρ (i.e., whose set of images is not empty) is called the *domain* of ρ. The set of all objects that are images under ρ of *some* object is called the *range* of ρ. If $\rho = \{(x, y) \mid x \rho y\}$, the domain consists of values of "x" and the range consists of values of "y." The domain consists of first components of pairs belonging to ρ and the range consists of second components.

(6) **Def.**   Dom $(\rho) = \{x \mid \exists y \; x \rho y\}$,

(7) **Def.**   Rge $(\rho) = \{y \mid \exists x \; x \rho y\}$.

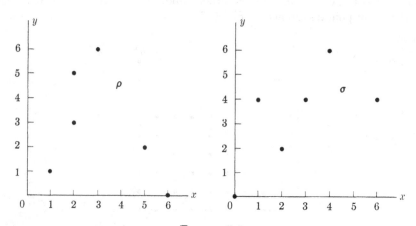

FIGURE 5–8

Let ρ be defined by a sentence $s(x, y)$ so that

$$s(x, y) \leftrightarrow x \; \rho \; y \leftrightarrow (x, y) \in \rho.$$

Then the domain of ρ is a subset of the range of significance of x in $s(x, y)$, and the range of ρ is a subset of the range of significance of y in $s(x, y)$.

(q) For ρ shown in Fig. 5–8 what are the domain and range? Find the images of each member of the domain.   (r) Answer the same questions for σ (sigma) in Fig. 5–8.   (s) Find the domain and range of Br defined by (4).   (t) Do the same for the relation in (1).   (u) Do the same for the relation in Exercise (o).

Familial relations are useful as illustrations of concepts discussed in this and following sections. Accordingly we add to the list begun with (4). The range of significance of the variables is understood to be persons born since 1800 A.D.

(8) $x \; \text{Si} \; y \; =_d y$ is a sister of x.

(9) $x \; \text{So} \; y \; =_d y$ is a son of x.

(10) $x \; \text{Da} \; y \; =_d y$ is a daughter of x.

(11) $x \; \text{Pa} \; y \; =_d y$ is a parent of x.

(12) $x \; \text{Fa} \; y \; =_d y$ is the father of x.

(13) $x \; \text{Mo} \; y \; =_d y$ is the mother of x.

(14) $x \; \text{Hu} \; y \; =_d y$ is a husband of x.

(15) $x \; \text{Wi} \; y \; =_d y$ is a wife of x.

(16) $x \; \text{Ch} \; y \; =_d y$ is a child of x.

(17) $x \; \text{Sb} \; y \; =_d y$ is a sibling of x.

(v) Define each relation in (8) through (17) as a set of pairs.   (w) Specify the domain and range of each.   (x) For each relation describe the images of an element of the domain.   (y) Why did we say "the" in (12) and (13), but "a" elsewhere?   (z) Define each relation by an idiomatic sentence whose subject is x instead of y.

PROBLEMS

Suppose we have two propositions called 0 and 1, the first false and the second true. In Problems 1 through 6 graph the relations in $\{0, 1\} \times \{0, 1\}$. Give their domains and ranges and find images of 0 and 1 under each one.

1. $\wedge$.   2. $\vee$.   3. $\rightarrow$.   4. $\leftrightarrow$.   5. $\underline{\vee}$.

★6. / (See Problem 23 in Section 2–5).

★7. What other relations are possible in $\{0, 1\} \times \{0, 1\}$? Treat each one as you did those in Problems 1 through 6. Suggest meanings in terms of logic.

In Problems 8 through 17 graph the relations in $A \times Re$, where $A = \{-9, -8, -7, -6, \ldots, -1, 0, 1, 2, \ldots, 9\}$.

8. I defined by $I = \{(x, y) \mid y = x\}$.
9. I^2 defined by $I^2 = \{(x, y) \mid y = x^2\}$.
10. $\{(x, y) \mid y = x + 1 \lor y = x + 2\}$.
11. $\{(x, y) \mid x + y = 8\}$.
12. $\{(x, y) \mid x^2 + y^2 - 2y - 5 = 0\}$.
13. $\{(x, y) \mid y = |x|\}$.
14. $\{(x, y) \mid x = |y|\}$.
★15. $\{(x, y) \mid y = \sin x\}$.
★16. $\{(x, y) \mid y = \tan x\}$.
★17. $\{(x, y) \mid y = \cos x\}$.

18. Describe the domain and range of each relation listed under Exercise (a).

ANSWERS TO EXERCISES

(a) $=, \land, \lor, \rightarrow, \leftrightarrow, \underline{\lor}, \in, \subset, \subseteq, //, \cancel{\in}, <, >, \leq, \geq, \neq.$ (b) For example, $=$ is the set of ordered pairs whose members are identical.   (c) Any verb, e.g. "loves," since "x loves y" defines the relation of loving.   (d) $(3, 4) \in <.$ (e) $(2, 2) \in \geq.$ (f) $(p, q) \in \rightarrow.$ (g) $(x, y) \in //.$ (h) All grid points not in Fig. 5–7.   (i) Like (h) but less points on diagonal.   (j) Fig. 5–7 plus points on diagonal.   (k) Ten points on diagonal.   (l) See Fig. 5–9.   (m) $\{4\}.$ (n) $\{y \mid y < 4\}.$ (o) $\{2, -2\}.$ (p) The empty set.   (q) Dom $(\rho) = \{1, 2, 3, 5, 6\}$, Rge $(\rho) = \{0, 1, 2, 3, 5, 6\}.$ (r) Dom $(\sigma) = \{0, 1, 2, 3, 4, 6\}$, Rge $(\sigma) = \{0, 2, 4, 6\}.$ (s) The domain is the set of all people with brothers, the range is the set of all males with siblings.   (t) Domain is the digits 1 through 9, range is G.   (u) Domain is non-negative reals, range is reals.

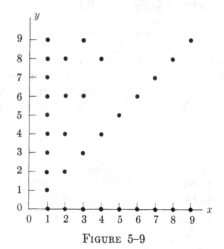

FIGURE 5–9

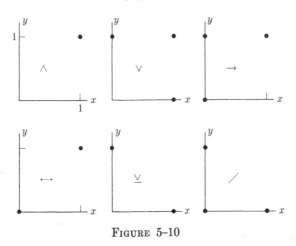

FIGURE 5–10

(v) For example, Si is the set of ordered pairs whose second components are sisters of their first components. (w) Rge (Da) = women, Dom (Da) = men and women who have at least one daughter. Rge (Pa) = parents, Dom (Pa) = people. Rge (Fa) = fathers, Dom (Fa) = people. Rge (Mo) = mothers, Dom (Mo) = people, Rge (Hu) = husbands, Dom (Hu) = wives. Rge (Wi) = wives, Dom (Wi) = husbands. (x) For example, the images of x under So are the sons of x. (y) "The" applies only where there is uniqueness. (z) For example, x Pa y = x is a child of y; and x So y = x is a parent of y and y is male.

ANSWERS TO PROBLEMS

1 through 6. See Fig. 5–10. 7. Ten others, including the empty relation (no points) and the universal relation (all four points). 9. Nineteen points. 11. Nineteen points. 13. Nineteen points arranged in a "V." 15. Remember, x is in radians!

5–4 Functions. We have seen that an object in the domain of a relation may have one or several images in the range. For example, in a monogamous society a person has at most one wife, so that each object in the domain of Wi has one and only one image under Wi. On the other hand, a person in the domain of Br (i.e., a person who has at least one brother) may have several images (brothers) under the relation Br. As another example, consider the relation defined by $y = x^2$ with domain the real numbers and range the non-negative real numbers. Each number in the domain has just one image, its unique square. On the other hand, consider the relation defined by $x = y^2$ with domain the non-negative real numbers and range the real numbers. For each nonzero value of x there are here two values of y that satisfy the equation, so that each object (except zero) in the domain of this relation has two images in the range. Finally, under the relation $<$ each real number has an infinite number of images, since for any x, the sentence $x < y$ has infinitely many solutions for y.

Relations such that every object in the domain has just one image are called *functions*. This means that a relation ρ is a function if and only if for any x in Dom (ρ), the set of images of x under ρ is a singleton. Letting "Fun" stand for the class of functions, we may write formally

(1) Def. $\qquad \rho \in \text{Fun} = \forall x\, [x \in \text{Dom}\, (\rho) \rightarrow \exists! y\; x\, \rho\, y].$

Since each object in the domain of a function has just one image, this unique image is determined as soon as the object in the domain is specified. Obviously, functional relations are of great importance whenever we have a situation in which one thing determines another. Indeed, historically the function concept preceded the more general concept of relation, and the idea of determination usually dominates the idea of relationship.

Since each object in the domain of a function F determines a unique image, and since we often wish to speak of the image of a particular object or of an unspecified object, it is convenient to have a special symbol for the image of the object x under the function F. This symbol is provided by the functional notation, which we have been using since Chapter 2. Formally, we may introduce this usage as follows.

(2) Def. $\qquad F(x) = y \ni x\, F\, y.$

Note that the right member of the definition makes no sense unless F is a function. It may be read "the y such that $x\, F\, y$," "the y that is an image of x under F," or "the image of x under F." This is a definition form that gives meaning to "$F(x)$" whenever F is a function and $x \in \text{Dom}\, (F)$. We have immediately

(3) $\qquad [y = F(x)] \leftrightarrow [x\, F\, y],$

(4) $\qquad F = \{(x, y)\,|\,y = F(x)\},$

(5) $\qquad x\ \ F\ \ F(x),$

(6) $\qquad (x, F(x)) \in F,$

(7) $\qquad F = \{(x, F(x))\,|\,x \in \text{Dom}\, (F)\}.$

Only functions are significant values of F in these laws. This is suggested by using the variable F, although any letter may be used.

We can now give meaning to the letter "f" in the functional notation "$f(x)$" introduced in Section 2–9. If "$f(x)$" is a formula, "$f(x)$" has a unique value corresponding to each value of "x," by our definition of the word "formula." Hence the sentence "$y = f(x)$" defines a function whose domain is the range of significance of "x" in "$f(x)$" and whose range is the range of "$f(x)$." Then by (4), "f" is a name of this function. Hence we were speaking quite precisely in Section 2–9 when we said that "the reader may think of it as standing for the relation between the values of x and the corresponding values of the formula."

Which of the following relations are functions?   (a) Mo,   (b) Da,   (c) Hu, (d) Pa,   (e) ρ in Fig. 5–8,   (f) σ in Fig. 5–8.
Read in words: (g) Mo (x),   (h) Fa (John),   (i) $\sigma(3)$ in Fig. 5–8.

We may visualize a function in a great variety of ways. All these go back to the fundamental fact that *a function is a set of ordered pairs in which no two different pairs have the same first component.* This is obvious if we recall the definition of a relation as a set of pairs and observe that in (1), $[\exists !y \; x \; \rho \; y] \leftrightarrow [\exists !y \; (x, y) \in \rho]$.

Which of the following are functions?   (j) $\{(0, 1), (1, 1)\}$;   (k) $\{(0, 0), (0, 1)(1, 0)\}$;   (l) $\{(0, 1), (1, 1), (2, 1)\}$.

Evidently a function can be specified by giving the roster of its members, provided there are only a few members. A more general way to specify a function is to give a *defining sentence* that defines the set of ordered pairs. (Compare Section 3–2.) This may be done *explicitly* by a sentence of the form $y = f(x)$ or *implicitly* by a sentence $s(x, y)$ of some other form. For example, $y = x^2$ and $x^2 - y = 0$ define the same function. The first gives an explicit formula for the image of each x. The second gives the image of each x implicitly, since it has to be solved for y in order to find the image in each case. When a function is defined by a sentence of the form $y = f(x)$, "f" is a name of the function as indicated by (4). If

$$\forall x \, [[\exists y \; s(x, y)] \rightarrow [\exists !y \; s(x, y)]],$$

then $\{(x, y) \,|\, s(x, y)\}$ is a function.

In each of the following, solve for y so as to get a sentence defining the function explicitly: (m) $y - x = 0$, and   (n) $y - 2x + 1 = 0$.

A third way to specify a function is to give a formula for the image of any object in the domain. Thus we could define the function discussed above by writing $f(x) = x^2$. Of course, it may not be convenient to write a single formula that gives the image of every object in the domain. All that is necessary is that some rule be given from which we can calculate the image of any object. For example, we might write $f(x) = x^2$ for $x > 0$ and $f(x) = x$ for $x < 0$. Then $f = \{(x, y) \,|\, (x \geq 0 \,\wedge\, y = x^2) \,\vee$

$(x < 0 \wedge y = x)\}$. We call $f(x)$ a *defining formula* of f. When a defining formula $f(x)$ is known, the function may be specified as in (7) without using a second variable.

(o) Write a defining formula for the function defined by the sentence $y^2 = x \wedge y \geq 0$. (p) Do the same for "x was sired by y."

Since functions are relations of a special type, the discussion of the previous section applies. For example, the domain of a function defined by $y = f(x)$ is the set of values of x for which $f(x)$ exists, and the range is the set of values of y for which $y = f(x)$ for some x. We have discussed functions by using x to stand for objects in the domain and y to stand for those in the range. Of course, other letters could be used. In any case, the variable used to stand for objects in the domain of a function is called the *independent variable,* and the variable used to stand for objects in the range is called the *dependent variable.* The terminology refers to the fact that the value of the dependent variable is determined by that of the independent variable.

Consider the function f defined by "$f = \{(x, y) \mid x$ has the army serial number $y\}$." (q) What is the domain? (r) What is the range? (s) Write an explicit formula for $f(x)$. (t) Which is the dependent and which the independent variable?

It is often convenient to visualize a function as the set of points corresponding to the pairs that make up the function. The *graph of a function* f is simply the graph of the sentence $y = f(x)$. Graphs of functions have one special feature not possessed by those of other relations: *no vertical line crosses the graph of a function in more than one point.* This is the geometrical counterpart of the fact that only one value of the dependent variable corresponds to each value of the independent variable. For example, Figs. 5–2 and 5–4 are graphs of functions, whereas Figs. 5–1 and 5–6 are not.

(u) Some relations in $Re \times Re$ are sketched in Fig. 5–11. Which are functions? Write a defining sentence for each. Indicate the range and domain.

Problems

1. If $F = \{(1, 0), (-3, 4), (0, 0), (2, 0), (3, -4)\}$, specify Dom (F), Rge (F), $F(-3)$, $F(0)$, $F(2)$. Sketch the function.

2. Do the same for $G = \{(-2, 4), (-1, 1), (0, 0), (1, 1), (2, 4)\}$. Suggest a defining formula and define G by using it.

3. Let $I^2 = \{(x, y) \mid y = x^2 \wedge x \in J\}$, where $J =$ the integers. Specify: (a) $I^2(1)$, $I^2(2)$, Dom (I^2), Rge (I^2); (b) three names for the function; (c) a defining formula for the function; (d) a defining sentence for the function. (e) What is the image of m under I^2? (f) What two numbers have the image 16

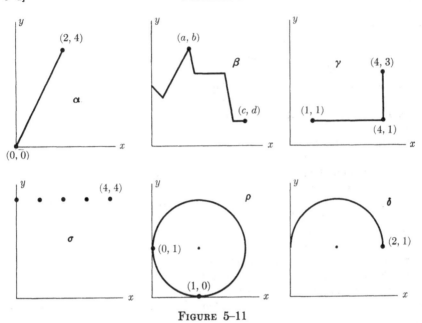

FIGURE 5–11

under I^2? (g) If the independent variable has value 3, what is the corresponding value of the dependent variable? (h) Sketch I^2.

4. Graph the following sentences. Which define functions? $x - y = 7$; $x^2 + y^2 = 4$; $y^2 = x \wedge y < 0$; $|y| = x$.

5. Is it legitimate to write "$P(x)$ = the point on the axis corresponding to the real number x," and to consider P a function? Explain.

6. Does "x is a name of y" define a set of pairs (x, y) that is a function? Explain.

7. Can any expression serve as a defining formula for a function? Explain.

8. What is the relation between the range of a formula involving just one variable and the range of the function it defines? between the range of the variable in such a formula and the domain of the function it defines?

9. If a function is defined by a propositional formula, what is the nature of its range? its domain?

10. If $T(x)$ = the truth value of x, what is the domain and range of T?

11. If $f(x)$ = the digit in the twenty-fifth decimal place of the decimal expansion of x, is f a function? What is its domain and range?

12. What are the domain and range of the function defined by $f(x) = \sim x$? Describe the image of x under f.

★13. Let $T(f)$ = the truth table of f. Is T a function? What is its range and domain?

★14. If $T(x)$ = the truth value of x, what can we say about the range of the function defined by $T(f(x))$ if $f(x)$ is a law?

15. Does "y dislikes x" define a function? Explain.

16. Does "y is married to x"? Explain.

★17. Mention some functions whose domains consist of sets. Give a defining formula for each and specify the domain and range.

★18. Specify a function whose domain is Re, the real numbers, and whose range consists of intervals.

19. Mention a function whose domain is a set of intervals of real numbers and whose range is a set of real numbers.

20. List several functions whose domain consists of ordered pairs of real numbers.

21. Do the same for the domain consisting of ordered pairs of sets and the range consisting of sets.

22. What is the domain of the function defined by $y = $ max (x)?

★23. In the theory of games a *strategy* is defined as a rule for choosing a move in any situation that may arise in play. Describe a strategy as a function, indicating its domain and range.

24. Which of the relations in Fig. 5–10 are functions?

25. Is any straight line the graph of a function?

26. Is Fig. 4–12 the graph of a function?

27. Comment on the following. "Let $\theta(\mathbf{u})$ be the direction angle of $\mathbf{u}$."

28. What is the domain of the function defined "by $y = $ the x-intercept of the straight line L," where L is the independent variable and y the dependent variable?

29. Does $y = \sin x$ define a function? What are its domain and range?

30. What are the domain and range of cosine? tangent?

★31. Let $Q\rho(x) = $ the images of x under ρ. (a) What are the domain and range of $Q\rho$? (b) Under what conditions does the range consist of singletons? (c) What is the union of all the members of the range of $Q\rho$? (d) If ρ is a function, is the range of $Q\rho$ the same as the range of ρ?

★32. Reformulate Problems 17 through 21, using the idea of cartesian product.

ANSWERS TO EXERCISES

(a), (c), and (f) are functions.   (g) The mother of x.   (h) The father of John.   (i) Four.   (j) and (l) are functions.   (m) $y = x$.   (n) $y = 2x - 1$.   (o) $\sqrt{x}$.   (p) Fa(x).   (q) Soldiers having serial numbers.   (r) A set of serial numbers.   (s) $y = $ the serial number of x.   (t) x independent, y dependent.   (u) $\alpha, \beta, \sigma, \delta$. Dom $(\beta) = (0\_\_c)$, Rge $(\beta) = (\underline{d\quad b})$. $\beta = \{(x, y) \mid (x, y)$ lies on the graph in Fig. 5–11$\}$.

ANSWERS TO PROBLEMS

1. Dom $= \{1, -3, 0, 2, 3\}$,   Rge $= \{0, 4, -4\}$, $F(-3) = 4$, $F(0) = 0$, $F(2) = 0$.   Graph consists of five points.   3. (a) $I^2(1) = 1$, $I^2(2) = 4$, Dom $(I^2) = J$, Rge $(I^2) = $ perfect squares; (b) "I^2," "$\{(x, y) \mid y = x^2 \wedge x \in J\}$," "$\{(x, x^2) \mid x \in J\}$"; (c) "$x^2$"; (d) "$y = x^2 \wedge x \in J$"; (e) m^2; (f) 4 and -4; (g) 9.   5. Yes, if an axis is assumed to have been specified, since just one point corresponds to each number.   7. No, the expression must be a formula, since it must be such as to have a unique value for each value of its variable.   8. Same

in both cases. 9. Range consists of propositions, domain is not limited by this information. 11. Yes; domain is real numbers, range is the digits. 13. Yes; domain is the set of logical formulas, range is the set of corresponding truth tables. 15. No, since often more than one person dislikes a given individual. 17. $F(x) = x'$; domain the power set of some set, range the corresponding complements (range = domain in this case). 19. $F(x)$ = the length of x; domain is intervals, range is non-negative reals. 21. $F(x, y) = x \cap y$. 23. The domain is the set of possible situations; the range the corresponding moves specified. 25. No, not vertical lines. 27. Nonsense, since a vector has many direction angles. 29. Yes, domain is real numbers, range is $(\underline{-1 \quad 1})$. 31. (a) Domain is the domain of ρ, range consists of sets of images. (b) When ρ is a function. (c) The range of ρ. (d) No, if ρ is a function the range of $Q\rho$ is the set of all singleton subsets of the range of ρ.

★5–5 Aspects of the function idea. The idea of one thing determining another is very old and very familiar. It is natural that this idea has been described in many different ways and appears under a variety of aliases in different fields. In Section 5–4 we have defined a function as a special kind of relation, that is, as a set of ordered pairs in which each first component is paired with only one second component. The cartesian graph of a function is a set of points that is met by any vertical line in at most one point. The function may be defined by listing the pairs, by giving a defining sentence or defining formula, or by the graph. In this section we discuss various other ways in which functions appear and terms in which they are described.

Tables. A function may be defined by a table that lists in some convenient fashion the images of members of the domain. Thus, the first two columns in Table I in the Appendix define the square-root function with domain the first 100 positive integers. Or we may think of Table I as giving merely a partial roster of the pairs belonging to $\{(x, y) \mid y = \sqrt{x}\}$ with domain the real numbers. Such partial tables can be used to find other images exactly or approximately by various devices. For example, we can find $\sqrt{2600}$ from Table I by noting that $\sqrt{2600} = 10\sqrt{26}$, and we can estimate $\sqrt{26.3}$ by linear interpolation as explained in Section 4–11.

(a) What function is defined by the first and third columns in Table I?
★(b) Use Table I to find $\sqrt[3]{86,000}$ and to estimate $\sqrt[3]{51.6}$.

Although we can tabulate only a finite number of pairs belonging to a function, we can think of a function as specified by an imaginary table giving all pairs. Each pair corresponds to a point on the graph of the function. Printed tables are then viewed as a selection, often (but not always) a selection of pairs with equally spaced values of the independent variable.

(c) What would be the appearance of a table of a nonfunctional relation? Have you ever seen such a table?

Mappings. When we construct a map of a portion of the earth's surface, each feature on the ground is represented by a corresponding symbol on the map. Letting $f(x) =$ the point on the map corresponding to the point x on the ground, f is a function whose domain is the mapped portion of the earth and whose range is the map. By analogy we describe any function as a mapping that maps its domain onto its range. Schematically, as suggested in Fig. 5-12, we may think of the domain and range of a function as sets of points, and of the function as mapping each point in the domain into its image in the range.

(d) Is it possible for a function to map two different points in its domain into the same point in its range?   (e) Sketch $\{(x, y) \mid y = x^2 \land x \in \{-2, -1, 0, 1, 2\}\}$ by representing the domain and range on two parallel axes and drawing an arrow from each object in the domain to its image in the range.

Correspondences. We have spoken of the image of an object in the domain as the object that corresponds to it. With this in mind, relations are often described as correspondences. If the relation is a function, each object in the domain corresponds to one and only one object in the range, though more than one object in the domain may correspond to the same object in the range. Hence relations are called *many-many correspondences*, and functions are called *many-one correspondences*.

(f) In the function defined by $y = x^2$, how many objects in the domain correspond to each object in the range?   (g) Answer the same question for the relation Pa defined by (5-3-11).

Operations. When we find the image of x under the function f, we perform an operation on x. A function f is often described as an operation, namely, the operation that yields $f(x)$ when it is applied to x. For example, the operation of squaring yields x^2 when it is applied to x. We may think of f as a kind of machine that produces $f(x)$ when x is fed into it, as suggested in Fig. 5-13. This way of thinking is not usually applied to

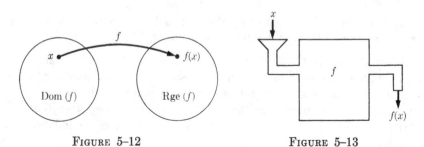

FIGURE 5-12   FIGURE 5-13

nonfunctional relations, but it could be by thinking of the output of the machine as the set of images of x under the relation.

(h) What is the operation defined by $y = x^2 - 3$?

When the domain of a function consists of ordered pairs from some set A, and its range of members of A, the function is often called a *binary operation on A*. For example, addition is a binary operation on numbers. When the domain of an operation on A consists of $A \times A$, that is, when every ordered pair of elements in A yields a member of A under the operation, the set A is said to be *closed* under the operation. For example, the real numbers are closed under addition, but not under division, since division by zero is undefined. Sometimes a function is called a *unary operation* on the members of its domain.

Transformations. Sometimes a function is described as a transformation that transforms each object in its domain into the corresponding object in its range. This can be visualized in terms of Fig. 5–12 or Fig. 5–13. When this terminology is used, the image of x is sometimes called the *transform* of x.

(i) What is the transform of x under Fa?

Traditional terminology. The above terms are really metaphorical ways of talking about functions. They are suggestive of the nature of functions, facilitate thinking about them, and provide alternative expressions for ideas that appear very frequently. They are picturesque ways of talking about the set of ordered pairs that *is* the function. However, there are many other common ways of talking about functions that are quite confusing, even though hallowed by long and continued usage.

Frequently the symbols used to define a function and to represent members of its domain and range are used as names of the function itself. Below is a table of some such usages, with the meaning of each.

Common usage	*Meaning*
The function $f(x)$.	The function defined by "$y = f(x)$."
The value of the function corresponding to x.	$f(x)$.
The value of the function at x.	$f(x)$.
y is a function of x.	$\exists f \; \forall z \; y(z) = f(x(z))$.
$f(x)$ is a single-valued function.	f is a function.
$f(x)$ is a multi-valued function.	f is a nonfunctional relation.

The expressions in the left column must be viewed as idioms. Taken literally, they are nonsense. For example, in the fourth entry y is a variable, not a name of a function (a set of ordered pairs). Very frequently in science one sees expressions such as, "The distance is a function of the time." This means that the time determines the distance, i.e., that there is a function f such that $y = f(x)$ gives the distance corresponding to the time x. It should be noted that when we say "determine" we are not assuming causality, but merely a correspondence (or pairing) of times and distances. The function idea does not imply causality. A function is a pairing, not an explanation of the pairing. We are perfectly free to pair members of any two sets.

The reader will not become confused with the variety of terms if he keeps in mind the essential features of a function as a set of ordered pairs in which no two second elements are paired with the same first element. Only when it comes to talking about functions in various contexts does the profusion of terms appear. We usually have an independent variable to refer to members of the domain of the relation, a dependent variable to refer to members of the range, and a formula, rule, sentence, table, or other means of specifying the second component corresponding to each first component. The function itself may be called a mapping, correspondence, operation, or transformation. The values of the variables may be referred to as points, objects, or components. The object in the range corresponding to a member of the domain may be called the image, the corresponding element, the transform, or the value of the function. Some of the terminology may suggest that the formula or rule *is* the function, but this is merely a way of speaking.

What is meant by the following? (j) "The function x^3." (k) "The work done in moving a given distance is a function of the force applied." (l) "The value of the function σ defined by Fig. 5–8 is 6 when $x = 4$."

PROBLEMS

1. Discuss photography in terms of the function idea.
2. Is a one-to-one correspondence a function?
3. Suppose we view a function in $Re \times Re$ as mapping the x-axis into the y-axis. If we are given the graph of the function, show how to determine the image on the y-axis of any point on the x-axis.

Graph the relations in Problems 4 through 12 by drawing parallel axes and using arrows as in Fig. 5–16. In the case of more than a finite number of pairs, give a few typical arrows.

4. ρ of Fig. 5–8.
5. σ of Fig. 5–8.
6. The relation of Fig. 5–9.
7. The relations of Fig. 5–10.
8. $\{(x, y) \mid y = x + 3\}$.
9. $\{(x, y) \mid y = 2x\}$.
10. $\{(x, y) \mid y = -x\}$.
11. $\{(x, y) \mid y = x\}$.
12. $\{(x, y) \mid y = 1 - x^2\}$.

13. Describe the relations in Problems 4 through 12 as mappings.

14. Describe them as correspondences.

15. Describe them as operations.

16. Describe them as transformations.

★17. "One quantity, or measurable thing y, is a function of another measurable thing x, if any change in x will produce or 'determine' a definite corresponding change in y. . . ." (*The Alphabet of Economic Science*, by Philip H. Wicksteed) Discuss.

★18. "Since the statement $y = f(x)$ implies a definite relation between the changes in y and the changes in x, it follows that a change in y will determine a corresponding change in x as well as *vice versa*. Hence if y is a function of x, it follows that x is also a function of y." (Same reference) Discuss.

19. Figure 5-14 graphs the function f. Sketch $y = -f(x)$, $y = f(-x)$, $y = |f(x)|$, $y = f(|x|)$.

20. Do the same for g, defined by Fig. 5-15.

★21. Consider the function partially tabulated in the first and fourth columns of Table I. Give a precise definition of the function. What are its domain and range? Find the image of 8.7 and of 87.2 under this function. Use the table to find 1/35 approximately.

ANSWERS TO EXERCISES

(a) Cube root with domain the first 100 positive integers.   (b) 44.14; 3.72. (c) More than one entry in the body of the table for some entry in the margin. (d) Yes.   (e) See Fig. 5-16.   (f) Two, except that only one corresponds to 0.

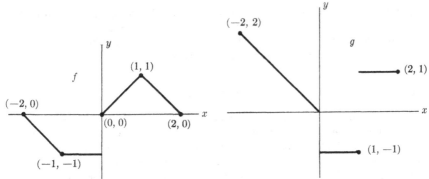

FIGURE 5-14   FIGURE 5-15

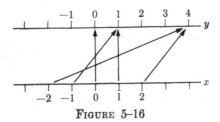

FIGURE 5-16

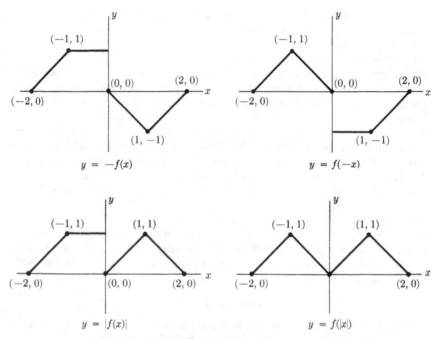

FIGURE 5-17

(g) For any y, as many as the children of y.   (h) Squaring x, then subtracting 3 from the result.   (i) The father of x.   (j) $\{(x, y) \mid y = x^3\}$.   (k) There is a function whose members are ordered pairs whose first components are forces and whose second components are the corresponding work done by them.   (l) $\sigma(4) = 6$.

Answers to Problems

1. The image of an object is the mark on the print corresponding to the object photographed.   3. From $(x, 0)$ on the x-axis go vertically to the graph, then go horizontally to $(0, f(x))$ on the y-axis.   13. For example, the function in Problem 8 maps each real number into one three greater; it shifts all points three units to the right.   15. The same function is the operation of adding three.   19. See Fig. 5–17.   21. Domain and range consist of all nonzero reals. $1000/8.7 \doteq 114.9$; $1000/87.2 \doteq 11.46$; $1/35 \doteq 0.02857$.

★5–6 Set algebra applied to relations. Since relations are sets of ordered pairs, all the concepts and laws of Chapter 3 are applicable. We may consider a cartesian product $A \times B$ and its power set, that is, the set of all relations in $A \times B$. Among these is the *null relation* (the relation that holds between no two objects chosen from A and B) and the *universal* *relation*, $A \times B$ itself, which holds between the members of any ordered pair in $A \times B$. Then the definitions and laws of Chapter 3 apply.

In particular, if ρ is a relation, we call ρ' the *complementary relation*. We have

(1) $$\rho' = \{(x, y) \mid \sim(x \, \rho \, y)\},$$

(2) $$(x \, \rho' \, y) \leftrightarrow \sim(x \, \rho \, y).$$

Often the complementary relation is indicated by drawing a diagonal line through a name for the relation. For example, we write $\neq$ for $='$ and $\notin$ for $\in'$.

(a) Read in idiomatic English: $3 <' 3$, $a \in' B$, $S \subseteq' T$, $a \, //' \, b$. (b) Identify the graph of the complement of $\rightarrow$ in your solution of Problem 3 in Section 5–3. (c) Does $\rightarrow'$ have another name? (d) Describe So′ where So is defined by (5–3–9). What is its domain? range?

The union of two relations ρ and σ is the relation $\rho \cup \sigma$ that holds whenever either or both hold; that is,

(3) $$[x \, (\rho \cup \sigma) \, y] \leftrightarrow [x \, \rho \, y \lor x \, \sigma \, y].$$

The intersection of two relations is that which holds for an ordered pair whenever they both hold; that is,

(4) $$[x \, (\rho \cap \sigma) \, y] \leftrightarrow [x \, \rho \, y \land x \, \sigma \, y].$$

One relation ρ is a *subrelation* of a second σ if and only if the second holds for any ordered pair for which the first does; that is,

(5) $$(\rho \subseteq \sigma) \leftrightarrow (\forall x, y \; x \, \rho \, y \rightarrow x \, \sigma \, y).$$

Two relations are disjoint if and only if they cannot hold for the same ordered pair.

(e) True or false? $> \subseteq \geq$; $\leq \subseteq =$; $= \subseteq \leq$. (f) Explain in words the meaning of $\subset$, $//$, $\not\backslash$, and $\underline{\cup}$ for relations. (g) Under what conditions is $\rho = \sigma$ where ρ and σ are relations? (h) Give examples of disjoint relations.

Consider the relation S defined by $y^2 = x$. It is not a function, since there are two images of each value of x unless $x = 0$. However, since

$$(y^2 = x) \leftrightarrow (y = \sqrt{x} \lor y = -\sqrt{x}),$$

$$S = \{(x, y) \mid y = \sqrt{x}\} \cup \{(x, y) \mid y = -\sqrt{x}\}.$$

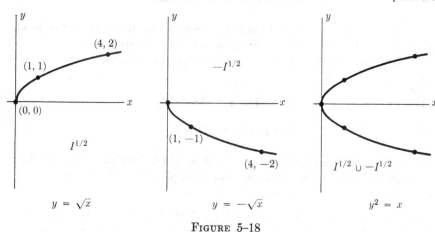

FIGURE 5–18

Hence S is the union of two functions. In Fig. 5–18 we have sketched $y = \sqrt{x}$, $y = -\sqrt{x}$, and $y^2 = x$.

(i) Show that $\{(x, y) \mid x^2 + y^2 = 4\} = \{(x, y) \mid y = \sqrt{4 - x^2}\} \cup$ $\{(x, y) \mid y = -\sqrt{4 - x^2}\}$. Sketch each function and their union as in Fig. 5–18. (j) Express the relation of Fig. 5–9 as a union of functions. (k) Argue that any relation is the union of functions. (l) When we wish to determine the relation $A \cap B$, where A is defined by $s(x, y)$ and B by $r(x, y)$, we refer to this as _____ the _____ sentences s and r. (Fill in the blanks.)

Problems

Complete the sentences in Problems 1 through 12.

1. Fa $\cup$ Mo = ?
2. Hu $\cup$ Wi = ?
3. Br$'$ = ?
4. ($<$ $\cup$ =) = ?
5. Hu $\cap$ Wi = ?
6. Fa $\cap$ Pa = ?
7. Br $\cup$ Si = ?
8. $<'$ = ?
9. So $\cup$ Da = ?
10. ($<$ $\cap$ $>$) = ?
11. ($\rightarrow$ $\cup$ $\leftrightarrow$) = ?
12. ($\vee$ $\cup$ $\wedge$) = ?

13. Graph $x^2 + y^2 = 9$ as the union of two functions.

14. Show that any circle is the graph of the union of two functions by showing that the equation of any circle is logically equivalent to the disjunction of two sentences defining functions.

15. Do the same for $(x - 3)^2 + (y + 2)^2 = 25$.

Graph 16 through 20 as the unions of functions.

16. $|y| = x$
17. $y^4 = x$.
18. $|y - 3| = 2x$.
19. $y^2 = x^3$.
20. $(y - 3)(y - x^2) = 0$.

21. Prove $\{(x, y) \mid f(x, y)g(x, y) = 0\}$
$$= \{(x, y) \mid f(x, y) = 0\} \cup \{(x, y) \mid g(x, y) = 0\}.$$

Graph 22 through 25.

22. $xy = 0$.   23. $(x - y)(x + y) = 0$.

24. $(x + 2y - 1)(3x + y) = 0$.   25. $(x^2 - y^2)(y + 2) = 0$.

26. Prove $\{(x, y) \mid [f(x, y)]^2 + [g(x, y)]^2 = 0\}$
    $= \{(x, y) \mid f(x, y) = 0\} \cap \{(x, y) \mid g(x, y) = 0\}$.

Graph 27 through 30.

27. $x^2 + y^2 = 0$.
28. $(x - y)^2 + (x + y)^2 = 0$.
29. $(x + 2y - 1)^2 + (3x + y)^2 = 0$.
30. $(x^2 - y^2)^2 + (y + 2)^2 = 0$.

31. Consider the relations $=$, $\neq$, $<$, $>$, $\leq$, and $\geq$ on the real numbers. Write as many true statements as you can about them, using the ideas of the set operations and relations. For example, $\neq \; = \; ='$, $< \; \subseteq \; \leq$, and $\leq \; \cap \; \geq \; = \; \dots$.

★32. Do the same for the binary relations on sets, $\subseteq$, $\subset$, $\not\subset$, $//$.
★33. Do the same for the logical relations, $\vee$, $\wedge$, $\rightarrow$, $\leftrightarrow$, $\underline{\vee}$.
★34. Do the same for the familial relations.

Answers to Exercises

(a) 3 is not less than 3, a is not a member of B, S is not a subset of T, a and b are not disjoint. (b) $\{(1, 0)\}$. (c) No. (d) Set of all ordered pairs of people in which the second component is not a son of the first; range = domain = set of all people. (e) T, F, T. (f) For example, $\rho \subset \sigma$ if and only if whenever $x \, \rho \, y$ then $x \, \sigma \, y$ but there is an (x, y) such that $x \, \sigma \, y$ but $x \, \rho' \, y$. (g) $x \, \rho \, y \leftrightarrow x \, \sigma \, y$. (h) $<$, $>$; σ, σ' for any σ; etc. (i) Use (2–8–21) $(x{:}y, a{:}1, b{:}0, c{:}x^2 - 4)$ or subtract x^2 from both members and use $a^2 = b^2 \leftrightarrow a = b \vee a = -b$. (j) $(y = 0x) \vee (y = x) \vee (y = 2x) \vee (y = 3x) \vee (y = 4x) \vee (y = 5x) \vee (y = 6x) \vee (y = 7x) \vee (y = 8x) \vee (y = 9x)$.

(k) We construct one of the functions by picking out one of the images of each object in the domain, a second by picking out another, etc. In this way we exhaust all images if the number is finite. If not, practical difficulties may arise, but we can still imagine the process as accomplished. For example, we can imagine $x > y$ as the conjunction of all sentences of the form $x = y + k$ where k takes all real positive values. (l) solving; simultaneous.

Answers to Problems

1. Pa. 2. Spouse. 3. Not brother. 4. $\leq$. 5. Empty. 7. Sb. 9. Ch. 11 $\rightarrow$. 13. Upper and lower halves of circle. 15. $y = -2 \pm \sqrt{25 - (x - 3)^2}$. 17. $y = \pm\sqrt{x}$. 19. $y = \sqrt{x^3} \vee y = -\sqrt{x^3}$. 21. (2–8–5).

5–7 Linear functions. A function that has a defining formula of the form $mx + b$ is called a *linear function*. We studied the geometric properties of the graphs of such functions in Section 4–5. We adopt $mI + b$ as a name for the function defined by $y = mx + b$; that is,

(1) **Def.** $mI + b = \{(x, y) \mid y = mx + b \wedge x \in Re\}$.

(a) Describe $2I + 3$.

A special class of linear functions consists of those for which $m = 0$. We define

(2) **Def.** $\underline{b} = 0I + b$.

The range of $\underline{b}$ consists of the singleton $\{b\}$, since the image of any x is just b. We call any function whose range is a singleton a *constant function*. The name suggests the idea that the image is the same for all objects in the domain.

(b) Describe $\underline{3}$.

Another special class consists of those linear functions for which $b = 0$. We define

(3) **Def.** $mI = mI + 0$.

We call these functions *direct variations*. Their graphs consist of straight lines through the origin. Among the direction variations is the *negative* defined by

(4) **Def.** $-I = (-1)I$.

(c) Describe the graphs of $-I, 2I, -3I$.

Note that "I" is not a variable. On the contrary,

(5) **Def.** $I = \{(x, y) \mid y = x \wedge x \in Re\}$.

It is a constant standing for the set of all ordered pairs of real numbers whose components are identical. We call I the *identity function*. The expressions of the form $mI + b$ used to designate linear functions have the advantage that a defining formula is obtained by substituting an independent variable for the constant I. We shall use this name of the identity function extensively for naming other functions.

A function whose graph consists of a straight line or of the union of portions of straight lines is called *piecewise linear*. Piecewise linear functions appear very frequently in science. As indicated in the problems following Section 4–5, a defining sentence of a straight line segment consists of the conjunction of a linear equation and an inequality requiring the independent variable to lie in an interval. Hence a defining sentence of a piecewise linear function consists of the disjunction of such sentences.

Graph the functions defined by the following: (d) $(y = x \wedge 0 \leq x \leq 2) \vee$ $(y = 3 \wedge 2 < x \leq 3) \vee (y = -x + 6 \wedge x \geq 3)$;   (e) $f(x) = 1$ for $0 \leq x \leq 2$, $f(x) = 2$ for $2 < x < 3$, $f(x) = 1$ for $4 \leq x \leq 5$, and $f(x) = 3$ for $6 \leq x \leq 8$. (f) Write Exercise (d) in the form of (e).   (g) Write Exercise (e) in the form of (d).

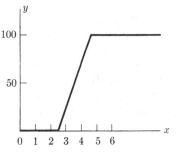

FIGURE 5–19

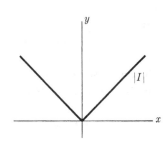

FIGURE 5–20

Figure 5–19 is similar to several appearing in "The Theory of the Neural Quantum in Discrimination of Loudness and Pitch," by S. S. Stevens, G. T. Morgan, and J. Volkmann in the *American Journal of Psychology*, July, 1941. The variable y represents the percentage of cases in which a person discriminates this pitch as different from a comparison pitch corresponding to $x = 0$. Estimating the position of the corners, we have the following defining sentence of this functional relation:

$$(x \leq 2.5 \wedge y = 0) \vee (2.5 \leq x \leq 4.5 \wedge y = 50x - 125)$$

$$\vee (x \geq 4.5 \wedge y = 100).$$

(h) Explain the equation $y = 50x - 125$.   (i) Find the defining sentence if the corner points are $(2.1, 0)$ and $(4.5, 100)$. Graph. From the graph and Fig. 5–19, discuss the reactions of the two individuals.

The norm. The *norm function* is defined by $y = |x|$. (See Section 3–9.) Its graph is suggested in Fig. 5–20. We designate it by $|I|$ so that $|I|(x) = |x|$.

(6) **Def.** $|I| = \{(x, y) \,|\, y = |x|\}.$

(j) Write a defining equation of the norm by completing the following:
$(y = x \land$ _____$) \lor (y = -x \land$ _____$)$.   (k) Graph $y = |2x|$.
(l) Show that $|I(x)| = |x|$.

The signum function. The *signum function*, sg, is defined by

$$(y = -1 \land x < 0) \lor (y = 0 \land x = 0) \lor (y = 1 \land x > 0).$$

It is graphed in Fig. 5-21.

(m) Find sg (0), sg (1), sg (-1), sg (10), sg (-3).   (n) Argue that
$x \cdot \text{sg}(x) = |x|$.

Step functions. *Step functions* are those whose graphs are made up of
horizontal line segments and, possibly, single isolated points. The signum
function is an example, since its graph consists of two infinite horizontal
segments and the origin. Step functions appear very frequently. For
example the postage charged on books is related to their weight by the
step function graphed partially in Fig. 5-22. Here y is the postage in
cents, and x is the weight in pounds. Note that the right-hand endpoints
are not included in each step.

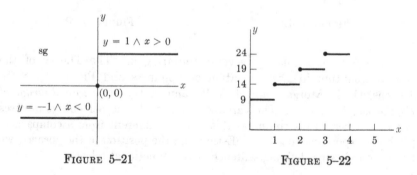

FIGURE 5-21   FIGURE 5-22

(o) What is the domain of the above function? Its range?   (p) Graph the
postage required on a first-class letter as a function of its weight.

A convenient function for dealing with step functions is the *unit step
function* $[I]$ defined by $y = [x]$, where $[x] = $ the largest integer less than
or equal to x. Formally,

(7) **Def.** $[x] = \max \{z \mid z \in J \land z \leq x\}$.

The unit step function is graphed in Fig. 5-23. Note that each segment is
open on the right and closed on the left.

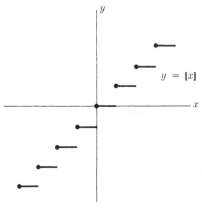

FIGURE 5–23

(q) Find $[1.5]$, $[1.7]$, $[2]$, $[0]$, $[-1.6]$, $[-3]$. (r) Graph $y = [2x]$. (s) Graph $y = 2[x]$. (t) Argue that Fig. 5–22 is the graph of $y = 9 + 5[x]$. (u) How can one tell whether "$[x]$" means the greatest integer less than or equal to x or whether the brackets are used to indicate grouping?

PROBLEMS

1. If $f = mI + b$, $f(x) = $?
2. What is the image of 3 under $3I - 2$?
3. Distinguish between 5 and $\underline{5}$.   4. Describe the graphs of $\underline{0}$, $\underline{1}$, $\underline{-1}$.
5. Show that f defined by "$f(x) = $ the place to which road x led in the days of the Roman empire" is a constant function. What is its domain? its range?
6. What function is both a direct variation and a constant function?
7. Show that if two ordered pairs ($\neq (0, 0)$) belong to the same direct variation, then their components are proportional; that is, $(x_1, y_1) \in mI \wedge (x_2, y_2) \in mI \rightarrow y_1/x_1 = y_2/x_2$.
8. Show that the conclusion of Problem 7 is equivalent to $y_1/y_2 = x_1/x_2$.
9. When in scientific discourse it is said that "y varies directly with x" or that "y is proportional to x," the meaning is that there exists a number m such that $(x, y) \in mI$ whenever x and y are corresponding values of the variables, i.e., that $y = mx$ defines the relation between the quantities involved, and we have a direct variation. For example, for constant speed, distance is proportional to time. Letting $y = $ distance and $x = $ time, $y = mx$ for some m. Here $m = $ the speed. If we know one pair of values, we can calculate m and so determine the function. We call m the *constant of proportionality*. Suppose that we know that when the time is 6 sec the distance is 25 ft. Find the function and determine the distance when time is 10 sec.
10. The revenue from a sale is proportional to the number of units sold. What do we call the constant of proportionality? Find it if 50 units yield a revenue of $225. What revenue would be yielded by 28 units?

11. The simple interest on a fixed deposit is proportional to the time. (a) What is the constant of proportionality called? If interest on $150 for 2 years is $3, (b) find the constant, and (c) determine the interest for 18 months.

12. If response varies directly as the stimulus and response is 25 when stimulus is 4, find the direct variation relating stimulus and response and determine the response when the stimulus is 8.

13. (a) If y varies directly with x, does it follow that y increases when x increases? (b) By how much does y change if x increases one unit and x and y are related by mI?

14. If y varies directly with x, does x vary directly with y?

15. The all-or-none law of response of a neuron to a stimulus is described as follows by Boring in *Foundations of Psychology*. "The magnitude of the activity in any single neural functional unit is as great as it can be in that unit at that time and is independent of the magnitude of the energy exciting it, provided only that the stimulating energy is sufficiently strong to excite the neuron at all." Letting y = the activity of the neuron, x = the energy exciting it, 10 = the minimum energy required for a response, and 20 = the maximum activity, graph the function, and write a defining sentence.

★16. Define sg (x) in terms of $|I|$.

Graph 17 through 19.

★17. $y = |x/2|$.   18. $y = |2x - 1|$.   19. $y = |x| + x$.

20. A car travels for 1 hr at 20 mi/hr, stops for 1/2 hr, then travels for 2 hr at 40 mi/hr. Letting s = the road distance from the starting point, and t = the time elapsed, graph the function relating t and s, considering t as the independent variable. Is t a function of s?

Graph 21 through 23.

★21. $y = [x]^2$.   ★22. $y = [x^2]$.   23. $y = x - [x]$.

24. According to *The Biology of Population Growth*, by R. Pearl, the density d of a fly population is defined by $d = N/v$, where N is the number of flies and v is the volume of effective free space. Experiments indicate that the density of the maximum population varies directly as the volume. Show that the size of the maximum population varies directly as the square of the volume of the space available.

★25. "For constant values of superthreshold reaction potential set up by massed practice, the number of unreinforced responses (n) producible by massed extinction procedure is a linearly decreasing function of the magnitude of the work (W) involved in operating the manipulanda, i.e., $n = 3.25(1.1476 - 0.00984W)$." (*A Behavior System*, by C. L. Hull) Graph the function. How is the "decreasing" character reflected in the formula and graph?

★26. A Finnish geographer who studied the pattern of geographical distribution of industries around the town of Turku in the 18th century found that the value of loads of lumber, rye, and flax varied as the square of the distance they were hauled to market. In particular he found that a load of lumber hauled a distance of 50 km had a value of 7 and a load of rye hauled a distance of

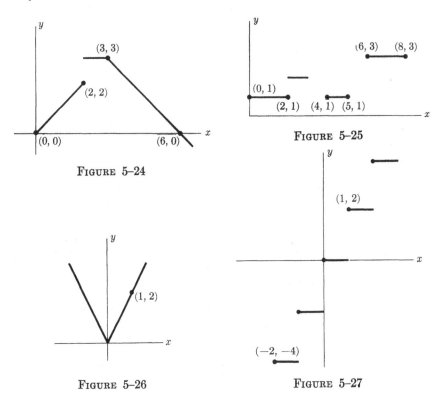

FIGURE 5–24

FIGURE 5–25

FIGURE 5–26

FIGURE 5–27

140 km had a value of 54. Find the constant of proportionality and verify his claim. If the value of a load of flax was 135, how far was it hauled?

★27. If V is the volume of a gas at constant pressure and t is the corresponding temperature, then $V = V_0(1 + b(t - t_0))$. Show that this defines a linear function. Describe V_0. Describe b. Sketch.

ANSWERS TO EXERCISES

(a) Set of ordered pairs whose second components are three more than twice their first; graph is a straight line with slope 2 and intercept $(0, 3)$.   (b) Set of pairs all of whose second components are 3.   (c) Lines through origin with slopes of -1, 2, -3.   (d) See Fig. 5–24.   (e) See Fig. 5–25.   (f) $f(x) = x$ for $0 \leq x \leq 2$, $f(x) = 3$ for $2 < x \leq 3$, and $f(x) = -x + 6$ for $x \geq 3$.   (g) $(y = 1 \wedge 0 \leq x \leq 2) \vee (y = 2 \wedge 2 < x < 3) \vee (y = 1 \wedge 4 \leq x \leq 5) \vee (y = 3 \wedge 6 \leq x \leq 8)$.   (h) (4–5–4) with the points $(2.5, 0)$ and $(4.5, 100)$.   (i) $y \doteq 42x - 88$.   (j) $x \geq 0$; $x < 0$.   (k) See Fig. 5–26.   (l) From (5) $I(x) = x$.   (m) 0, 1, -1, 1, -1.   (n) Cases.   (o) Positive reals. $\{9, 14, 19, \ldots\}$ (p) Like Fig. 5–22 but with first step at $y = 4$ and risers of height 4.   (q) 1, 1, 2, 0, -2, -3.   (r) Similar to Fig. 5–27 but with key points at $(-1, -2)$, $(-1/2, -1)$, $(0, 0)$, $(1/2, 1)$, etc.   (s) See Fig. 5–27.   (t) Check cases.   (u) Context.

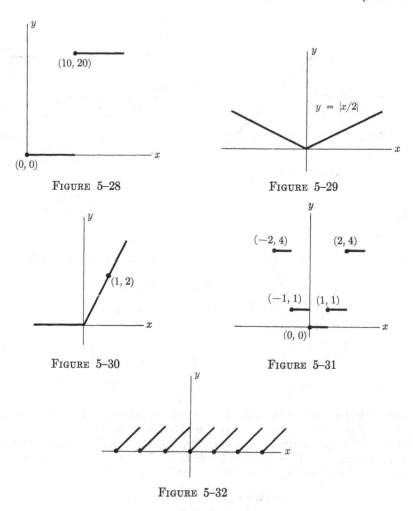

FIGURE 5–28

FIGURE 5–29

FIGURE 5–30

FIGURE 5–31

FIGURE 5–32

ANSWERS TO PROBLEMS

1. $mx + b$. 3. A real number; the set of all ordered pairs whose first component is a real number and whose second component is 5. 5. Since all roads led to Rome, the image of x is always Rome. Domain is set of all roads at that time, range is the singleton {Rome}. 7. $y_1 = mx_1 \wedge y_2 = mx_2 \rightarrow y_1/x_1 = m \wedge y_2/x_2 = m$. 9. $25 = 6m$. Hence $m = 25/6$ and the function is $(25/6)I$. When $x = 10$, $y = (25/6)(10) = 250/6 = 125/3$. 11. (a) The interest per period, (b) \$1.50, (c) \$2.25. 13. (a) No, only if $m > 0$, (b) m units. 15. See Fig. 5–28. 17. See Fig. 5–29. 19. See Fig. 5–30. 21. See Fig. 5–31. 22. Not the same as Problem 21! 23. See Fig. 5–32. 25. Negative slope. 27. V_0 is the image of t_0; b is the change in volume per unit of original volume per unit change in temperature.

5–8 Quadratic functions. A function with defining formula of the form $ax^2 + bx + c$ with $a \neq 0$ is called a *quadratic function*. We define

(1) **Def.** $aI^2 + bI + c = \{(x, y) \mid y = ax^2 + bx + c \wedge x \in Re\}$.

Consider, for example, the function $2I^2 - 3I - 4$ defined by $y = 2x^2 - 3x - 4$. By making a table for a few integral values of x and drawing a smooth curve through the corresponding points, we find the graph of Fig. 5–33. The graph appears to extend indefinitely upward but to have a lowest point. By completing the square, as in Section 4–8, we find

(2) $$2x^2 - 3x - 4 = 2(x - 3/4)^2 - 41/8.$$

Since the first term on the right is never negative and is zero for $x = 3/4$, we see that the smallest value of y is $-41/8$ and that the lowest point on the graph is $(3/4, -41/8)$. This means that the range of $2x^2 - 3x - 4$ is the interval $(-41/8 \ \infty)$.

More generally,

(3) $$ax^2 + bx + c = a(x + b/2a)^2 + (4ac - b^2)/4a.$$

If $a > 0$, the first term is never negative and is zero for $x = -b/2a$.

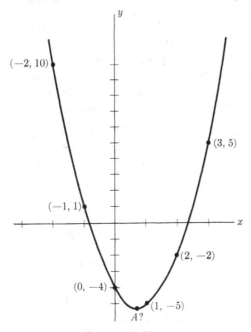

FIGURE 5–33

Hence the minimum is $(4ac - b^2)/4a$. On the other hand, if $a < 0$, the first term is never positive and we have a maximum at the point

$$(-b/2a, (4ac - b^2)/4a).$$

(a) Sketch $-2I^2 - 3I + 4$ by plotting selected points. Find the highest point on the graph by completing the square. Find the range of the function.

We can summarize the above results as follows:

(4) $f = aI^2 + bI + c \land a > 0$
$$\to \min \text{Rge}(f) = f(-b/2a) = (4ac - b^2)/4a,$$

(5) $f = aI^2 + bI + c \land a < 0$
$$\to \max \text{Rge}(f) = f(-b/2a) = (4ac - b^2)/4a.$$

We call the high or low point the *vertex* of the curve. Evidently in every case the vertex is the point corresponding to $x = -b/2a$, a result that is easy to remember.

(b) What is the vertex of the graph in Exercise (a)?

Let us imagine a company producing and selling x units of output at a cost C given by $C = x^2 + 100x + 100$. Suppose that sales x and price p are related by $x = -p + 200$, that is, $p = 200 - x$, so that revenue R from sales is given by $R = px = 200x - x^2$. Then profit y is given by $y = R - C = -2x^2 + 100x - 100$. This equation defines the function $-2I^2 + 100I - 100$. For what output does the company maximize its profit y? Using $(5)(a{:}-2, b{:}100, c{:}-100)$, we find that for an output of 25 the company has a maximum profit.

(c) Plot the profit function.   (d) The path of a projectile fired from the origin in the xy-plane is given by $y = 2x - x^2$. How high does it rise and where does it again reach the level from which it was fired?

Looking again at Fig. 5–33, we observe that we have not yet found the points where the graph crosses the x-axis. These, of course, are the points where $y = 0$, that is, where $ax^2 + bx + c = 0$. They are easily found by (2-8-21) to be $-b/2a - \sqrt{b^2 - 4ac}/2a$ and $-b/2a + \sqrt{b^2 - 4ac}/2a$.

(e) Find the coordinates of these points in Fig. 5–33.

The above results show that the two x-intercepts of a quadratic function lie at equal distances to the right and left of the line $x = -b/2a$. The appearance of Fig. 5–33 suggests that if we draw any horizontal line meeting the curve, its points of intersection will be equidistant from the line

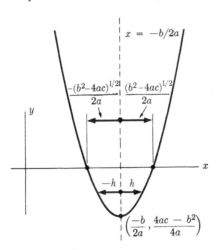

FIGURE 5–34   FIGURE 5–35

$x = -b/2a$. To prove this let $f(x) = ax^2 + bx + c$. Then $f(-b/2a + h) =$ $a(-b/2a + h)^2 + b(-b/2a + h) + c = a(b^2/4a^2 - bh/a + h^2) +$ $b(-b/2a + h) + c = b^2/4a - bh + ah^2 - b^2/2a + bh + c = ah^2 -$ $b^2/4a + c$. Similarly, $f(-b/2a - h) = ah^2 - b^2/4a + c$.

This means that if we move h units to the right from the vertex we get the same y-coordinate for a point on the curve as if we moved h units to the left. The situation is sketched in Fig. 5–34. We describe this property by saying that the curve is *symmetric* about the line $x = -b/2a$, and we call the line $x = -b/2a$ the *axis of symmetry*. (Symmetry is discussed systematically in Sections 5–14 and 5–15.)

(f) Sketch $I^2 - 2I + 2$ by finding the vertex, axis of symmetry, and a few other points.

From the discussion in Section 2–8 it is evident that the graph of the quadratic function may fail to meet the x-axis at all. Indeed, it will meet the x-axis if and only if the discriminant $b^2 - 4ac$ is non-negative. If $a > 0$, we have the three cases sketched in Fig. 5–35. When the discriminant $b^2 - 4ac$ is positive, we have two points of intersection; when it is zero, the graph is tangent to the x-axis; when it is negative, the graph lies entirely above the x-axis.

(g) Discuss the case $a < 0$.   (h) Obtain results found in the above paragraph and in Exercise (g) from (4) and (5) without using results concerning the roots of quadratic equations.

In Section 3–5 we gave some methods of solving quadratic inequalities. The results of the present section make it easy to solve any inequality of the form $ax^2 + bx + c < (>, \leq, \geq) \, 0$. We simply sketch the function

$aI^2 + bI + c$ and observe where it is above or below the x-axis. The solution is either an interval between the roots of $ax^2 + bx + c = 0$ or the complement of this interval. If there are no real roots, the inequality is satisfied by all real numbers or none.

Verify the previous comments by solving the following inequalities:
(i) $2x^2 + 3x - 1 < 0$,   (j) $2x^2 + 3x - 1 > 0$,   (k) $x^2 - 4x + 5 \geq 0$,
(l) $3x^2 - x + 1 \leq 0$.

Since we are often concerned with the maximum and minimum values of the dependent variable and with the values of the independent variable for which the dependent variable is zero, it is convenient to have some special terminology. We call max Rge (f) the *maximum of the function* f and min Rge (f) the *minimum of the function* f. If $f(x_0) = $ max Rge (f), we say that f takes its maximum at x_0. Similarly, if $f(x_0) = $ min Rge (f), we say that f takes its minimum at x_0. If $f(x_0) = 0$, we call x_0 a *zero of the function*. A zero of a function f is thus simply a root of the equation $f(x) = 0$.

(m) Give an example of a function that has neither a maximum nor a minimum.   (n) Give an example of one that has a maximum but no minimum. (o) Assuming that a, b, and c are real, write a necessary and sufficient condition that the zeros of $aI^2 + bI + c$ are imaginary,   (p) equal,   (q) unequal.

PROBLEMS

Graph the functions in Problems 1 through 6. Show the axis of symmetry, vertex, and intercepts. Find the zeros, maxima, and minima. For each one find the subset of the domain for which the images are positive and the subset for which they are negative.

1. $I^2 - I$.   2. $I^2 + 3$.   3. $2I^2 + 3I - 1$.
4. $9I^2 + 6I + 1$.   5. $9I^2 + 6I - 1$.   6. $-3I^2 + 2I + 2$.

7. Solve for m: $10m^2 - 136m\sqrt{N}/R + N/R^2 = 0$.
8. Solve for α: $\alpha(\alpha - 1) + p_0\alpha + q_0 = 0$.
★9. Find a quadratic function whose zeros are the squares of the zeros of $aI^2 + bI + c$.
★10. The following formula appears in *Fertility and Reproduction*, by R. R. Kuczynski:

$$T = (1/2)[R_1/R_0 + \sqrt{(R_1/R_0)^2 - 2(R_2/R_0 - (R_1/R_0)^2) \log_e R_0}].$$

Do you think that this formula was derived by solving a quadratic equation? Why or why not? If so, what is the other root? Write an equation of which this is a root.

11. Solve for r and for p: $(r - p)^2 = p(1 - p)/b$.
12. For what values of k will the graph of the function $kI^2 + 3kI - 4$ lie entirely below the x-axis?

★13. A company has a cost function given by $K = Ax^2 + Bx + C$ and a demand function given by $x = ap + b$. Find the output and price for maximum profit. Argue that $A > 0$ and $a < 0$, and show that the result must actually be a maximum and not a minimum.

14. When a projectile is launched with muzzle velocity of v_0 in a direction making an angle α with the horizontal, its path is given by $x = (v_0 \cos \alpha)t$, $y = (v_0 \sin \alpha)t - (1/2)gt^2$, where t is the time. What is the highest point on its path? Where does it strike the ground? Show that its path is the graph of a quadratic function and find this function. For what angle α will it go the farthest?

Graph 15 and 16.

★15. $y = x^2 + x + 1 \wedge x \in (\underline{-2 \quad 3})$.

★16. $y \geq 2x^2 - 3x - 1 \wedge y \leq 2 \wedge -1 \leq x \leq 1$.

Answers to Exercises

(a) Range is $(-\infty \quad \underline{41/8})$.   (b) $(-3/4, 41/8)$.   (d) To $(1, 1)$ and at $(2, 0)$.
(e) $(-0.85, 0)$, $(2.3, 0)$.   (f) Vertex is $(1, 1)$, axis of symmetry is the line $x = 1$.   (g) Same, except that all curves open downward and curve is entirely below x-axis if discriminant is negative.   (h) The sign of the minimum or maximum of the range determines whether the curve crosses the axis; it does so if this sign is opposite to that of a.   (i) Open interval.   (j) Union of two infinite intervals.   (k) $(-\infty \quad \underline{\quad} \infty)$.   (l) Null set.   (m) Any nonconstant linear function.   (n) Quadratic with $a < 0$.   (o) Discriminant negative,   (p) zero, (q) nonzero.

Answers to Problems

1. Positive on $(-\infty \underline{\quad} 0) \cup (1 \underline{\quad} \infty)$.   7. $(68 \pm \sqrt{4614})\sqrt{N}/10R$.   9. Check by substitution.   13. $x = (b + aB)/2(1 - aA)$ (see *Mathematical Introduction to Economics*, by G. C. Evans, for more problems of this kind).

5–9 Composition. Suppose a ship P is sailing due east on a line 50 mi north of an observation point O. We are interested in the distance OP in Fig. 5–36, where x is the distance to the ship from the point B on its course nearest to O. Clearly $y = \sqrt{50^2 + x^2}$. If the ship is maintaining

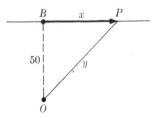

Figure 5–36

a speed of 20 mi/hr, and if we measure time t beginning at the point B, we have $x = 20t$. Then $y = \sqrt{50^2 + (20t)^2}$. Here we have three functions, one relating t and x, one relating x and y, and one relating t and y. We say that the third is the composite of the first two. To find the image of t under the composite, we find the image x of t under the first function and then the image y of x under the second function; that is, we go from t to x and then from x to y.

Again, suppose that y is the father of z and z is the father of x. We have $z = \text{Fa}\,(x)$ and $y = \text{Fa}\,(z)$. Then $y = \text{Fa}\,[\text{Fa}\,(x)]$. We call the function defined by this last equation the composite of Fa and Fa. Suppose that z is a parent of x and that y is a parent of z; that is, x Pa z and z Pa y. Then y is a grandparent of x, or x Gp y. We call Gp the composite of Pa and Pa. Now suppose we are given x and y and wish to show that x Gp y. We should have to show that there was some individual who was a parent of x and a child of y; that is, $\exists z\, x$ Pa $z \wedge z$ Pa y.

We use this idea to define the *composite* $\rho \circ \sigma$ of the relations ρ and σ as follows:

(1) **Def.** $\rho \circ \sigma = \{(x, y)\,|\,\exists z\, x\, \rho\, z \wedge z\, \sigma\, y\},$

(2) $x\, (\rho \circ \sigma)\, y \leftrightarrow \exists z\, x\, \rho\, z \wedge z\, \sigma\, y,$

(3) $(x\, \rho\, z \wedge z\, \sigma\, y) \rightarrow x\, (\rho \circ \sigma)\, y.$

From (3) we see that to find the images of x under $\rho \circ \sigma$, we find the images of x under ρ and then the images of these images under σ. For example, to find the grandparents of x we find the parents of x and then the parents of these. When the relations are functions, it is easy to find a defining formula for the composite from those of the two functions. If F and G are functions, (2) becomes $y = (F \circ G)(x) \leftrightarrow \exists z\, z = F(x) \wedge y = G(z) \leftrightarrow y = G[F(x)]$. The last equation says simply that the image under $(F \circ G)$ is found by finding the image under F, namely $F(x)$, and then the image of this under G, namely $G[F(x)]$. This was exactly the procedure we followed in the example at the beginning of the section, where we have $F(t) = 20t$, $G(x) = \sqrt{50^2 + x^2}$, and $G[F(t)] = \sqrt{50^2 + (20t)^2}$. We have, then,

(4) $(F \circ G)(x) = G[F(x)].$

Note the reversal of order. If we think in terms of operations, $F \circ G$ means the operation F followed by the operation G, and so does $G[F(x)]$. It seems natural to use $G[F]$ to stand for the composite of F and G, when F and G are functions. Hence we adopt the notation,

(5) **Def.** $G[F] = F \circ G.$

The notation on the left could be used also for nonfunctional relations, but its convenience comes from its similarity to $G[F(x)]$. With this notation, (4) becomes

(4') $$(G[F])(x) = G[F(x)].$$

(a) Suppose $y = z^2$ and $z = 2x$. What is the relation between x and y? Find a defining formula of: (b) $(2I) \circ (3I)$, (c) $(I^2) \circ (2I + 1)$, (d) $(2I + 1) \circ I^2$. (e) What is the meaning of $y = \text{Mo} (\text{Fa} (x))$? (f) If $F(x) = 1 - x$ and $G(x) = 3x^2$, find $G[F(x)]$ and $F[G(x)]$.

We see from Exercise (f) that composition of relations is not commutative. However, a little reflection shows that it is associative; that is,

(6) $$\rho \circ (\sigma \circ \tau) = (\rho \circ \sigma) \circ \tau.$$

Indeed, the images of x under the left side are found by determining the images under ρ, then the images of these under $\sigma \circ \tau$, that is, the images of these under σ and then the images of these under τ. But the right side calls for the same operations.

Since composition is associative, it is natural to ask whether we can derive other results analogous to the laws of the algebra of numbers. For example, is there a relation that plays a role like that of the number 1 under multiplication and has the property similar to that given in the law $1 \cdot a = a \cdot 1 = a$? We define

(7) **Def.** $$I_A = \{(x, y) \,|\, x = y\} \cap (A \times A).$$

We call I_A the *identity relation on* A. It is evidently the set of all ordered pairs with identical members belonging to A. Then,

(8) $$\rho \subseteq A \times A \rightarrow (I_A \circ \rho = \rho \circ I_A = \rho),$$

(9) $$\rho \subseteq A \times B \rightarrow (I_A \circ \rho = \rho),$$

(10) $$\rho \subseteq A \times B \rightarrow (\rho \circ I_B = \rho).$$

Let ρ and σ be defined by Fig. 5–8. Graph: (g) $\rho \circ \sigma$, (h) $\sigma \circ \rho$, (i) I_A where $A = \{0, 1, 2, 3, 4, 5, 6\}$. (j) Verify (8) for these relations. (k) What is the relation between I_A and I defined by (5–7–5)?

The above results are fairly obvious, since I_A maps any x into itself. Hence $I_A \circ \rho$ maps x into x and then into images of x under ρ, and $\rho \circ I_A$ maps each x into its images under ρ and then these into themselves.

When ρ is a function, (8) becomes

(11) $F \subseteq A \times A \rightarrow (F[I_A] = I_A[F] = F)$.

PROBLEMS

In Problems 1 through 12, defining formulas for F and G are given in that order. Find a defining formula for $G[F]$.

1. $2x$, $5x$.
2. $2x$, $3x + 1$.
3. x^2, $x + 1$.
4. $x + 1$, x^2.
5. $1/x$, $1/x$.
6. $3 + x$, $3 - x$.
7. x^2, x^2.
8. 2, $5x$.
9. $5x$, 2.
10. $-x$, $--x$.
11. $\sin x$, $2x$.
12. $3x$, $\cos x$.

Determine 13 through 18 by using defining formulas of the functions.

13. $(2I) \circ (5I)$.
14. $(I^2 + 2I) \circ (3I)$.
15. $4I \circ \sin$.
16. $5 \circ I^2$.
17. $(2I + 3) \circ (-I + 1)$.
18. $(mI + b) \circ (m'I + b')$.

★19. Suppose that a point moves so that the vector to it from a fixed point moves through an angle of ω units per unit time. Then we call ω (omega) the *angular velocity* about the fixed point. If the point moves in a circle of radius r about the fixed point, it covers an arc of $r\omega$ per unit time. We call $r\omega$ its *linear velocity*. Suppose the angular velocity varies with time, so that $\omega = 3t$. Find the function that maps the time into the linear velocity if the radius is 25 ft.

★20. The radius of a circle is increasing at the rate of 100 yd/min. Find the function that gives the area of the circle in terms of time t if the radius is zero when $t = 0$.

★21. Show that the composite of two linear functions is linear.

★22. Show that composition of linear functions is not commutative.

★23. Show that the composite of a linear and quadratic function in either order is a quadratic function.

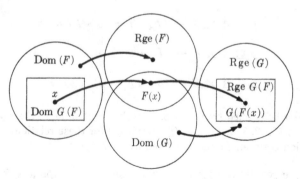

FIGURE 5-37

★24. With the aid of Fig. 5–37, justify

(12)   Dom $(F \circ G) \subseteq$ Dom (F),

(13)   Rge $(F \circ G) \subseteq$ Rge (G).

★25. Draw a figure similar to Fig. 5–13 to suggest the idea of composition.
★26. Find the domain and the range of Hu, Mo, and Hu ∘ Mo. Verify (12) and (13). Note how Rge (Hu) and Dom (Mo) overlap.
★27. Of what set is Rge $(F \circ G)$ the map?
★28. Repeat Problem 26 for Mo ∘ Hu.
★29. Do the same for $F = \{(x, y) \,|\, y = x^2 \wedge x \in J\}$, $G = \{(x, y) \,|\, y = 2x \wedge x$ is odd$\}$.
★30. Translate into words: $x(\text{Pa} \circ \text{Mo})y$; $x(\text{Mo} \circ \text{Pa})\,y$; $x(\text{So} \circ \text{Da}) \circ \text{So}\,y$; $x(\text{Wi} \circ \text{Hu}) \circ \text{Mo}\,y$.
★31. Identify: $> \circ >$; $= \circ =$; $\neq \circ \neq$.
★32. Is it possible for the composite of two nonfunctional relations to be a function?

Answers to Exercises

(a) It is defined by $y = 4x^2$; that is, it is $4I^2$.   (b) $6x$.   (c) $2x^2 + 1$.
(d) $4x^2 + 4x + 1$.   (e) y is the mother of the father of x.   (f) $3(1 - x)^2$ and $1 - 3x^2$.   (g) through (i) See Fig. 5–38.   (k) $I = I_{Re}$, where $Re =$ the real numbers.

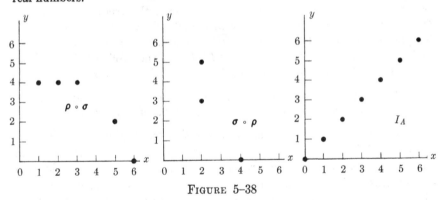

FIGURE 5–38

Answers to Problems

1. $10x$. 3. $x^2 + 1$. 5. x. 7. x^4. 9. 2. 11. $2 \sin x$. 13. $10I$. 15. $y = \sin 4x$.
17. $-2I - 2$. 18. $m'mI + m'b + b'$. 19. Letting $y =$ the linear velocity, $y = 25\omega$ and $\omega = 3t$; hence $y = 75t$ and the function is $75I$. 21. See Problem 18. 23. $(mI + b) \circ (AI^2 + BI + C) = Am^2I^2 + (2Amb + mB)I + Ab^2 + Bb + C$; $(AI^2 + BI + C) \circ (mI + b) = mAI^2 + mBI + mC + b$. 25. Two machines, one feeding into the other. 27. Rge $(F) \cap$ Dom (G).

5-10 Converse relations. To introduce the concept of the converse of a relation, we ask: if x is greater than y, so that $(x, y) \in >$, what is the relation that y bears to x? Obviously y is less than x, so that $(y, x) \in <$. We call each of $>$ and $<$ the converse of the other. The *converse* ρ^* of a relation ρ is the relation such that $y \,\rho^*\, x$ if and only if $x \,\rho\, y$; that is,

(1) Def. $\rho^* = \{(x, y) \mid (y, x) \in \rho\}$,

(2) $[(b, a) \in \rho^*] \leftrightarrow [(a, b) \in \rho]$,

(3) $(b \,\rho^*\, a) \leftrightarrow (a \,\rho\, b)$.

These laws mean that the converse of a relation is obtained by interchanging components in each pair belonging to the relation.

(a) Graph ρ^* and σ^*, where ρ and σ are defined by Fig. 5-8. (b) Graph $\{(1, 1), (3, 1), (-2, -5), (1, -1)\}$ and its converse on the same drawing, using circles for points belonging to the original relation and crosses for those belonging to the converse. With a straight line connect each point with the corresponding point in the graph of the converse.

As suggested by the solution to Exercise (b), the points (a, b) and (b, a) are always placed so that the graph of I (the line defined by $y = x$) is the perpendicular bisector of the segment joining them. This can easily be proved by reference to Fig. 5-39. Indeed, the triangles OAB and $OA'B'$ are congruent because they are right triangles whose sides have lengths $|a|$ and $|b|$. Hence $\overline{OB} = \overline{OB'}$. Also, since $\angle AOB = \angle A'OB'$ and $\angle AOC = \angle A'OC$, $\angle BOC = \angle B'OC$. It then follows that $\overline{B'C} = \overline{BC}$ and that I is perpendicular to BB', as we wished to prove.

(c) Justify the last statement.   (d) Make a new drawing and check the proof for a point in the second quadrant.   (e) What if (a, b) lies in the line I?

From the above it follows that the cartesian graph of ρ^* is obtained by "reflecting" the graph of ρ in the line defined by $y = x$. We use the word "reflecting," because an object reflected in a mirror appears to be located on the other side of the mirror a distance equal to its distance in front of the mirror, just as (b, a) is located in relation to the line $y = x$ and the point (a, b). Another way of describing the relation is to say that the graph of ρ^* may be obtained by rotating the graph of ρ up out of the plane around the line $y = x$ through $180°$.

Use the above idea to graph the converse of: (f) $2I$,   (g) I^2.

(4) $\{(x, y) \mid s(x, y)\}^* = \{(x, y) \mid s(y, x)\}$.

This law, which follows immediately from (2), indicates explicitly that a defining sentence of the converse of a relation is obtained by inter-

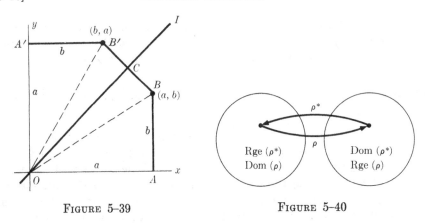

FIGURE 5–39   FIGURE 5–40

changing the variables in a defining sentence of the original relation. For example, $2I$ is defined by $y = 2x$. Hence $(2I)*$ is defined by $x = 2y$ or $y = x/2$. Hence $(2I)* = (I/2)$, which is confirmed by Exercise (f). Similarly, I^2 is defined by $y = x^2$. Hence $(I^2)*$ is defined by $y^2 = x$.

(h) What is $(2I + 1)*$? Find a defining formula and graph $2I + 1$ and its converse.

It may be helpful to visualize a relation and its converse as mappings. A relation ρ maps each element in its domain into the corresponding images in its range. The converse $\rho*$ maps each element in its domain (the range of ρ) into its images under $\rho*$. The images of x under $\rho*$ are precisely the objects one of whose images under ρ is x. This is suggested in Fig. 5–40, where, for simplicity, we indicate only one image under both relation and converse.

(i) Make a graph similar to Fig. 5–16 of the converse of the function graphed there.   (j) Formulate a rule for making this kind of a representation of a converse.

We see from Exercise (g) that the converse of a function may not be a function. However, if both F and $F*$ are functions, each establishes a one-to-one correspondence. When F and $F*$ are functions we call each the *inverse* of the other. The following hold for inverses.

(5) $F* = \{(x, y) \mid x = F(y)\}$,

(6) $F*(x) = [y \ni x = F(y)]$,

(7) $[y = F*(x)] \leftrightarrow [x = F(y)]$,

(8) $F*[F(x)] = x$,

(9) $F[F*(y)] = y$.

Theorem (5) is simply (4) applied to a relation with a defining sentence of the form $y = F(x)$. The others follow from the discussion in Section 5–4. In (9) we use y to remind the reader that the range of the variable is the domain of $F*$, i.e., the range of F.

(k) Is $F*[F(x)] = F[F*(x)] = x$ a law?   (l) Argue for (8) and (9) in terms of mappings.

We now have procedures for investigating the converse of a known relation. To find the graph, we rotate about the line $y = x$. To find a defining sentence, we interchange variables. If the converse of a function defined by $y = F(x)$ is also a function, we can find its defining formula by solving $x = F(y)$ for y so as to get $y = F*(x)$. For example, the inverse of $2I + 3$, defined by $y = 2x + 3$, is obtained by solving $x = 2y + 3$ for y. We find $y = x/2 - 3/2$, so that $(2I + 3)* = I/2 - 3/2$.

(m) Find $(mI + b)*$.   (n) Is there any linear function whose converse is not a function?   (o) Describe $mI + b$ and its inverse as operations. (p) Which linear functions are their own inverses?   (q) Graph the relation defined by $y = \sqrt{x}$, and graph its converse.

PROBLEMS

In Problems 1 through 12 graph the relation indicated by a defining sentence or name. Find a defining sentence for the converse, and a defining formula if the converse is a function.

1. $y = -x + 1$.   2. $-2I - 5$.   3. $y = 1 - x^2$.
4. $y = x^2$.   5. $y = x^2 \wedge x \geq 0$.   6. $y = x^2 \wedge x < 0$.
7. $y = x^3$.   8. $y = -x$.   9. $x^2 + y^2 = 4$.
10. $y = 1/x$.   11. $I^2 - 1$.   12. $I^2 + 2I$.

For Problems 13 through 16 verify (8) and (9).

13. $F = -I - 7$.   14. $F = 10I + 13$.
15. $F = \{(x, y) \,|\, y = x^2 \wedge x \geq 1\}$.   16. $F = \{(1, 1), (2, 7), (3, -1)\}$.

Identify the relations in 17 through 26.

17. $<*$.   18. $>*$.   19. $\leq*$.   20. $\geq*$ 21. $I*$.
22. $(-I)*$.   ★23. Hu*.   ★24. Wi*.   ★25. Sb*.   ★26. Br*.

27. Find a defining sentence for the converse of the function $AI^2 + BI + C$. Express this converse as a union of two functions.

28. Graph $2I^2 + I - 1$ and $(2I^2 + I - 1)*$ on the same drawing.

★29. Defining x Ch $y =_d y$ is a child of x, define Ch as the converse of an already defined relation.

★30. Derive the following from the definition of the converse.

(10)   [Dom $(R*)$ = Rge (R)] $\wedge$ [Rge $(R*)$ = Dom (R)],

(11) $(R*)* = R,$

(12) $(R \cup S)* = R* \cup S*,$

(13) $(R \cap S)* = R* \cap S*,$

(14) $R'* = R*',$

(15) $I* = I.$

31. Argue for the following law in terms of images and mappings:

(16) $(\rho \circ \sigma)* = \sigma* \circ \rho*.$

★32. Noting that (y is a parent of the wife of x) = (x is the husband of a child of y), illustrate (16) for various familial relations.

33. Argue for (16) in terms of operations.

34. Verify (16) for $\rho = 3I + 2$ and $\sigma = -I + 5$.

★35. Prove (16).

★36. Show that $\rho \subseteq \sigma \rightarrow \rho* \subseteq \sigma*.$

★37. What happens if we take the composite of a relation and its converse? Show the following and illustrate them for familial relations.

(17) $x \,(\rho* \circ \rho)\, y \leftrightarrow \exists z\, z\, \rho\, x \wedge z\, \rho\, y,$

(18) $x \,(\rho \circ \rho*)\, y \leftrightarrow \exists z\, x\, \rho\, z \wedge y\, \rho\, z.$

★38. Show that, if F and $F*$ are functions,

(19) $F*[F] = I_{\text{Dom}(F)},$

(20) $F[F*] = I_{\text{Rge}(F)}.$

★39. Illustrate (19) and (20) with familial relations.

★40. In the sentence $x \in A$, what is the subject? the verb? the object? When names of relations are used as verbs, what is the grammatical role of a name of the converse? the complement? Give the converse and complement of "dominates."

ANSWERS TO EXERCISES

(a) Check by later discussion.   (b) $\{(1, 1),\ (1, 3),\ (-5,\ -2),\ (-1,\ 1)\}.$ (c) OC is bisector of vertex angle of an isosceles triangle; hence it is also a median and an altitude.   (e) Then $(a, b) = (b, a)$, and the segment reduces to a point.   (f) See Fig. 5–41.   (g) See Fig. 5–42.   (h) $(1/2)I - 1/2.$ (i) Reverse arrows.   (j) See Exercise (i).   (k) Yes, but the range of x is Dom $(F) \cap$ Dom $(F*)$.   (l) F carries x into its image, and $F*$ carries this image back into x.   (m) $(1/m)(I - b).$ (n) The constant functions.

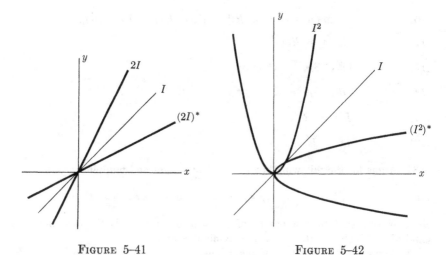

FIGURE 5–41 FIGURE 5–42

(o) $mI + b$ multiplies by m and adds b; $(mI + b)*$ subtracts b and divides by m. (p) I and $-I + b$. (q) The graph of the inverse is just the right half of the graph of I^2.

ANSWERS TO PROBLEMS

1. $y = -x + 1$ (own inverse); 3. $y^2 = 1 - x$. 5. $y = \sqrt{x}$. 7. $y = \sqrt[3]{x}$. 9. $x^2 + y^2 = 4$ (own inverse). 11. $y^2 = x + 1$. 13. $F* = -I - 7$; $F*(F(x)) = -(-x - 7) - 7 = x$. 15. $F* = \{(x, y) \mid y = \sqrt{x} \wedge x \geq 1\}$; $F*(F(x)) = \sqrt{x^2} = x$; $F(F*(x)) = (\sqrt{x})^2 = x$. Note that without some limitation on x, the inverse of the function defined by $y = x^2$ is not a function. 17. $>$. 19. $\geq$. 21. I. 23. Wi. 25. Sb. 27. $Ay^2 + By + C = x$.

$$(AI^2 + BI + C)* = \left\{(x, y) \,\middle|\, y = \frac{-B + \sqrt{B^2 - 4A(C - x)}}{2A}\right\}$$

$$\cup \left\{(x, y) \,\middle|\, y = \frac{-B - \sqrt{B^2 - 4A(C - x)}}{2A}\right\}.$$

29. Ch = Pa*. 33. We perform ρ then σ in $\rho \circ \sigma$. To undo this, we start by undoing the last operation, i.e., by performing $\sigma*$ then $\rho*$. 35. $y (\rho \circ \sigma)* x \leftrightarrow x \rho \circ \sigma y \leftrightarrow \exists z\, x\, \rho\, z \wedge z\, \sigma\, y \leftrightarrow \exists z\, z\, \rho*\, x \wedge y\, \sigma*\, z \leftrightarrow \exists z\, y\, \sigma*\, z \wedge z\, \rho*\, x \leftrightarrow y\, \sigma* \circ \rho*\, x$.

37. $y(\text{Pa}* \circ \text{Pa})x \rightarrow y$ and x are parents of the same child. 39. $x(\text{Hu} \circ \text{Hu}*)y = x(\text{Hu} \circ \text{Wi})y \rightarrow (y = x)$ in our society. 40. x is the subject, $\in$ is the verb, A is the object. The converse is represented by the passive if the original verb was in active form, or the active if the original was in passive form. The complement is represented by a negated verb. (dominates)* = is dominated by; dominates$'$ = does not dominate.

5–11 Circular functions. Figure 5–43 shows the monthly load y of active cases of family welfare agencies in a certain city in relation to time x during the years 1926 to 1929, where 100 = the average load in 1927. The function relating y and x has the property of practically repeating itself in successive intervals. A function f is said to be *periodic* if there exists a positive number T such that $f(x + T) = f(x)$ for all x in Dom (f). Any such number is called *a period*, and the smallest period (if it exists) is called *the period*.

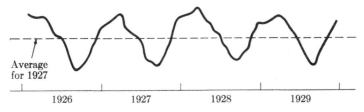

Average
for 1927

| 1926 | 1927 | 1928 | 1929 |

FIG. 5–43. Adapted from *Statistics in Social Studies*, by S. A. Rice, 1930, p. 52.

(a) State the definition of "periodic," using quantifiers.   (b) Cite examples of periodic functions.   (c) Show that if T is a period, so is nT for $n \in J^+$. (d) Why "positive number" in the definition?   (e) What approximately is the period of the function of Fig. 5–43?

Because of the cyclic, repetitive character of many natural and social phenomena (orbits of planets and atomic particles, motions in machines, biological cycles, waves, economic fluctuations, etc.), periodic functions are very common. Perhaps the most useful periodic functions, and the only ones we consider here, are the *circular functions* sin, cos, tan, cot, sec, and csc. For example, sin is defined by $y = \sin x$, and is periodic according to (4–12–19). Here we use x instead of θ as the independent variable. The geometric picture of the relation between x and $\sin x$ is suggested in Fig. 5–44, where x is the directed arc length from (1, 0) measured along the unit circle. The graph of sin in Fig. 5–45 may be obtained by

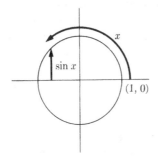

x

$\sin x$

$(1, 0)$

FIGURE 5–44

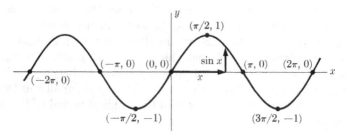

FIGURE 5–45

measuring directed arcs x along the horizontal axis and erecting verticals of directed length sin x measured on Fig. 5–44, that is, by "unwrapping" Fig. 5–44.

(f) What are the domain and range of sin? What is the period? (g) Review Sections 4–11, 4–12, and 4–13. Why do you think we use the term "circular" here? (h) From (4–12–9) prove $\forall x \cos x = \sin(x + \pi/2)$.

Because of the identity proved in Exercise (h), the graph of cos may be obtained by shifting the graph of sin a distance of $\pi/2$ toward the left. (i) Explain this. (j) Make a large graph of cos. What is the period of cos?

A comparison of Figures 5–43 and 5–45 suggests that the latter could be made to "fit" the former by only moderate distortion. Consider $A \sin[mI + b]$ defined by

(1) $$y = A \sin(mx + b).$$

Its domain is the real numbers, as for the sin, but its range is $(-|A| \quad |A|)$. We call $|A|$ the *amplitude* of the function. Since $\sin(m(x + 2\pi/m) + b) = \sin(mx + b + 2\pi) = \sin(mx + b)$, a period is $2\pi/m$. Since this is clearly the smallest positive number with this property, $2\pi/m$ is the *period*. If we think of x as the time, the period is the time required to repeat. If we take its reciprocal (divide one unit of time by the time per repetition), we get the *frequency* f defined by $f = 1/T$. For (1) the frequency is $m/2\pi$.

To see the significance of b, let us compare (1) with the function defined by $y = A \sin mx$, where $b = 0$. Noting that

$$\sin(mx + b) = \sin m(x + b/m),$$

we see that $\sin[mI]$ and $\sin[mI + b]$ have the same range and domain, but that the graph of the latter may be obtained by shifting that of the former a distance b/m to the left. The number b/m is called the *phase*.

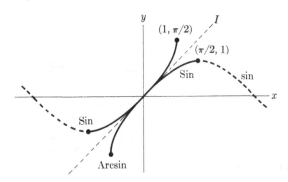

FIGURE 5–46

Graph each of the following, showing a range including several periods: (k) $2 \sin$, (l) $\sin [2I]$, (m) $\sin [I/2]$, (n) $\sin [I + \pi/3]$.

Obviously, the converse of a periodic function is not a function. In particular, $x = \sin y$ has infinitely many solutions for y corresponding to each value of x so that sin* is not a function.

(o) Sketch sin*.

However, by limiting the domain of sin, its converse becomes a function.

(2) **Def.** $\mathrm{Sin} = \{(x, y) \,|\, y = \sin x \,\wedge\, -\pi/2 \leq x \leq \pi/2\}$.

Then Sin* is a function, as suggested in Fig. 5–46. We call Sin* the *Arcsin*, and read "Arcsin x" as "the arc (or angle) whose sine is x."

(3) **Def.**   Arcsin $= \mathrm{Sin*}$.

Similarly,

(4) **Def.** $\mathrm{Cos} = \{(x, y) \,|\, y = \cos x \,\wedge\, 0 \leq x \leq \pi\}$.

(5) **Def.**   Arccos $= \mathrm{Cos*}$.

(p) Sketch Sin. What are its domain and range? (q) Sketch Cos. (r) Sketch Arccos. What are its domain and range?

Our choice of domain for Sin and Cos was dictated by a desire to have functions whose graphs are unbroken, have 0 in the domain, and contain points in the first quadrant.

(s) Suggest an alternative for Sin that would violate only the first requirement. (t) Suggest one that would violate only the second. (u) Suggest an alternative for Cos that would violate only the third requirement. (v) Show that the criteria completely determine the choice.

PROBLEMS

1. Prove that 2π is the period of sin.

Graph each function in Problems 2 through 10, indicating the amplitude and period of each.

2. $\sin [3I]$.

3. $\cos [2I]$.

4. $-\sin [I]$.

5. $\sin [I + \pi/6]$.

6. $\sin [I + \pi/3]$.

7. $\sin [I + \pi/2]$.

8. $\sin [I - \pi/3]$.

9. $3 \sin [I/2 + \pi/2]$.

10. $2 \sin [2I - \pi/6]$.

11. We wish to construct a sine curve that crosses the x-axis at $x = \pi/4$, has a period of $\pi/2$, and has a maximum of 10. What is its equation?

12. Draw a sketch of tan. (*Suggestion:* Since tan x is undefined when x is an odd multiple of $\pi/2$, draw vertical lines at such points on the x-axis. Make use of the identities of Section 4–12.) What is the period of tan?

13. Sketch tan*.

14. Using the criteria of the section, define "Tan" in such a way that Tan* is a function. Define "Arctan" and sketch Arctan.

15. Arcsin $0 = ?$

16. Arcsin $1 = ?$

17. Arcsin $2 = ?$

18. Arcsin $(-1) = ?$

19. Arcsin $(1/\sqrt{2}) = ?$

20. Arcsin $(0.89) \doteq ?$

21. Arccos $(-1) = ?$

22. Arccos $0 = ?$

23. Arccos $(0.89) \doteq ?$

24. Arctan $(-1) = ?$

25. Arcsin $(-0.71) \doteq ?$

26. Arctan $(5.8) \doteq ?$

Prove 27 through 30.

27. $\sin (\text{Arcsin } x) = x$.

28. $\sin (\text{Arccos } x) = \sqrt{1 - x^2}$.

29. $\cot (\text{Arctan } x) = 1/x$.

30. $\sec (\text{Arccos } x) = 1/x$.

Find all solutions of Problems 31 through 39.

31. $\sin x = 1$.

32. $\sin x = 0.5$.

33. $\sin x \doteq -0.3$.

34. $\sin 2x = 1$.

35. $\sin 5x \doteq 0.3$.

36. $\sin^2 x = 1$.

37. $\sin^2 x + 1 = 0$.

38. $\sin^2 x + 2 \sin x - 3 = 0$.

39. $2 \cos x + \sin 2x = 0$.

★40. Show that all solutions of $\sin y = x$ are given by $y = (-1)^k \text{Arcsin } x + k\pi$, where k is an integer. Define the set of images of x under sin*.

★41. Treat cos* and tan* as we treated sin* in Problem 40.

★42. Show that the criteria by which we chose a definition for Sin, Cos, and Tan can be applied to cot but not to sec and csc.

★43. If a body of mass m moves along a straight line so that its acceleration is always toward a fixed point and proportional to its distance from that point with a constant of proportionality equal to k, its position is given by $x = A \cos \sqrt{k/mt}$. (Sears and Zemansky, *University Physics*, p. 195) What is A?

What is the period? What is the frequency? Define the same function as a composite of a linear function and sin.

★44. The phenomenon of wave motion is a familiar one. Examples are waves in water and the waves observed in a rope whose ends are moved periodically. Such waves appear to have a sinusoidal shape and to move (along the water or the rope, for example). Such a moving wave may be represented by an equation of the form $y = A \cos 2\pi(t/T - x/\lambda)$, where t is the time and x is the distance from some fixed point. (Sears and Zemansky, *University Physics*, p. 372) For fixed t, the equation gives the shape of the moving wave at that instant. For fixed x, the equation gives the motion of a particle (of water or of the rope) at that point as a function of time. What is the period for t? for x? With what speed is the wave moving and in which direction? Graph the equation for $T = 1, \lambda = 0.5, t = 1$. Do the same for $t = 1.25$.

★45. Show that $\pi/4 = 4 \operatorname{Arctan}(1/5) - \operatorname{Arctan}(1/239)$.

ANSWERS TO EXERCISES

(a) $\exists T \ T > 0 \land \forall x \ f(x + T) = f(x)$.   (b) Dates, the days of the week, positions of point on a rotating wheel, repeated daily activities, etc. (c) $f(x + 2T) = f(x + T + T) = F(x + T) = f(x)$, etc.   (d) The definition holds for *any* function for $T = 0$; permitting negative T adds nothing. (e) 1 year.   (f) Dom sin = Re, Rge sin = $(\underline{-1 \quad 1})$.   (g) Because of the definition in terms of arcs on unit circles.   (h) Immediate, since $\sin \pi/2 = 1$ and $\cos \pi/2 = 0$.   (i) For a fixed x, $\cos x$ has the same value as $\sin (x + \pi/2)$, i.e., the value that $\sin x$ has $\pi/2$ toward the right.   (j) Looks like Fig. 5–45 with y-axis shifted $\pi/2$ to right; 2π.

(k) Like Fig. 5–45 but with high and low points at $(\pi/2, 2)$, $(3\pi/2, -2)$, etc. (l) Like Fig. 5–45 but with zeros at $\pi/2, \pi, 3\pi/2$, etc.; curve is compressed horizontally.   (m) Fig. 5–45 with curve expanded horizontally, zeros twice as far apart.   (n) Fig. 5–45 moved $\pi/3$ to the left.   (o) Use Section 5–10.   (p) Portion of Fig. 5–45 between $(-\pi/2, -1)$ and $(\pi/2, 1)$; domain is $(\underline{-\pi/2 \quad \pi/2})$, range is $(\underline{-1 \quad 1})$.   (q) See Fig. 5–47.   (r) See Fig. 5–47; domain is $(\underline{-1 \quad 1})$, range is $(\underline{0 \quad \pi})$.   (s) Let domain be $(\underline{0 \quad \pi/2}) \cup (\underline{3\pi/2 \quad 2\pi})$.   (t) Domain $(\underline{\pi/2 \quad 3\pi/2})$.   (u) $(\underline{-\pi \quad 0})$.   (v) Start at $x = 0$ and go into the first quadrant; the inverse must be a function.

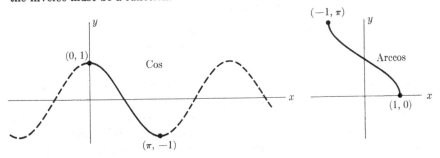

FIGURE 5–47

ANSWERS TO PROBLEMS

1. Letting $x = 0$, we have $\sin T = \sin 0$ for any period T. The only value
of T that satisfies this and that is smaller than 2π is π. But π is not a period,
since $\sin (\pi/2 + \pi) = -1 \neq 1 = \sin \pi/2$. 3. $1, \pi$. 5. $1, 2\pi$. Shift is $\pi/6$
to left. 7. $1, 2\pi$. Same as cos. 9. $3, 4\pi$. 11. $y = 10 \sin (4x - \pi)$. 13. See
Fig. 5–48. 15. 0. 17. Undefined. 19. $\pi/4$. 21. π. 23. 0.48. 25. -0.79.
    27. (5–10–9). 29. $\tan (\text{Arctan } x) = x$, but $\cot y = 1/\tan y$. 31. $\{x \mid \exists k\, k \in J \wedge x = \pi/2 + 2k\pi\}$. 33. $\{x \mid \exists k\, k \in J \wedge [x = -0.3 + 2k\pi \vee x = 3.44 + 2k\pi]\}$.   35. $\{x \mid \exists k\, k \in J \wedge [x = 0.06 + 2k\pi/5 \vee x = 0.57 + 2k\pi/5]\}$.
37. Null set. 39. $\{x \mid \exists k\, k \in J \wedge [x = \pi/2 + k\pi]\}$.

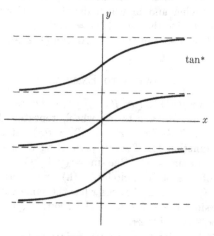

FIGURE 5–48

5–12 Arithmetic operations on functions. If f and g are functions
whose ranges are subsets of the real numbers, we define the sum of f and
$g, f + g$, as the function that maps x into $f(x) + g(x)$; that is,

(1) **Def.** $f + g = \{(x, y) \mid y = f(x) + g(x)\}$.

From this definition we have immediately $(f + g)(x) = f(x) + g(x)$.

Complete the following: (a) The image of x under $f + g$ is the _____
of the images of x under _____ and under _____.   (b) $f + g = h \ni h(x) = $ _____.   (c) Dom $(f + g) = $ _____.

In Fig. 5–49, we graph f, g, and $f + g$ where $f = 0.5I + 1$ and $g = -I + 2$. Note how we find $f(x) + g(x)$ for a given x by going to $(x, f(x))$
and then a vertical directed distance $g(x)$. Since these vectors can be
measured on the graph with a ruler or straight edge and compasses, once
f and g are sketched, $f + g$ can be drawn without further calculations.
For a given x, we find $f(x)$, $g(x)$, and $f(x) + g(x)$ from the drawing.

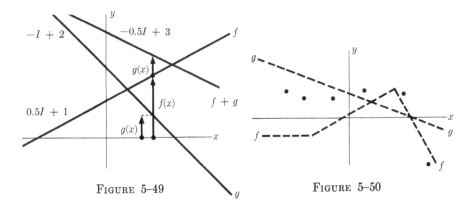

FIGURE 5–49   FIGURE 5–50

Complete the following: (d) The graph of f meets the graph of $f + g$ at the values of x where _____.   (e) The graph of g meets the graph of $f + g$ where _____.   (f) At values of x where the graphs of f and g cross, $(f + g)(x) =$ _____.   (g) The graph of $f + g$ will be below that of f where _____.   (h) The graph of $f + g$ crosses the x-axis where _____.   (i) Sketch $f + g$ in Fig. 5–50.

The above procedure of graphing $f + g$ is called *addition of ordinates.* (The y-coordinate of a point is sometimes called its *ordinate.*) Addition of ordinates is particularly useful when the nature of the graph of a sum is not obvious.   For example, Fig. 5–51 shows the graph of $y = \sin x$, $y = \sin 2x$, and $y = \sin x + \sin 2x$.   The graph of the sum is found rather quickly from the drawing by addition of ordinates at key points.

Graph by addition of ordinates: (j) $\sin + I$,   (k) $I + \text{sg}$ (see Section 5–7).

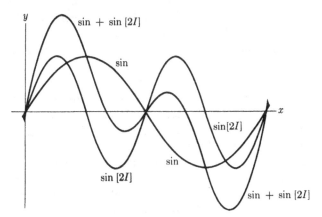

FIGURE 5–51

We define the other arithmetic operations on functions as follows:

(2) Def. $-f = \{(x, y) \mid y = -f(x)\}.$

(3) Def. $f - g = f + (-g).$

(4) Def. $f \cdot g = \{(x, y) \mid y = f(x)g(x)\}.$

(5) Def. $1/f = \{(x, y) \mid y = 1/f(x)\}.$

(6) Def. $f/g = f \cdot (1/g).$

(7) Def. $f^n = \{(x, y) \mid y = (f(x))^n\}.$

(l) Suggest a quick way to graph $-f$ from a sketch of f, and sketch $-f$ and $-g$ where f and g are defined by Fig. 5–50.   (m) Complete: $(-f)(x) =$ _____; $(f - g)(x) =$ _____; $(fg)(x) =$ _____(note that the dot is omitted when no confusion is likely); $(1/f)(x) =$ _____; $(f/g)(x) =$ _____; $(f^n)(x) =$ _____.   (n) Graph $f - g$ in Fig. 5–50. Suggest two ways to graph $f - g$ from the graphs of f and g.   (o) What are the domains of $-f$, $1/f$, and f^n?   (p) What are the domains of $f - g$, $f \cdot g$, and f/g?

From the definitions it is easy to show that functions follow laws similar to those for numbers. For example, $f + g = g + f$ is true for any functions, since

$$f + g = \{(x, y) \mid y = f(x) + g(x)\} = \{(x, y) \mid y = g(x) + f(x)\} = g + f,$$

the second equality being justified by the commutative law of addition. Indeed, since every operation on functions is defined by the "same" operation on the images, each law for numbers has a counterpart for functions which can be proved by applying the corresponding law for numbers to the images. Indeed, if we have a set of functions closed under addition, negation, and multiplication, axioms (1–16–1) through (1–16–9) are satisfied, where the constant functions $\underline{0}$ and $\underline{1}$ play the roles of the numbers 0 and 1. Complications arise for the reciprocal since the domain of $1/f$ does not contain the zeros of f. Hence (1–16–10) does not hold as it stands. (Incidentally, we do not use f^{-1} as a synonym for $1/f$ since f^{-1} is sometimes used to stand for the inverse $f*$.) It follows that we can manipulate the names of functions just as we would symbols standing for numbers, provided we take into account the domains of all functions involved and keep in mind that equality of functions means identical range and domain as well as defining formula. By taking advantage of this, we can state many identities without the explicit use of variables. For example, $\sin^2 + \cos^2 = \underline{1} \leftrightarrow \forall x \, \sin^2 x + \cos^2 x = 1$.

(q) Complete the following: $\sin/\cos =$ _____; $(I^2 - 1)/(I + 1) =$ _____; $\underline{1} + \cot^2 =$ _____; $2\sin\cos =$ _____. (Note that we sometimes use a numeral without the bar under it to stand for a constant function where no confusion is likely.)

Since $I(x) = x$, we can construct familiar functions by arithmetic operations applied to I and constant functions. In particular, we have $(\underline{a}I)(x) = ax$, $I^n(x) = x^n$, $\underline{a}I^n(x) = ax^n$, and $(1/I)(x) = 1/x$. If we form an expression from "I" and symbols standing for constant functions by indicating arithmetic operations, we have a name of a function whose defining formula is obtained by changing "I" to "x" and changing names of constant functions to names of the corresponding numbers. We have already used this idea in naming linear and quadratic functions. Thus a defining formula for $\underline{A}I^2 + \underline{B}I + \underline{C}$ is "$Ax^2 + Bx + C$," since $(\underline{A}I^2 + \underline{B}I + \underline{C})(x) = (\underline{A}I^2)(x) + (\underline{B}I)(x) + \underline{C}(x) = Ax^2 + Bx + C$.

(r) Find a defining formula for: I^3, $1/I^2$, $(2 + I)I^4$, $2\sin[3I^2]$.

A function obtained from I and constant functions by the arithmetic operations exclusive of division is called a *polynomial*. A polynomial $\underline{A}_nI^n + \underline{A}_{n-1}I^{n-1} + \ldots + \underline{A}_2I^2 + \underline{A}_1I + \underline{A}_0$ with $A_n \neq 0$ is called a polynomial of the nth *degree*. A function that is the quotient of two polynomials is called a *rational function*.

(s) Write a typical defining formula for a polynomial; for a rational function. (t) Sketch $I^3 - I$ by addition of ordinates. (u) Sketch the same function by drawing I and $I^2 - 1$ and noting that $I^3 - I = I(I^2 - 1)$. (v) Sketch $1/(I - 1)$.

It is not true that we get a defining formula for a function by substituting an independent variable for "I" in *any* name of the function. For example, "$\sin + x^2$" is not a defining formula for $\sin + I^2$. However, $\sin + I^2 = \sin[I] + I^2$, and "$\sin x + x^2$" is a defining formula obtained from the second name by $(I:x)$. Evidently, if we have a defining formula for a function we can always construct a name for it by putting "I" in place of the independent variable. For example, we did this when we adopted the name "$|I|$" for the function defined by "$y = |x|$." Also, given a name of a function, we can always insert "I" by using the fact that $F = F[I]$, so that the resulting name will yield a defining formula by the substitution $(I:x)$. For example, $\sin[I] = \sin$. Note that the brackets, indicating composition, are required here, since "$\sin I$" might be read as "$\sin \cdot I$." When possible, we shall usually follow the practice of naming functions so that the substitution $(I:x)$ yields a defining formula. Of course this may not always be convenient, for example when different defining formulas are required in different subsets of the domain.

(w) Write a name for the function defined by each of the following: $y = 2/x$; $y^2 = x \wedge y \geq 0$; $x^3 - y = 0$; $y =$ the largest integer less than or equal to x.

This method of naming functions has a considerable advantage for the purpose of quickly finding composites. For any functions F and G, $F = F[I]$, $G = G[I]$, and $G[F] = G[F[I]]$. Defining formulas are $F(x)$, $G(x)$, and $G(F(x))$. [See (5-9-4) and (5-9-5).] Note that a name for the composite is obtained by substituting $F(I)$ for I in $G(I)$, just as a defining formula for the composite is obtained by substituting $F(x)$ for x in $G(x)$. So long as we use names obtained from defining formulas by $(x:I)$, we may as well work with the names directly instead of going to the defining formulas, finding a defining formula of the composite, then going back to the names. For example, in Exercise (c) of Section 5-9 we found $(2I + 1)[I^2]$ by noting that $(2I + 1)(x) = 2x + 1$, $I^2(x) = x^2$, then using (5-9-4') to get $((2I + 1)[I^2])(x) = 2x^2 + 1$. Finally, we concluded that $(2I + 1)$ $[I^2] = 2I^2 + 1$. This final result could be obtained immediately by substituting I^2 for I in $2I + 1$.

(x) Do Exercises (d) and (f) in Section 5-9 by the above method.   (y) Complete: $I^3[1 - I] =$ _____; $I^2[\sin [I]] =$ _____; $I^2[\sin] =$ _____.

We see from the last example that a name for $G[F]$ may be found by substituting a name for F in place of I in a name for G, provided only that the name for G is such that a defining formula of G is obtained by the substitution $(I:x)$.

(z) Is "I" a variable? Is substitution for "I" justified by the Rule of Substitution? What is the justification for the device discussed above?

PROBLEMS

1. How would you construct a table of $f + g$, $-f$, $f - g$, fg, $1/g$, f/g, and f^n from tables of f and g? (Such procedures are used to construct tables of complicated functions from those of simpler ones.)

2. Assuming that you can build machines that perform arithmetic operations on numbers, draw pictures of machines along the lines of Fig. 5-13 to illustrate the definitions of this section.

3. Into what does $I^3 + I - \underline{3}$ map z?

4. What is the transform of -3 under $f + 2g$?

5. What corresponds to -1 under $\text{Arcsin} + I^2$?

6. Into what does $(\underline{1} - I)^2$ transform 3?

In Problems 7 through 12 sketch f, g, $-f$, $1/f$, $f + g$, $f - g$, $f \cdot g$, f/g, and $g[f]$. Name and give a defining formula of each function.

7. $f = \underline{2}, g = I^2$.

8. $f = I^3, g = -I$.

9. $f = \sin, g = I$.

10. $f = \sin^2, g = \cos^2$.

11. $f = \sin, g = \sin [3I]$.

12. $f = \sin, g = \sin [4I]$.

★13. Simple musical sounds are representable by functions of the type sin $[nI]$, where $n \in J$ and the pitch is determined by n. Show that the sum of any two such functions is periodic, and find the period.

14. Show that $A \sin x + B \cos x = C \sin (x + D)$, where $C = \sqrt{A^2 + B^2}$ and D satisfies $\cos D = A/\sqrt{A^2 + B^2} \wedge \sin D = B/\sqrt{A^2 + B^2}$. Illustrate your result by graphing sin + cos by addition of ordinates and comparing it with the graph of $\sqrt{2} \sin [I + \pi/4]$.

In Problems 15 through 18 repeat Problem 14 for the given sum.

15. $\sqrt{3} \sin + \cos.$ 16. $3 \sin + 4 \cos.$

17. $\sin - 2 \cos.$ 18. $0.3 \sin + 0.7 \cos.$

In Problems 19 through 24 find the composite $F \circ G$, that is, $G[F]$, directly from the names by substituting the name of F in the name of G in place of I. Find it also by using the defining formulas, and compare.

19. $F = 2I^2 - I, G = 1 - I^3.$ 20. $F = 3 \sin, G = 5I^2.$

21. $F = 5I^2, G = 3 \operatorname{Sin}.$ 22. $F = \operatorname{Sin}, G = |I|.$

23. $F = \operatorname{Arccos}, G = \operatorname{Sin}.$ 24. $F = \operatorname{sg}, G = \operatorname{Arcsin}.$

25. Prove that $f/f = \underline{1}$ is not a law for functions, but that $f/f = \underline{1} \cap \{(x, y) \,|\, f(x) = 0 \wedge y = 1\}'$ is a law.

★26. $f = \{(x, y) \,|\, y = 1/x \wedge -1 \le x \le 1\}, g = \{(x, y) \,|\, y = x^2 \wedge 1 \le x \le 2\}$. Find and sketch $f + g, f \cdot g, g/f,$ and $f[g]$.

★27. Sketch $\operatorname{sg} + (I - I^2)$, $\operatorname{sg} \cdot (I - I^2)$, and $\operatorname{sg} [I - I^2]$.

★28. $f(x) = x^2 - 1, (f + g)(x) = (x + 1)^2; g(x) = ?$

★29. Solve for F: $F^2 - I = 0.$

★30. Suggest a definition of $\sqrt{f}$, where f is a function. What is its domain?

ANSWERS TO EXERCISES

(a) sum; f; g.   (b) $f(x) + g(x)$.   (c) Dom $(f) \cap$ Dom (g).   (d) $g(x) = 0$, i.e., at the zeros of g.   (e) $f(x) = 0$.   (f) $2f(x)$.   (g) $g(x) < 0$.   (h) $f(x) = -g(x)$.   (i) Three segments through dots in Fig. 5–50.   (j) A wavy curve

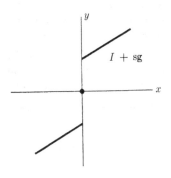

FIGURE 5–52

oscillating about the line $y = x$, varying from a distance of 1 vertically above to a distance of 1 vertically below; *not* the same as just swinging whole sine curve around the origin by an angle of $\pi/4$. (k) See Fig. 5–52. (l) "Reflect" curve in x-axis; in Fig. 5–50 reflect corners and connect.

(m) $-f(x)$; $f(x) - g(x)$; $f(x)g(x)$; $1/f(x)$; $f(x)/g(x)$; $(f(x))^n$. (n) Graph f and $-g$ and add ordinates; graph f and g and subtract $g(x)$ from $f(x)$ at key points from graph. (o) Dom $(-f)$ = Dom (f^n) = Dom (f); Dom $(1/f)$ = Dom $(f) \cap \{x|f(x) = 0\}'$. (p) Dom $(f - g)$ = Dom (fg) = Dom $(f) \cap$ Dom (g); Dom (f/g) = Dom $(f) \cap$ Dom $(1/g)$. (q) tan; $(I - 1) \cap \{(-1, -2)\}'$; $\csc^2$; $\sin [2I]$. (r) x^3; $1/x^2$; $(x + 2)x^4$; $2 \sin 3x^2$. (s) $A_n x^n + A_{n-1}x^{n-1} + \cdots$; $(A_n x^n + A_{n-1}x^{n-1} + \cdots + A_0)/(B_m x^m + B_{m-1}x^{m-1} + \cdots + B_0)$.

(t) Crosses x-axis at $(0, 0)$, $(1, 0)$, $(-1, 0)$. (v) Domain does not contain 1; curve does not cross line $x = 1$. (w) $2/I$; $\sqrt{I}$; I^3; $[I]$. (y) $(1 - I)^3$; $\sin^2$; $\sin^2$. (z) No. No. It works! It does so because we use names for functions so that it does work!

ANSWERS TO PROBLEMS

1. Include columns for $f(x)$, $g(x)$, $f(x) + g(x)$, etc. 3. $z^3 + z - 3$. 5. $-\pi/2 + 1$. 7. Defining formulas: 2, x^2, -2, $1/2$, $2 + x^2$, $2 - x^2$, $2x^2$, $2/x^2$, 4. 9. $\sin x$, x, $-\sin x$, $1/\sin x$, $x + \sin x$, $\sin x - x$, $x \sin x$, $(\sin x)/x$, $\sin x$. 11. $\sin x$, $\sin 3x$, $-\sin x$, $1/\sin x$, $\sin x + \sin 3x$, $\sin x - \sin 3x$, $(\sin x)(\sin 3x)$, $(\sin x)/(\sin 3x)$, $\sin (3 \sin x)$. 15. $2 \sin (x + \pi/6)$. 17. $\sqrt{5} \sin (x - 1.11)$. 25. $\underline{1}$ is the set of all ordered pairs of real numbers whose second component is 1, but f/f is the set of all such pairs except those pairs whose first component is a zero of f.

5–13 The conic sections. Mathematicians of ancient Greece obtained a very interesting class of curves by imagining a plane cutting a right circular cone. As suggested in Fig. 5–53, a right circular cone is generated by choosing a point V (called the *vertex*) on a line (called the *axis*) perpendicular to the plane of a circle at its center, and then letting a line through the vertex move around the circumference of the circle. Any plane perpendicular to the axis meets the cone in a circle. If the plane is rotated, the intersection first becomes a closed curve called an *ellipse*. If the rotation continues until the plane is parallel to a straight line on the cone, the curve of intersection is called a *parabola*.

(a) Prove indirectly that a parabola is not a closed curve.

If the plane is rotated past the position for a parabola, it meets both halves of the cone. The intersection is called a *hyperbola* and evidently has two branches.

(b) What happens if we let the plane pass through the vertex?

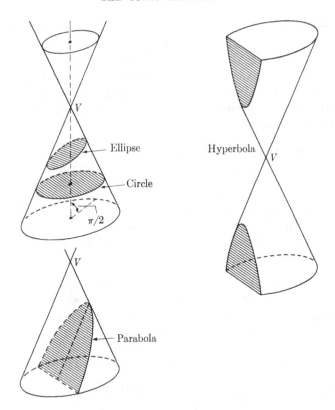

FIGURE 5–53

Though originally studied for their own sake, conic sections have been found to have many important applications, for they appear as paths of projectiles, orbits of planets, shapes of mirrors and lenses, patterns of points in statistical scatter diagrams, and architectural designs.

By using theorems of plane and solid Euclidean geometry, mathematicians discovered many interesting properties of conics. However, the methods of analytic geometry yield the same results more easily. It turns out that the conic sections all have equations of the form

$$(1) \qquad Ax^2 + Bxy + Cy^2 + Dx + Ey + F = 0,$$

where $\qquad \sim(A = B = C = 0).$

Conversely, any equation of this form (except those equations that define two parallel lines) is an equation of a curve that can be obtained by letting a plane cut a right circular cone. (Proof may be found in some textbooks on analytic geometry, e.g., in that by W. A. Wilson and J. L. Tracey.) Since all the properties of conics can be derived from such equations without reference to their historical origin as sections of a cone, it is customary today to define a *conic* as a curve with an equation of the form (1). Accordingly, by "conic" we mean a curve with an equation of the form (1). When the locus is the empty set, a single point, a single line, or two lines, it is called a *degenerate conic*.

(c) From the definition show that a circle is a conic. (d) Write an equation of a single point in the form (1). How could we get a single point by letting a plane cut a cone? (e) Write an equation of the form (1) that defines the null set of points in the cartesian plane. (f) Write an equation of two intersecting lines in the form (1). (g) of two parallel lines. (h) Show that the graph of a quadratic function is a conic. What kind? (i) Are all conics graphs of functions? (j) Show that the converse of a relation whose graph is a conic is a relation whose graph is a conic.

Some properties of conics are peculiar to one of the ellipse, parabola, or hyperbola. We consider these in the next three sections. Others are common to all conics and will be introduced here. An obvious property that follows immediately from the definition is

(2)   *A straight line meets a conic in at most two points.*

(k) Prove (2) by considering how one would find the intersection of (1) with a linear equation. (l) Find the points of intersection of $2x^2 + xy + y^2 - x + 4y - 6 = 0$ with $y - x = 1$.

A less obvious property is the following:

(3)

A plane curve is a conic if and only if there is a point (called the focus) *and a line (called the* directrix) *and a number (called the* eccentricity), *such that for every point on the curve the ratio of the distance between that point and the focus to the distance between that point and the directrix is equal to the eccentricity.*

Because of the "if and only if," theorem (3) asserts two implications: a plane curve is a conic if it has the geometric property; a plane curve is a conic only if it has the geometric property. The first implication is equivalent to "if a curve has the geometric property, then it is a conic." The second implication is equivalent to "if a plane curve is a conic, then it has the geometric property." It is easy to prove the first implication. Suppose that a curve has the indicated property. Then by the Rule of Choice (2–10–9), let the eccentricity be e, the focus be (h, k), and the line be defined by $A_1x + B_1y + C_1 = 0$. Then the locus is the set of points (x, y) satisfying

(4) $$e \frac{|A_1 x + B_1 y + C_1|}{\sqrt{A_1^2 + B_1^2}} = \sqrt{(x - h)^2 + (y - k)^2}.$$

(m) Justify (4). Show that $e \geq 0$.

Now (4) is not in the form (1), but squaring both members yields an equivalent equation of the form (1), which proves that every curve with the geometric property is a conic.

(n) Find an equation in the form (1) of the locus of points such that the distance from $(2, 3)$ is twice the distance from $x + y - 1 = 0$.   (o) Justify the last step in the above proof by carrying out and explaining the algebra.

The proof of the second part of (4) is not so easy. We shall accomplish it by showing in Sections 5–14 through 5–16 that it holds for the special cases of ellipse, parabola, and hyperbola, and then showing in Section 5–17 that every conic is one of these or a degenerate case. We shall find that the second part of the theorem does not hold for the circle and degenerate cases if we interpret it literally. However, by considering these special conics as "limiting cases," we squeeze them into a liberal interpretation of the theorem.

It is possible to tell from the equation of a conic whether it is an ellipse, parabola, or hyperbola. Indeed, we have

(5)   *A conic whose equation is in the form* (1) *is an ellipse, parabola, or hyperbola according as* $B^2 - 4AC$ *is less than, equal to, or greater than* 0.

The type of curve is determined also by the eccentricity according to the following:

(6)   *A conic is an ellipse, parabola, or hyperbola according as the eccentricity is less than, equal to, or greater than one.*

These theorems are proved in the following sections.

PROBLEMS

1. Show that (1) is the equation of a circle if $B = 0$ and $A = C$.
2. Show that any circle has an equation of the form (1) with $B = 0$ and $A = C$.
3. Show that $B^2 - 4AC < 0$ for a circle, in conformity with (5).

4. Calculate $B^2 - 4AC$ for the quadratic function defined by $y = ax^2 + bx + c$ after putting the equation in the form (1). Does your result conform to (5)?

Graph Problems 5 through 10, show that they are conics, and identify the type from the graph.

5. $xy = 2$.

6. $2x^2 + y^2 = 1$.

7. $x^2 - y^2 = 2$.

8. $x^2 - y^2 = 0$.

9. $x^2 + 2xy + y^2 = 0$.

10. $(2x + 3y - 4)(x - 5y + 10) = 0$.

Find the members of the sets defined by the simultaneous equations in Problems 11 and 12.

11. $xy = 1; x + y = 4$.

12. $x^2 + 4xy - 3y^2 + 2x - 5y + 7 = 0; x - 2y - 3 = 0$.

For Problems 13 through 16 find the equation in the form (1) of the conic with the indicated eccentricity, focus, and directrix. Graph.

13. $1, (1, 0), x = -1$.

14. $0.5, (1, 0), x = -1$.

15. $2, (1, 0), x = -1$.

16. $0, (1, 0), x = -1$.

17. Verify that your results in Problems 5 through 10 conform to (5).

18. Verify that your results in Problems 13 through 16 conform to (6).

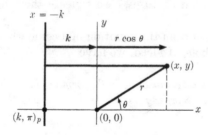

FIGURE 5–54

19. The equation of a conic takes a very simple form in polar coordinates if the focus is placed at the origin, and the directrix parallel to one of the axes. Figure 5–54 shows a focus at $(0, 0)$ and a directrix parallel to the y-axis through the point $(k, \pi)_p$ with $k > 0$. Show that an equation of the conic with eccentricity e is

(7) $$r = ke/(1 - e \cos \theta).$$

Graph the conics in 20 through 25 by plotting selected points. Sketch the directrix in each case, and indicate the eccentricity.

20. $r = 1/(1 - \cos \theta)$.

21. $r = 2/(1 - 0.5 \cos \theta)$.

22. $r = 2/(1 - 2 \cos \theta)$.

23. $r = 3/(3 - \cos \theta)$.

24. $r(2 - 5 \cos \theta) = 0$.

25. $r(1 + \cos \theta) = 0$.

★26. Find the form of the equation if the directrix is parallel to the y-axis but to the right, parallel to the x-axis and above, and parallel to the x-axis and below.

★27. Construct and graph examples illustrating your results in Problem 26.

ANSWERS TO EXERCISES

(a) If it were closed it would have to meet the line in the cone to which the cutting plane is parallel.   (b) Two lines, one line (plane tangent to cone), or a point.   (c) See Section 4–8.   (d) $(x - h)^2 + (y - k)^2 = 0$ defines $\{(h, k)\}$.   (e) $x^2 + y^2 = -1$.   (f) $x^2 - 4y^2 = 0$.   (g) Expand $(x + y + 1)$ $(x + y + 2) = 0$.   See Problem 21 in Section 5–6.   (h) In form (1) with $(B:0, C:0)$; parabola.   (i) No; example: two intersecting straight lines.   (j) Interchanging variables does not change form (1).   (k) Solving linear equation for one variable and substituting in (1) yields a quadratic, which has at most two roots.   (l) $(1/4)(-3 + \sqrt{13}, 1 + \sqrt{13})$ and $(1/4)(-3 - \sqrt{13}, 1 - \sqrt{13})$.

(m) (4–7–10) and (4–4–14); we must have $e \geq 0$ since the other terms are non-negative.   (n) $x^2 + 4xy + y^2 + 2y - 11 = 0$.   (o) First step: $e^2(A_1x + B_1y + C_1)^2 = (A_1^2 + B_1^2)((x - h)^2 + (y - k)^2)$.   Then let $A_1^2 + B_1^2 = A^2$, and carry on bravely.   The equations are equivalent to (4) since $a^2 = b^2 \leftrightarrow |a| = |b|$ is a law.   [Prove it from (3–9–10) and (2–8–38)].

ANSWERS TO PROBLEMS

1. See Section 4–8.   3. $B = 0$, $A = C$.   Hence $AC > 0$ and $-4AC < 0$.   5. Hyperbola similar to Fig. 5–5.   7. Hyperbola, Fig. 5–5, rotated through $-45°$.   9. Single line, $y = -x$.   11. $(a, 1/a)$ where $a = 2 \pm \sqrt{3}$.   13. $y^2 - 4x = 0$.   Parabola.   15. $3x^2 - y^2 + 10x + 3 = 0$.   Hyperbola.   17. 6 is ellipse; 8 and 10 two lines (a degenerate hyperbola).   19. Equation before simplifying is $r/(k + r \cos \theta) = e$.   21. Ellipse.   23. Divide numerator and denominator by 3 to put in form (7).

5–14 The ellipse.   The equation

$$(1) \qquad \frac{x^2}{a^2} + \frac{y^2}{b^2} = 1$$

is in the form (5–13–1) with $(A:1/a^2,\ B:0,\ C:1/b^2,\ D:0,\ E:0,\ F:-1)$. We let a and b be positive.   Solving for y and x we find

$$(2) \qquad y = \pm(b/a)\sqrt{a^2 - x^2},$$

$$(3) \qquad x = \pm(a/b)\sqrt{b^2 - y^2}.$$

Since we are considering only real values of the variables, (2) shows that

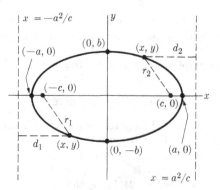

FIGURE 5–55

we must have $a^2 - x^2 \geq 0$, that is, $|x| \leq a$. Similarly, from (3), $|y| \leq b$. This shows that the locus is entirely within or on the boundary of the rectangle defined by $-b \leq y \leq b \wedge -a \leq x \leq a$.

(a) Carry through the manipulations to obtain (2) and (3).   (b) Justify the conclusions drawn from them.

Substitution shows that the points $(a, 0)$, $(-a, 0)$, $(0, b)$, $(0, -b)$ lie on the locus. They are called *vertices* of the curve. By plotting a few more points, the reader can convince himself that the curve is shaped as indicated in Fig. 5–55. Equation (1) is called the *standard form* of the equation of an ellipse. The segments joining $(a, 0)$ to $(-a, 0)$ and $(0, b)$ to $(0, -b)$ are called the *axes* of the ellipse, and their intersection is called the *center*. The longer one is called the *major axis*, the smaller one the *minor axis*.

Graph the following by plotting numerous points: (c) $x^2/9 + y^2/4 = 1$, and (d) $x^2 + y^2/4 = 1$.   (e) Show that $x = a \cos t$, $y = b \sin t$ are parametric equations of the ellipse defined by (1), i.e., that by letting t take all real values we get precisely the points satisfying (1).

It is convenient to use the concept of symmetry to describe the ellipse. We say that *two points are symmetric with respect to a line* if the line is the perpendicular bisector of the segment joining them.

(f) We have shown that the points (x, y) and (y, x) are symmetric to the line defined by $y = x$. Where?   (g) Show that the points (x, y) and $(-x, y)$ are symmetric with respect to the y-axis.   (h) Show that the points (x, y) and $(x, -y)$ are symmetric with respect to the x-axis.

We say that *a set of points is symmetric with respect to a line* if, corresponding to each point in the set and not on the line, there is another point in the set such that the pair of points is symmetric with respect to the line. The line is called *an axis of symmetry* of the set of points. From Fig. 5–55, it appears that an ellipse is symmetric with respect to each of its axes. To prove this we note that if (x, y) satisfies (1), then so do $(-x, y)$ and $(x, -y)$. The result follows from Exercises (g) and (h).

We now show that (5–13–3) holds for the ellipse defined by (1). For this purpose we suppose that $a > b$, and let $c = \sqrt{a^2 - b^2}$. Then we prove that $(-c, 0)$ is a focus and that the line defined by $x = -a^2/c$ is the corresponding directrix of the curve. Letting the eccentricity be c/a, equation (5–13–4) becomes

$$(4) \qquad (c/a)|x + a^2/c| = \sqrt{(x + c)^2 + y^2}.$$

This equation defines a set of points whose distances from $(-c, 0)$ and the line defined by $x = -a^2/c$ are in the ratio c/a. Squaring the equation, simplifying, and making use of the fact that $a^2 - c^2 = b^2$, we find (1). Hence a focus of the ellipse defined by (1) is $(-c, 0)$, the corresponding directrix is the line defined by $x = -a^2/c$, and the eccentricity is c/a. Figure 5–55 shows the focus and directrix and the distances r_1 and d_1 for a typical point (x, y).

(i) Carry through the algebra to show that (4) is equivalent to (1). (j) Show that $(c, 0)$ and $x = a^2/c$ are an alternative focus and directrix. (k) Show that in conformity with (5–13–5), $B^2 - 4AC < 0$ in (1). (l) Show that in conformity with (5–13–6), the eccentricity of (1) is less than one. (m) What happens if $a = b$? What about the foci, the eccentricity, the directrices? (n) What is the relation between a, b, and c if $a < b$? Where are the foci? the directrices?

PROBLEMS

In Problems 1 through 6 sketch the graph of the equation, indicate the vertices, the lengths of the major and minor axes, the foci, the directrices, and the domain and range of the relation defined by the equation.

1. $x^2/4 + y^2/16 = 1$.   2. $x^2/9 + y^2/25 = 1$.
3. $4x^2 + 9y^2 = 36$.   4. $100x^2 + y^2 = 100$.
5. $16x^2 + y^2 = 4$.   6. $4x^2 + 9y^2 = 1$.

In Problems 7 through 12 sketch the ellipse determined by the given conditions, and write the equation of each one.

7. Focus at $(3, 0)$, vertex at $(5, 0)$, center at origin.
8. Center at origin, vertices at $(1, 0)$ and $(0, 2)$.
9. Center at origin, focus at $(3, 0)$, eccentricity 0.75.
10. Vertices at $(0, -3)$, $(2, 0)$, $(0, 3)$, $(-2, 0)$.
11. Focus at $(-3, 0)$, corresponding directrix at $x = -16/3$, center at $(0, 0)$.
12. Eccentricity 0.5, major axis of length 10 and along y-axis, center at origin.
★13. Half the length of the major axis is sometimes called the *semimajor axis*.

The earth's path is elliptical with the sun at one focus. Taking the semimajor axis as 92.8 million miles and the eccentricity as 0.017, write an equation of the path.

★14. Do the same for the path of Mercury, whose eccentricity is 0.2 and semimajor axis is 0.39 times that of the earth's orbit.

★15. Why did we assume a and b positive in (1)?

★16. Prove that (1) is an equation of the set of points the sum of whose distances from $(-c, 0)$ and $(c, 0)$ is $2a$, where $c^2 = a^2 - b^2$ and $a > b$. (This is sometimes taken as a defining property of the ellipse.)

★17. Formulate and prove a similar result for the case where $a < b$.

★18. Show that the parametric equations $x = 2a \sin t \cos t$, $y = b \cos^2 t - b \sin^2 t$ define an ellipse.

★19. Suggest a way of constructing an ellipse by using the property of Problem 16.

★20. What can you say about the family of ellipses given by $(a^2 - c^2)x^2 + a^2 y^2 = a^2(a^2 - c^2)$ for different values of a and c?

Answers to Exercises

(a) See Section 2-8.   (c) Major axis horizontal.   (d) Major axis vertical, vertices at $(1, 0)$, $(-1, 0)$, $(0, 2)$, $(0, -2)$.   (e) Substitute in (1) to show that any point given by them lies on locus; also, all values of x in $(-a \quad a)$ are given by $x = a \cos t$. [Compare the discussion of (4–5–1) and (4–5–5).]   (f) In Section 5–10.   (g) The line joining the points is obviously perpendicular to the x-axis, and the points are equidistant from the x-axis.   (h) Similarly.

(i) Square and simplify.   (j) Equation (4) is the same except that the first two plus signs are changed to minus signs.   (k) Since here $B = 0$, and A and C are greater than zero.   (l) $c^2 = a^2 - b^2$, hence $c < a$, and $c/a < 1$.

(m) A circle, zero eccentricity, foci at center, directrices "at infinity"; the circle is a limiting case of an ellipse in which c approaches zero.   (n) $c^2 = b^2 - a^2$. In all cases $c^2 = |a^2 - b^2|$. The foci at $(0, c)$, $(0, -c)$. The directrices are given by $y = b^2/c$ and $y = -b^2/c$.

Answers to Problems

3. Divide both members by 36.   5. Divide both members by 4, then divide numerator and denominator of first term by 4.   7. $x^2/25 + y^2/16 = 1$. 9. $x^2/16 + y^2/7 = 1$.   11. $x^2/16 + y^2/7 = 1$.   13. $x^2/(92.8)^2 + y^2/(0.99)$ $(92.8)^2 = 1$.   15. For convenience, so that a and b would be half the lengths of the axes.   17. Sum of distances from $(0, c)$ and $(0, -c)$ is $2b$, where $c^2 = b^2 - a^2$. Use distance formula to express distances. Set sum equal to $2a$. Manipulate so that one square root is alone on one side and square. Simplify and repeat.

5–15 The hyperbola. The standard form of the equation of a hyperbola is

$$(1) \qquad \frac{x^2}{a^2} - \frac{y^2}{b^2} = 1.$$

A discussion of this equation is naturally very similar to that of the standard form of the equation of an ellipse in the previous section. Accordingly we follow such a pattern, leaving it to the reader to fill in the details.

Equation (1) is in the form (5–13–1) with _____

(a) Complete the previous sentence and equations (2) and (3).

As before we let a and b be positive. Solving for y and x, we find

(2) $y = \pm$ _____

(3) $x = \pm$ _____

Since we are considering only real values of the variables, (2) shows that

we must have $x^2 - a^2 \geq 0$, that is, _____

(b) Complete the previous sentence.

However, since $y^2 + b^2 \geq 0$ for all real y, the range of y is the real numbers.

Substitution shows that $(a, 0)$ and $(-a, 0)$ lie on the curve. We call these the *vertices* of the hyperbola. By plotting a few more points, the reader can convince himself that the curve is shaped as indicated in Fig. 5–56. The segment joining the vertices is called the *principal axis*. Its

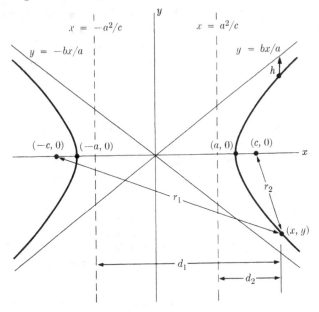

FIGURE 5–56

midpoint is called the *center* of the hyperbola. The perpendicular bisector of the principal axis is called the *conjugate axis*.

Graph the following by plotting numerous points: (c) $x^2/9 - y^2/4 = 1$, and   (d) $x^2 - y^2/4 = 1$.   (e) Show that $x = a \sec t$, $y = a \tan t$ are parametric equations of the hyperbola defined by (1).   (f) Show that (1) is symmetric with respect to the x-axis and with respect to the y-axis.

Symmetry with respect to a line is called *axial symmetry*. Another type of symmetry, called *central symmetry*, is defined with respect to a point. Two points are said to be *symmetric with respect to a point* if the point is the midpoint of the segment joining them.

(g) Show that the points (x, y) and $(-x, -y)$ are symmetric with respect to the origin.

A figure is said to be *symmetric with respect to a point* called the *center* if, corresponding to each point of the figure not at the center, there is another point of the figure such that the pair of points is symmetric with respect to the center. Both the ellipse and the hyperbola appear to possess central symmetry. To prove this it is sufficient to observe that their equations are equivalent to the equations we get by the substitution $(x{:}{-}x, y{:}{-}y)$.

(h) Why is this sufficient?

We now show that (5-13-3) holds for the hyperbola defined by (1). Letting $c = \sqrt{a^2 + b^2}$, we prove that $(-c, 0)$ is a focus and that the line defined by $x = -a^2/c$ is the corresponding directrix of the hyperbola. Letting the eccentricity be c/a, equation (5-13-4) becomes

(4) $$(c/a)|x + a^2/c| = \sqrt{(x + c)^2 + y^2}.$$

This equation defines a set of points whose distances from $(-c, 0)$ and the line defined by $x = -a^2/c$ are in the ratio c/a. Squaring the equation, simplifying, and making use of the fact that $c^2 = a^2 + b^2$, we find (1). Figure 5-56 shows the focus and directrix and the distances r_1 and d_1 for the typical point (x, y).

(i) Carry through the algebra to derive the above results.   (j) Show that $(c, 0)$ and $x = a^2/c$ are an alternative focus and directrix.   (k) Show that in conformity with (5-13-5), $B^2 - 4AC > 0$ in (1).   (l) Show that in conformity with (5-13-6), the eccentricity of (1) is greater than one.   (m) Must the discussion be altered according to whether $a < b$ or not?

Figure 5-56 shows two lines with equations $y = bx/a$ and $y = -bx/a$. It appears that the hyperbola approaches these lines.

(n) Draw the lines on your graphs of Exercises (c) and (d).

To show that the curves actually do approach these lines, we consider h, the directed distance parallel to the y-axis from the hyperbola to the line. We consider first the branch of the hyperbola in the first quadrant and the corresponding line there. We have

(5) $h = bx/a - (b/a)\sqrt{x^2 - a^2} = (b/a)(x - \sqrt{x^2 - a^2})$.

Now we multiply and divide the right member by $x + \sqrt{x^2 - a^2}$. The result is

(6) $h = (b/a)(x^2 - (x^2 - a^2))/(x + \sqrt{x^2 - a^2})$

(7) $= (ab)/(x + \sqrt{x^2 - a^2})$.

(o) Why is this multiplication and division justified?   (p) Explain (6) and (7).

Now consider what happens to h when x gets very large. Certainly h is positive, since both numerator and denominator are positive. Hence the hyperbola remains below the line. However, as x gets large, the denominator gets large without limit. Hence h gets small. Indeed, by going out far enough (i.e., by taking x large enough) we can make h as small as we like. We describe this by saying that the hyperbola approaches the line, and we called the line an *asymptote* of the curve.

(q) Show that in the fourth quadrant the hyperbola approaches the line $y = -bx/a$ as x increases.

A quick way to sketch a hyperbola is to draw the asymptotes and vertices, then draw a curve through each vertex approaching the asymptotes. An easy way to remember the equations of the asymptotes is to note that they are obtained from the equation of the hyperbola by putting 0 in place of 1 in the right member of (1), for we have

(8) $\dfrac{x^2}{a^2} - \dfrac{y^2}{b^2} = 0 \leftrightarrow (x/a + y/b)(x/a - y/b) = 0$

(9) $\leftrightarrow x/a + y/b = 0 \vee x/a - y/b = 0$

(10) $\leftrightarrow y = -bx/a \vee y = bx/a$.

Hence (8) is the equation of the union of the two asymptotes.

(r) Justify (8) through (10).

PROBLEMS

In Problems 1 through 6 sketch the graph, indicate the vertices, the foci, the directrices, the asymptotes, and the domain and range of the relation defined by the equation.

1. $x^2/4 - y^2/16 = 1$.
2. $x^2/9 - y^2/25 = 1$.
3. $4x^2 - 9y^2 = 36$.
4. $100x^2 - y^2 = 100$.
5. $16x^2 - y^2 = 4$.
6. $4x^2 - 9y^2 = 1$.

In Problems 7 through 12 sketch the hyperbola determined by the given conditions and write equations for each one.

7. Focus at $(5, 0)$, vertex at $(3, 0)$, center at $(0, 0)$.
8. Center at origin, vertex at $(2, 0)$, focus at $(3, 0)$.
9. Center at origin, focus at $(3, 0)$, eccentricity 2.
10. Center at origin, focus at $(0, 6)$, vertex at $(0, 4)$.
11. Focus $(-3, 0)$, corresponding directrix at $x = -1$.
12. Eccentricity 1.5, principal axis of length 10 along x-axis, center at origin.

13. Graph $x^2/9 - y^2/4 = -1$.
★14. Discuss the relation defined by (1). What is its domain? its range? Express it as the union of two functions. What is the appearance of its converse? Illustrate by sketching a particular hyperbola and its inverse.
★15. Treat the ellipse as in Problem 14.
16. Prove that (1) is an equation of the set of points the difference of whose distances from $(-c, 0)$ and $(c, 0)$ is $2a$, where $c^2 = a^2 + b^2$.
★17. Graph $x^2 - y^2 = 1$ and compare with Fig. 5–5.
★18. Prove that if a figure is symmetric with respect to two perpendicular lines, it is symmetric with respect to their intersection.

ANSWERS TO EXERCISES

(a) $(A:1/a^2, B:0, C:-1/b^2, D:0, E:0, F:-1)$;   (2) $(b/a)\sqrt{x^2 - a^2}$;   (3) $(a/b)\sqrt{y^2 + b^2}$.   (b) $|x| \geq a$.   (c) Vertices at $(3, 0)$, $(-3, 0)$.   (d) Vertices at $(1, 0)$, $(-1, 0)$.   (e) Like Exercise (e) in Section 5–14. See (4–12–7). (f) $(1) \leftrightarrow (1)(y:-y) \leftrightarrow (1)(x:-x)$.   (g) The two triangles formed by the component vectors are congruent.   (h) Since this shows that (x, y) lies on the graph if and only if $(-x, -y)$ does, and each such pair is symmetric.   (k) Since here $A > 0$ and $C < 0$.   (l) Since $c^2 = a^2 + b^2$, $c > a$.   (m) The treatment is the same regardless of the relation between a and b.   (n) $y = \pm 2x/3$ in Exercise (c) and $y = \pm 2x$ in (d).   (o) (1–14–2).   (p) (1–12–11). (q) Here $h < 0$; but manipulations are similar.   (r) (1–12–11), (2–8–5).

ANSWERS TO PROBLEMS

7. $x^2/9 - y^2/16 = 1$.   9. $x^2/9 - y^2/27 = 1/4$.   11. $x^2/3 - y^2/6 = 1$.
13. Vertices $(0, 2)$ and $(0, -2)$. Asymptotes, $y = \pm 2x/3$.

5–16 The parabola. A *parabola* is a conic with eccentricity equal to 1. Let the focus be at $(c, 0)$ and the directrix be defined by $x = -c$, where $c > 0$ as in Fig. 5–57. Then the condition that the ratio of the distance r from the focus to the distance d from the directrix be 1 is $r/d = 1$ or $r = d$. Since $r = \sqrt{(x - c)^2 + y^2}$ and $d = |x + c|$, we have, after squaring and simplifying,

(1) $$y^2 = 4cx.$$

The curve is easily sketched by substituting values of y. Indeed, (1) is equivalent to $x = y^2/4c$, which defines the converse of the quadratic function $I^2/4c$ defined by $y = x^2/4c$.

Graph the following parabolas, indicating focus and directrix: (a) $y^2 = 4x$, (b) $y^2 = x$,   (c) $y^2 - 7x = 0$.

(d) Express the relation defined by (1) as the union of two functions whose intersection is the origin.   (e) Show that (1) is of the form (5–13–1) when it is written $y^2 - 4cx = 0$.

The graph in Fig. 5–57 looks very much like the graphs of the quadratic functions in Section 5–8. Is the graph of any quadratic function a parabola? To show that the answer to this question is affirmative, we first show that the equation of the graph of any quadratic function can be put in the form $y = ax^2$ by appropriately placing the axes. From Section 5–8 we know that the graph of $y = ax^2 + bx + c$ has vertex $(-b/2a, (4ac - b^2)/4a)$. We are going to show that if we move the axes so that the origin is at this vertex, the equation of the curve becomes $y' = ax'^2$, where (x', y') are coordinates of points with respect to the new axes.

Let us imagine, as suggested in Fig. 5–58, that the axes are moved without any rotation so that the coordinates of the new origin O' with respect to the old origin O are (x_0, y_0), and the new axes are parallel to the old.

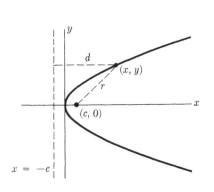

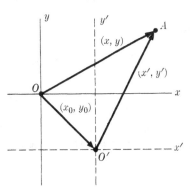

FIGURE 5–57   FIGURE 5–58

Consider a point A whose coordinates with respect to the old axes are (x, y) and with respect to the new axes are (x', y'). These coordinates are components of the vectors OO', OA, and $O'A$ respectively, as indicated in the figure. Evidently

$$(2) \qquad (x, y) = (x_0, y_0) + (x', y').$$

(f) Justify (2) by an argument not depending on a figure.

Hence we have $(x, y) = (x' + x_0, y' + y_0)$, which is equivalent to each of the following two pairs of equations.

$$(3) \qquad x = x' + x_0, y = y' + y_0,$$

(Equations of translation)

$$(4) \qquad x' = x - x_0, y' = y - y_0.$$

The equations give us the relationship between the coordinates of a fixed point with respect to two different axes. When the axes are moved without rotation, we speak of a *translation*. Equations (2) and (4) refer to a translation through the vector (x_0, y_0).

Let us consider now the quadratic function $2I^2 - 3I - 4$ defined by $y = 2x^2 - 3x - 4$ and the sketches in Fig. 5–33. In (5–8–2) we rewrote the defining formula as $2(x - 3/4)^2 - 41/8$. Hence the same graph is defined by the equation $y + 41/8 = 2(x - 3/4)^2$. Now suppose we let $x' \doteq x - 3/4$, $y' = y + 41/8$. Then the equation becomes $y' = 2x'^2$. But these equations are $(4)(x_0:3/4, y_0:-41/8)$. Hence they correspond to a translation of the axes to the vertex $(3/4, -41/8)$. We see that in relation to the new axes, the equation is $y' = 2x'^2$.

(g) Draw a sketch showing the curve and the two pairs of axes.

More generally, for any quadratic function $aI^2 + bI + c$ defined by

$$(5) \qquad y = ax^2 + bx + c,$$

we have from (5–8–3) the equivalent defining equation,

$$(6) \qquad y - (4ac - b^2)/4a = a(x - (-b/2a))^2.$$

A translation to the point $(-b/2a, (4ac - b^2)/4a)$, that is, to the vertex of the curve, yields the new equation

$$(7) \qquad y' = ax'^2,$$

as we wished to prove.

(h) Graph $y = 3x^2 - 4x + 2$ by finding the vertex, translating the origin there, and graphing with respect to the new axes.

We have shown that the graph of an equation of the form $y = ax^2 + bx + c$ has an equation $y' = ax'^2$ with respect to new axes obtained by a translation. Now, interchanging x' and y' is equivalent to rotating the curve about the line $y = x$, and such a rotation certainly does not change its shape or size in any way. But the interchange yields the equation $x' = ay'^2$ or $y'^2 = (1/a)x'$, which is of the form (1) and hence is a parabola. Hence the graph of any quadratic function is a parabola.

(i) Show that the graph of any equation of the form $x = ay^2 + by + c$, where $a \neq 0$, is a parabola. Indicate its focus and directrix.   (k) Find the focus and directrix of the graph of $y = ax^2 + bx + c$.   (l) Consider the graph of (5–13–1) when $A = B = D = 0$ or $B = C = E = 0$.   (m) Verify (5–13–5) and (5–13–6) for parabolas.

PROBLEMS

In Problems 1 through 8 sketch the curve and find its focus and directrix.

1. $y^2 = 5x$.

2. $2y^2 = 3x$.

3. $y^2 = -4x$.

4. $y^2 = 4x - 2$.

5. $y^2 + 3y = 4x$.

6. $x = 2y^2 - 3y + 7$.

7. $3y^2 + 2x + 4y - 5 = 0$.

8. $x^2 - 2y + 4 = 0$.

9. The axes are translated through the vector $(2, -3)$. The old coordinates of a point are $(-8, 3)$. What are its new coordinates?

10. In the same situation, what are old coordinates of a point whose new coordinates are $(0, -2)$?

11. Under a translation, the point whose old coordinates are $(3, -7)$ has coordinates $(-1, -6)$ in the new system. To what point was the origin translated?

★12. Sketch the graph of $y = 3x + 2$. Translate the origin to its y-intercept, and find the equation with respect to the new axes.

In Problems 13 through 16 find an equation of a conic with respect to focus and directrix placed so that the equations are not in standard form. Use (5–13–4) to find the equation and put it in the form (5–13–1).

★13. Eccentricity 1, focus $(3, 2)$, directrix $x + y = 1$.

★14. Eccentricity 2, focus $(3, 2)$, directrix $x + y = 1$.

★15. Eccentricity $\frac{1}{2}$, focus $(-3, 1)$, directrix $y = 0$.

★16. Eccentricity $\frac{1}{2}$, focus $(-3, 1)$, directrix $y = 4x + 3$.

★17. Show that the equation of any circle can be put in the form $x^2 + y^2 = r^2$ by translating the origin appropriately. Illustrate by discussing Problem 1 in Section 4–8.

ANSWERS TO EXERCISES

(a) $(1, 0)$, $x = -1$.   (b) $(1/4, 0)$, $x = -1/4$.   (c) $(7/4, 0)$, $x = -7/4$.

(d) $\{(x, y) \mid y^2 = 4cx\} = \{(x, y) \mid y = 2\sqrt{cx}\} \cup \{(x, y) \mid y = -2\sqrt{cx}\}$.

(e) $(A:0, B:0, C:1, D:-4c, E:0, F:0)$.   (f) Since the two sets of axes are parallel and have the same scales, the components of a vector are the same with respect to them. To arrive at a point A, we may go from the old origin via the vector (x, y) or from the old to the new origin via the vector (x_0, y_0) then via the vector from the new origin to the point (x', y'). Since the two paths have a common beginning and ending, we have (2) immediately regardless of the position of the axes or the point.   (g) New axes with origin at $(3/4, -41/8)$.

(h) Vertex is $(2/3, 2/3)$.   (i) Complete the square on y.   Procedure is as before with $(x:y, y:x)$.   Focus: $((4ac - b^2 + 1)/4a, -b/2a)$, directrix: $x = (4ac - b^2 - 1)/4a$.   (k) $(-b/2a, (4ac - b^2 + 1)/4a)$ and $y = (4ac - b^2 - 1)/4a$.   (l) In the first case the equation takes the form $Cy^2 + Ey + F = 0$, which is an equation of two lines parallel to the x-axis, one such line, or the empty set according as $E^2 - 4CF > 0$, $=0$, or <0.

ANSWERS TO PROBLEMS

1. $(5/4, 0)$, $x = -5/4$.   3. $(-1, 0)$, $x = 1$.   (Note that the parabola is open to the left and the range of x is the nonpositive reals.)   5. $(7/16, -3/2)$, $x = -25/16$.   7. $(3, -2/3)$, $x = 10/3$.   9. $(-10, 6)$.   11. $(4, -1)$.

13. $(x + y - 1)/\sqrt{2} = \sqrt{(x - 3)^2 + (y - 2)^2}$, which reduces to $x^2 - 2xy + y^2 - 10x - 6y + 25 = 0$.

★**5–17 The general case.** In the previous section we showed by appropriate translations of the axes that an equation of the form

(1) $$Ax^2 + Bxy + Cy^2 + Dx + Ey + F = 0,$$

with

$$A = B = 0 \wedge C \neq 0 \wedge D \neq 0$$

or with

$$B = C = 0 \wedge A \neq 0 \wedge E \neq 0,$$

is an equation of a parabola. By similar methods we now show that (1) with $B = 0$ and neither A nor C zero is an equation of an ellipse or hyperbola.

Suppose we translate the axes through the vector (h, k). Using (5–16–3) $(x_0:h, y_0:k)$, we replace x by $x' + h$ and y by $y' + k$ in

(2) $Ax^2 + Cy^2 + Dx + Ey + F = 0.$

Simplifying, we find

(3) $Ax'^2 + Cy'^2 + (2Ah + D)x' + (2Ck + E)y'$
$$= -(Ah^2 + Ck^2 + Dh + Ek + F).$$

Clearly, if we let $h = -D/2A$ and $k = -E/2C$, the equation takes the form

(4) $Ax'^2 + Cy'^2 = K,$

where $K = -(Ah^2 + Ck^2 + Dh + Ek + F)$. Hence, by an appropriate translation we can eliminate the linear terms in an equation of the form (2).

(a) Carry through the substitution to reduce $2x^2 + 5y^2 + 2x - 8y - 10 = 0$ to the form (4). Graph the result in relation to the new axes. (b) Suggest a method of finding K that is easy to remember.

Assuming that $K \neq 0$, we may divide both members of (4) by it and rewrite (4) in the form

(5) $$\frac{x'^2}{K/A} + \frac{y'^2}{K/C} = 1.$$

We may assume $A > 0$, since if it is not we could multiply both members of (4) by -1. If $C > 0$ also and $K > 0$, then $K/A > 0$ and $K/C > 0$. Hence there are real numbers a and b such that $K/A = a^2$ and $K/C = b^2$, and (5) takes the standard form of the equation of an ellipse. If $C > 0$ and $K < 0$, then K/A and K/C are negative, and the locus contains no points. In this case we call the locus an *imaginary ellipse*. Now if $C < 0$ and $K > 0$, we have $K/A = a^2$ and $K/C = -b^2$, and (5) reduces to the standard form of the equation of a hyperbola. If $C < 0$ and $K < 0$, $K/A = -a^2$ and $K/C = b^2$. An interchange of x and y then puts the equation in the standard form for the hyperbola. Finally, when $K = 0$, (4) is an equation of the single point given by $x' = y' = 0$ if $A/C > 0$, or of a pair of straight lines through the origin if $A/C < 0$.

(c) Justify the last statement.

We have now considered every case for (1) with $\sim(A = B = C = 0)$ for which $B = 0$. We found that the equation defined a parabola, ellipse, or hyperbola or else the null set, a single point, a line, or a pair of lines. We refer to these last four cases as *degenerate conics*. With this understanding, we can say that (1) with $B = 0$ and $\sim(A = C = 0)$ always defines a conic.

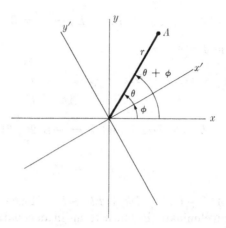

FIGURE 5–59

To complete our proof that any equation of the form (1) with
$\sim(A = B = C = 0)$ defines a conic, we must consider the case in which
$B \neq 0$. We shall show that the xy-term may always be eliminated by an
appropriate rotation of the axes about the origin. Before we can do this,
we need the equations giving the relation between the old and new coordinates of a point when the axes are rotated.

In Fig. 5–59 we show the point A, the original axes, the new axes, and
the angle ϕ (Greek "phi") of rotation. In polar coordinates the point A is
given by $(r, \theta)_p$ with respect to the new axes and by $(r, \theta + \phi)_p$ with
respect to the old axes. According to (4–11–2), we have $x/r = \cos(\theta + \phi)$,
$y/r = \sin(\theta + \phi)$, $x'/r = \cos\theta$, $y'/r = \sin\theta$. Now we apply (4–12–10)
to the expression for x/r to get

$$(6) \qquad x/r = \cos\theta\cos\phi - \sin\theta\sin\phi$$

$$(7) \qquad = (x'/r)\cos\phi - (y'/r)\sin\phi.$$

Multiplying both members by r we find the first formula of (8) giving the
old coordinates in terms of the new.

$$
(8) \qquad
\begin{aligned}
x &= x'\cos\phi - y'\sin\phi, \\
y &= x'\sin\phi + y'\cos\phi.
\end{aligned}
\qquad \text{(Equations of rotation)}
$$

(d) Derive the second equation in (8) by considering the formula for y/r.

If in (1) we replace x and y by their synonyms from (8), we find

(9) $(A \cos^2\phi + B \sin\phi \cos\phi + C \sin^2\phi)x'^2$

$+ ((C - A) \sin 2\phi + B \cos 2\phi)x'y'$

$+ (A \sin^2\phi - B \sin\phi \cos\phi + C \cos^2\phi)y'^2$

$+ (D \cos\phi + E \sin\phi)x'$

$+ (-D \sin\phi + E \cos\phi)y' + F = 0.$

(e) Verify (9) by making the replacement and simplifying.

Evidently we can eliminate the term involving the product $x'y'$ if we can so choose ϕ that $(C - A) \sin 2\phi + B \cos 2\phi = 0$. But this is equivalent to $\cot 2\phi = (A - C)/B$. Since the range of cot is the real numbers, there is a value of ϕ such that this equation is satisfied, and for this rotation, equation (1) reduces to the form (2), of which every case has already been considered. Hence we have proved that every equation of the form (1) with $\sim(A = B = C = 0)$ defines a conic.

(f) Carry through the rotation to eliminate the xy-term for the equation $xy = 1$. Graph it in relation to the old and new axes.

PROBLEMS

The derivation of equation (4) shows that when $A \neq 0 \wedge C \neq 0$ the center of the graph of (2) is at $(-D/2A, -E/2C)$ and that the equation with respect to axes translated to the center is (4). Use this procedure to graph 1 through 8, showing both old and new axes.

1. $3x^2 + 3y^2 + 6x - 12y - 2 = 0.$
2. $16x^2 + 2y^2 - 32x + 16y + 16 = 0.$
3. $x^2 - 4y^2 - 6x - 16y - 11 = 0.$
4. $4x^2 - 9y^2 + 8x + 72y - 264 = 0.$
5. $2x^2 + y^2 - 8x + 6y + 17 = 0.$
6. $2x^2 + y^2 - 8x + 6y + 20 = 0.$
7. $2x^2 - y^2 - 8x - 6y - 1 = 0.$
8. $2x^2 - y^2 - 8x - 6y + 1 = 0.$

9. An alternative way of deriving (4) and of simplifying an equation in the form (2) in any particular case is to complete the square on the terms in each variable separately, as we did for one variable in Sections 5–16 and 4–8. Deal with Problems 1 through 8 in this way.

★10. Derive formulas giving the new coordinates in terms of the old after rotating the axes through an angle ϕ. Do this in two ways: first by solving (8) for x' and y', second by relabeling Fig. 5–59 so that θ is the angle of A in the old system and $\theta - \phi$ is its angle in the new.

★11. Find approximately the angle ϕ to eliminate the xy-term in $5x^2 - 4xy + 2y^2 = 0$. Find $\sin \phi$ and $\cos \phi$ exactly.

★12. Letting A', B', C' be the coefficients of x'^2, $x'y'$, and y'^2, respectively, show that $A' + C' = A + C$ and $B'^2 - 4A'C' = B^2 - 4AC$.

★13. The previous problem shows that rotation does not change the value of the expression formed by squaring the coefficient of xy and subtracting four times the product of the coefficients of x^2 and y^2. Show that this expression is not changed by a translation.

★14. Explain how the results of Problems 12 and 13 suffice to demonstrate (5-13-5).

★15. By rewriting (5-13-7) in rectangular coordinates, prove (5-13-6).

ANSWERS TO EXERCISES

(a) $x = x' - 1/2$, $y = y' + 4/5$, $2x'^2 + 5y'^2 = 137/10$, $x'^2/(137/20) + y'^2/(137/50) = 1$, $a \doteq 2.6$, $b \doteq 1.7$. (b) Substitute the coordinates of the new origin in the left member and take the negative of the result. (c) If $A/C > 0$, $C > 0$, and we have $Ax'^2 + Cy'^2 = 0$, which is satisfied only for $x' = y' = 0$. However, if $A/C < 0$, that is, $C < 0$, then we have $Ax'^2 - (-C)y'^2 = 0$, whose left member factors to give $(\sqrt{A}x' + \sqrt{-C}y') = 0 \vee (\sqrt{A}x' - \sqrt{-C}y') = 0$. (d) $y/r = \sin(\theta + \phi) = \sin\theta\cos\phi + \cos\theta\sin\phi = (y'/r)\cos\phi + (x'/r)\sin\phi$. (e) We need to use (4-12-22) and (4-12-23). (f) Result is $x'^2 - y'^2 = 2$.

ANSWERS TO PROBLEMS

1. Center: $(-1, 2)$; new equation: $x'^2 + y'^2 = 17/3$. 3. $(3, -2)$, $x'^2/4 - y'^2 = 1$. 5. $(2, -3)$, $2x'^2 + y'^2 = 0$, point ellipse. 7. $y'^2 - 2x'^2 = 0$. 9. For example, after completing the square in Problem 1, we find $3(x + 1)^2 + 3(y - 2)^2 = 17$. Then for $x' = x + 1$, we have $x_0 = -1$ from (5-16-4).

CHAPTER 6

NUMBERS

★6-1 What are numbers? Everyone knows what numbers are. A person begins to learn about them almost as soon as he can speak, has many experiences with them as he grows older, and seldom passes a day of his life without using them in some way. Our store of knowledge about them is so vast and so rapidly growing that only a few specialists have a clear picture of it. Yet the sentence with which this paragraph begins is hardly true if it is taken to mean that everyone can say precisely what numbers are.

It would be more accurate to say that everyone knows something about numbers. Similarly, everyone knows something about electricity, and some specialists know a great deal about it. But who knows what electricity "really is"? If we attempt to define numbers (or electricity) we must do so in terms of other undefined terms, as pointed out in Section 1–13. Then we may formulate the properties of numbers by means of axioms and other laws. Such a theory comes as close as is possible to telling what numbers "really are."

There are many satisfactory theories describing our number system. Although they differ in their undefined terms, their definitions, and their axioms, they exhibit the same laws, for the obvious reason that they all have been designed to formalize the properties of numbers as they are known and used.

The purposes of this chapter are (1) to suggest a way in which cardinal numbers (including the natural numbers and the transfinite numbers) can be defined in terms of sets, (2) to familiarize the student with the basic properties of the natural numbers, particularly the principle of finite induction, (3) to outline the way in which the real and complex number systems can be built upon a few axioms about the natural numbers, and (4) to give the student further insight into the nature of numbers.

Some of the sections of this chapter are marked as optional because, in spite of their considerable mathematical interest, they are not essential to later chapters. The treatment is mostly informal, proofs and many details being omitted where only a bird's-eye view of the ideas is given.

★6-2 The cardinal numbers. The Mohave indians of California used to ally themselves with the Yumas to make joint raids on other tribes. To coordinate an attack they would send the Yuma warriors a knotted string, keeping a similar one for themselves. Each tribe would untie one knot

321

each morning, then attack on the morning of the last knot. (R. F. Heizer and M. A. Whipple, *The California Indians*, 1931, p. 377)   We should now describe their action by saying that they tied the same number of knots in each string.   But their technique did not require numbers or counting.   It depended solely on the recognition that two sets (knots, days, mornings) with quite different members may be similar in that they can be paired off (knot for knot, knot for morning, etc.) so as to exhaust all members of each set.   That the Mohaves were pairing off sets in this way shows that they were near the idea of number, for the next natural step would have been to assign a name to each kind of string of knots and use the name to describe any set that could be paired with it.

This historical example suggests that natural numbers are properties of sets.   Indeed, we ordinarily say that two sets have the same number of members when we can pair them off.   Counting a set merely amounts to pairing off the members of the set with a subset of the natural numbers. We shall call two sets similar if they can be paired off.   Then we shall define cardinal numbers as classes of similar sets. We use the word "class" as a synonym for "set" to avoid phrases such as "a set of sets."

To give a precise meaning to the concept of pairing off two sets, we must see clearly what the process involves.   When we pair off two finite sets, we pick out a member of one and a member of the other, and put them aside.   We repeat the process until all members of both sets are exhausted. If any members of either set are left over, we conclude that the sets cannot be paired.   In short, we attempt to form a set of ordered pairs whose first components are from one set, whose second components are from the other set, and such that every member of the first set and every member of the second set belongs to just one ordered pair.   Such a class of pairs is a relation whose domain is the first set and whose range is the second. More than that, such a class is a function and so is its converse, since each first component and each second component belongs to only one pair. In the language of correspondences, such a pairing is called a one-to-one correspondence.

(a) Review the discussion of correspondences in Section 5–5 and the remarks just before (5–10–5).

We say that the set A is *similar* to the set B, and we write $A \approx B$ if a one-to-one correspondence can be set up between A and B.   Formally,

(1) **Def.** $A \approx B$

$$= \exists F \; F \in \text{Fun} \wedge F* \in \text{Fun} \wedge \text{Dom}\,(F) = A \wedge \text{Rge}\,(F) = B.$$

To show that two sets are similar we may exhibit a function required by (1).   [See (2–10–5) and the paragraphs that follow it.]   For example, to show that $\{a, b\} \approx \{c, d\}$, we exhibit $\{(a, c), (b, d)\}$, which is a function of the required type. Evidently if one one-to-one correspondence exists between two sets, there may be many others, but it is sufficient to show the existence of at least one. Where the sets have many or an infinite number of members, it is not convenient to tabulate the one-to-one correspondence. In such cases, the function may be specified by giving a defining sentence.

Show that the following are true by exhibiting a one-to-one correspondence. (b) $\{0, 1, 2\} \approx \{1, 2, 3\}$.   (c) Married men $\approx$ married women (monogamy is assumed).   (d) The digits $\approx$ the even numbers from 2 to 20.   (e) $E \approx D$, where $E =$ the even positive integers, and $D =$ the odd positive integers.

The following fundamental properties of similarity are easily proved.

(2) $A \approx A$,

(3) $(A \approx B) \to (B \approx A)$,

(4) $[A \approx B \wedge B \approx C] \to A \approx C$.

To prove (2) we note that the identity relation, I_A, defined in (5-9-7), is a function satisfying the definition. Theorem (3) follows from the observation that the right member of (1) is logically equivalent to the result of interchanging A and B. To prove (4), suppose by hypothesis that there is a one-to-one correspondence F between A and B, and a one-to-one correspondence G between B and C. Then $F \circ G$ is a one-to-one correspondence between A and C, and $G(F(x))$ in C corresponds to x in A.

(f) Illustrate the proof of (3) by exhibiting a one-to-one correspondence between $\{1, 3\}$ and $\{8, 15\}$ and the converse of this correspondence.   (g) Similarly illustrate the proof of (4) by exhibiting a one-to-one correspondence between $\{1, 3\}$ and $\{8, 15\}$, between $\{8, 15\}$ and $\{0, 5\}$, and between $\{1, 3\}$ and $\{0, 5\}$.   (h) How did we use (2-10-9) in proving (4)?

We now define the *cardinal number of a set* (roughly speaking, the number of members in a set) as the class of all sets similar to it.

(5) **Def.** $\mathfrak{N}(S) = \{x \mid x \approx S\}$,

(6) $[\mathfrak{N}(A) = \mathfrak{N}(B)] \leftrightarrow (A \approx B)$.

Definition (5) defines a function $\mathfrak{N}$, whose domain is the class of all sets under consideration and whose range is a set whose members are classes of similar sets. $\mathfrak{N}$ maps each set A onto the class of all sets similar to A. These classes of similar sets are disjoint, since if $\mathfrak{N}(A)$ and $\mathfrak{N}(B)$ have a set in common, then because of (4) and (5) they each contain all sets similar to it and are therefore identical. Hence every set has a unique cardinal number associated with it.

(i) List several members of $\mathfrak{N}(\{a, b\})$, where $a \neq b$.

When a child is learning the meaning of numbers, he associates each number with various sets having the appropriate number of members. Thus the child associates 2 with a set of two ducks, two children, two toys, etc. All such sets associated with 2 are similar. They are members of the class of all sets similar to any one of them. We call this class of similar sets the cardinal number of any one of them. We define 2 to be this cardinal number. That is the idea behind the following definitions.

(7) **Def.** $$0 = \mathfrak{N}(\emptyset).$$

(8) **Def.** $$1 = \mathfrak{N}(\{0\}).$$

(9) **Def.** $$2 = \mathfrak{N}(\{0, 1\}).$$

Thus we define the number 0 to be the cardinal of the null set, i.e., the class of all sets similar to the null set. We define the number 1 to be the cardinal of a singleton. Any singleton could be used in place of $\{0\}$, but in the sequence of definitions the most obvious procedure is to use the symbols already defined to make each additional definition. In this way we use the symbols "0" and "1," defined in (7) and (8), to designate the pair $\{0, 1\}$, whose cardinal is 2.

(j) How many members does 0 have? (k) How many members does 1 have? (l) Define the cardinal 3 similarly. (m) List several members of 1. (n) Define 4 and list several of its members. (o) For (9) to be a satisfactory definition of 2, we need to prove that $0 \neq 1$, since otherwise $\{0, 1\}$ might be a singleton. Suggest an idea on which such a proof might be based.

The cardinals 1, 2, 3, ... obtained by continuing the definitions (7) through (9) are called *natural numbers*. We let N stand for the set of natural numbers. Each successive natural number is defined as the cardinal of a set whose members are just the previously defined cardinals, so that, roughly speaking, the sets belonging to each successive natural number have one more member than those belonging to the preceding natural number. It would be "natural" to call 0 a natural number, but actually zero was understood only recently and the term "natural numbers" refers to the sequence beginning with 1.

(p) Assuming that 0 through 100 have been defined, define 101.

If the cardinal of a set is 0 or is a member of N, i.e., a natural number, we say that the set is *finite*. In the contrary case we say that the set is *infinite*. Cardinals of finite sets are called *finite cardinals*, and those of infinite sets are called *transfinite*.

The set of natural numbers N has a cardinal, $\mathfrak{N}(N)$, which is called *aleph null*, $\aleph_0$. Any set that can be put into one-to-one correspondence with the natural numbers has the cardinal $\aleph_0$ and is said to be *denumer-*

able. For example, the set of even numbers is denumerable, since the function $2I$ is a function satisfying the requirements of the definition (1). Under it, each natural number n corresponds to the even number that is its double. The correspondence may be suggested by a partial tabulation: $\{(1, 2), (2, 4), (3, 6), \ldots, (17, 34), \ldots\}$.

(q) Define this one-to-one correspondence explicitly and formally.   (r) Show that the odd numbers are denumerable.

The cardinal of the natural numbers, $\aleph_0$, is the "smallest" transfinite cardinal. It seems obvious that it is not finite and fairly obvious that any "smaller" cardinal is finite, but the proof is not trivial. The interested reader is referred to *Theory of Sets,* by E. Kamke, and to the classic *Contributions to the Founding of the Theory of Transfinite Numbers,* by Georg Cantor.

We see from the examples that an infinite set may be put into one-to-one correspondence with one of its own proper subsets. This is a characteristic of all infinite sets and is sometimes taken as a definition of them.

PROBLEMS

1. Exhibit as many one-to-one correspondences as you can between $\{1, 2\}$ and $\{3, 4\}$.

2. Do the same for $\{1, 2, 3\}$ and $\{0, 1, 2\}$.

3. Do the same for $\{1, 2, 3\}$ and $\{1, 2, 3\}$.

★4. A one-to-one correspondence of a set with itself is called a *permutation.* How many permutations are there of a set of 3 elements? 4 elements? n elements?

5. What conclusion can you draw from $A \approx \emptyset$?

6. Show that $A \approx B \rightarrow A = B$ is not a law.

7. Show that the converse is a law.

8. Show that $A \approx B \wedge C \approx D \rightarrow (A \cup C) \approx (B \cup D)$ is not a law.

9. Show that with the added hypothesis $A//C \wedge B//D$, it is a law.

10. Show that the perfect squares 1, 4, 9, 16, ... are denumerable.

11. Show that the set of finite cardinals $\{0, 1, 2, 3, \ldots\}$ belongs to $\aleph_0$.

12. Show that the integers are denumerable.

13. Show that the unit fractions 1/1, 1/2, 1/3, ... are denumerable.

14. Show that $\mathfrak{N}(\{1\}) = 1$, $\mathfrak{N}(\{1, 2\}) = 2$, $\mathfrak{N}(\{1, 2, 3\}) = 3$. What does this have to do with the usual process of counting?

★15. Show that the positive rational numbers are denumerable by making use of the schema in Fig. 6–1. Note that every rational appears in the array more than once. However, by following the indicated path and using each rational only when it is first named, we may set up the required one-to-one correspondence.

★16. Use the result of Problem 15 to show that the set of all rationals is denumerable.

★17. Show $A \in \aleph_0 \wedge B \in \aleph_0 \rightarrow (A \cup B) \in \aleph_0$.

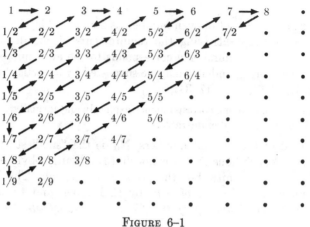

FIGURE 6–1

★18. Show that the union of a finite set and a denumerable set is denumerable.

★19. Show that the union of a finite number of denumerable sets is denumerable.

★20. Show that the union of a denumerable number of denumerable sets is denumerable!

★21. A set that is either finite or denumerable is called *countable*. Are all sets countable? To show that the answer is no, it is sufficient to exhibit an infinite set that is not denumberable. We consider the set of all real numbers between 0 and 1, and use a proof by contradiction. Suppose, then, that this set is denumerable and let $F(i)$ be the real number corresponding to the natural number i. Then we imagine the real numbers between 0 and 1 arranged as follows:

$$F(1) = 0.a_{11}a_{12}a_{13} \ldots a_{1i} \ldots,$$
$$F(2) = 0.a_{21}a_{22}a_{23} \ldots a_{2i} \ldots,$$
$$F(3) = 0.a_{31}a_{32}a_{33} \ldots a_{3i} \ldots,$$
(10) $$F(4) = 0.a_{41}a_{42}a_{43} \ldots a_{4i} \ldots,$$
$$F(5) = 0.a_{51}a_{52}a_{53} \ldots a_{5i} \ldots,$$
$$\ldots\ldots\ldots\ldots\ldots\ldots\ldots\ldots\ldots\ldots\ldots,$$
$$F(i) = 0.a_{i1}a_{i2}a_{i3} \ldots a_{ii} \ldots.$$

Here a_{11} is the first digit in the decimal expansion of $F(1)$, a_{54} is the fourth digit in the expansion of $F(5)$, and so on. Now establish the contradiction by exhibiting a decimal between 0 and 1 that is not among the decimals in (10)!

★22. Show that the set of all points in the interval (0_1) is similar to the set of all points on the entire infinite straight line!

ANSWERS TO EXERCISES

(b) $\{(0, 1), (1, 2), (2, 3)\}$.   (c) $\{(x, y) \mid y =$ the husband of $x\}$, that is, Hu of Section 5–3.   (d) $\{(x, y) \mid y = 2(x + 1) \wedge x$ is a digit$\}$.   (e) $\{(x, y) \mid y = x + 1 \wedge x$ is an even number$\}$.   (f) $\{(1, 8), (3, 15)\}, \{(8, 1), (15, 3)\}$.   (g) $\{(1, 8), (3, 15)\}, \{(8, 0), (15, 5)\}, \{(1, 0), (3, 5)\}$.   (h) **Knowing the existence of some** one-to-one correspondences, we were permitted to introduce "F" and "G" with the desired properties. (i) (1, 2), (0, 1), ($\emptyset$, U), ... (j) One. (k) $\mathfrak{N}$(the set of all singletons). (l) $\mathfrak{N}(\{0, 1, 2\})$. (m) Any singletons. (n) $\mathfrak{N}(\{0, 1, 2, 3\})$, (1, 2, 3, 4), (Hu, Wi, Mo, Fa), ... (o) See Exercises (j) and (k). (p) $\mathfrak{N}(\{0, 1, 2, \ldots, 99, 100\})$. (q) $\{(x, y) \mid y = 2x$ and x is a natural number$\}$. (r) $F = 2I - 1$.

ANSWERS TO PROBLEMS

1. There are two. 3. There are six. 5. $A = \emptyset$. 7. By (2). 9. A one-to-one correspondence for the conclusion is just the union of those for the two similarities. 11. $F = I + 1$. Note that "belongs to aleph null" is synonymous with "is denumerable." 13. They are similar to the natural numbers by $1/I$. 15. The rational corresponding to n is found by counting along the path, omitting duplications, to the nth rational. Conversely, the natural number corresponding to a rational is found by counting to find the position of the first occurrence of the rational. 17. In setting up the correspondence, take elements alternately from A and B. 21. Form a decimal by choosing for the ith digit a digit different from a_{ii} and from 9!

★6–3 **Arithmetic of the cardinals.** When a child first learns to add numbers, he does so in terms of the union of sets of objects. He has four objects in one hand and two different objects in the other. He puts all the objects in one hand (forms the union) and finds that he has six objects. The number of members in the union of two disjoint sets is the sum of the numbers of members in each. We use this idea to define formally the sum of any two cardinals.

(1) **Def.** $m + n$

$$= z \ni \forall x \, \forall y \, [[x//y \wedge m = \mathfrak{N}(x) \wedge n = \mathfrak{N}(y)] \to z = \mathfrak{N}(x \cup y)],$$

(2) $$A//B \to [\mathfrak{N}(A \cup B) = \mathfrak{N}(A) + \mathfrak{N}(B)].$$

(a) Translate (1) literally into ordinary English. (b) Explain how (1) defines $m + n$ as the cardinal of any two disjoint sets having m and n members. (c) How does (2) follow from (1)?

From (1) it is not too difficult to prove the laws of addition for natural numbers. For example, the commutative law $m + n = n + m$ follows immediately from (2) and (3–7–3).

(d) Why? (e) Argue for (1–16–7) for natural numbers.

Now, from (1–14–19) the rules for subtraction of natural numbers follow, and we can derive the following law, which has considerable use in probability and statistics.

(3) $$\mathfrak{N}(A \cup B) = \mathfrak{N}(A) + \mathfrak{N}(B) - \mathfrak{N}(A \cap B).$$

This follows from (2) and the observation that $A \cup B = A \cup (A' \cap B)$ and that $B = (A' \cap B) \cup (A \cap B)$. Since the terms in the right member of each of these equations are disjoint, by (2) $\mathfrak{N}(A \cup B) = \mathfrak{N}(A) + \mathfrak{N}(A' \cap B)$ and $\mathfrak{N}(B) = \mathfrak{N}(A' \cap B) + \mathfrak{N}(A \cap B)$. Subtraction of the second equation from the first yields (3).

(f) Show that $A \cup B = A \cup (A' \cap B)$ and that $A//(A' \cap B)$. (g) Show that $B = (A' \cap B) \cup (A \cap B)$ and that $(A' \cap B)//(A \cap B)$. (h) Argue for (3) with a Venn diagram. (i) In 1953 the membership of the American Mathematical Society was 4,473 and that of the Mathematical Association was 5,289. In a joint directory, 7,590 names were listed. How many persons were members of both organizations?

Multiplication of finite cardinals is sometimes defined in terms of repeated addition, but it can be defined very neatly in terms of the idea of a cartesian product.

(j) Glance again at Section 5–2.

Suppose we have two finite sets A and B with $m = \mathfrak{N}(A)$ and $n = \mathfrak{N}(B)$. The cartesian product $A \times B$ can be visualized as in (4). There is a row in this array for each element of A and a column for each element of B; i.e., there are m rows of n pairs each, or mn elements altogether.

$$(a_1, b_1)(a_1, b_2) \ldots (a_1, b_n),$$

$$(a_2, b_1)(a_2, b_2) \ldots (a_2, b_n),$$

(4)
$$\vdots$$

$$(a_m, b_1)(a_m, b_2) \ldots (a_m, b_n).$$

The above discussion suggests defining the product of any two cardinals m and n as the cardinal of the cartesian product of any two sets with m and n members; that is,

(5) **Def.**

$$m \cdot n = z \ni \forall x \, \forall y \, ([m = \mathfrak{N}(x) \wedge n = \mathfrak{N}(y)] \rightarrow z = \mathfrak{N}(x \times y)),$$

(6) $$\mathfrak{N}(A \times B) = \mathfrak{N}(A) \cdot \mathfrak{N}(B).$$

(k) Why does (6) follow from (5)? (l) Derive the commutative law of multiplication for natural numbers from (5) and (6). (m) Why did we not require that A and B be disjoint here?

From (5) the familiar laws of multiplication can be proved. Then division can be defined as in (1–14–1), and the laws of division proved.

Of course, subtraction and division may not always have answers if we limit ourselves to the natural numbers, so that the laws hold only where the result of an operation is defined.

There is nothing in the definitions (1) and (5) that limits them to finite cardinals. For example, we can use (1) to find the sum of aleph null and itself, that is, $\aleph_0 + \aleph_0$. According to (1), we may do so by choosing any two disjoint sets with aleph null members (i.e., any two disjoint denumerable sets), forming their union, and determining the cardinal of the result. But by Problem 17 in Section 6-2, the cardinal of the union is also denumerable. Hence $\aleph_0 + \aleph_0 = \aleph_0$. Here we see an example of the fact that the arithmetic of transfinite numbers is not the same as that of finite numbers, even though the laws that follow from (1) and (5) hold for all cardinals. The reader who wishes to pursue this fascinating branch of mathematics further may consult the references given at the end of Section 6-2.

PROBLEMS

1. A has 6 members, B has 25, and the two have 3 in common. How many members in their union?

2. A has 20 members, B has 100 members, and their union has 100 members. What is the situation?

3. A sociologist added together the estimates of different types of pathological deviants (alcoholics, criminals, mathematicians, etc.) in the United States as found in various representative textbooks. The total was about 104,000,000. Does this result discredit the figures from which it was derived?

4. Derive the associative law of addition for cardinals.

Verify (6) for the sets in Problems 5 through 8.

5. $A = \{1, 2, 3\}, B = \{4, 5\}$.
6. $A = \{1, 2, 3\}, B = \{2, 3\}$.
7. $A = \{a\}, B = \{1, 2, \ldots, 10\}$.
8. $A = \{a\}, B = \aleph_0$.

9. Derive the associative law of multiplication for cardinals.
10. Derive (1-16-8).
11. Which axioms of Section 1-16 do not hold for cardinals?
12. Derive the distributive law for cardinals.
13. Show directly from (5) that $0 \cdot m = m \cdot 0 = 0$.
14. Show from the definitions of this and the previous section that $1 + 1 = 2$, $2 + 1 = 3$, and $0 + 1 = 1$.
15. Show that $\aleph_0 \cdot \aleph_0 = \aleph_0$.

ANSWERS TO EXERCISES

(a) m plus n is the z such that for every x and y if x is disjoint with y (see 3-8-18) and m is the cardinal of x and n is the cardinal of y, then z is the cardinal of the union of x and y.   (b) The right member says that z, that is, $m + n$, is just the cardinal of the union of any x and y that are disjoint and have cardinals m and n respectively.   (c) Let $m = \mathfrak{N}(A)$ and $n = \mathfrak{N}(B)$, then use (3–9–1).   (d) By (3–7–3) $\mathfrak{N}(A \cup B) = \mathfrak{N}(B \cup A)$. Hence $\mathfrak{N}(A) + \mathfrak{N}(B) = \mathfrak{N}(B) + \mathfrak{N}(A)$ for any A and B. Or we may use (1) directly by noting that its right member is not changed by interchanging the dummies x and y. Interchanging m and n and the dummies, we find that the right members are identical except for differences that can be eliminated by use of the commutative laws of logical conjunction and of union.   (e) Use (6–2–7) and (3–7–15).

(f) Expand right member, using (3–7–7), (3–7–19), and (3–7–18). To show disjointness, use (3–8–18).   (g) Similarly.   (h) When we add $\mathfrak{N}(A)$ and $\mathfrak{N}(B)$ we count $\mathfrak{N}(A \cap B)$ twice.   (i) 2172.   (k) $m \cdot n$ is defined as the cardinal of the cartesian product of any two sets whose cardinals are A and B. Or similarly to Exercise (c).   (l) Similar to Exercise (d).   (m) Because we are dealing with ordered pairs of members.

Answers to Problems

1. 28. 3. No. 5. 6 members. 7. 10 members. 9. Use (5) twice, then laws of logic and cartesian product. 11. The last three. 13. An informal argument: Since the only set with cardinal 0 is the null set, we need the cardinal of the cartesian product of the null set with another set. But this cartesian product can have no members, since each of its members must have a first element from the null set. 15. See Problem 20 in Section 6–2.

★6–4 Axioms for the natural numbers.

In the last two sections we have sketched one way of defining the natural numbers and the operations on them. The familiar properties of natural numbers can be derived from those definitions. An alternative procedure is to take the natural numbers as undefined, characterize them by a few axioms, define the operations upon them, and make use of these axioms and definitions to derive other laws. The way of doing so that we outline in this section is due to G. Peano, an Italian mathematician who lived from 1858 to 1932. Of course, no matter how we lay the foundations of a theory of the natural numbers we arrive at the same laws and properties. The axioms that we state in this section, known as the *Peano postulates* (though our axioms here are not identical with his), could be derived from the theory sketched in the previous sections.

We now take as undefined three constants: "N" (which we consider as a name for the set of all natural numbers), "1" (which we consider a name for one), and "$\mathcal{S}$" (a name for a function whose domain is N). We call $\mathcal{S}(x)$ the *successor* of x, and $\mathcal{S}$ the *successor function*. Our axioms will characterize the natural numbers by utilizing the fact that they form a sequence in which each number has an immediate successor.

Our first axiom merely asserts that 1 is a natural number.

(1) **Ax.** $1 \in N.$

We are familiar with the fact that every natural number has a unique successor and that this successor is also a natural number. This is our second axiom.

(2) **Ax.** $n \in N \rightarrow S(n) \in N.$

(a) How is the uniqueness included in the theory?   (b) Graph S in the cartesian plane for the first few values in its domain.

We now wish to introduce the special position of 1 as the first natural number. We do this by specifying that no number precedes it, i.e., there is no natural number whose successor is 1. This is our third axiom.

(3) **Ax.** $\sim \exists x \, x \in N \wedge S(x) = 1.$

(c) Show that (3) $\leftrightarrow \forall x \, x \in N \rightarrow S(x) \neq 1.$ (d) If we have in mind that $S(n) = n + 1$, is there a number whose successor is 1? Does this contradict (3)?

We note now that every natural number except 1 has a unique predecessor, i.e., if the successor of m is the same as that of n, then $m = n$. This is our fourth axiom.

(4) **Ax.** $[S(m) = S(n)] \rightarrow (m = n).$

(e) Show from (4) that $m = n \rightarrow S*(m) = S*(n)$.   (f) Argue that (4) means that each natural number has no more than one predecessor.   (g) Argue that (4) means simply that $S*$ is a function.

The preceding axioms are not sufficient to characterize the natural numbers. We need an axiom to indicate that N contains *just* the numbers obtainable by continuing to take successors. According to (1) and (2), N contains 1 and the successor of each of its members. Now we wish to specify that N consists of only those numbers that can be obtained from 1 by successive application of S, that is, that N consists *only* of 1 and numbers that are successors of natural numbers. A convenient way of stating this is to assert that if M is a set that contains 1 and that contains the successor of each of its members, then M must contain all the natural numbers; that is, $N \subseteq M$. Formally,

(5) **Ax.** $[1 \in M \wedge \forall n \, (n \in M \rightarrow S(n) \in M)] \rightarrow N \subseteq M.$

In other words, N is the smallest set of numbers that have the properties (1) and (2). Still another formulation is that we can reach any natural number by starting with 1 and finding successors.

(h) Consider the set consisting of all the natural numbers and the numbers 1.6, 2.6, 3.6, . . . , where $S(n) = n + 1$. Show that this set satisfies all the axioms except (5).   (i) What is the range of S?   What are the domain and range of $S*$?   (j) Recalling that 1, N, and $S(n)$ are undefined above, let 1 be the usual number -1, N be the negative integers, and $S(n) = n - 1$. Show that *all* the axioms are satisfied!

The example in Exercise (j) indicates that the Peano postulates apply to sets other than the natural numbers.   Indeed, any denumerable set arranged in one-to-one correspondence with the natural numbers satisfies the axioms.   For let F be any function whose domain is N and for which $F*$ is a function.   Then let $F(1)$ play the role of 1 in the axioms, let $F(n)$ play the role of n, and let $S(F(n)) = F(S(n))$.   Then all axioms are satisfied by the range of F.   In Exercise (j), $F(n) = -n$. F serves simply to rename the members of N.

Show that the following satisfy the Peano axioms, and indicate F in each case: (k) the even integers,   (l) the odd integers.

To complete our picture of the natural numbers we must introduce their properties associated with addition and multiplication.   We do so by defining "$+$" and "$\cdot$" in terms of S.   We begin by defining $m + 1$ as $S(m)$.   Then we define $m + S(n)$ as $S(m + n)$; that is,

(6) Def. $m + 1 = S(m)$, $m + S(n) = S(m + n)$.

For convenience we adopt the usual decimal notation for natural numbers, introduced formally by definitions such as the following.

(7) Def. $2 = S(1)$, $3 = S(2)$, $4 = S(3)$, $5 = S(4)$, ...

(m) Rewrite (1) through (5) with $S(n)$ replaced by $n + 1$.

Then we can use Eqs. (6) and (7) to find the sum of any two natural numbers.   For example, to find $3 + 2$, we note that $3 + 2 = 3 + S(1) = S(3 + 1) = S(S(3)) = S(4) = 5$.   The easiest way to derive the addition tables is to begin by finding $n + 2$ for values of n as we just did for $n = 3$. Having found these we calculate $n + 3$ for values of n.   For example, knowing that $3 + 2 = 5$, we have $3 + 3 = 3 + S(2) = S(3 + 2) = S(5) = 6$.   Next we find the values of $n + 4$, and so on.   In this way we can build up an addition table from (6) alone.

(n) Show that $2 \in N$.

The following definition of multiplication is similar to (6).

(8) Def. $m \cdot 1 = m$, $m \cdot S(n) = m \cdot n + m$.

(o) What is the scope of the multiplication dot in the right member of the right equation in (8)?

We should like to prove

(9) $1 \cdot m = m$.

Now certainly (9) is true for $(m{:}1)$, since this is just $1 \cdot 1 = 1$, which is just $(m \cdot 1 = m)(m{:}1)$. To prove that it is true for m any natural number, we are going to adopt what seems like a rather artificial device. Let M be the set of values of m for which (9) is true; that is,

$$M = \{m \,|\, 1 \cdot m = m\}.$$

Now we shall prove that $N \subseteq M$, that is, that all natural numbers belong to M. This is the same as saying that (9) is true for m any natural number. But $N \subseteq M$ is the conclusion of axiom (5). Hence our strategy will be to prove the hypotheses of (5). Evidently, $1 \in M$, since we have shown above that (9) holds for $(m{:}1)$. To prove $\forall n\, n \in M \rightarrow S(n) \in M$, we assume $n \in M$ by the Rule of Hypothesis. That is, we assume $1 \cdot n = n$, and hope to derive $1 \cdot S(n) = S(n)$. Now, by definition (8), $1 \cdot S(n) = 1 \cdot n + 1$. By hypothesis, $1 \cdot n = n$. Hence $1 \cdot S(n) = n + 1 = S(n)$, which is what we wished to prove. Hence the implication holds by Q.E.D. Since both parts of the hypothesis of (5) hold, the conclusion follows by the Rule of Inference.

(p) Rewrite the proof in your own words. Review Section 2–6.

With the aid of the above results, it is now easy to derive the multiplication tables step by step along the lines described for addition. Then, by methods like those in the proof of (9), we can prove all the laws of arithmetic for natural numbers. Our discussion here has been very brief and incomplete. For one thing, we have failed to show that the definitions (6) and (8) are really adequate to define addition and multiplication of natural numbers. The matter is not as simple as it seems. For further details, the reader should consult *Foundations of Analysis*, by Edmund Landau.

PROBLEMS

1. Argue that (5) implies the following:

(5′) $(M \subseteq N \wedge 1 \in M \wedge \forall n[n \in M \rightarrow S(n) \in M]) \rightarrow M = N$.

2. Read each of (1) through (5) in words.
3. Show that N is the intersection of all sets to which 1 belongs and which have the property that if n belongs, then $S(n)$ belongs.
4. Derive the addition table for $2 + n$ with $n = 1, 2, 3, \ldots, 9$.
5. Derive the multiplication table for $2 \cdot n$.
6. Show that $m \cdot 2 = m + m$.
7. We may define $<$ for natural numbers as follows:

(10) Def. $m < n = \exists x\, x \in N \wedge m + x = n.$

Show that

(11) $m < m + 1.$

8. Argue that the natural numbers as defined in Section 6–2 satisfy the axioms and that $S(n) = \Re(\{x \mid x = 0 \vee (x \in N \wedge x \leq n)\}) = \{x \mid \exists y\, \exists z\, y \in n \wedge z \in 1 \wedge y//z \wedge x = y \cup z\} = \{x \mid \forall y\, y \in x \leftrightarrow (x - \{y\}) \in n\}.$

9. Show that the natural numbers greater than 100 satisfy the Peano postulates.

10. Show the same for the natural numbers and zero.

ANSWERS TO EXERCISES

(a) Since S is a function. (b) Discrete points lying on $y = x + 1$. (c) (2–10–3), (2–5–24). (d) Yes, 0, but 0 is not a natural number. Here we are speaking only of natural numbers. (e) (4) $= [m \neq n \rightarrow S(m) \neq S(n)]$ by (2–5–28). In this $(m:S^*(m),\ n:S^*(n))$ yields $S^*(m) \neq S^*(n) \rightarrow S(S^*(m)) \neq S(S^*(n))$. But by (5–10–9) the conclusion is the same as $m \neq n$. Hence $S^*(m) \neq S^*(n) \rightarrow m \neq n$, which is the same as what we wish to prove by (2–5–28). (f) 1 has none. (g) (5–4–1). (h) Here N in the axioms is the set of all natural numbers together with those of the form $n + 6/10$, where n is a natural number. However, now N is not the smallest set satisfying (1) and (2). (i) $N - \{1\}$; $N - \{1\}, N$. (j) Check each one. (k) $F = 2I$. (l) $2I - 1$. (n) (2), (7).

ANSWERS TO PROBLEMS

1. (3–8–5) 3. Since it is included in every set with these properties. 7. (2–10–5). 9. 101 plays the role of 1. 10. 0 plays the role of 1.

6–5 Finite induction. Let us imagine a ladder extending into the sky without end. Suppose, first, that we are able to reach the bottom rung, and, second, that we are able to move from *any* rung to the next one. It seems reasonable to conclude that we could reach any rung. This is an intuitive image illustrating the idea of finite induction.

Suppose we have a sentence $s(x)$ in which the range of the variable includes the natural numbers. Suppose we prove that $s(1)$ is true (we can reach the bottom rung) and that for any x, *if* $s(x)$ is true, *then* $s(x + 1)$ is true (we can get from any rung to the next). Then it seems reasonable to conclude that $s(x)$ is true for all natural numbers (we can get to any rung).

(a) Imagine a long line of toy soldiers so arranged in a line that if any one falls it will knock over the next one. What happens if the first one falls? (b) In the ladder image, what is $s(x)$? (c) What is $s(x)$ in the toy-soldier image?

These ideas can be stated in a theorem as follows.

(1) $[s(1) \wedge \forall x\, [s(x) \rightarrow s(x+1)]] \rightarrow \forall n\, [n \in N \rightarrow s(n)]$.

The letter n is used on the right to suggest that its values are natural numbers.

(d) Could we use n as a dummy on the left in (1)?   (e) Let $M = \{x \mid s(x)\}$, and state (1) in terms of set membership and set inclusion.

To prove (1) we make use of the ideas of Exercise (e) and of the proof of (6–4–9). (Readers who have omitted Section 6–4 may omit this and the next paragraph.) Let M be the set of values of x for which $s(x)$ is true; that is, $M = \{x \mid s(x)\}$. Since $s(1)$ by hypothesis, $1 \in M$. Since $\forall x\, s(x) \rightarrow s(x+1)$ by hypothesis, $\forall x\, x \in M \rightarrow (x+1) \in M$. Hence by (6–4–5), $N \subseteq M$; that is, all natural numbers belong to M. In other words, $s(x)$ is true for all natural numbers!

We illustrate the way in which we can use (1) by proving

(2) $1 + m = m + 1$

from the axioms and results of the previous section. Here $s(x) = [1 + m = m + 1]$, $s(1) = [1 + 1 = 1 + 1]$, $s(x+1) = [1 + (x+1) = (x+1) + 1]$. We have, therefore, to show two things: first that $1 + 1 = 1 + 1$, and, second, that $\forall x [1 + x = x + 1 \rightarrow 1 + (x+1) = (x+1) + 1]$. The first is true by (2–5–1). To prove the second we use the Rule of Hypothesis by assuming $1 + x = x + 1$. To derive $1 + (x+1) = (x+1) + 1$ from this assumption, we note that from (6–4–6)$(m{:}1, n{:}x)$ we have $1 + (x+1) = \mathsf{S}(1+x)$. But by hypothesis, $1 + x = x + 1$. Hence $1 + (x+1) = \mathsf{S}(x+1)$. And since $\mathsf{S}(x+1) = (x+1) + 1$, we have $1 + (x+1) = (x+1) + 1$, as desired. Now we have shown both hypotheses of (1) and so can conclude that $\forall n\, n \in N \rightarrow s(n)$; that is, that (2) holds for all natural numbers m.

It is important to notice the pattern of the above proof. First we show that the sentence holds for 1. Next we show that *if* it holds for a natural number, then it holds for the next one. Then from (1) we conclude that the sentence holds for all natural numbers.

The idea of induction can be generalized. Suppose that we are able to reach the tenth rung of a ladder and able to get from the tenth, or any later rung, to the next one. Then it seems reasonable to suppose that we can get to *any* rung past the tenth. The idea could be stated formally and proved, but we prefer to embody it in a new rule of proof.

RULE OF INDUCTION: *To prove that a sentence involving a variable is true for all natural numbers greater than or equal to a given one, it is sufficient to prove that the sentence holds for the given natural number and that if it holds for a natural number not smaller than this, then it holds for the succeeding natural number.*

(3)

Schematically,

$$s(n_1)$$

(4) $$\frac{s(x) \rightarrow s(x + 1) \text{ for } x \geq n_1}{\therefore s(n) \text{ for } n \geq n_1}.$$

(f) Write out a law embodying (3) and (4). Include the quantifiers omitted in (4).

When the Rule of Induction is used, the proof of the sentence for a particular case [n_1 in (4)] is called the *verification*. The proof that $s(x) \rightarrow s(x + 1)$ is called the *induction*. To carry out the induction, it is usual to use the Rule of Hypothesis to assume $s(x)$ and derive $s(x + 1)$ from it. In that case, $s(x)$ is called the *induction hypothesis*.

(g) Review Section 2–6, especially the Rule of Hypothesis. Does assuming $s(x)$ in the induction amount to assuming what we are trying to prove? Can we use substitution for x in $s(x)$ when it is so assumed?

Students often have difficulty understanding induction, but understanding comes with familiarity. The procedure is basically simple. To construct an induction proof, the following steps are recommended.

I. *Before beginning to write the proof, write the theorem in the form $n \in N \rightarrow s(n)$, and write out $s(n_1)$, $s(x)$, and $s(x + 1)$. Usually n_1 is 1.*

(5)   II. *Begin by proving $s(n_1)$. This is usually easy.*

III. *Complete the proof by proving that $\forall x \, s(x) \rightarrow s(x + 1)$. This is usually most easily done by assuming $s(x)$ and deriving $s(x + 1)$ from it.*

We follow this pattern to prove that for any natural number n, $\sin (\theta + 2n\pi) = \sin \theta$. The theorem is $n \in N \rightarrow [\sin (\theta + 2n\pi) = \sin \theta]$. Here $s(1) = [\sin \theta = \sin (\theta + 2\pi)]$, $s(x) = [\sin (\theta + 2x\pi) = \sin \theta]$, and $s(x + 1) = [\sin (\theta + 2(x + 1)\pi) = \sin \theta]$. Now for the proof: $s(1)$ is true by (4–12–19). Our induction hypothesis is $s(x)$, or $\sin (\theta + 2x\pi) = \sin \theta$. We begin with the left member of $s(x + 1)$ and try to use the induction hypothesis and known laws to show that it is equal to $\sin \theta$. We have

(6) $\sin (\theta + 2(x + 1)\pi) = \sin (\theta + 2x\pi + 2\pi)$

(7) $= \sin (\theta + 2x\pi)$ (4-12-19)

(8) $= \sin \theta$ (Induction hypothesis).

But (8) is just $s(x + 1)$. So the proof is complete.

(h) Prove by induction that $\cos (\theta + 2n\pi) = \cos \theta$ for all natural numbers n.

Both parts (the verification and the induction) of an induction proof are necessary. It may happen that $s(1)$ [or even $s(n)$ for many values of n] is true, and yet that $s(n)$ is not true for all natural numbers. For example, if $s(n) = [n^2 - 1{,}000{,}000 < 0]$, then $s(n)$ is true for the first 999 values of n, but not for all natural numbers. Here we could perform the verification but not the induction. On the other hand, if $s(n) = [n = n + 1]$, then $s(n)$ is always false, yet it is easy to show that $s(n) \rightarrow s(n + 1)$. Here the verification is impossible and the induction easy.

(i) Show that $[n = n + 1] \rightarrow [n + 1 = (n + 1) + 1]$.   (j) Construct examples in which $s(1)$ is true but $n \in N \rightarrow s(n)$ is not a law.   (k) Construct examples in which $s(n) \rightarrow s(n + 1)$ is true, but $n \in N \rightarrow s(n)$ is not a law.

It should be clear from the above discussion that (1) is a theorem of mathematics and of deductive logic. It is called the *principle of finite induction* to indicate that it deals with only *finite numbers*. Although the word "induction" is involved in its name, it must not be confused with the process of induction by which we guess general results from particular cases. Induction in that sense is a process of heuristic, a method of discovery. Induction in mathematics refers to the use of the rigorous rule of proof (3).

By using induction we can prove the commutative, associative, distributive, and other laws for natural numbers. The process is too long to give here, and it can be found in the reference given at the end of Section 6-4. When these proofs have been carried out, (1-16-1) through (1-16-8) are theorems about the natural numbers instead of axioms.

To illustrate once more the use of induction, we prove that the natural numbers are closed under addition and multiplication; that is,

(9) $(m \in N \land n \in N) \rightarrow m + n \in N$,

(10) $(m \in N \land n \in N) \rightarrow m \cdot n \in N$.

To prove (9), we follow (5). Step I: Let $s(x) = [m + x \in N]$. We could write (9) as $m \in N \rightarrow [n \in N \rightarrow s(n)]$, then assume $m \in N$ by Hyp and derive $n \in N \rightarrow s(n)$. Hence it is sufficient to consider m as though it were a constant, and to carry through an induction proof on n,

as follows: $s(1) = [m + 1 \in N]$; $s(x + 1) = [m + (x + 1) \in N]$. Step
II: $s(1)$ is true by (6–4–2). Step III: The induction hypothesis is
$(m + x) \in N$. From this, (6–4–2)$(n:m + x)$ and (6–4–6), it follows by
Inf that $(m + x) + 1 \in N$. But by (6–4–6) this is the same as
$m + (x + 1) \in N$, which is just $s(x + 1)$. Hence $s(x) \to s(x + 1)$ by
Q.E.D., and the proof is complete by (3).

(l) Prove (10) similarly.   (m) Why does (9) $= [m \in N \to (n \in N \to m + n \in N)]$?

PROBLEMS

We shall use induction many times later in the book. The following problems
will help the reader become familiar with the method, but are not supposed to
be sufficient to master it. In every case, follow the outline (5). Note that Step I
is a matter of "tooling up" for the proof. The proof consists of Steps II and III.

1. $\tan (\theta + n\pi) = \tan \theta$.
2. In a polygon of n sides there are $n - 3$ diagonals issuing from each vertex.
★3. Show that every nonempty subset of the natural numbers has a minimum.
(*Hint:* Assume the contrary, that is, that $M \subseteq N$ and M has no minimum.
Then prove that $M' = N$, that is, $M = \emptyset$. It follows that M has no minimum $\to M = \emptyset$, or $M \neq \emptyset \to M$ has a least member. This property is
described by saying that the natural numbers are *well ordered*.)
4. $n \in N \to n + 1 \neq n$.
★5. From the Peano axioms and other results obtained from them, prove
the commutative, associative, and distributive laws for natural numbers.
★6. The sum of the interior angles of a polygon of n sides is $(n - 2)\pi$. (Note
here that $n_1 = 3$. To perform the induction, imagine a polygon of x sides and
see what happens when one more side is added.)
★7. Prove that a polygon of n sides has $(n^2 - 3n)/2$ diagonals.

ANSWERS TO EXERCISES

(a) They all fall.   (b) $s(x) =$ we can reach the xth rung.   (c) $s(x) =$ the xth
soldier falls.   (d) Yes.   (e) It is just (6–4–5).   (f) $[s(n_1) \wedge \forall x \, x \geq n_1 \to$
$(s(x) \to s(x + 1))] \to [\forall n \, n \in N \wedge n \geq n_1 \to s(n)]$.   (g) No, it is merely a
device for proving an implication. No, since we are not assuming that $s(x)$ is
true for all x (which would be assuming what we are trying to prove!), but merely
introducing it to prove an implication.   (h) The same proof with "cos" for
"sin."   (i) $n = n + 1 \to S(n) = S(n + 1)$, but $S(x) = x + 1$. Or we may think
of assuming $n = n + 1$ and adding 1 to both members.   (l) $s(x) = mx \in N$,
$s(1) = m1 \in N$, $s(x + 1) = m(x + 1) \in N$. $s(1)$ is synonymous with $1 \in N$.
By (6–4–8), $m(x + 1) = mx + m$. The induction hypothesis is $mx \in N$.
Hence by (9), $mx + m \in N$, and we have derived $s(x + 1)$.   (m) (2–5–30).

ANSWERS TO PROBLEMS

1. Like (6) through (8). 3. Begin by showing that 1 belongs to M'. Then
show that if x belongs to M', so does $x + 1$.

6–6 Sequences. A function whose domain is the natural numbers is called a *sequence*. It is customary to use letters near the middle of the alphabet, such as n, i, and j, as independent variables for such functions and to use those near the end of the alphabet as dependent variables. Thus, if f is defined by $y = f(x) \wedge x \in N$, we usually write $y = f(n)$. Also, the image of n under f is usually written y_n, so that $y_n = f(n)$. In this section the range of n is always N, so that we need not indicate this explicitly in each case.

The identity function on N, I_N, maps each natural number into itself. Here $y_n = n$. A linear function on the natural numbers, $mI_N + b$, maps n into $mn + b$, so that $y_n = mn + b$, $y_1 = m + b$, $y_2 = 2m + b$, $y_3 = 3m + b$, and $y_n = y_1 + (n - 1)m$. We have $y_{n+1} - y_n = m(n + 1) + b - mn - b = m$, so that the difference between two successive images is always the same. We call a sequence defined by a linear function an *arithmetic progression*. Part of the graph of an arithmetic progression is sketched in Fig. 6–2. It consists of equally spaced discrete points on a straight line. In this context, the slope is called the *common difference*.

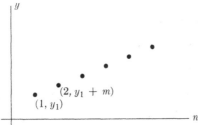

FIGURE 6–2

(a) An employee begins at a salary of $5000 a year and receives an increase of $250 a year. Define the sequence relating the years to his salaries and sketch it. What is his salary after five years?

It is natural to think of a sequence as an infinite list of images: y_1, y_2, $\ldots$, y_{n-1}, y_n, y_{n+1}, $\ldots$. We call these values of the dependent variable the *terms* of the sequence. We call y_1 the *first term* and y_n the *n*th *term*. We call y_n and y_{n+1} *successive terms*.

We have seen that a sequence may be defined by giving a defining formula for the function. One may also be defined by specifying the initial term and giving a rule for finding the successor of any term. Thus we could define an arithmetic progression by writing

$$y_1 = b \wedge y_{n+1} = y_n + m.$$

(b) Show by induction that if $y_{n+1} = y_n + m$, then $y_n = y_1 + (n - 1)m$.

When we define a sequence by giving the initial term and a formula for finding the successor of any term, the sequence is said to be defined *recursively*. The equation giving y_{n+1} in terms of y_n is called a *recursion relation*. Just above we defined an arithmetic progression recursively with the recursion relation $y_{n+1} = y_n + m$.

(c) Where previously have we used a recursive definition?

In Section 1–13 we defined a^2 and a^3, but we did not give a general definition of a^n for any natural number n. This is most conveniently done recursively, as follows.

(1) **Def.** $$a^1 = a, \qquad a^{n+1} = a \cdot a^n.$$

(d) Use definition (1) to find a^2, a^3, a^4, and a^5.

Definition (1) defines a^n for any value of a for which multiplication is defined. We can use it very conveniently to prove the following laws when the exponents are natural numbers.

(2) $$a^n \cdot a^m = a^{n+m},$$

(3) $$(a^m)^n = a^{mn},$$

(4) $$(ab)^n = a^n b^n,$$

(5) $$(a/b)^n = a^n/b^n,$$

(6) $$a^n/a^m = a^{n-m} \qquad \text{(when } n > m\text{)},$$

(7) $$a^n/a^m = 1/a^{m-n} \qquad \text{(when } n < m\text{)}.$$

The simplest procedure is to prove (2) through (5) by induction and then derive the other two from them. We illustrate by proving (2) and (6). We prove (2) by induction on n. (Compare the beginning of the proof of (6–5–9)). Here $s(x) = [a^x a^m = a^{x+m}]$, $s(1) = [a^1 a^m = a^{1+m}]$, $s(x+1) = [a^{x+1} a^m = a^{x+1+m}]$. By (1), $s(1)$ is true. For the induction we assume $a^x a^m = a^{x+m}$. Now $a^{x+1} a^m = aa^{x+m}$. Using (1) again, we find $a^{x+1} a^m = a^{x+m+1}$, which is seen to be synonymous with $s(x+1)$. Having proved $s(1)$ and derived $s(x+1)$ from $s(x)$, we may assert (2) by the Rule of Induction.

(e) Similarly prove (3) by induction on n.

To prove (6) we use the definition (1–14–1) to get the equivalent equation $a^n = a^m a^{n-m}$. But by (2) this is equivalent to $a^n = a^n$.

(f) Why $n > m$ in (6)? (g) Prove (7).

It would be natural to extend the meaning of a^n to the case of $n = 0$. We are free to define a^0 as we wish, but we should like to do so in such a way that laws (3) through (8) still hold. For example, we wish $a^0 a^m = a^{0+m}$. But this is the same as $a^0 a^m = a^m$, which implies $a^0 = 1$. Accordingly, we define.

(8) Def. $a^0 = 1$ $(a \neq 0)$.

The limitation $a \neq 0$ is dictated by the fact that $0^1 = 0$ and $0^n = 0$ by (1), so that we might expect both $0^0 = 0$ and $0^0 = 1$. Accordingly, we take "0^0" as undefined.

(h) Characterize the sequence defined by (1) when $a = 0$.   (i) Show now that (2) holds when m or n is zero.

The function defined by $y = cr^x$ is called an *exponential function*. We designate it by cr^I. So far cr^x is defined only for natural numbers (and zero), so that we are dealing with an exponential function on the natural numbers. The sequence cr^{I-1} is called a *geometric progression*. We have $y_1 = c$, $y_2 = cr$, $y_3 = cr^2$, $y_n = cr^{n-1}$, and $y_{n+1}/y_n = r$. The number r is called the *common ratio* of the geometric progression. We also call r^n the nth power of r, and call r the *base*.

(j) Sketch the geometrical progression $2 \cdot (1.5)^{I_N}$.   (k) Show by induction that if $y_{n+1} = ry_n$, then $y_n = y_1 r^{n-1}$.

We define $n!$, called *n-factorial*, by the recursive definition.

(9) Def. $0! = 1$, $(n + 1)! = (n + 1) \cdot n!$.

(l) From (9) find $1!$, $2!$, $3!$, $4!$, $5!$, $10!$.   (m) Argue that $n!$ = the product of the first n natural numbers.   (n) Argue that the number of ways four objects can be arranged in a row is $4!$. (*Hint:* Any one of the four can be placed first, then....)   (o) Argue that $n!$ = the number of ways n objects can be arranged in a row.

Problems

In Problems 1 through 8 give the first few terms of the sequences and sketch them.

1. $y_n = 3 + n^2$.
2. $y_n = 2 - n$.
3. $y_n = 10(0.5)^n$.
4. $y_n = 1$.
5. $y_1 = 0$, $y_{n+1} = y_n - 1/10$.
6. $y_1 = -1$, $y_{n+1} = 2y_n$.
7. $-I_N$.
8. $2(-1)^{I_N}$.

9. The first term of an arithmetic progression is 5, and the common difference is 2. Find the 500th term.

10. Suppose that money doubles every ten years if left in the bank to draw interest. If one dollar is left for 1000 years, how much will be on deposit?

11. In how many orders can three people be seated around a table? (We are interested here only in their relative positions!)   four people? n people?

12. Prove (4) by induction.

13. Derive (5).

14. Prove that (3) and (4) hold when m or n is zero.

15. Prove that (5) through (7) hold when m or n is zero.

16. Can a sequence ever be both an arithmetic and a geometric progression?

17. The first term of a geometric progression is 3 and the common ratio is -1. Find the 25th term.

★18. Prove by induction that $a^{2n} > 0$.

★19. Prove by induction that the number of ways of arranging n objects along a straight line is $n!$

★20. Let a sum of money P be deposited at simple interest rate i, so that the interest per period is iP. Let y_n be the total of original deposit and accumulated interest at the end of n periods. Show that the mapping n into y_n is an arithmetic progression.

★21. Let a sum P be deposited at compound interest i, so that the interest for the nth period is iy_n, where y_n is the amount on deposit during the nth period, and $y_{n+1} = y_n + iy_n$. Show that the function that maps n into y_n is a geometric progression, and find r.

★22. In his *An Essay on the Principles of Population* (1798), Malthus wrote, "I think I may fairly make two postulates. First, That food is necessary to the existence of man. Secondly, That the passion between the sexes is necessary, and will remain nearly in its present state.... Assuming, then, my postulata as granted, I say, that the power of population is indefinitely greater than the power in the earth to produce subsistence for man. Population, when unchecked, increases in a geometrical ratio. Subsistence only increases in an arithmetical ratio. A slight acquaintance with numbers will show the immensity of the first power in comparison with the second." Illustrate the last remark by sketching on the same graph the arithmetic progression with $y_1 = 1$, $m = 2$ and the geometric progression with $y_1 = 1$, $r = 2$. Similarly sketch $y_1 = 10$, $m = 3$; and $y_1 = 1$, $r = 1.5$.

★23. Suppose that for each 1000 living people, an average of 30 more are born than die in each year. Show that the populations in successive years are the terms of geometric progression.

★24. Show that population increases in geometric progression if and only if the yearly increase in population is a fixed percentage of the population in that year.

★25. How do you explain the fact that population has not increased in geometric progression in the last century?

★26. Under what conditions does subsistence increase only in arithmetic progression? Do you think these conditions are fulfilled?

★27. Find the first few terms of the sequence defined by $y_0 = 0$, $y_1 = 1$, $y_{n+1} = y_n + y_{n-1}$. (This sequence is due to Fibonacci, an Italian mathematician of the twelfth century; for some of its applications, see *Mathematics for Science and Engineering*, by P. L. Alger, McGraw-Hill, 1957.)

★28. The following recurrence relation is adapted from *Business Cycles in the United States of America 1919–1932*, by J. Tinbergen: $y_n = ay_{n-1} - by_{n-2}$. For each of the following cases, calculate enough terms of the sequence to guess its character: $a = 1.6$, $b = 1$, $y_1 = 0$, $y_2 = 10$; $a = 2.5$, $b = 1$, $y_1 = 0$, $y_2 = 10$; $a = 1.8$, $b = 0.1$, $y_1 = 0$, $y_2 = 10$.

ANSWERS TO EXERCISES

(a) $y_1 = 5000$, $y_n = 250(n - 1) + 5000$. Sequence is $250I_N + 4750$. $y_6 =$ 6250. (b) $s(x) = (y_x = y_1 + (x - 1)m)$. $s(1) = (y_1 = y_1 + 0 \cdot m)$. By hypothesis, $y_{x+1} = y_x + m$. From the induction hypothesis, $y_{x+1} = y_1 + (x - 1)m + m = y_1 + (x + 1 - 1)m$, which completes the proof, since $s(x + 1) = (y_x = y_1 + (x + 1 - 1)m$. (c) To define addition and multiplication in Section 6-4. (d) $a^5 = aaaaa$. (e) $s(x) = [(a^m)^x = a^{mx}]$. The induction goes like this: $(a^m)^{x+1} = a^m(a^m)^x = a^m a^{mx} = a^{m+mx} = a^{m(x+1)}$. These steps are justified respectively by (1), the induction assumption, (2), and the distributive law. But $s(x + 1) = [(a^m)^{x+1} = a^{m(x+1)}]$.

(f) If $n \leq m$, $n - m$ is not a natural number, and so the expression would as yet be undefined. (g) It is equivalent to $a^n a^{m-n} = a^m$. (h) All zero. (i) We have done so for $n = 0$, but $a^n a^0 = a^0 a^n$ by the commutative law, so it follows immediately for $m = 0$. (j) Discrete points lying on the curve defined by $y = 2(3/2)^x$. (k) Use an inductive argument. (l) 1, 2, 6, 24, 120, 3,628,800. (m) Use induction. (n) We can place any one of the objects in first place (in four ways), then any of the others in the next (in three ways), then either of the remaining in the next place (two ways), and the last number in the last place (one way). Hence the total number of ways is $4 \cdot 3 \cdot 2 \cdot 1$. (o) Induction.

ANSWERS TO PROBLEMS

1. 4, 7, 12, 19, 28,.... 3. 5, 2.5, 1.25,.... 5. 0, $-1/10$, $-2/10$, $-3/10$,.... 7. -1, -2, -3, -4,.... 9. $y_{500} = 5 + 499 \cdot 2$. 11. 2; 6; $(n - 1)!$. 17. 3.

6-7 The summation notation. Let f be a sequence whose range consists of numbers. We are often interested in the sum of several successive terms. For example, let $y_n = 2n - 1$, where $n \in N$, so that we are dealing with the sequence whose terms are the odd numbers. We note that

$$(1) \qquad y_1 = 1 = 1^2,$$

$$(2) \qquad y_1 + y_2 = 1 + 3 = 4 = 2^2,$$

$$(3) \qquad y_1 + y_2 + y_3 = 1 + 3 + 5 = 9 = 3^2,$$

$$(4) \qquad y_1 + y_2 + y_3 + y_4 = 1 + 3 + 5 + 7 = 16 = 4^2.$$

It begins to look as though the sum of the first n odd natural numbers is just n^2. But if we wish to prove this, or even to talk about it precisely, we need a notation to indicate the sum of n terms.

Let us indicate the sum of the first n terms of a sequence by s_n. Then the values of s_n are terms of a sequence. Furthermore,

$$(5) \qquad s_1 = y_1, \qquad s_{n+1} = s_n + y_{n+1}.$$

It appears that s_n can be found from (5) for any value of n. Indeed, (5) amounts to a recursive definition of s_n. (For a proof, see Landau's *Foundations of Analysis*.)

The notation s_n has the disadvantage of not indicating explicitly the sequence that is being summed. The following notation, called the *summation notation*,

(6)
$$\sum_{i=1}^{i=n} y_i \quad \text{or} \quad \sum_{i=1}^{i=n} f(i),$$

stands for s_n and includes an expression for the terms to be added. The summation sign indicates that we add together the results of substituting for i, successively, names of each integer from 1 to n inclusive. Now we can write our conjecture as follows.

(7)
$$\sum_{i=1}^{i=n} (2i - 1) = n^2.$$

(a) Argue that i is a dummy in (6) and (7). (b) Write, using the summation notation, and evaluate numerically, s_5 where $y_n = 2n - 1$, and (c) s_5 where $y_n = (1/2)^n$.

We get a formal definition of the summation notation by simply replacing s_n in (5) with either expression in (6).

(8) **Def.**
$$\sum_{i=1}^{i=1} f(i) = f(1),$$

(8') **Def.**
$$\sum_{i=1}^{i=n+1} f(i) = \left[\sum_{i=1}^{i=n} f(i) \right] + f(n + 1).$$

(d) Rewrite (8) and (8'), using the other expression in (6).

In (6) we call $f(i)$ or y_i the *summand*, the dummy i the *variable of summation*, and 1 and n the *lower and upper limits of summation*. The notation indicates the sum of the values of the summand for all integral values of the variable of summation between the lower and upper limits of summation, inclusive. For example,

(9)
$$\sum_{i=1}^{i=4} (-1)^i = (-1)^1 + (-1)^2 + (-1)^3 + (-1)^4,$$

(10)
$$\sum_{i=1}^{i=4} i = 1 + 2 + 3 + 4.$$

Write out the following summations, as we have (9) and (10), with summation on the left and expanded form on the right: (e) $f(i) = i^2$, $n = 4$; (f) $y_i = ar^{i-1}$, $n = 3$; (g) $y_i = 3i$, $n = 5$.

Now we can prove (7) by induction. Before starting the proof, we note that

(11) $$s(x) = \left[\sum_{i=1}^{i=x} (2i - 1) = x^2 \right],$$

(12) $$s(1) = \left[\sum_{i=1}^{i=1} (2i - 1) = 1^2 \right],$$

(13) $$s(x + 1) = \left[\sum_{i=1}^{i=x+1} (2i - 1) = (x + 1)^2 \right].$$

Now $s(1)$ is immediate from (8). Also from (8), we have

(14) $$\sum_{i=1}^{i=x+1} (2i - 1) = \left[\sum_{i=1}^{i=x} (2i - 1) \right] + 2(x + 1) - 1.$$

Using the induction assumption $s(x)$, we have

(15) $$\sum_{i=1}^{i=x+1} (2i - 1) = x^2 + 2x + 1 = (x + 1)^2.$$

But (15) is just $s(x + 1)$, and the proof is complete.

(h) Similarly prove that the sum of the first n even positive natural numbers is $n(n + 1)$.

We may wish to indicate the sum of successive terms of a sequence beginning with some term other than the first. We allow for this by the following definition.

(16) **Def.** $$\sum_{i=m}^{i=n} f(i) = \sum_{j=1}^{j=n-m+1} f(m - 1 + j).$$

Note that in the right member $f(m - 1 + 1) = f(m)$, $f(m - 1 + 2) = f(m + 1)$, and so on, until $f(m - 1 + n - m + 1) = f(n)$. Hence the summation is just the sum of all values of $f(i)$ for i between m and n inclusive. Note that there are $n - m + 1$ terms in the sum. This definition also has meaning for m and/or n zero or negative integers, provided that $m < n$ and that $f(i)$ is defined for $m \leq i \leq n$.

Expand the following summations: (i) $\sum_{i=2}^{i=5} (5i)$; (j) $\sum_{i=3}^{i=7} y_i$; (k) $\sum_{i=5}^{i=10} y_{2i}$; (l) $\sum_{i=n}^{i=n+3} i^2$.

Sometimes, to avoid the summation notation, a summation is expressed by writing the first few terms, the last term, and an expression for the ith term, with dots to indicate the missing terms. The notation is defined as follows.

(17) **Def** $f(1) + f(2) + \cdots + f(i) + \cdots + f(n) = \sum\limits_{i=1}^{i=n} f(i).$

(m) Define $f(m) + f(m + 1) + \cdots + f(i) + \cdots + f(n).$

The following are fairly obvious, but can be proved from (8) and (8').

(18) $\sum\limits_{i=m}^{i=n} cf(i) = c \sum\limits_{i=m}^{i=n} f(i).$

(19) $\sum\limits_{i=m}^{i=n} c = (n - m + 1)c.$

(20) $\sum\limits_{i=m}^{i=n} (f(i) + g(i)) = \sum\limits_{i=m}^{i=n} f(i) + \sum\limits_{i=m}^{i=n} g(i).$

(21) $\sum\limits_{i=m}^{i=n} f(i) = \sum\limits_{i=m}^{i=p} f(i) + \sum\limits_{i=p+1}^{i=n} f(i) \qquad (m \le p \le n).$

(n) Illustrate (18) through (21) for $f(i) = y_i$, $g(i) = x_i$, $m = 1$, $n = 3$, $p = 2$.   (o) Of what law is (18) a generalization?   (p) Of what law is (21) a generalization?

Problems

1. Use the summation notation to rewrite the left member of (22) and then prove it by induction.

(22) $1 + 2 + \cdots + i + \cdots + n = n(n + 1)/2.$

2. State (22) in words.
3. Write (23) with the summation notation, and then prove it.

(23) $1 + 4 + \cdots + i^2 + \cdots + n^2 = n(n + 1)(2n + 1)/6.$

4. Prove that if $y_n = 1/n(n + 1)$, then $s_n = n/(n + 1)$.
5. Prove (18) by induction on n. (*Hint:* Let $n_1 = m + 1$.)
6. Prove (19).
★7. Prove (20) and (21)
8. Use (18) through (21) to prove that the sum of the first n terms of an arithmetic progression is half the number of terms times the sum of the first and nth terms; that is,

(24) $\sum\limits_{i=1}^{i=n} [y_1 + (i - 1)m] = (n/2)[2y_1 + (n - 1)m]$

$$= (n/2)(y_1 + y_n).$$

9. Prove by induction that the sum of the first n terms of geometric progression is given by

(25)
$$\sum_{i=1}^{i=n} y_1 r^{i-1} = y_1(1 - r^n)/(1 - r).$$

10. Find the sum of the first 1000 positive integers.
11. Find the sum of the first ten terms of the sequence 1, 2, 4, 8, ...
12. Find the sum of the first 100 terms of the sequence $(1/2)^{I_N-1}$.
★13. Prove $[\sum_{i=1}^{i=n} i]^2 = \sum_{i=1}^{i=n} i^3$.
★14. The summation notation is used to indicate the application of addition to successive terms of a sequence. Similar notations are used to indicate other operations. For example,

(26)
$$\prod_{i=m}^{i=n} y_i$$

means the product of all values of y_i for $m \le i \le n$. What is $\prod_{i=1}^{i=n} i$?
★15. Define (26) along the lines of the definition of the summation notation, or in some other way of your choice.
★16. Give a formal definition of $y_1 \cdot y_2 \cdot y_3 \cdot \ldots \cdot y_n$.
★17. State and justify laws for $\prod$ analogous to (18) through (21).
★18. Prove that $\prod_{i=1}^{i=n} i = n!$.

ANSWERS TO EXERCISES

(a) See Section 2–11.   (b) $\sum_{i=1}^{i=5} (2i - 1) = 1 + 3 + 5 + 7 + 9 = 25$.
(c) $\sum_{i=1}^{i=5} (1/2)^i = 1/2 + 1/4 + 1/8 + 1/16 + 1/32 = 31/32$.
(d) $\sum_{i=1}^{i=1} y_i = y_1$, $\sum_{i=1}^{i=n+1} y_i = [\sum_{i=1}^{i=n} y_i] + y_{n+1}$.
(e) $\sum_{i=1}^{i=4} i^2 = 1 + 4 + 9 + 16$.   (f) $\sum_{i=1}^{i=3} ar^{i-1} = a + ar + ar^2$.
(g) $\sum_{i=1}^{i=5} 3i = 3 + 6 + 9 + 12 + 15$.
(h) Here $s(x) = [s_x = x(x + 1)]$. For $x = 1$, we have $2 = 1(1 + 1)$. $s_{x+1} = s_x + 2(x + 1)$. Using the induction assumption, $s_x = x(x + 1)$, we find $s_{x+1} = x(x + 1) + 2(x + 1) = (x + 1)(x + 1 + 1)$, which is just $s(x + 1)$.
(i) $10 + 15 + 20 + 25$.   (j) $y_3 + y_4 + y_5 + y_6 + y_7$.   (k) $y_{10} + y_{12} + y_{14} + y_{16} + y_{18} + y_{20}$.   (l) $n^2 + (n + 1)^2 + (n + 2)^2 + (n + 3)^2$.
(m) $\sum_{i=m}^{i=n} f(i)$.   (n) $cy_1 + cy_2 + cy_3 = c(y_1 + y_2 + y_3)$, $c + c + c = 3c$, $(y_1 + x_1) + (y_2 + x_2) + (y_3 + x_3) = (y_1 + y_2 + y_3) + (x_1 + x_2 + x_3)$, $y_1 + y_2 + y_3 = (y_1 + y_2) + y_3$.   (o) Distributive law.   (p) Associative law.

ANSWERS TO PROBLEMS

1. $\sum_{i=1}^{i=n} i$. 3. $\sum_{i=1}^{i=n} i^2$. 9. Assuming $s_x = y_1(1 - r^x)/(1 - r)$ for the induction, we have $s_{x+1} = s_x + y_{x+1} = y_1(1 - r^x)/(1 - r) + y_1 r^x = y_1(1 - r^{x+1})/(1 - r)$. 11. $2^{10} - 1$. 13. Use (22).

★6–8 Extension of the number system. The natural numbers have an origin lost in the very first cultural steps of mankind. Other kinds of numbers are relatively much more recent. The extensions of the number system have been dictated by inadequacies of existing numbers for dealing with practical problems and by the urge for mathematical completeness. Both motives may be summarized by saying that it is desirable for practical and aesthetic reasons to have a number system that is closed under the operations we wish to perform and also closed under their inverses. In this section we sketch in outline the history and logical structure of the system of rational numbers.

(a) What does "closed" mean in this context?

The natural numbers, N, are suitable for counting, and they are closed under addition and multiplication, which are essentially methods of counting. But they are not closed under the inverse operations of subtraction and division. Geometrically, this means that they are suitable for measuring undirected distances equal to an exact number of units of measurement, but inadequate for indicating direction or fractional parts of distances. Algebraically, it means that they are insufficient for solving equations of the form $a + x = b$ when $a \geq b$, or of the form $ax = b$ when a does not divide b.

Historically, the inadequacy of natural numbers to represent fractional parts was felt first. Accordingly, the positive rationals were invented in very ancient times, with the numerical and geometric intepretation indicated in Section 1–2. There are many ways of introducing the positive rationals into a formal theory of numbers. A natural one is to consider the *rational number* m/n as a relation (a ratio) to which (x, y) belong if and only if $x/y = m/n$, that is, $xn = my$. Formally,

(1) Def. $m/n = \{(x, y) \,|\, my = nx \,\wedge\, (x, y) \in N \times N\}.$

According to (1) the rational $3/4$ consists of all those ordered pairs of members of N whose components are in the indicated ratio, that is, $(3, 4)$, $(6, 8)$, $(9, 12), \ldots$. It is then easy to prove

(2) $(a/b = c/d) \leftrightarrow (ad = bc).$

From this (1–14–2) follows. Then if we *define* addition and multiplication of rationals by (1–14–3) and (1–14–8), and subtraction and division, as usual, by (1–14–19) and (1–14–1), we can prove all the laws concerning positive fractions. Finally, if we define $<$ by

(3) Def. $(a/b < c/d) = (ad < bc),$

the rules of inequalities can be proved for the positive rationals.

(b) Derive (2) from (1).   (c) By relying on laws assumed already proved for natural numbers and on (2), show from (1–14–3) as a definition of addition that $a/b + c/d = (ad + bc)/bd$.   (d) Show similarly that addition of positive rationals is commutative.   (e) Do the same for multiplication.

Among the positive rationals Ra^+, let us consider just those of the form $n/1$ where $n \in N$, and let us call the set of all such rationals K. Note that $n/1 = \{(n, 1), (2n, 2), (3n, 3), \ldots\}$, so that $n/1$ and n *are not identical!* However, we note that $K \approx N$, since the function F defined by $F(n) = n/1$ maps N onto K in a one-to-one correspondence. Moreover, this correspondence has the following interesting properties.

$$(4) \qquad\qquad F(n + m) = F(n) + F(m),$$

$$(5) \qquad\qquad F(n \cdot m) = F(n) \cdot F(m),$$

$$(6) \qquad\qquad F(m) < F(n) \leftrightarrow m < n.$$

In words, the image of a sum is the sum of the images, and the image of a product is the product of the images. These laws follow immediately from the definitions of F and the operations. For example, $F(n + m) = (m + n)/1 = m/1 + n/1 = F(m) + F(n)$.

(f) Similarly justify (5).   (g) What is $F*$?   (h) Argue that (4) and (5) hold also for $F*$.   (i) Complete the following verbal interpretation of (6): The images of two natural numbers are in the relation of less than _____. (j) Show that the positive rationals are closed under addition, multiplication, and division.

The results of the last two paragraphs may be verbalized by saying that there exists between N and K a one-to-one correspondence that preserves sums, products, and order. In other words, the natural numbers and the positive rationals with denominator 1 *behave exactly the same under the operations and relations in which we are interested.* It follows that in any calculation it makes no difference whether we use the symbol n or $n/1$. It is for this reason that we can safely treat n and $n/1$ as the same in arithmetic.

The relation existing between N and K above is called *isomorphism.* Two sets are said to be isomorphic with respect to certain operations and relations if there exists a one-to-one correspondence between them that preserves these operations and relations. More precisely, suppose we have a set A on which there is a set S_O of operations and a set S_R of relations. Similarly, suppose we have a set A' with operations S_O' and relations S_R'. Then we say that A and A' are *isomorphic* with respect to these operations and relations provided there exists a one-to-one correspondence F between $A \cup S_O \cup S_R$ and $A' \cup S_O' \cup S_R'$ that maps a in A into a'

in A', o in S_O into o' in S_O', and r in S_R into r' in S_R', and such that $(a \ o \ b)' = a' \ o' \ b'$ and $a \ r \ b \leftrightarrow a' \ r' \ b'$ are laws.

When two sets are isomorphic with respect to certain operations and relations, every law about one has a corresponding law in the other. The two theories differ only in the symbols used and are *formally* the same. The concept is a powerful one, because it often happens that theories about quite different things turn out to be isomorphic. This means that the things under consideration do not differ in any expressible way *with respect to the operations and relations being considered.* Either may be used as an accurate picture of the other, and a single formal theory suffices to tell us everything about them both *as far as these relations and operations are concerned.*

We may summarize the discussion of the positive rationals by saying that they include a proper subset isomorphic to the natural numbers with respect to the operations of addition and multiplication, and the relation $<$.

The negative numbers and zero were added to the number system only very recently. Until a few hundred years ago, $a - b$ was considered absurd if $a < b$, and zero is not mentioned in European manuscripts until the fourteenth century, although the Hindus and Arabs had used zero before that time. Zero and negative numbers may be introduced by definitions analogous to those for the rationals. We may define the difference $a - b$ as the relation

$$\{(x, y) | a + y = b + x \land (x, y) \in N \times N\},$$

define 0 as $a - a$, and then write $-b$ for $0 - b$ and a for $a - 0$. We then get all the *integers* J defined as differences of natural numbers. The subset J^+ of *positive integers* (given by $a - 0$ with $a \in N$) is then isomorphic to N.

Having defined the integers J as relations on the natural numbers N, we may then define rationals as ratios of integers by $(1)(N : J)$ and so arrive at the rationals Ra. Or we may define the difference $a - b$ as $\{(x, y) | a + y = b + x \land (x, y) \in Ra^+ \times Ra^+\}$. This gives us all the rationals as differences of positive rationals. The rationals defined in this way are not identical with Ra, but they are isomorphic to Ra. The integers and rationals can be introduced in many other ways but, barring errors, they will always be isomorphic to the numbers as defined here. Also, the subset of those differences $a - b$ for which $a > b$, is isomorphic to Ra^+. The isomorphisms hold with respect to the arithmetical operations and the relation of order. On this account, so long as we are interested in only these operations and relations, it makes no difference in which way our numbers have been defined. In the same context we may ignore the difference between the natural numbers and the positive integers.

(k) Show from the definition of $a - b$ that $c - c = d - d$, so that 0 is uniquely defined by $a - a$. (l) Define $<$ for integers. (m) Show that the set of integers of the form $a - 0$ with $a \in N$ is isomorphic to N with respect to addition, multiplication, and $<$.

The rationals are closed under addition, multiplication, and their inverses (except division by zero). However, they are not closed under the inverse of taking powers, i.e., root extraction. To prove this we need just one counterexample:

(7)   *There is no rational number whose square is 2.*

(n) Restate (7), using quantifiers and other logical symbols. State it using the word "irrational."

To show (7) we need to use the fact that a natural number is even if its square is even.

(8) $$m^2 \in E \to m \in E.$$

This is easily proved indirectly by assuming $\sim(m \in E)$. Then m is odd and can be written in the form $2n + 1$ where $n \in N$. Then $m^2 = 4n^2 + 4n + 1 = 2(2n^2 + 2n) + 1$, which is clearly odd. We have shown $\sim(m \in E) \to \sim(m^2 \in E)$, which is equivalent to (8).

Now we show (7) indirectly by assuming its negative. Since (7) $= \sim \exists m \, \exists n \, m \in N \wedge n \in N \wedge (m/n)^2 = 2$, its negation is the assumption that there do exist natural numbers m and n such that $(m/n)^2 = 2$. Moreover, we can assume that m and n are not both even, since a common factor of 2 could be divided out by (1–14–2). We have, then, $m^2/n^2 = 2$ and $m^2 = 2n^2$. We see that m^2 must be even, since it has 2 as a factor. Hence by (8) m is even; that is, $\exists p \, p \in N \wedge m = 2p$. Replacing m by $2p$, we find $(2p)^2 = 2n^2$ or $2p = n^2$. It follows that n^2 is even and by (8) that n is even. But this means that m and n are both even, which contradicts our assumption. Since the negation of (7) leads to a contradiction, it is true by (2–7–43).

The discovery of (7) is attributed to the Pythagoreans, a society of mathematicians which flourished during the sixth century B.C. Since one can easily construct a segment whose length is $\sqrt{2}$ (the diagonal of a square with side 1), the discovery meant that rational numbers are inadequate to represent distances. It was one cause of the divorce of arithmetic and geometry, which lasted until very recent times, for the extension of the number system to include irrationals is not as easy as the previous extensions. Whereas rationals can be defined as simple relations on the natural numbers, it requires infinite sets of rational numbers to define a real number. We postpone discussion of this extension to the next section. However, its pattern is the same as before: definitions in terms of previous numbers, then proof of the laws of algebra.

(o) Prove that $\sqrt{3}$ is not rational.

Problems

1. Show that the even numbers are closed under addition and multiplication.

2. Show that the odd numbers are closed under multiplication but not under addition.

3. If we defined rationals as ratios of integers, instead of just natural numbers, what additional limitation would we have to add to (1)?

4. Show that multiplication of rationals defined as ratios of natural numbers is commutative.

5. Show that the set of vectors (x, y) whose first component is a natural number and whose second component is 0 is isomorphic to the natural numbers under the operation of addition.

6. Prove (1–14–2) from (2).

7. Once negatives and rationals have been introduced, it is natural to try to extend the meaning of a^n to cases where n is any rational number. We do so by the following definitions.

(9) Def. $a^{m/n} = x \ni x^n = a^m \wedge [n \in E \rightarrow x \geq 0]$,

(10) Def. $a^{-n} = 1/a^n$.

Since we have previously defined only non-negative integral exponents, (9) applies only to m and n belonging to N or $m = 0$ and defines non-negative rational exponents. Then (10) has meaning for all non-negative rational values of n and defines rational exponents. Read (9) in words. What is the range of significance of a in (9)?

8. Use (9) to show that $a^{1/2} = \sqrt{a}$ by reference to (1–13–16).

9. Use (9) to show that non-negative rational exponents follow the laws (6–6–2) to (6–6–7).

10. Assuming that the proofs of Problem 9 have been accomplished, use (10) to show that rational exponents satisfy these laws.

11. The number $a^{1/n}$, defined by (9)$(m{:}1)$, is called the *principal nth root of a* and is usually designated by $\sqrt[n]{a}$. Any number x such that $x^n = a$ is called an *nth root of a*. How many square roots $(n = 2)$ does a non-negative real number have? a negative real number?

12. Show that $a^{m/n} = \sqrt[n]{a^m}$.

13. Show that for $a \geq 0$, $b \geq 0$,

(11) $$\sqrt[n]{a}\sqrt[n]{b} = \sqrt[n]{ab},$$

(12) $$\sqrt[k]{\sqrt[n]{a}} = \sqrt[kn]{a},$$

(13) $$(\sqrt[n]{a})^k = \sqrt[n]{a^k}.$$

14. Show that (13) is not a law if the limitation on a and b is removed.

★15. If $a \in C \wedge b \in C \wedge a < b \wedge \sim \exists x\, x \in C \wedge a < x < b$, we call a and b *adjacent* members of C, a the *immediate predecessor* of b, and b the *immediate successor* of a. A set in which every member (except the smallest, if any) has an immediate predecessor and every member (except the largest, if any) has an immediate successor, is called *discrete*. Explain why the natural numbers are discrete.

★16. Show that the rationals are not discrete.

★17. Show that the negative integers are discrete.

★18. Show that any subset of the integers is discrete.

★19. A set in which there are no adjacent members is called *dense*. Show that the rationals are dense.

★20. Show that we can find a rational number arbitrarily close to any given rational number.

Answers to Exercises

(a) See Section 5-5.   (b) a/b consists of all pairs (x, y) such that $ay = bx$, and c/d consists of those for which $cy = dx$. If $a/b = c/d$, then these sets must be the same. That is, (x, y) satisfies $ay = bx$ if and only if it satisfies $cy = dx$. Alternatively, we note first that $(m, n) \in m/n$ by (1), since $mn = nm$. Hence if $a/b = c/d$, then $(c, d) \in a/b$ or $cb = da$. Conversely, if $ad = bc$, then $(x, y) \in a/b \leftrightarrow ay = bx \leftrightarrow acdy = bcdx \leftrightarrow cy = dx \leftrightarrow (x, y) \in c/d$, so that $a/b = c/d$ by (3-3-4).   (c) $a/b + c/d = ad/bd + cb/bd = (ad + cb)/bd$.   (d) Since $ad + cb = cb + ad$ by commutativity of addition of natural numbers.   (e) Using the commutativity of multiplication for natural numbers and the definition (1-14-8), $(a/b)(c/d) = ac/bd = ca/db = (c/d)(a/b)$.   (f) $F(nm) = nm/1 = (n/1)(m/1) = F(n)F(m)$.   (g) $F^*(n/1) = n$.   (h) Since the correspondence is one-to-one and (4) and (5) hold for F.   (i) if and only if the corresponding natural numbers are in the same relation.   (j) Since in each case the definition yields a ratio of positive natural numbers.   (k) $a + y = a + x \leftrightarrow y = x \leftrightarrow b + y = b + x$.   (l) $[a - b < c - d] = [a + d < b + c]$.   (m) $F(a) = a - 0$. $F(a + b) = a + b - 0 = a - 0 + b - 0 = F(a) + F(b)$.   (n) $\sim \exists x\, x \in \mathrm{Ra} \wedge x^2 = 2$. $\sqrt{2}$ is irrational.   (o) $\sqrt{3} = m/n \rightarrow [3n^2 = m^2] \rightarrow [\exists p\, p \in N \wedge m = 3p] \rightarrow 3n^2 = 9p^2 \rightarrow n^2 = 3p^2 \rightarrow \exists q\, q \in N \wedge n = 3q$.

Answers to Problems

1. $2m + 2n = 2(m + n)$, $(2m)(2n) = 2(2mn)$. 3. $n \neq 0$. 5. Since $(x, 0) + (x', 0) = (x + x', 0)$, $F(x) = (x, 0)$. 7. Non-negative reals when n is even, reals when n is odd if we wish x to be real. Of course, if we were being thoroughly rigorous, a would be restricted to rational values since real numbers have not been defined as yet.

6–9 Real numbers. There are many ways in which the real numbers can be defined satisfactorily. All methods have the following features in common.

(1)   *The real numbers are defined in terms of rational numbers.*

(2)   *The arithmetic operations and the relation of inequality are defined in terms of these relations for rational numbers.*

(3)   *With these definitions, the real numbers are proved to satisfy the fundamental algebraic identities (1–16–1) through (1–16–10) and the laws of inequalities of Section 3–5.*

(4)   *Included in the real numbers is a subset isomorphic to the rationals.*

(5)   *The real numbers are isomorphic to the points on a straight line with respect to the operations of arithmetic and order.*

(a) If you did not study Section 6–8, read now the paragraphs on isomorphism following Exercise (e).   (b) Why does it follow from (4) that the reals include subsets isomorphic to the integers, natural numbers, evens, etc.? (c) What is the one-to-one correspondence under which (5) holds? What operations on points correspond to the operations of arithmetic? What is the image of $<$?

To carry out (1) through (5) in detail is much too time-consuming for our purpose here. Instead we mention some ways in which the definitions can be formulated. In Section 1–15 we described real numbers in terms of infinite decimals, an idea which can be used to formulate rigorous definitions and proofs. For details the reader is referred to the first pages of *The Theory of Functions*, by J. F. Ritt. The real numbers may be defined as cuts of rationals. A *cut* is defined in Edmund Landau's *Foundations of Analysis* as a subset of rationals that contains at least one but not all the rationals, that contains no greatest member, and every one of whose members is less than every rational not belonging to it. Speaking in rough geometric terms, a cut is the set of all rationals to the left of a given point on the axis, and the point (real number) is defined by specifying the cut. For details, the reader is referred to Chapter III of Landau's book. Also, real numbers may be defined in terms of sets of nested intervals. A *sequence of nested intervals* is defined as a sequence of closed intervals I_n, each consisting of the rational numbers between two rationals, each one included in the previous one, and the length of the intervals approaching zero as n increases. An informal discussion of this approach may be found in Chapter II of *What is Mathematics?*, by R. Courant and H. Robbins.

(d) Describe an infinite decimal as a special kind of sequence.   (e) Review Section 1–15 and suggest a connection between the definitions in terms of nested intervals and in terms of infinite decimals.   (f) Suggest a definition of < for infinite decimals.   (g) Show that $\{x \mid x < \sqrt{2} \wedge x \in Ra\}$ is a cut.

It is not hard to show that between any two rational numbers there is another, since if a and b are rational so is $(a + b)/2$. It follows that there are an infinite number of rationals between any two of them. We describe this property by saying that the rationals are *dense*.

(h) Why does $a \in Ra \wedge b \in Ra \rightarrow (a + b)/2 \in Ra$?   (i) Show that $a < b \rightarrow a < (a + b)/2 < b$.   (j) Show that the number of rationals between two rationals is not finite (i.e., is infinite).   (k) Show that our definition of "dense" here is equivalent to that given in Problem 19 of Section 6–8.

The property of denseness of the rationals means that they are very "tightly packed" on the line. On the other hand, from Problem 21 in Section 6–2 we know that in some sense there are "more" reals than rationals, so that the real numbers must be even more tightly packed on the line than the rationals! We close our sketch of the real numbers by discussing this idea more carefully.

We call x an *upper bound* of a set B if all members of B are no greater than x; that is, $\forall z \, z \in B \rightarrow z \leq x$. Similarly we call y a *lower bound* of B if $\forall z \, z \in B \rightarrow z \geq y$. A set that has an upper (lower) bound is said to be *bounded from above* (*below*). One that has both is said to be *bounded*. One that has neither is said to be *unbounded*.

(l) Give examples of each category.

A bounded set has infinitely many bounds, since if x is an upper bound, any number greater than x is an upper bound, and similarly for lower bounds. For example, the set of all numbers between 0 and 1 is bounded. Among its upper bounds are 1, 2, 1.001, $\sqrt{2}$, 1000, and 1.3. Among its lower bounds are -99, -0.00001, and 0. Is there a smallest among its upper bounds? Yes, 1 is the smallest upper bound, for if $x < 1$, then the denseness of the reals implies that there is a number between x and 1, and hence a member of the set greater than x. Hence x cannot be an upper bound.

If there exists a number that is smaller than all other upper bounds of a set, we call it the *least upper bound* (lub). Similarly, the *greatest lower bound* (glb) is the largest of the lower bounds. For the open interval between 0 and 1, the lub is 1 and the glb is 0. Note that in this case they are not members of the set.

Let us consider the set A of positive rationals whose squares are less than 2, $A = \{x \mid x^2 < 2 \wedge x \in Ra \wedge x > 0\}$.   This set is certainly bounded from above. Among its upper bounds are 2, 3, 2.4, 1.5, 1.42,

1.415, etc. But does it have a least upper bound? Not if we limit ourselves to rational numbers, for if x is any rational greater than $\sqrt{2}$, we
find another rational between x and $\sqrt{2}$. Hence A has no least upper
bound among the rationals. However, if we include all reals, A has the
least upper bound $\sqrt{2}$.

Find the lub and glb, if any, of the following: (m) N,   (n) J,   (o) Re^+,
(p) $\{x \mid x^2 < 4\}$,   (q) $\{x \mid x^2 < 3\}$,   (r) $\{x \mid x^2 \leq 3\}$

We have seen that unless we include the irrational points on the line
it may happen that a bounded set has no least upper or greatest lower
bound. However, if the irrationals are included, then it can be proved
from the definitions of real numbers that

(6)   *Every set of real numbers that is bounded from above (below)
has a lub (glb) that is a real number.*

A set B such that every subset bounded from above (below) has a lub
(glb) that is a member of B is called *continuous*. The real numbers are
continuous, whereas the rationals are not. The real numbers are sometimes
called the *continuum*. It is the continuousness of the reals that enables
them to completely fill in the line without any jumps (such as exist between the integers) or gaps (such as exist among the rationals).

(s) Argue that every real number is the lub of the set of all the approximations s_i formed by replacing all digits after the ith by zeros in its decimal expansion. Note that this defines each real as the lub of a set of rationals—an
approach alternative to infinite decimals.   (t) Show that every non-empty
subset of N contains its own glb.   ★(u) Prove from (6) that there is a real
number whose square is 2.

The continuousness of the real numbers, expressed by (6), is useful in
many ways. As an illustration, it enables us to extend the definitions of
exponents to all real values. When $a > 0$, a^x is defined for all rational
values of x by (6–8–9) and (6–8–10). Then for $a > 1$ we can define a^c for
any real number c as the least upper bound of all the numbers a^x for rational
values of $x < c$; that is,

(7) **Def.** $a^c = \text{lub } \{z \mid \exists x\, x \in Ra \land x < c \land z = a^x\}.$

Roughly speaking, a^c is a real number that is approximated arbitrarily
closely from below by taking rational powers with exponents smaller
than c. With a similar definition for $0 < a < 1$ all real exponents can be
proved to follow the familiar laws.

(v) Suggest a definition of the product of two positive reals in terms of least
upper bounds.

PROJECTS

1. Study the discussion of the number system in Chapter II, sections 1 and 2, of *What is Mathematics?*, by R. Courant and H. Robbins.
2. Study the theory of infinite decimals in *The Theory of Functions*, by J. F. Ritt. Try to reformulate the discussion as symbolically as possible.
3. Study Edmund Landau's book, *Foundations of Analysis*. (This is suitable for a term project.)
4. Study *Theory of Sets*, by E. Kamke (Dover, 1950), for further details on continuity, denseness, etc.

ANSWERS TO EXERCISES

(b) Because the rationals include such subsets, and hence any isomorphic set must also have such subsets.   (c) $F(x)$ = the point corresponding to x. Dom (F) = real numbers, Rge (F) = axis. The image of $<$ is the relation (to the left of).   (d) $F(0)$ = the integral part, $F(1)$ = the first digit after the decimal point, $F(i)$ = the ith digit.   (e) The infinite decimal specifies a sequence of nested intervals in which the ith has length 10^{-i}.   (f) The greater is the one that has a greater number in the first position in which they differ. (g) A subset of Ra, since $x \in Ra$ is part of the defining sentence; non-empty, since $0^2 < 2$; does not contain all Ra, since $3^2 > 2$; contains no greatest member, since, if $x < \sqrt{2}$, we can find a rational y such that $x < y < \sqrt{2}$; and satisfies the final condition, since if $y \geq \sqrt{2}$ and $x < \sqrt{2}$, then $x < y$.   (h) Because rationals are closed under addition and division.   (i) $a + (b - a)/2 = (a + b)/2 = b - (b - a)/2$.

(j) Suppose the number is finite, then choose the largest. There will be another between it and the bigger of the two given rationals, thus contradicting the assumption.   (k) No adjacent members means that there is always another between any two.   (m) No lub, glb = 1.   (n) Not bounded.   (o) No lub, glb = 0.   (p) lub = 2, glb = −2.   (q) lub = $\sqrt{3}$, glb = $-\sqrt{3}$. (r) Same as (q).   (s) See Section 1–15.   (t) See Problem 3 in Section 6–5. (v) The least upper bound of the set of all products of positive rationals less respectively, than the two real factors.

★6–10 **Systems of numeration.** Our decimal system of expressing numbers is called a *positional system*, because the meaning of each digit in a numeral depends on its position relative to the others. Thus in 9375, 9 stands for 9000, 3 for 300, 7 for 70, 5 for 5, and 9375 for the sum of these.

(1) $9375 = 9 \cdot 10^3 + 3 \cdot 10^2 + 7 \cdot 10^1 + 5 \cdot 10^0.$

The number 10 is called the *base* of the system. Every number is expressed as a sum of terms each of which is the product of a number less than the base times a power of the base.

Different civilizations have had distinctive systems of numeration, and it is only recently that the decimal system, based on the number 10, has come into general world-wide use. There is nothing essential about the use of 10. Other civilizations have used other bases, or even no base at all, and 10 is not the best base for all purposes.

The system of numeration using the base 2, called the *binary system*, requires just two digits, 0 and 1. It has great practical importance today because it is the system used by most of the new electronic "brains."

To distinguish numbers written in the binary system we place the subscript b after binary numerals whenever they appear with common decimals. Then we have

(2)

$$1_b = 1 \qquad 101_b = 5 \qquad 1001_b = 9 \qquad 1101_b = 13$$

$$10_b = 2 \qquad 110_b = 6 \qquad 1010_b = 10 \qquad 1110_b = 14$$

$$11_b = 3 \qquad 111_b = 7 \qquad 1011_b = 11 \qquad 1111_b = 15$$

$$100_b = 4 \qquad 1000_b = 8 \qquad 1100_b = 12 \qquad 10000_b = 16$$

Note that $1_b = 2^0$, $10_b = 2^1$, $100_b = 2^2$, $1000_b = 2^3$, and $10000_b = 2^4$, just as $1 = 10^0$, $10 = 10^1$, $100 = 10^2$, $1000 = 10^3$, and $10000 = 10^4$. Note also that any binary numeral expresses a number as a sum of terms each of which is either 0 or 1 times a power of 2. For example,

(3) $$101101_b = 1 \cdot 2^5 + 0 \cdot 2^4 + 1 \cdot 2^3 + 1 \cdot 2^2 + 0 \cdot 2^1 + 1 \cdot 2^0.$$

Write as in (3) and then evaluate in the common decimal notation: (a) 101_b, (b) 1001_b,   (c) 10001_b,   (d) 100001_b,   (e) 101011_b.

Express in the binary notation: (f) 9,   (g) 17,   (h) 23,   (i) 34,   (j) 50.

The reader may wonder why it is advantageous to use a notation in which so many digits have to be written to represent a number. The answer lies in the following addition and multiplication tables for the binary system.

Addition   *Multiplication*

(4)

+	0	1
0	0	1
1	1	10

·	0	1
0	0	0
1	0	1

With the use of the tables, binary arithmetic follows the usual rules. Since (4) is so simple, very little strain is put on the memory and the operations are very easy. This is particularly important for electronic brains, since they perform routine operations very quickly, but are relatively poor in memory and lack initiative. To add two binary numerals, we place them in the usual way. Then each column adds to zero or one, with one carried when both digits in the column are 1's. For example,

(5)
$$
\begin{array}{r}
1011011 \\
+\ 101110101 \\
\hline
111010000
\end{array}
$$

Multiplication is even easier, since to multiply by 1 we simply recopy and to multiply by 0 we do not recopy! For example,

(6)
$$
\begin{array}{r}
1011101 \\
\times\ 10110 \\
\hline
1011101 \\
1011101 \\
1011101 \\
\hline
11111111110
\end{array}
$$

Perform the following operations, where all numerals are binary, by using (4) and the usual rules of arithmetic: (k) $101111 + 101001$, (l) $1011 + 1011 + 10101$, (m) $11101 - 1011$, (n) $11011 \cdot 11001$, (o) $11011/11$.

Decimals less than 1 can be represented in the binary system as easily as in the decimal system. For example, $1.101_b = 1 + 1/2 + 1/8 = 1 + 2^{-1} + 2^{-3}$.

(p) Find in the usual decimal notation 11.011_b and 0.101011_b. (q) Find, by using the binary notation only, $(1.101_b)(0.101011_b)$ and check your result by changing to the usual decimal notation.

The number system based on twelve is called the *duodecimal system*. It requires two additional digits. Let $e_d = 10$, $f_d = 11$, where the subscript d refers to duodecimal numerals.

(r) Construct addition and multiplication tables for duodecimal arithmetic. (s) Transform to common decimals: 11_d, 23_d, 30_d. (t) Transform to duodecimals: 13, 95, 144.

PROBLEMS

1. Find the decimal expansion in the duodecimal system of $1/3$, $1/4$, $1/6$.

2. Can you think of any advantages of the duodecimal system over the decimal system?

3. Investigate the system with base 3. Make addition and multiplication tables, transform numbers between this system and the common decimal system, do sample computations, and evaluate fractions.

4. Could we represent all real numbers as binary decimals? Would this simplify the arguments of Section 1–15? Carry through the argument of that section with this change.

5. Show that if one carries out the computation of Problem 11 in Section 1–10, the last two digits in the final answer give the change in one's pocket and the first two one's age. (*Suggestion:* Let $A = 10x + y$, $C = 10a + b$.)

6. Why do you suppose that the binary system is convenient for machines that operate electrically? (*Hint:* The current is either off or on in a circuit, so one could represent 0 and 1 by....)

★7. On the face of a nuclear scaler used to count cosmic radiation particles, there appears a row of eight little lights labeled 1, 2, 4, 8, 16, 32, 64, 128. When the machine is operating, the lights flash on and off. Immediately after all the lights are on at once, they all go off at once, and at the same time the number showing in a little window increases by one. Then the lights begin flashing, seeming to move from left to right. What is the explanation?

★8. Use simple examples to justify the usual procedures for doing arithmetic by reference to laws of algebra.

★9. Show that a number is divisible by four if and only if the number represented by its last two digits in the decimal system is divisible by four.

★10. Show that a number written in decimal notation leaves the same remainder when divided by nine as does the sum of its digits.

★11. Show the same for a number divided by eleven when the number is written in duodecimal notation.

ANSWERS TO EXERCISES

(a) 5.   (b) 9.   (c) 17.   (d) 33.   (e) 43.   (f) 1001.   (g) 10001.
(h) 10111.   (i) 100010.   (j) 110010.   (k) 1011000.   (l) 101011.
(m) 10010.   (n) 1010100011.   (o) 1001.   (p) 27/8 and 43/64.   (s) 13, 27, 36.   (t) 11_d, $7f_d$, 100_d.

ANSWERS TO PROBLEMS

1. 0.4_d, 0.3_d, 0.2_d. 2. More simple fractions are terminating decimals. 4. Each interval would be 1/2 the size of the preceding one. 5. $50(2(10x + y) + 5) + (10a + b) - 365 + 115 = 1000x + 100y + 10a + b$.

★6–11 **Approximation.** We assume that the reader is able to carry out arithmetic operations with decimals. Effective numerical calculation involves more than this. Indeed, because of the power of modern calculating machines, the ability to do arithmetic is less important than an understanding of the nature of arithmetical operations and the errors inevitably involved. Error is an essential part of most numerical work for several reasons. In the first place, in practical computations we are usually working with numbers obtained by measurements that cannot be exact. Secondly, we have seen that most numbers cannot be expressed exactly by terminating decimals, and irrational numbers cannot be expressed as repeating decimals or ratios of integers. Thirdly, the capacities of machines and the time and patience of human beings set limits to the accuracy of our work. Even if we knew two numbers exactly to 100 decimal places, we would not care to find their product exactly, since this would require

200 decimal places! The purpose of this section is to familiarize the reader with standard terminology for talking about approximation.

Errors. Suppose A is an approximation to the true value T. Then we define the *error* E by

(1) $$E =_d A - T.$$

For example, let $A = 0.67$ and $T = 2/3 = 0.\dot{6}$. Then $E = 0.67 - 0.\dot{6} = 0.67 - 0.666\ldots = 0.00333\ldots = 0.00\dot{3}$.

(a) Complete: The error is the _____ minus the _____. The error is negative when _____. (b) Verify the result in the example in the previous paragraph by calculating $E = 67/100 - 2/3$.

The *relative error* R is defined by

(2) $$R =_d E/T.$$

(c) Complete: the relative error is the _____ divided by the _____, or the ratio of the _____ to the _____. (d) What is the relative error when 0.26 is used as an approximation to 0.264?

Precision. Usually we do not know the error when we are using an approximation, but we almost always have some idea of how large it may be. We may know, for example, that the digits in a decimal are reliable from the first digit on the left up to a certain one, but unreliable after that. If the last reliable digit is to the left of the decimal point, we say that the number is precise to units, tens, hundreds, etc., according to the position of the last reliable digit. If the last reliable digit is the nth one at the right of the decimal point, we say that the number is precise to n decimal places. When we say that a certain digit is *reliable*, we mean that the absolute value of the error is not greater than the number represented by 5 in the *next* decimal place. We use the word *precision* to refer to reliability measured in this way.

(e) How large could the error be if a decimal is known to three decimal places? (f) if it is known to units? (g) if it is known to hundreds?

Accuracy. The reliable digits in a decimal, not including initial zeros, are called *significant digits*. For example, there are 4 significant digits in 0.03105, if we assume that all digits are reliable. The concept of significant digit is not applied to an expression for zero. The reliability of a decimal may be indicated by giving the number of significant digits. We use *accuracy* to refer to reliability measured in this way.

(h) Suppose 3.149 is accurate to three significant digits. Find the largest possible error and largest possible relative error. (i) Do the same for 31.49.

Rounding off. It is often desirable to *round off* a decimal either to eliminate unreliable digits or to simplify calculations. The following rule is generally used.

(3)
If the digits to be dropped represent less than half a unit in the last digit retained, drop them and make no other change. If they represent more than half a unit, drop them and increase the last retained digit by one. If they represent exactly half a unit, the last retained digit is left unchanged if it is even and increased by one if it is odd.

In brief, round off to the closer approximation, choosing the even possibility if they are equally close. For example, we round off 3.14159 to five significant digits (four decimal places) by writing 3.1416, which is closer to 3.14159 than is 3.1415. But we round off 3.1415 to 3.142, since 3.141 and 3.142 are equally close. This rule covers rounding off to as few as one significant digit.

(j) Round off to 3 significant digits: 3.14159; 0.03841; 99.47; 99.45; 37,251; 63140.

Scientific notation. It is customary in science not to write unreliable digits, except sometimes during calculations. When this convention is followed, we know that all digits are reliable. But if we try to round off a decimal accurate to, say, hundreds, we find that we must write unreliable digits in the usual notation. For example, suppose 582,937 is accurate only to hundreds. We may round it off to 582,900, but we must keep the zeros even though they are unreliable. The difficulty is avoided if we shift the decimal point until it lies to the right of the first significant digit (i.e., the first nonzero digit from the left), and compensate by multiplying by some power of 10. Thus $582,937 = 58293.7 \times 10 = 5829.37 \times 10^2 = 582.937 \times 10^3 = 5.82937 \times 10^5$. Now we round off to 5.829×10^5, where we no longer need the zeros. In scientific notation a numeral is in the form $A \times 10^n$, where $1 \leq A < 10$ and n is an integer. The number 0 is not written in scientific notation.

A rule for writing numerals in this way is:

(4)
Move the decimal point to the position immediately to the right of the first significant digit and multiply by 10^n, where n is the number of places from this position to the original position of the decimal point, counting positive to the right, negative to the left.

Numerals expressed in this way are said to be in *scientific notation*.

(k) Write in scientific notation: 48.2; 951.6; 2.97; 0.199; 0.03846; 30.016; 0.00074; 3,001,309; 1,000,000,000.   (l) Write in the usual decimal notation: 8.91×10^2, 3×10^6, -4.397×10^3, 1.31×10^{-1}, 4.985×10^{-6}.

The scientific notation has other advantages. It saves space in writing decimals representing very large or very small numbers. It permits the quick comparison of numbers. It indicates at a glance the accuracy of a decimal, since the number of digits that appear is the number of significant digits.

Order of magnitude. If we express a decimal in scientific notation in the form $A \times 10^n$ with $1 \leq A < 10$, then replace A by 1 or 10 according to whether $A \leq 5$ or $A > 5$, the result is a power of ten that we call the *order of magnitude* of the original number. It is roughly the nearest power of ten. We may think of it as rounding off to no significant digits. For example, the order of magnitude of 4.92 is 1, of 5.78 is 10, of 68 is 100, of 0.3 is 10^{-1}, etc. This definition gives the order of magnitude of any non-zero number. We define the order of magnitude of zero as zero.

Possible errors. Suppose that a decimal (not in scientific notation) has been rounded off to n decimal places and is known to be reliable to the nth place. Then the absolute value of the error cannot exceed $1/2$ unit in the nth place or 5 in the next place to the right. Thus precision gives us an upper bound to possible absolute values of the error, which we call the *possible error*. The *possible relative error* is the possible error divided by the true value.

Suppose that a decimal has been rounded off to n significant digits, where all are known to be reliable. We cannot estimate the error, since the number of significant digits does not tell us about the position of the decimal point. But we can estimate the relative error. Consider 3.19, where we assume all digits significant. The possible error is 0.005, the possible relative error is $0.005/3.19 = 5/3190 \doteq 0.0016 = 1.6 \times 10^{-3}$. For 31.9 we have $0.05/31.9 = 5/3190$, for 319 we have $0.5/319 = 5/3190$, for 0.319 we have $0.0005/0.319 = 5/3190$, and so on. The possible relative error evidently is independent of the position of the decimal point. Hence, to get at the relation between significant digits and possible relative error, we need consider only numbers between 1 and 10.

Consider the following examples:

(5)	Decimal	1	2	5	9	10
(6)	Number of significant digits	1	1	1	1	2
(7)	Possible error	0.5	0.5	0.5	0.5	0.5
(8)	Possible relative error (approximate)	0.5	0.25	0.1	0.06	0.05
(9)	Order of magnitude of (8)	10^{-1}	10^{-1}	10^{-1}	10^{-1}	10^{-2}

It appears that any decimal accurate to one significant digit has a possible relative error whose order of magnitude is 10^{-1}. For two significant digits the relative errors would be $1/10$ as great, and so on. Hence we may state that if a decimal has n significant digits, the order of magnitude of its possible relative error is 10^{-n}.

(m) Find the order of magnitude of the possible relative errors of 10, 20, 30, 50, 70, 90, 12, 99, 100, assuming two significant digits.   (n) Precision gives us an estimate of the _____, accuracy an estimate of the _____.

PROBLEMS

In Problems 1 through 10 an approximation and true value are given, in that order. In each case find the error, the relative error, the precision, and the accuracy. Round off so that all digits are significant, and express in scientific notation.

1. 3.14; 3.13.   2. 31.4; 31.3.
3. 0.314; 0.313.   4. 3,339,412; 3,000,000.
5. 256; 259.   6. 100.01; 100.05.
7. 0.0003818; 0.0003812.   8. 20, 10.
9. 10, 20.   10. 0.001; 0.

11. Express in scientific notation: 8.912; 81.92; 0.8192; 819.2; 0.08192; 0.00008192; 1,300,000; 750,000,000; 1,000,000,000; 0.100,000,000; 8.100,000,001; 358.2×10^{-6}.

12. Express without scientific notation: 3×10; 3.48×10^2; 3.48×10^{-2}; 3×10^{-6}; 1.8901×10^{-3}; 10.000195×10^2; 3.00123×10^{10}.

★13. What is the order of magnitude of the U. S. national debt?

★14. Of the U. S. population?

★15. Of the population of China?

★16. Of the average family income in the U. S.?

★17. Of the prison population in the U. S.?

18. Find the order of magnitude of $16.35 \cdot 10^2$, 3,895,113, and 0.00345.

★19. Argue that the order of magnitude of the ratio of two numbers is approximately the ratio of their orders of magnitude, and illustrate with examples.

★20. Argue by examples that the order of magnitude of the difference of two numbers is not equal to the difference of their orders of magnitude.

★21. Show that there is one number for which "order of magnitude" has no meaning.

ANSWERS TO EXERCISES

(a) approximation; true value. the approximation is less than the true value.   (b) 1/300.   (c) error; true value; error; true value.   (d) $-1/66$ or approximately 0.017.   (e) 0.0005.   (f) 0.5.   (g) 50.   (h) 0.005; 5/3144 or approximately 0.0016.   (i) 0.05; same as in Exercise (h).   (j) 3.14; 0.0384; 99.5; 99.4; 37,300; 63100.   (k) 4.82×10; 9.516×10^2; 2.97×10^0; 1.99×10^{-1}; 3.846×10^{-2}; 3.0016×10; 7.4×10^{-4}; 3.001309×10^6; 10^9.   (l) 891; 3,000,000; -4397; 0.131; 0.000004985.   (m) All 10^{-2}.   (n) error; relative error.

ANSWERS TO PROBLEMS

1. 0.01; approx. 0.0032; precision is to 1 decimal place; accuracy is to two significant figures; 3.1; 3.1×10^0. 3. 0.001; approx. 0.0032; precision is to 2 decimal places; accuracy is to two significant figures; 3.1×10^{-1}. 5. -3; approx. -0.012; tens; two significant digits; 2.6×10^2. 7. 0.0000006; approx. 0.0016; 5 decimal places; two significant digits; 3.8×10^{-4}. 9. -10; -0.5;

hundreds; no digits are significant; the order of magnitude, 10, is all that is known. 11. 8.912×10^0; 8.192×10^1; 8.192×10^{-1}, etc. 13. 10^{11}. Note that all we need here is to know that it is in the hundreds of billions and less than 500 billion. 15. 10^8 or 10^9. 17. 10^6 is a good guess.

★6–12 **Numerical calculations.** We are concerned in this section with the problems that arise as a result of the fact that the numbers we usually work with in applied mathematics are approximations. Suppose for example that $T = T_1 + T_2$, $T_1 \doteq 2.0 = A_1$, $T_2 \doteq 10.12 = A_2$. Assuming that the digits written are reliable, which is the usual convention followed, how reliable is the sum $A_1 + A_2 = 12.12$ as an approximation to T? We have $1.95 \leq T_1 \leq 2.05$, $10.115 \leq T_2 \leq 10.125$, from which it follows that $12.065 \leq T \leq 12.175$.

(a) Why?

We see that in 12.12, the second digit after the decimal point is certainly not reliable. At most we can say that the result is very nearly reliable to one decimal place.

Let us try to give a more general answer to questions of this kind. We have $E_1 = A_1 - T_1$, $E_2 = A_2 - T_2$, $E = A - T = (A_1 + A_2) - (T_1 + T_2) = (A_1 - T_1) + (A_2 - T_2) = E_1 + E_2$. That is,

(1)   *The error in a sum is the sum of the errors.*

Suppose now that we are adding two numbers, one of which is precise to more decimal places than the other. The possible error in the less precise number is at least ten times that of the more precise number. Accordingly, we could not expect the sum to be more precise than the number of lesser precision. If both errors are positive, as in the case above, the sum may actually not have even the precision of the least precise number. However, usually we do not know the actual errors. They may reinforce or cancel one another in any particular case. Evidently, in estimating the precision of a sum there is bound to be uncertainty. However, considerable experience and theoretical considerations too advanced to be considered here suggest the following rule of thumb.

(2)   *The precision of a sum is that of the least precise of the numbers being added.*

This rule is applied no matter how many numbers are added, and it includes subtraction.

(b) Do you think rule (2) would be more or less satisfactory when applied to more than two numbers?   (c) Why does (2) apply also to subtractions?   (d) Suppose $T_1 = 3.965$, $T_2 = 134.20813$, $T_3 = 0.0295$. Find the exact sum and round it off to units. Round off each number to units and find the sum. To what position do they agree? (e) Round off to one decimal place and add. Compare the result with the sum rounded off to one decimal place. (f) Do the same for two decimal places. (g) Round off the first to units and add to the other two. To what place does the result agree with the true sum? (h) Round off the second to units and add to T_1 and T_3. Compare with the true sum.

Can we construct a similar rule of thumb to cover multiplication and division? We have $E = A_1A_2 - T_1T_2$, but there seems to be no easy way to relate this to the errors $A_1 - T_1$ and $A_2 - T_2$. However, we may introduce the errors by noting that $A_1 = T_1 + E_1$ and $A_2 = T_2 + E_2$. Then $E = (T_1 + E_1)(T_2 + E_2) - T_1T_2 = E_1T_2 + E_2T_1 + E_1E_2$. We now have the errors in the equation, but it is still hard to see what the relation is. However, if we divide both members by $T = T_1T_2$, we find $E/T = E_1/T_1 + E_2/T_2 + (E_1/T_1)(E_2/T_2)$. Now we see that the relative error in the product is the sum of the relative errors plus their product. But if the relative errors are small, the additional term will be of small consequence. Accordingly,

(3)   *The relative error in a product is approximately the sum of the relative errors.*

(i) Suppose $E_1/T_1 = 0.01$ and $E_2/T_2 = 0.04$. Find E/T exactly and compare it with the approximate relative error given by (3).

As in the case of addition, errors may reinforce or cancel each other in multiplications. Since each additional significant digit reduces the possible relative error by a factor of 10, it appears likely that if several numbers are multiplied

(4)   *The accuracy of a product is that of the least accurate of the numbers being multiplied.*

Rules (2) and (4) serve as satisfactory rough guides as to what extent decimals should be rounded off during computations and in final answers. It is desirable to round off decimals as much as possible before computations to avoid needless work, and if answers are not rounded properly, we may either lose information or give an illusion of greater accuracy than exists. Frequently computers keep one more digit during computations than is required by (2) and (4), then round it off in the final result. Such a practice is a hedge against possible reinforcement of small errors. The reader should note that precision (number of decimal places) applies to addition and subtraction, while accuracy (number of significant digits) applies to multiplications and divisions. When both are involved, it may be desirable to keep more digits than would be indicated otherwise.

Often we desire a rough estimate of the result of a numerical computation, either as a check, or to locate the decimal point in a result, or simply because we are interested only in getting a rough idea of the answer. Consider for example,

(5) $A = (848.31)(14.831)^{1/2}(0.0183)/(25.167).$

About how big is the result? To make an estimate, we round off the numbers, making use of scientific notation, to the number of significant digits in which we are interested, and then carry out the arithmetic, rounding off at each step. Thus, rounding off to one significant digit, we find

(6) $A \doteq 8 \times 10^2 \times 15^{1/2} \times 2 \times 10^{-2}/3 \times 10$

(7) $\doteq 8 \times 4 \times 2 \times 3^{-1} \times 10^{-1} \doteq 2 \times 10^0.$

Here to find $8 \times 4 \times 2 \times 3^{-1}$ we thought "32, 64, 21, 20."
The order of magnitude of A is 1, and 2 is an accurate estimate to one significant digit, since $A \doteq 2.3755$ to 5 significant digits.

Estimate the following to one significant digit and compare with complete answer properly rounded off: (j) $(2894)(0.0139)$,   (k) $(18.35)(-0.003)^2$.

There are many aids and short cuts in numerical calculations. Among these are logarithms, the slide rule, desk computers, and electronic computers. All but the last are accessible to most students and suitable for problems of modest complication. Tables and handbooks give simple instructions for using such devices, and we shall not discuss them here. However, the rules about reliability and the methods of rough approximation described above are helpful when these devices are used. The rough methods are used for estimation and checking.

If we are to calculate $A = (0.39517)(4.865)(.00913)(3835)(1.3)$,   (l) how many significant digits should appear in the final result?   (m) How many significant digits should be carried in the computations?   (n) Rewrite the factors rounded-off as you would round them before beginning computation. (o) Rewrite A with the factors rounded to one digit.   (p) What is the order of magnitude of A?   (q) If the numbers were to be added, how many decimal places would you keep in the result?   (r) how many significant digits?   (s) Round off the numbers as you would before adding.

PROBLEMS

In Problems 1 through 5 estimate to one significant digit and compare with the complete rounded answer.

1. $((387.19)(2.8321)(0.489))^{1/3}$.   2. $(285)^4$.
3. 51.38% of $4,892,831.16$.   4. $(1.01)^{10}$.
5. $(253.8)(0.9514)/(5.935)$.

6. Estimate the order of magnitude of $2^{100,000,000}$ (the number of different states of the human eye's 100,000,000 photoreceptors).

7. Estimate the result in Problem 5 to two significant digits and compare with the complete answer.

8. The approximation $1/(1 + x) \doteq 1 - x$ is sometimes used. What is the error? Under what conditions will it be small? Show that the relative error is $-x^2$.

9. Discuss the approximation $(1 + x)/(1 + y) \doteq 1 + x - y$.

10. Discuss the approximation $(1 + z)^{1/2} \doteq 1 + z/2$. When is it good?

11. To calculate $(50.0004/49.9993)^{1/2}$, a chemist rewrote the expression as follows: $((50 + 0.0004)/(50 - 0.0007))^{1/2}$; $((1 + 0.000008)/(1 - 0.000014))^{1/2}$; $(1.000022)^{1/2}$; 1.000011. Explain the manipulations.

★12. The human egg has a diameter of about $1/2000$ inch. How many could be put in a quart bottle without crowding?

★13. The light-year is the distance traveled by light in one year. If the speed of light is 186,339 mi/sec, find the number of miles in a light year to one significant digit.

★14. Find the following ratios to two significant digits: $81/80$; $125/128$; $531441/524288$. (For their significance, see *Introduction to Musicology*, by Glen Haydon, p. 33)

★15. When is $1 + 2a$ a good approximation for $(1 + a)^2$? What is the error? the relative error?

★16. Assume that one digit occupies $\frac{1}{8}$ inch. How long a strip of paper is required to write a decimal with 10^{100} digits? Suppose that it takes $\frac{1}{5}$ sec to write a digit. How long would it take to write it out? Compare the length of the strip with the distance to the sun (93,000,000 miles) and the time to the age of the earth (about 3 billion years).

★17. In *Statistical Methods*, by G. W. Snedecor, the following equation appears on page 203: $1,310x^3 - 113,303x^2 + 20,540,068x - 317,722,720 = 0$. He writes, "A first approximation may be got by ignoring the first two terms." Find the approximation.

★18. With the aid of a figure, show that when $\theta \doteq 0$, $(\sin \theta)/\theta \doteq 1$.

★19. The following equations appeared in a paper on the variability of awards in damage cases tried by juries and by judges: $12(1.3574)/n = 1.625$; $2.328/m = 1.3574$. Solve for m and n to the nearest integer.

★20. A. A. Eddington's "cosmical number," which he believed to be the number of protons and electrons in the universe, is $2 \times 136 \times 2^{256}$, or $2 \times 15,747,$ $724,136,275,002,577,605,653,961,181,555,468,044,717,914,527,116,709,366,231,$ $425,076,185,631,031,296$. Show that these numbers are the same to three significant digits.

★21. Discuss the approximation $(1 - B^2)^{-1/2} \doteq 1 + (1/2)B^2$. When is it good?

★22. Suppose $\tan u = h/(s + \delta)$. If u is small and δ is small compared with s, justify $\tan u \doteq h/s$.

★23. From $1/\lambda = R(1/2^2 - 1/n^2)$, $R = 1.097 \times 10^{-7}$, $n = 3$, find λ.

★24. Suppose $\sin[(A + a)/2] = n \sin(A/2)$. If A and a are small, justify $a = (n - 1)A$.

ANSWERS TO EXERCISES

(a) (3-5-42).   (b) It depends on circumstances.   (c) $a - b = a + (-b)$.
Subtraction merely changes the sign of the error.   (d) 138.20263; 138; 4, 134, 0,
138; to units.   (e) 4.0, 134.2, 0.0; 138.2; 138.2. They agree to one decimal place.
(f) 3.96; 134.21, 0.03; 138.20. They agree to two decimal places.   (g) 138.23763.
To one decimal place.   (h) 137.9945. Agrees only to units.   (i) 0.0504; 0.05.
(j) If the decimals are rounded to one digit, we get an estimate of 3×10^1,
which does not agree with the complete result, 4.02×10^1. Rounding off the
decimals to two digits yields 4×10^1, as does the estimate.   (k) 2×10^{-4};
1.65×10^{-4}.   (l) Two.   (m) Three.   (n) $3.95 \times 10^{-1} \times 4.86 \times 9.13 \times 10^{-3} \times$
$3.84 \times 10^3 \times 1.3$.   (o) $(0.4)(5)(0.009)(4 \times 10^3)(1)$.   (p) 100.   (q) None.
(r) Four.   (s) $0.4 + 4.9 + 0 + 3835 + 1.3$.

ANSWERS TO PROBLEMS

1. "The complete rounded answer" means the result of following rules (2)
and (4). Here one should round off to three significant digits before beginning. Use cube root tables. 3. $51\% = 51/100$. 5. Rounding off numbers to
one digit yields 5×10^1, compared with the complete answer of 4.068×10^1.
7. 4.1×10^1 and 4.07×10^1. 9. $(1 + x)/(1 + y) = (1 + x)(1/(1 + y))$, and
use Problem 8. Good when both x and y are small. 11. Multiply numerator
and denominator by $1/50$, then use approximations of Problems 9 and 10.
13. 6×10^{12} mi. 15. When a is small; $-a^2$; $-a^2/(1 + a)^2$. 17. 15. 19. 2, 10.

★6-13 Complex numbers.   In Section 4–3 we defined the addition of
vectors, and the multiplication of vectors by real numbers in such a way
that the geometric interpretation of these operations was the same as for
real numbers. We now wish to define the multiplication of two plane
vectors so as to satisfy the same conditions. When we do so we shall find
that we have solved the mystery of imaginary and complex numbers!

(a) Review the definitions of Section 4–3 and the discussion of definitions
in Section 1–13.

Our task now is to decide on an appropriate definition of multiplication
of plane vectors (ordered pairs of reals). Of course, we are free to choose
this definition as we wish, but, if possible, we desire that addition and
multiplication of vectors follow the usual laws (such as the commutative,
associative, and distributive laws) and that multiplication has the same
vector interpretation as for real numbers. We defined the sum of two vectors as the vector obtained by adding their components. Suppose we defined the product of two vectors as the vector obtained by multiplying
their components, that is, $(a, b)(c, d) = (ac, bd)$? It is easy to show that
this gives a unique product which is also a vector and that the commuta-

tive, associative, and distributive laws are satisfied. Moreover, $(0, 0)$ plays the role of 0, and $(1, 1)$ plays the role of 1, so that (1–16–1) through (1–16–9) are satisfied with "vector" in place of "real number." But (1–16–10) is not satisfied, since, for example, the vector $(2, 0)$ is not $(0, 0)$ and yet it has no reciprocal.

(b) What do we mean by saying that $(1, 1)$ plays the role of 1 with this definition of multiplication? (c) Show that with this definition of multiplication $\sim \exists x \, \exists y \, (2, 0)(x, y) = (1, 1)$.

Since the situation seems more complicated than for addition, we may take another approach. Let us imagine that we have somehow defined multiplication so that the laws hold and see what conclusions we can draw. Then we have

(1) $(a, b)(c, d) = [(a, 0) + (0, b)] \cdot [(c, 0) + (0, d)]$ (4–3–2),

(2) $= [a(1, 0) + b(0, 1)][c(1, 0) + d(0, 1)]$ (4–3–7),

(3) $= ac(1, 0)(1, 0) + ad(1, 0)(0, 1)$
           $+ \, bc(0, 1)(1, 0) + bd(0, 1)(0, 1)$ (1–12–7).

Our problem is reduced to deciding on how to multiply together the vectors $(1, 0)$ and $(0, 1)$. That is, we can solve our problem by deciding on definitions for $(1, 0)(1, 0)$, $(1, 0)(0, 1)$, $(0, 1)(1, 0)$, and $(0, 1)(0, 1)$. Here the concept of isomorphism will be helpful. The vectors with second component zero are isomorphic to the real numbers with respect to addition under the correspondence $F(x) = (x, 0)$, for we have $F(a + b) = (a + b, 0) = (a, 0) + (b, 0) = F(a) + F(b)$. In other words, the vectors with second component zero behave exactly like the real numbers under addition. Moreover, $F(0) = (0, 0)$ and $F(1) = (1, 0)$. This is to be expected, since we began by interpreting real numbers as vectors on an axis.

(d) Review the discussion of isomorphism in Section 6–8.

It would be natural, then, to define multiplication of vectors so that the vectors $(x, 0)$ were isomorphic to the reals under multiplication also. This would require $F(ab) = (ab, 0) = F(a)F(b) = (a, 0)(b, 0)$. In particular, we would have $(1, 0)(1, 0) = (1, 0)$. From this isomorphism we see that $(1, 0)$ plays the role for vectors that 1 does for real numbers, for $(1, 0)(x, 0) = (x, 0)$, and there can be only one number with this property in any number system satisfying the axioms of Section 1–16. Hence we would expect $(1, 0)(x, y) = (x, y)$ for any x and y. In particular we have $(1, 0)(0, 1) = (0, 1)$ and, by commutativity, $(0, 1)(1, 0) = (0, 1)$.

(e) How is Problem 30 in Section 2–11 related to the above?

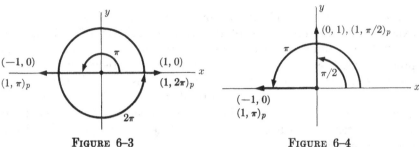

FIGURE 6-3   FIGURE 6-4

There remains only $(0, 1)(0, 1)$ or $(0, 1)^2$. How shall we define it? An appropriate decision is suggested by the following considerations. According to (3–9–13), the absolute value (length of the corresponding vector) of the product of two real numbers is the product of their absolute values. Accordingly we might hope that the length of the product of two plane vectors would be the product of their lengths. This is indeed the case in the three products considered so far. For example, the lengths of $(1, 0)$, $(0, 1)$, and their product $(0, 1)$ are all 1. Hence we would expect the length of $(0, 1)^2$ to be 1. Using polar coordinates, we have $(0, 1)^2 = (1, \pi/2)_p^2 = (1, \pi)_p$.

(f) Review polar coordinates in Section 4–10.

Our problem now appears to be one of deciding on the relation between the direction angle of a vector and the direction angle of its square. The square $(-1, 0)^2$ is suggestive. Because of the isomorphism between vectors of the form $(x, 0)$ and the corresponding real numbers x, we expect that $(-1, 0)^2 = (1, 0)$, as sketched in Fig. 6–3. In polar coordinates we have $(1, \pi)_p^2 = (1, 2\pi)_p$. This suggests that the direction angle of a square is found by doubling the original direction angle. This conjecture (backed at present by only one example!) suggests $(0, 1)^2 = (1, \pi/2)_p^2 = (1, \pi)_p = (-1, 0)$. (See Fig. 6–4.) Replacing the products in (3) by the expressions found in the last few paragraphs, we find

(4) $(a, b)(c, d) = ac(1, 0) + ad(0, 1) + bc(0, 1) + bd(-1, 0),$

(5) $= (ac, 0) + (0, ad) + (0, bc) + (-bd, 0),$

(6) $= (ac - bd, ad + bc).$

This suggests the following definition.

(7) **Def.** $(a, b)(c, d) = (ac - bd, ad + bc).$

(g) Use definition (7) to find $(0, 1)(0, 1)$, $(2, 1)(3, -2)$, $(-1, 0)(5, -1)$, and $(0, -1)^2$.

From (7) it is not difficult to prove that plane vectors satisfy all the axioms (1-16-1) through (1-16-10) with "plane vector," "(0, 0)," and "(1, 0)" substituted for "real number," "0," and "1" throughout. The proof is based on the assumption that the reals satisfy the axioms. It follows that plane vectors (ordered pairs of reals) with these definitions of addition and multiplication satisfy all the laws of algebra derivable from these axioms, that is, all the laws except those related to order. In such a context it is customary to call plane vectors (ordered pairs of reals) *complex numbers*.

The relation between the complex numbers and the reals is that there exists a subset of the complex numbers, namely those with second component 0, that is isomorphic to the real numbers with respect to the operations of addition and multiplication, under the correspondence $F(x) = (x, 0)$. Thus $F(1) = (1, 0)$, $F(-1) = (-1, 0)$, $F(0) = (0, 0)$, and so on. Because of this isomorphism, it is customary to write simply "1" for "(1, 0)," "−1" for "(−1, 0)," and generally "x" for "$(x, 0)$."

(h) Why does this practice cause no difficulty even though, strictly speaking, 1 and (1, 0) are not identical?

From (7) it is now easy to prove the key conjecture we used in deciding on (7), namely

$$(8) \qquad\qquad (0, 1)^2 = (-1, 0).$$

But $(-1, 0)$ corresponds to -1 under the above isomorphism. Hence this may be written

$$(9) \qquad\qquad (0, 1)^2 = -1.$$

It is convenient to represent complex numbers as sums of their component vectors. According to (4-3-9), $(a, b) = a(1, 0) + b(0, 1)$. Using "1" in place of "(1, 0)" and adopting the letter "i" as a name for "(0, 1)," we have

(10) **Def.** $\qquad\qquad i = (0, 1),$

(11) $\qquad\qquad\qquad (a, b) = a + bi.$

This expresses a complex number (ordered pair of reals) as the sum of a real number (a) times the unit real vector (1 or (1, 0)) and a real number (b) times a unit vector (i or (0, 1)) called the *imaginary unit vector*.

(i) Prove (12).

$$(12) \qquad\qquad i^2 = -1.$$

(j) What is the other square root of -1?

In the complex number (a, b) we called a the *real part* and b the *imaginary part*. Note that the so-called imaginary part is as real as the real part! Complex numbers whose imaginary parts are zero are called *real numbers*, though actually they are merely isomorphic to the real numbers. They may be written in the form $a + 0i$, $(a, 0)$, or a. Complex numbers whose imaginary parts are not zero are called *imaginaries*. Those whose real parts are zero and whose imaginary parts are nonzero are called *pure imaginaries*. They may be written in the form $0 + bi$, $(0, b)$, or bi. These terms have only historical significance, since an ordered pair of real numbers (a complex number, a point in the plane, a plane vector) is as real (possibly even twice as real!) as a single real number.

(k) Plot the following points: 2, $2i$, i, $-i$, $2 + 4i$, $-1 + i$.   (l) Give four interpretations of "$(2, 3)$," indicating in each case the terms applied to 2 and 3.

Since all the laws of algebra apply to complex numbers, we may operate with their names in the form $a + ib$ with the usual rules of algebra, keeping in mind, of course, the law (12). For example, using the usual manipulative rules we find $(a + bi)(c + di) = ac + adi + bci + bdi^2 = ac - bd + (ad + bc)i$, as we would expect according to (7).

Carry out the following operations, plotting the complex numbers involved. Repeat with numbers written in the form $a + bi$.   (m) $(2, 0)(3, 4)$,   (n) $(1, 1)(1, -1)$,   (o) $(3, 4)(2, 1)$.

We can now prove our conjecture that the direction angle of a square is twice the original direction angle. Let $(x, y) = (r, \theta)_p$. Then by (4–11–2), $x = r \cos \theta$ and $y = r \sin \theta$. Hence

(13) $[(x, y) = (r, \theta)_p] \leftrightarrow [x + yi = r(\cos \theta + i \sin \theta)]$.

The right member of the second equation is said to be the *polar form* of writing complex numbers, while its left member is said to be in *rectangular form*. Using the polar form, we have

(14) $(r, \theta)_p^2 = (r(\cos \theta + i \sin \theta))^2$

(15) $= r^2(\cos^2\theta - \sin^2\theta + i2 \sin \theta \cos \theta)$

(16) $= r^2(\cos 2\theta + i \sin 2\theta)$

(17) $= (r^2, 2\theta)_p$.

We see from this that the length of the square is the square of the length and the angle of the square is twice the angle.

(p) Explain the transition from (14) to (15),   (q) from (15) to (16).   (r) What is the reason for (17)?

The above result can be generalized in two ways. First we may generalize by taking the product of two different complex numbers and prove that

$$(18) \qquad (r_1\theta_1)_p(r_2, \theta_2)_p = (r_1r_2, \theta_1 + \theta_2)_p.$$

In words, the length of the product is the product of the lengths and the angle of the product is the sum of the angles.

(s) Prove (18) by continuing the following: $(r_1, \theta_1)(r_2, \theta_2) = (r_1(\cos\theta_1 + i\sin\theta_1)(r_2(\cos\theta_2 + i\sin\theta_2)) = \cdots$

Second, we may generalize the exponent by writing

$$(19) \qquad (r, \theta)_p^n = (r^n, n\theta)_p.$$

This is known as *DeMoivre's theorem* and can be proved to hold for all numerical values of n, including complex exponents when these have been defined properly. The proof of (19) for n a natural number is by induction. The verification is immediate. For the induction we have

$$(20) \qquad (r, \theta)_p^{n+1} = (r, \theta)_p(r, \theta)_p^n$$

$$(21) \qquad = (r, \theta)_p(r^n, n\theta)_p \qquad \text{(Induction assumption)}$$

$$(22) \qquad = (r^{n+1}, \theta + n\theta)_p \qquad (18)$$

$$(23) \qquad = (r^{n+1}, (n+1)\theta)_p.$$

But the last equation is just $(19)(n{:}n + 1)$. Hence the induction is complete and the theorem is proved.

(t) Does it matter that we have used the variable n in the induction instead of x as in Section 6–5? (u) Write out (17), (18), (19) with the complex numbers in polar form. (v) Complete the following: The natural numbers are closed under addition and multiplication but not under _____ and _____. The integers are closed under addition, subtraction, and _____, but not closed under _____. The rationals are closed under _____ but not under _____. The reals are closed under _____ but they are not closed under _____ since _____.

PROBLEMS

In Problems 1 through 10 carry through the operations by using the definitions. Then rewrite in rectangular form and do the operations by the usual algebraic rules and compare your results. Sketch the numbers involved and verify (18).

1. $(2, -1)(3, 0)$.   2. $(0, -3)(1, 2)$.   3. $(3, 4)^2$
4. $(1, 1)(1, -1)$.   5. $(-1, 2)^2$.   6. $(1, -2)^2$
7. $(0, 1)(0, -1)$.   8. $(1, 1)^3$.   9. $(-0.5, 0.5\sqrt{3})^3$.
10. $(a, b)(a, -b)$.

11. With definition (24) prove that $1/(a, b)$ is the reciprocal of (a, b) called for by (1–16–10).

(24) **Def.** $1/(a, b) = (a, -b)/(a^2 + b^2)$.

12. Find $1/(a, b)$ by multiplying numerator and denominator of $1/(a + bi)$ by $a - bi$.

13. Show that the length of the quotient is the quotient of the lengths and the angle of the quotient is the difference of the angles. State and prove a rule for reciprocals.

★14. As for real numbers, we call any complex number z such that $z^n = w$ an nth *root* of w. The root with smallest non-negative direction angle is called the *principal nth root*. For example, $i^2 = -1$ and $(-i)^2 = (-1)^2 i^2 = -1$, so that i and $-i$ are square roots of -1, but i is the principal square root. Sketch the two square roots of 1 and the four fourth roots of 1.

★15. Show that 1, $-0.5 + 0.5\sqrt{3}i$, and $-0.5 - 0.5\sqrt{3}i$ are cube roots of 1.

★16. Find the cube roots of -1.

★17. Show that when $x > 0$, $i\sqrt{x}$ and $-i\sqrt{x}$ are the square roots of $-x$.

★18. We now show that any complex number (except 0) has just n distinct nth roots. Suppose we wish $(s, \phi)_p$ to be an nth root of $(r, \theta)_p$. We have $(s, \phi)^n = (s^n, n\phi) = (r, \theta)$, where we omit the p since we shall be using only polar co-ordinates. From this we have $s^n = r$, where s and r are real and positive since they are lengths and we are excluding the null vector. Hence $s = r^{1/n}$, the principal nth root of r. By Problem 12 in Section 4–10 we have $n\phi = \theta + 2k\pi$, where k is some integer, or $\phi = \theta/n + 2k\pi/n$. Hence $(r^{1/n}, \theta/n + 2k\pi/n)_p$ is an nth root of $(r, \theta)_p$ for any such k. All these roots have the same length, and therefore the corresponding points lie on a circle of radius $r^{1/n}$ and center at the origin. If θ is the smallest non-negative direction angle of $(r, \theta)_p$, then the principal root is $(r^{1/n}, \theta/n)$. The others are then obtained by going counter-clockwise around the circle through successive angles of $2\pi/n$. Taking $k = 0, 1, 2, \ldots, n - 1$, we get all possible distinct points, since beginning with $k = n$, we reach the same points as before. Hence

$$(25) \qquad \begin{aligned} (s, \phi)_p^n &= (r, \theta)_p \\ &\leftrightarrow \exists k\, k \in J \wedge (s, \phi) = (r^{1/n}, \theta/n + 2k\pi/n). \end{aligned}$$

Verify that the formula gives the three cube roots of 1 found in Problem 15 and the four fourth roots of 1.

★19. Find the five fifth roots of 1 and sketch them. (*Suggestion:* Express them in polar form.)

★20. Find the sixth roots of 1 by (25) and also by factoring $x^6 - 1 = 0$.

★21. The *complex conjugate* of $a + bi$ is defined as $a - bi$. The complex conjugate of the complex number z is written $\bar{z}$. What is the geometric relation between a number and its conjugate? When does $z = \bar{z}$? $z = -\bar{z}$?

★22. Show that $\overline{z + w} = \bar{z} + \bar{w}$, $\overline{z - w} = \bar{z} - \bar{w}$, $\overline{zw} = \bar{z}\,\bar{w}$, $\overline{(z/w)} = \bar{z}/\bar{w}$, $\overline{z^n} = \bar{z}^n$, and $z\bar{z} = |z|^2$.

★23. By expanding $(\cos \theta + i \sin \theta)^3$ and using DeMoivre's theorem, find formulas for $\cos 3\theta$ and $\sin 3\theta$.

★24. Under what operations are the complex numbers closed?

★25. The closure of the complex numbers is even broader than is suggested by the answer to Problem 24. Indeed, any algebraic equation whose coefficients are complex numbers (one of the form $a_n x^n + a_{n-1} x^{n-1} + \cdots + a_1 x + a_0 = 0$, where $a_n \neq 0$ and where the a's are complex) has at least one root that is a complex number. (It may have as many as n distinct roots.) Study the proof of this result, known as the *Fundamental Theorem of Algebra*, given in *What Is Mathematics?*, by R. Courant and H. Robbins, Chapter II, Section 5.

Answers to Exercises

(b) $(x, y)(1, 1) = (x, y)$ is a law. (c) With this definition, $(2, 0)(x, y) = (2x, 0y) = (2x, 0) \neq (1, 1)$ no matter what values of x and y we choose. (e) Problem 30 in Section 2–11 states that there is one and only one number with the property $(1-16-8)$. Hence if we wish the complex numbers to satisfy the same laws as the reals, this must hold also. (g) $(-1, 0)$, $(8, -1)$, $(-5, 1)$, $(-1, 0)$. (h) We are using "x" as an abbreviation for "$(x, 0)$," but we will never get anything contradicting what we would expect for "x" representing a single real number, because the isomorphism means that $(x, 0)$ and x have identical properties with respect to the operations of algebra. (i) Find $(0, 1)(0, 1)$ by (7). (j) $-i$. (k) $(2, 0)$, $(0, 2)$, $(0, 1)$, $(0, -1)$, $(2, 4)$, $(-1, 1)$. (l) The point with coordinates 2 and 3; the vector with components 2 and 3; the complex number with real and imaginary parts 2 and 3; the ordered pair with first member 2 and second member 3. (m) $(2 \cdot 3 - 0 \cdot 4, 2 \cdot 4 + 0 \cdot 3) = (6, 8)$; $2(3 + 4i) = 6 + 8i$. (n) $(2, 0)$, 2. (o) $(2, 11)$, $2 + 11i$. (p) $(1-12-9)$ and $i^2 = -1$. (q) $(4-12-22)$, $(4-12-23)$. (r) $(13)(r{:}r^2, \theta{:}2\theta)$. (s) Multiply, collect terms involving i together, and use $(4-12-9)$ and $(4-12-10)$. (t) No. It is a dummy in the induction. See Sections 2–6 and 2–11. (u) DeMoivre's theorem becomes $(r \cos \theta + ir \sin \theta)^n = r^n (\cos n\theta + i \sin n\theta)$.

Answers to Problems

11. Multiply by (a, b) and show that the product is $(1, 0)$. 13. Use (24), $(4-12-17)$, and $(4-12-18)$.

CHAPTER 7

CALCULUS

7–1 What is calculus? Calculus had its origins in a number of problems that have troubled scientists and philosophers since ancient times. At first sight these problems appear to be quite different, but it turns out that they have a common character and can be attacked by very similar methods.

First, consider the problem of finding the area of an arbitrarily given region in a plane. This is easy if the region happens to be bounded by a rectangle, parallelogram, triangle, trapezoid, or polygon. Indeed, by using a formula for the area of a triangle we can easily find the area of any region that is the union of triangular regions.

(a) How does one find the area of a polygon?

But how can we find the areas of regions that cannot be divided into triangular regions? What about regions bounded by circular arcs, ellipses, sine curves, and so on? In very ancient times, mathematicians already took the correct path to the solution of this problem. They approximated areas within curves by replacing the curves with numerous straight line segments joining points on the curves. For example, they approximated the area of a circle by inscribing regular polygons whose areas could be calculated. As is evident in Fig. 7–1, the larger the number of sides, the closer the approximation. None of these approximations is the exact area, but it appears reasonable to say that as the number of sides increases, the area of the polygons approaches the exact area of the circle. We may think of the area within a closed curve as a number that is approximated arbitrarily closely by the area of inscribed polygons.

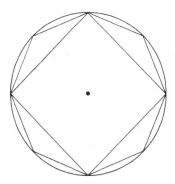

FIGURE 7–1

377

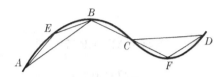

A somewhat similar problem is that of length. We know how to find the length of a straight line segment or of a broken line (curve made up of connected straight line segments). But what of an arc on a circle, ellipse, hyperbola, or other curve not made of segments? It seems reasonable here to approximate the curve by a broken line and consider the total length of such a broken line as an approximation to the length of the curve. As suggested in Fig. 7-2, by choosing the endpoints of the segments close enough together on the curve, we may make the broken line arbitrarily close to the curve.

A third problem, which appears to be quite different from the first two, is that of finding the exact velocity of a moving body. Of course this is easy if the body is known to be moving at a fixed velocity, for we have velocity = distance/elapsed time. But this relation does not hold when velocity is changing. Imagine a body moving along a straight line, and let y be its directed distance from a fixed point on this line. Let x be the corresponding time. We assume a function f such that $y = f(x)$. Suppose at time x_1 the body is at y_1 and at time x_2 the body is at y_2. In an elapsed time of $x_2 - x_1$ the body has moved a directed distance of $y_2 - y_1$. If we know that it moves at fixed velocity, this velocity is evidently $(y_2 - y_1)/(x_2 - x_1)$ or $(f(x_2) - f(x_1))/(x_2 - x_1)$. But if the velocity is changing, the most we can say is that this ratio is the *average velocity* during the time interval $(x_1 \quad x_2)$. What do we mean by the exact velocity at the time x_1? It seems natural to say that we could get a very close approximation to it by choosing x_2 very close to x_1. We may define the *exact velocity* (or *instantaneous velocity*) as the limit of the average velocity as the time interval is shortened toward zero.

(b) Why not let $x_2 = x_1$?   (c) What is the interpretation if $x_2 > x_1$ and $y_2 < y_1$?

To visualize the problem of velocity graphically, we may graph the function f relating time to distance y as in Fig. 7-3. The average velocity $(y_2 - y_1)/(x_2 - x_1)$ is just the slope of the line joining the two points on the graph of f. Thus the average velocity appears as the slope of a secant line. As we let the point (x_2, y_2) move closer to (x_1, y_1), the secant line swings around and approaches the tangent line T. Its slope, then, must approach the slope of the tangent line, which is the geometric counterpart

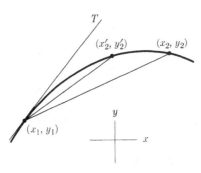

FIGURE 7-3

of the exact velocity. Hence finding the exact velocity is equivalent to finding the slope of a tangent line.

We can generalize the above. Let us imagine a function f given by $y = f(x)$. If x changes from x_1 to x_2, it has an *increment* of $x_2 - x_1$. At the same time, y changes from $f(x_1)$ to $f(x_2)$ by an increment $f(x_2) - f(x_1)$. The ratio of the change in y to the change in x is $(f(x_2) - f(x_1))/(x_2 - x_1)$, which we call the *average rate of change of y* (*or $f(x)$*) *with respect to x in the interval* $(x_1 \; x_2)$. If we hold x_1 fixed and let x_2 change so that it approaches x_1, the average rate of change may approach a number. If it does so, we call this number the *exact rate of change of y with respect to x and x_1*. Geometrically it is the *slope of the tangent line* to the graph of f at the point (x_1, y_1).

(d) To determine the tangent line at a given point on a curve is it sufficient to know the slope?

As a final example of problems giving rise to calculus, we mention one of Zeno's famous paradoxes, that of the runner in a race. Some twenty-five centuries ago Zeno argued that a runner can never reach the finish line. For, Zeno reasoned, the runner must first go half way. Then he must go half of the remaining distance. Having arrived at the three-quarter mark, he must go half of the rest, and so on. In this way he must cover an infinite number of smaller and smaller distances without ever arriving at the finish. The situation is illustrated in Fig. 7–4, where we take the total distance to be unity.

Let $y_i =$ the length of ith interval covered by the runner. Then $y_1 = 1/2$, $y_2 = 1/4$, $y_3 = 1/8$, $y_4 = 1/16$, and for any i, $y = 2^{-i}$. The y_i

FIGURE 7–4

are terms of a geometric progression with common ratio 2. Let $s_n =$ the sum of the first n terms. Then $s_n = 1 - 2^{-n}$. Let us assume that the runner moves with a constant velocity equal to 1. Then let $t_i =$ the time required to cover the ith interval. It is easy to see that $t_i = y_i$ so that if we let $T_n =$ the time required to cover the first n intervals, we have $T_n = 1 - 2^{-n}$. Evidently there is here no question of the runner requiring an infinite time to reach his goal, for the T_i are bounded from above by 1.

(e) Justify these results by reference to Section 6–7.

But there is a real difficulty here. Zeno has divided the segment of unit length into an infinite number of subdivisions. How can an infinite number of segments add up to a finite segment? No one of the s_n is equal to 1. Does it make sense to speak of the sum of *all* the y_i? Does $1/2 + 1/4 + 1/8 + 1/16 + 1/32 + \ldots$ ad infinitum $= 1$? If we can answer these two questions in the affirmative, we shall have a complete answer to Zeno. We can say to him that even when the unit interval is sliced into an infinite number of subintervals, their sum is still unity. If he wishes to think of the runner passing over an infinite number of intervals of decreasing length, the time required for each interval decreases also, and the sum of all the times is still one.

To achieve these results we have to define what is meant by adding together an infinite number of numbers. To do so we use the idea of approximation. The sum of the first n times is given by $T_n = \sum_{i=1}^{i=n} t_i = 1 - 2^{-n}$. As n gets larger this clearly gets closer and closer to 1. We then *define* the sum of all the t_i, that is, of all the terms of the sequence, as 1. More generally, we define the sum of all the terms y_i of any sequence as the limit (if any) approached by s_n (the sum of the first n terms) as n gets large.

(f) Why does T_n approach 1 as n gets large?   (g) Study the infinite geometric series $1/3 + 1/9 + 1/27 + \ldots$. Find a formula for the sum of the first n terms. What would you guess as the sum of all the terms?

The reader must have observed that these problems, though different in origin, have several things in common. In each case we have a concept that is easy to handle in simple cases (area of rectilinear figures, length of a segment, constant velocity, secant lines, average rate of change, sum of a finite number of terms of a sequence), but which does not yield to the same methods when we try to generalize (area of a region, length of a curve, changing velocity, tangent lines, exact rate of change, sum of an infinite number of terms). The difficulty appears to lie in our seeming inability to do more than approximate what we seek. Indeed, in each case it is not even clear that we know what we are looking for until we can formulate a clear definition (of area, length, rate of change, sum) that will apply to the general case. The way out appears to lie in making use of the idea that a number may be defined in some way as a limit of approximations.

The characteristic feature of calculus is the use of the concepts of function and limit to attack problems along the lines suggested here. The purposes of this chapter are (1) to give the student an intuitive appreciation of the fundamental ideas of calculus, (2) to suggest how a rigorous theory could be constructed, (3) to familiarize the student with the techniques and applications of calculus in simple situations.

Answers to Exercises

(a) It can always be done by adding the area of triangles.   (b) The ratio is then undefined.   (c) y is decreasing, the velocity is negative.   (d) See Section 4–5, Exercise (q).   (e) (6–7–25)   (f) $1/2^n$ clearly can be made arbitrarily small by taking n large enough.   (g) Sum is $1/2$.

7–2 Rates of change. We begin by considering the problem of exact rate of change and the related problem of tangent lines. We have already treated (Section 4–5) the very simplest case in which x and y are related by a linear function $mI + b$ so that $y = mx + b$. There we found that the change in y divided by the change in x in the interval $(x_1 \quad x_2)$, $(y_2 - y_1)/(x_2 - x_1)$, is m, the slope of the straight line graph of $mI + b$. This ratio is the average rate of change of y with respect to x in the interval. It is also the exact rate of change, since the ratio is the same for any choice of the two points.

(a) Suppose a body moves with constant speed m along a path in which its distance from a fixed point along the path is y and the corresponding time is x. What function relates time to distance if its distance from the fixed point is b when $x = 0$?   (b) Review Section 4–5.   (c) What is the significance in terms of rate of change and slope of $m < 0$, $m = 0$, $m > 0$?

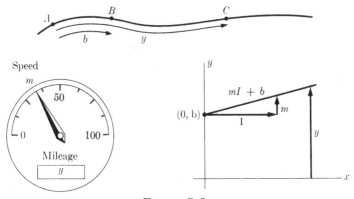

FIGURE 7–5

Figure 7–5 illustrates the discussion of Exercise (a). We imagine an automobile moving along a road with its position given by y and the time by x. Distances are measured from the point A. The car is at point B, a distance of b from A, when $x = 0$. Point C suggests an unspecified time x and the corresponding distance y. On the speedometer the arrow points to the speed m and the dial shows the distance y (assuming the mileage was set to zero at A). The graph shows the function relating x and y. Here m appears as a slope.

(d) At what time is the car at A, assuming always a maintained speed of m?

The idea that a car can maintain a fixed speed for more than a very short time is certainly not very realistic. Suppose, then, that our car moves at various speeds during different time intervals but maintains in each interval a uniform speed. Let the car start at the point B at time $x = 0$. Assume that it goes 60 mi/hr for $\frac{1}{2}$ hr, then 20 mi/hr for 12 min, stands still for 24 min, and finally goes at 40 mi/hr for 54 min. Measuring time x in hours and distance y in miles, we see that when $x = 0.5$, $y = 30$, when $x = 0.5 + 0.2 = 0.7$, $y = 30 + 20(0.2) = 30 + 4 = 34$, when $x = 0.7 + 0.4 = 1.1$, $y = 34 + 0 = 34$, and when $x = 1.1 + 0.9 = 2$, $y = 34 + 40(0.9) = 70$. These points are sketched in Fig. 7–6. Since the car is traveling at a constant speed in these intervals, its position is given by the piecewise linear function joining these points. Moreover, the slope of each segment is the speed during that interval. For example, in the time interval (0__30) the speed is 60. The function f relating time and distance is graphed as a solid line. The function Df, giving the slope of the graph of f (or the rate of change, or the velocity), is graphed with dashes.

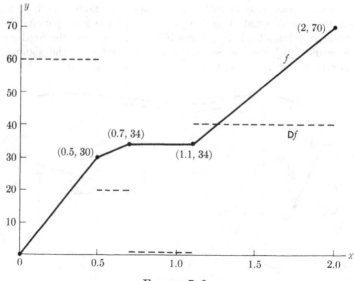

FIGURE 7–6

(e) What is the slope of f in the interval $(0.5\_\_0.7)$? How is this reflected in the graph of Df?   (f) Why does the graph of Df lie along the x-axis in the interval $(0.7\_\_1.1)$?   (g) What is the speed at the time $x = 0.5$?   (h) In Fig. 7–6 what is $Df(0.6)$? $Df(1.5)$? $Df(1.1)$?

The problem just discussed is not very realistic because of the situation at the corner points of the piecewise linear function. What is the speed at the time $x = 1.1$? What does the speedometer show at this instant? Actually, if a car could really change instantaneously from 0 to 40 mi/hr, the speedometer would be shifting *instantaneously* from 0 to 40 and would have no defined position. In terms of the graph, there is no defined slope at $(1.1, 34)$, since at any point (however close) toward the left of $(1.1, 34)$ the rate of change (slope) is 0, but at any point toward the right it is 40. We see then that the domain of Df does not contain the values of x at the corner points 0.5, 0.7, 1.1. We say that the rate of change is not defined there.

Let f be the piecewise linear function with corner points $(0, 0)$, $(0.5, 25)$, $(0.8, 31)$, $(1.2, 31)$ and $(2.2, 80)$.   (i) Graph it.   (j) On the same axes, graph Df, the function giving the slope of the graph of f.   (k) Interpret your results in terms of an automobile as we did.   (l) What are the domain and range of f? (m) of Df?   (n) What is $Df(1)$?

Let us now try to be still more realistic in describing an automobile whose movements are approximated in Fig. 7–6. Even the most extreme hot-rod driver cannot "jack rabbit" his car from a standing position to 40 mi/hr instantaneously, as Fig. 7–6 indicates at the point $(1.1, 34)$. It actually takes *some* time for a car to change its speed. Figure 7–7 sug-

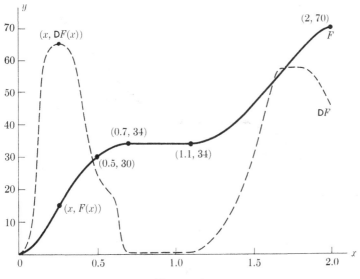

FIGURE 7–7

gests what actually happens. For any value of x, $F(x)$ represents the total distance traveled and $Df(x)$ represents the speed at the time x (the reading on the speedometer). When $x = 0$, both distance and speed are zero and the graphs of f and Df are together. The speed, given by the dotted line, gradually rises and then falls during the interval (0 0.5), so that its average is 60. Note that where the graph of F is steepest, the graph of DF has its highest point. This is the time when the speed is maximum. As the driver slows down before the second interval, his speedometer reading falls, so that during the second time interval it averages 20. When the car comes to a stop, the graph of DF reaches the x-axis where $DF(x) = 0$ and stays there until the driver again accelerates so as to achieve an average speed of 40 in the last interval. We have assumed that he is again slowing down toward the time $x = 2$.

(o) Why is the graph of F steepest where the speed is greatest?   (p) How does the graph show the driver slowing down toward the end?

Figure 7–7 is a fairly realistic picture of a possible situation. For each value of x, $DF(x)$ is the exact speed as given by the speedometer reading. There are no points where $DF(x)$ fails to exist. To see more clearly the relation between F and DF in Fig. 7–7, we approximate the smooth curve F by a piecewise linear function g consisting of small segments. This is suggested in Fig. 7–8, where we make the approximating segments rather long, since short ones can hardly be distinguished from the smooth curve. Then Dg is the step function shown in the figure. In any interval where

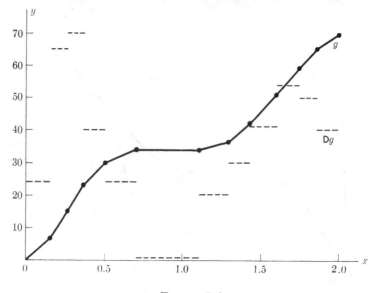

FIGURE 7–8

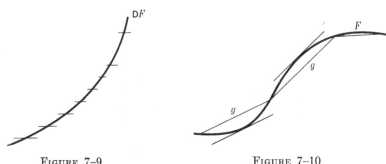

FIGURE 7-9   FIGURE 7-10

the graph of g is a segment, $Dg(x)$ is the slope of that segment. Now let us imagine the number of segments to be very large so that g differs from F very little. Then Dg will consist of many little steps approximating DF, as suggested for a part of Dg in Fig. 7-9. The segments making up g practically coincide with the curve F, and their slopes are practically the same as the slopes of the tangent lines to F in each interval, as suggested in Fig. 7-10. Hence it seems reasonable to say that $DF(x)$ is the rate of change of $F(x)$ at the point $(x, F(x))$, that is, the slope of the line tangent to F at the point $(x, F(x))$.

(q) Trace Fig. 7-7 on a piece of paper. Join the points on the curve for which $x = 0, 0.2, 0.4, 0.6, 0.8, 1.0, 1.2, 1.4, 1.6, 1.8, 2.0$ by straight line segments, and call this graph G. Estimate the slopes of each segment by estimating the coordinates of the endpoints. In each interval draw the graph of DG such that $DG(x) = $ the slope of the segment in the interval in which x lies. Compare with Figs. 7-6 and 7-8. Why is $DG(x)$ undefined at endpoints of segments?

The procedure we used for finding DF approximately when F is given is often useful. From the graph of F we can estimate DF by observing that $DF(x)$ is approximately the slope of a short segment joining two points on F near $(x, F(x))$. However, it is desirable to have more accurate methods of finding rates of change. For this purpose we introduce precise definitions of "rate of change" and "tangent" in the following sections. Here we consider a particular simple case in order to familiarize the reader with our intentions.

Consider the function defined by $y = x^2$. Figure 7-11 shows a very much enlarged portion of its graph in the neighborhood of $(3, 9)$. We label the tangent line at $(3, 9)$ T. The secant lines joining $(3, 9)$ to three other nearby points are labeled $S_1, S_2,$ and S_3. By (4-5-11) with $(x_1, y_1) = (3, 9)$, we find that the slopes of $S_1, S_2,$ and S_3 are $7, 6.5,$ and 6.1. Following is a table showing the slope of secants for points still closer to $(3, 9)$. The first column gives the x-coordinate of the second point on the secant, the second column the corresponding value of y, the last three columns the

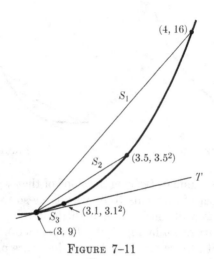

FIGURE 7–11

differences of the coordinates and the slope. Slopes of secants determined by letting the second point lie at the left are included also. The table shows that as we let the second point approach the first, the slope of the secant appears to approach the value 6.

x_2	y_2	$y_2 - y_1$	$x_2 - x_1$	$(y_2 - y_1)/(x_2 - x_1)$
3.01	9.0601	0.0601	0.01	6.01
3.001	9.006001	0.006001	0.001	6.001
3.0001	9.00060001	0.00060001	0.0001	6.0001
2.5	6.25	−2.75	−0.5	5.5
2.9	8.41	−0.59	−0.1	5.9
2.99	8.9401	−0.0599	−0.01	5.99
2.999	8.994001	−0.005999	−0.001	5.999
2.9999	8.99940001	−0.00059999	−0.0001	5.9999

It appears that if we take a sequence of secant lines, S_1, S_2, S_3,... passing through (3, 9) and another point that moves toward (3, 9), the slope of the secants approaches 6. Since the sequence of secants approaches coincidence with the tangent line at (3, 9), we suppose that the slope of this tangent line is 6. In other words, if $y = x^2$, the instantaneous rate of change of y with respect to x at (3, 9) is 6.

(r) Consider the graph of $y = x^2$ at $x = 2$. Find the slopes of the secant lines joining (2, 4) to other points as we did above. Make a table for $x_2 = 3$, 2.5, 2.1, 2.01, 2.001, 1.5, 1.9, and 1.99. What is the slope of the tangent line?

This sort of procedure could be used whenever we know a defining formula for a function. We could calculate slopes of secants for a sequence of values of the variable getting closer to the value in which we are interested, then guess the limit. However, by using algebra instead of arithmetic, we can get more precise and general results.

Consider f defined by $y = x^2$. Let (x, y) be a point at which we wish the instantaneous rate of change of y with respect to x, that is, $Df(x)$. Let $(x + h, (x + h)^2)$ be another point on f, where h may be > 0 or < 0. Here $x_1 = x$, $x_2 = x + h$, and $x_2 - x_1 = h$. The slope of the secant is given by

$$(1) \qquad m = \frac{(x + h)^2 - x^2}{(x + h) - x} = \frac{2hx + h^2}{h} = 2x + h.$$

It is obvious that as h approaches zero (i.e., the secant approaches the tangent), m approaches $2x$. In brief, as $h \to 0$, $m \to 2x$, where we read the arrow as "approaches" or "goes to." For example, when $x = 3$, we have $m \to 2 \cdot 3$ or 6, as we found above.*

(s) Verify that (1) holds in the table above.   (t) Why not let $h = 0$ in (1)?

For any value of x the formula $2x$ gives the slope of the line tangent to the graph of $y = x^2$ at the point (x, x^2). For example, when $x = -1$, the slope is $2(-1)$ or -2. Hence the line tangent at the point $(-1, 1)$ to the graph of I^2 is $y - 1 = -2(x + 1)$.

(u) Sketch I^2 and tangent lines at $x = 0$, $x = 3$, $x = -1$. Write the equations of the tangent lines there. Do their slopes seem consistent with your drawing?

The formula $2x$, giving the slopes of tangent lines to the graph of I^2, is a defining formula for the function $2I$. We call $2I$ the derivative of I^2. More generally, a function under which the image of x is the slope of the line tangent to the graph of f at the point $(x, f(x))$ is called the *derivative* of f and is designated by Df. $Df(x)$ is the slope of the tangent line to the graph of f at $(x, f(x))$ and also the exact rate of change of $f(x)$ with respect to x. We may summarize our discussion in the law

$$(2) \qquad\qquad DI^2 = 2I.$$

(v) Graph I^2 and $2I$ on the same plane. Explain the relationship between the two graphs in terms of rate of change and tangents.   (w) Suppose that

---

* Although this symbol is the same as that used for implication in this book, confusion is unlikely because of the different contexts in which the symbols are used. In particular, the implication arrow stands *only* between sentences. In order to avoid the double use of symbols, mathematicians often write longer arrows or double-shafted arrows for implication.

$Df(x) > 0$ for a certain x. What conclusion could you draw? (x) Answer the same question for $Df(x) < 0$, and for $Df(x) = 0$. (y) Use the above methods to find $D(3I^2)$. (z) Justify

$$(3) \qquad D(aI^2) = 2aI.$$

PROBLEMS

In Problems 1 through 10 find the slope of the line tangent to the graph of the indicated function at the point with given x-coordinate, write the equation of the tangent line, and sketch.

1. $y = x^2$ at $x = -2$.
2. $y = x^2$ at $x = 2$.
3. I^2 at $x = 0.5$
4. $2I^2$ at $x = -1$.
5. $-3I^2$ at $x = -1$.
6. I^3 at $x = 0$.
7. I^3 at $x = -1$.
8. I^3 at $x = 1$.
9. $2I^3$ at $x = x_0$.
10. $-I^3$ at $y = 8$.

11. Justify (4).

$$(4) \qquad D(aI^3) = 3aI^2.$$

12. Justify (5) and (6) in terms of rate of change.

$$(5) \qquad D(I) = \underline{1}.$$

$$(6) \qquad D(\underline{c}) = \underline{0}.$$

13. A freely falling body falls a distance y given by $y = 16x^2$, where x is time measured in seconds from the time when the body was at rest and y is in feet. Find a formula giving the velocity of the body at any time.

14. Recalling that acceleration is the rate of change of velocity with respect to time, find a formula for the acceleration of the body in Problem 13.

15. Let x be the output of a factory and y the total cost of the product, and imagine Fig. 7–6 as the graph of the function $y = f(x)$. What is the interpretation of Df? Do you think that a cost function might have this appearance? Would Fig. 7–7 be more realistic?

16. Let x be the time in years and y be the population in millions, and imagine Fig. 7–6 as the graph of the function $y = f(x)$. Answer the questions of Problem 15.

17. Let x be time spent in memorizing and y be the amount of material memorized, and imagine Fig. 7–6 to represent the graph of $y = f(x)$. Answer the questions of Problem 15.

18. V. O. Key (*Primer of Statistics for Political Scientists*, pp. 74–81) finds that the relation between the New York State presidential vote Y and the New York City registration X in millions was given approximately by $Y = 976{,}710 + 1.6026X$ during the years 1924–1948. Indicate what this means in terms of slope and rate of change.

19. In each of the following cases discuss the rate of change of y with respect to x.

	y	x
(a)	Distance	Time
(b)	Wealth	Time
(c)	Total cost	Amount produced
(d)	Income from sales	Amount sold
(e)	Amount of gasoline in a tank	Time
(f)	Words memorized	Time
(g)	Height of person	Time
(h)	Circumference of circle	Diameter
(i)	Value of $a(b + c)$	$b + c$

★20. Suppose y is the distance of a car from New York City and $y = -35x + 550$. What is going on? Sketch the function.

In Problems 21 through 23 we assume that the variables are known to be related by linear functions.

★21. Suppose $m = 2$. List several possible functions with this property and sketch them on the same graph.

★22. Suppose that $m = 2$ and $y = 3$ when $x = 1$. Find the function and graph it.

★23. Suppose the interest rate is 10% and that the amount of money on deposit after 5 years is $150. Find the function and determine the amount on deposit after 6 years and also at the beginning.

ANSWERS TO EXERCISES

(a) $mI + b$.　(c) y decreasing; not changing; increasing as x increases. (d) $-b/m$.　(e) 20 mi/hr.　(f) The slope of f is zero there.　(g) Undefined. (h) 20 mi/hr; 40 mi/hr; undefined.　(l) (0__2.2), (0__80).　(m) (0__0.5) ∪ (0.5__0.8) ∪ (0.8__1.2) ∪ (1.2__2.2), {50, 20, 0, 49}.　(n) 0.　(o) The steeper the graph, the faster y is increasing.　(p) DF turns downward, i.e., decreases. (q) No unique rate of change there.　(r) 4.　(t) m is undefined there; $(2hx + h^2)/h = 2x + h$ only for $h \neq 0$.　(u) $y = 0$; $y - 9 = 6(x - 3)$; $y - 1 = -2(x + 1)$.　(v) Consider the intervals where $2x < 0$, $= 0$, > 0. (w) $f(x)$ is increasing at that point.　(x) $f(x)$ is decreasing; stationary.　(y) $6I$. (z) $[a(x + h)^2 - ax^2]/h = 2ax + ah \to 2ax$.

ANSWERS TO PROBLEMS

1. $y - 4 = -4(x + 2)$.　3. $y - 0.25 = (x - 0.5)$.　5. $y + 3 = 6(x + 1)$. 7. $DI^3 = 3I^2; y + 1 = 3(x + 1)$.　9. $y - 2x_0^3 = 6x_0^2(x - x_0)$.　11. $a(x + h)^3 -$

$ax^3 = 3ax^2h + 3axh^2 + ah^3$.   **13.** $D16I^2 = 32I$; $v = 32x$.   **19.** (a) Velocity,
(b) Net rate of income, (c) Marginal cost, rate of change of cost per unit change
in output, (d) Price, (e) Net flow of gasoline into tank, negative when driving,
positive when filling tank, (f) Rate of memorization, (g) Rate of growth,
(h) π, (i) a.

7-3 Limits and continuity. From the previous section it appears that
in order to define precisely the notions of "exact rate of change" and
"slope of the tangent line" we need a definition of the notions of "approach"
and "limit." We wish to define "$f(x) \to L$ as $x \to a$" (f of x approaches L
as x approaches a) and the synonymous expression "$\lim_{x \to a} f(x) = L$"
(the limit of $f(x)$ as x approaches a is L). Then we can derive rules for
finding limits with precision and certainty. We introduce the definition
by considering typical examples.

To begin with a very simple case, consider $2I$ defined by $f(x) = 2x$.
We have $f(a) = 2a$. Also it seems plausible that $f(x) = 2x \to 2a$ as
$x \to a$. Since $|f(x) - 2a| = |2x - 2a| = 2|x - a|$, $f(x)$ differs from $2a$
by an amount that becomes smaller as $|x - a|$ becomes smaller. Indeed by
taking x close enough to a we can make $|f(x) - 2a|$ as small as we like.
Suppose, for example, that we wish to make $|f(x) - 2a| < \epsilon$, where ϵ
is any positive number. It is sufficient for this purpose to make $|x - a| <
\epsilon/2$. Indeed for any x such that $|x - a| < \epsilon/2$, $|f(x) - 2a| < 2(\epsilon/2) = \epsilon$.
In particular $f(x) \to 0$ as $x \to 0$.

Now consider $2I$ with the point $(0, 0)$ deleted (see Fig. 7-12). Here
$f(x) = 2x$ for $x \neq 0$, and $f(0)$ is not defined. Nevertheless it seems rea-
sonable to say that $f(x) \to 0$ as $x \to 0$, since $|f(x) - 0| = |2x - 0| =
2|x|$, and this is less than any positive ϵ, however small, provided $0 < |x| <
\epsilon/2$. Note that we had to exclude $x = 0$, since $f(x)$ is undefined there.

Now consider $2I$ with $(0, 0)$ deleted and $(0, 2)$ added, that is, $f =
(2I \cap \{(0, 0)\}') \cup \{(0, 2)\}$. Here $f(0) = 2$, yet we feel that $f(x) \to 0$
as $x \to 0$, since $|f(x)|$ is arbitrarily small provided $|x|$ is small and not zero.

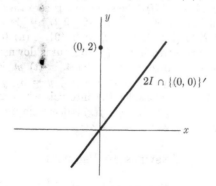

FIGURE 7-12

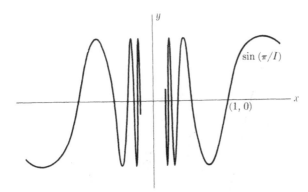

FIGURE 7-13

We see from these examples that the limit of $f(x)$ as $x \rightarrow a$ does not depend on $f(a)$, which may be undefined or different from the limit. Hence our definition should not put any conditions on the function at $x = a$.

(a) In this last example, what is $|f(x) - 0|$ when $x = 0$?

To see the necessity of requiring that we can make $|f(x) - L|$ *arbitrarily* small by taking $|x - a|$ *sufficiently* small, consider the function $I^2 + 10^{-10}$. As $x \rightarrow 0$, $x^2 + 10^{-10}$ always gets closer to 0 and the difference becomes very small. But $\lim_{x \to 0}(x^2 + 10^{-10}) \neq 0$, for $x^2 + 10^{-10}$ cannot be made closer to 0 than 10^{-10}.

Now consider the function $\sin [\pi/I]$ defined by $f(x) = \sin (\pi/x)$ and sketched in Fig. 7-13. For $x \in (\underline{1 \quad} \infty)$, π/x decreases from π toward zero, and $\sin (\pi/x)$ runs from 0 to 1 and back toward zero. However, as x goes from 1 toward 0, π/x increases from π without bound. When $x = 1/n$ for n integral, $\sin (\pi/x) = \sin n\pi = 0$. When $x = 2/(4n + 1)$ with n integral, $\sin (\pi/x) = 1$. The function is not defined for $x = 0$.

(b) For what values of x is $\sin (\pi/x) = -1$?

Evidently the graph oscillates an infinite number of times as $x \rightarrow 0$. In *any* interval about the origin, no matter how small, the function oscillates between -1 and 1 an infinite number of times. We say that $f(x)$ has no limit as $x \rightarrow 0$ in this case. Note, however, that we can make $f(x)$ arbitrarily small. In fact $f(x) = 0$ at infinitely many points. This emphasizes again that for $f(x) \rightarrow L$ as $x \rightarrow a$, we must require that $|f(x) - L|$ *remains* arbitrarily small for *all* values of x *sufficiently* close to a.

Finally, consider $I \sin (\pi/I)$, sketched in Fig. 7-14. We may imagine it constructed by distorting the graph of Fig. 7-13 so that it oscillates between the lines $y = x$ and $y = -x$. Here $f(0)$ is undefined, but the curve seems to be approaching the origin. Since $|\sin (\pi/x)| \leq 1$, we have

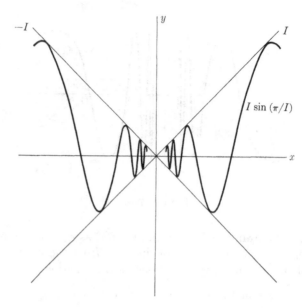

$I \sin (\pi/I)$

FIGURE 7-14

$|x \sin (\pi/x)| \leq |x| < \epsilon$ for all x such that $|x| < \epsilon$. Note that here $f(x)$ is not always getting nearer its limit, but in spite of the oscillations, it stays within an arbitrarily small distance of its limit provided x is close enough to its limiting value.

It is convenient to visualize the situation in terms of a function f mapping points on the axis of x into points on the axis of $f(x)$, as in Fig. 7–15. We call the open interval $(L - \epsilon\_L + \epsilon)$ defined by $|f(x) - L| < \epsilon$ an *epsilon neighborhood of L*. Similarly, we call $(a - \delta\_a + \delta)$ a *delta neighborhood of a*. We call $(a - \delta\_a + \delta) \cap \{a\}'$ defined by $0 < |x - a| < \delta$, a *deleted delta neighborhood of a*. If we are to have $\lim_{x \to a} f(x) = L$, we require that for x sufficiently near a, $f(x)$ is arbitrarily near L. That is, given any ϵ-neighborhood of L (that is, given any ϵ), there must be a deleted δ-neighborhood of a (determined by a δ) such that all points in the deleted neighborhood of a map into points in the neighborhood of L. Nothing is said here about the manner in which the points in the deleted δ-neighborhood map into the ϵ-neighborhood. If we imagine x as a point moving toward a, the corresponding point $f(x)$ may not always be moving toward L. But there is *some* deleted δ-neighborhood of a all of whose points map into the ϵ-neighborhood of L.

We can now give a precise definition. We require that for *any* given ϵ greater than 0, there must *exist* a δ such that if x is in the deleted δ-neighborhood of a, then $f(x)$ is in the ϵ-neighborhood of L. Without the geomet-

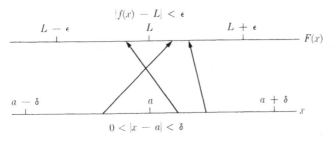

$$|f(x) - L| < \epsilon$$

FIGURE 7-15

ric imagery we have: given any positive ϵ there exists a δ such that $0 < |x - a| < \delta \rightarrow |f(x) - L| < \epsilon$. In symbols,

(1) **Def.** $[\lim_{x \to a} f(x) = L] = [f(x) \to L \text{ as } x \to a]$

$= [\forall \epsilon \, \epsilon > 0 \to \exists \delta \, \delta > 0 \wedge \forall x \, 0 < |x - a| < \delta \to |f(x) - L| < \epsilon].$

This definition is the outcome of the work of several generations of mathematicians. It may seem rather complicated, but all its parts are necessary, as is suggested by the preceding discussion. The idea is most easily grasped in terms of neighborhoods, and (1) is just a reformulation in terms of the sentences defining the neighborhoods.

(c) In the first example in which we argued that $2x \to 2a$ as $x \to a$, the definition is satisfied since we showed that $\delta = \epsilon/2$ is such that $0 < |x - a| < \delta \to |2x - 2a| < \epsilon$. Find another value for δ such that this sentence is satisfied. (d) Rephrase this situation in terms of neighborhoods. (e) In the second example where we argued that $2x \to 0$ as $x \to 0$, give two values for δ that satisfy the condition. (f) Show that for $\delta = \epsilon/10, 0 < |x| < \delta \to |x \sin (\pi/x)| < \epsilon$. (g) Prove that $x^2 \to 0$ as $x \to 0$ by finding a formula that gives the required δ in terms of ϵ. (h) Show that $2ax + h \to 2ax$ as $h \to 0$ by exhibiting the δ corresponding to any ϵ. (i) Justify (2) and (3) by reference to the definition.

(2) $$\lim_{x \to a} x = a.$$

(3) $$\lim_{x \to a} \underline{c}(x) = c.$$

From the definition can be derived all the laws we need for calculating limits. In particular,

(4) $[\lim_{x \to a} f(x) = A \wedge \lim_{x \to a} g(x) = B] \to [\lim_{x \to a} (f(x) + g(x)) = A + B]$

(5) $\to [\lim_{x \to a} (f(x)g(x)) = AB]$

(6) $[\lim_{x \to a} f(x) = A \ne 0] \to [\lim_{x \to a} (1/f(x)) = 1/A]$.

(j) Why $A \ne 0$ in (6)?   (k) State (4), (5), and (6) in words.

The plausibility of (4) through (6) is easy to establish. For example, the hypothesis of (4) tells us that we can make $f(x)$ arbitrarily close to A and $g(x)$ artitrarily close to B by requiring x to be sufficiently close to a. Clearly, by making x close enough to a, we can then make $f(x) + g(x)$ arbitrarily close to $A + B$. This suggests the way in which a proof could be constructed. Applying (1) to the hypothesis of (4) with $(L{:}A, \epsilon{:}\epsilon/2)$ and $(L{:}B, \epsilon{:}\epsilon/2)$, we see that there is a δ (call it δ_1) such that $0 < |x - a| < \delta_1 \to |f(x) - A| < \epsilon/2$ and a δ (call it δ_2) such that $0 < |x - a| < \delta_2 \to |f(x) - B| < \epsilon/2$. Let $\delta_3 = \min\{\delta_1, \delta_2\}$. Then for $0 < |x - a| < \delta_3$, both conclusions hold and hence $|f(x) - A| + |f(x) - B| < \epsilon/2 + \epsilon/2$. But $|f(x) + g(x) - (A + B)| = |f(x) - A + g(x) - B| \le |f(x) - A| + |g(x) - B|$. Hence $0 < |x - a| < \delta_3 \to |f(x) + g(x) - (A + B)| < \epsilon$, which is equivalent to the conclusion of (4) according to the definition (1). The other proofs we leave to optional problems.

(l) What laws and rules about existential quantifiers have we used here? (m) What law of inequalities did we use?   (n) Derive (7), (8), and (9) from (2) through (6).

(7) $[\lim_{x \to a} f(x) = A] \to [\lim_{x \to a} cf(x) = cA]$.

(8) $[\lim_{x \to a} f(x) = A \land \lim_{x \to a} g(x) = B] \to [\lim_{x \to a} [f(x) - g(x)] = A - B]$.

(9) $[\lim_{x \to a} f(x) = A \land \lim_{x \to a} g(x) = B \ne 0] \to [\lim_{x \to a} [f(x)/g(x)] = A/B]$.

We have not required that $f(a) = L$ whenever $f(x) \to L$ as $x \to a$. Indeed, we have seen that $f(a)$ may be different from L or undefined. However, when $f(x)$ approaches a limit that is equal to $f(a)$, we say that f is continuous at a. More precisely, f *is continuous at a* if and only if $\lim_{x \to a} f(x) = f(a)$. Note that f may fail to be continuous at a point for any one or a combination of the following reasons: $f(a)$ may not be defined, that is, a may not be in the domain of f; $f(x)$ may not have a limit as $x \to a$; or $f(a)$ and $\lim f(x)$ may exist but be different. When a function is not continuous at a we say that it is *discontinuous* or that a is a *point of discontinuity*. The simplest kinds of discontinuities are at points where the graph of a function has a jump.

(o) Cite examples of discontinuities from previous sections.   (p) Is $I \sin (\pi/I)$ discontinuous at the origin?   Answer the same question for the function $I \sin (\pi/I) \cup \{(0, 0)\}$.

When a function is continuous at a point, its limit there can be evaluated very easily by the substitution $(x{:}a)$ in $f(x)$. Hence it is important to know the continuity properties of functions. From (2) and (3) we see that the constant functions and the identity function are continuous at all points. By applying (4), (5), (7), (8), and (9) it is easy to show the following.

(10)   *If f and g are continuous at a, then cf, $f + g$, $f - g$, and fg are continuous at a. In addition, if $g(a) \neq 0$, f/g is continuous at a.*

(q) Show that all linear functions are continuous everywhere.   (r) Show that I^2 is continuous at all points.

When a function is continuous at all points of an interval it is said to be *continuous on the interval*. A function that is continuous on $(-\infty\_\infty)$ is said to be *continuous everywhere*.

(s) Show by induction that I^n is continuous everywhere for n a natural number.   (t) Show that all quadratic functions are continuous everywhere. (u) Use the theorems of this section to justify $2ax + h \to 2ax$ as $h \to 0$.

PROBLEMS

1. Sketch in the intervals $(-1.5 \quad -1/7)$ and $(1/7 \quad 1.5)$ the function defined by $f(x) = (-1)^{[1/x]}$, where $[1/x]$ is the largest integer less than or equal to $1/x$. Does $\lim f(x)$ exist as $x \to 0$? Is $f(x)$ defined for $x = 0$?
2. Do the same for the function defined by $g(x) = xf(x)$ where f is defined in 1. What is $\lim_{x\to 0} g(x)$? Is the function continuous at the origin? Could it be made so by adding a point to the graph? With the addition of this point, is there any interval containing the origin on which the function is continuous?
3. Find $\lim_{x\to 0} (mx + b)/(cx + d)$ assuming that $d \neq 0$ and justify your conclusion.
4. Find $\lim_{x\to 0} (x^2/x)$.
5. State (11) in words and justify it.

(11) $[\exists \delta\ \delta > 0 \wedge \forall x\ 0 < |x - a| < \delta \to f(x) = g(x)] \to [\lim_{x\to a} f(x) = \lim_{x\to a} g(x)]$.

6. Use (11) to find $\lim_{x\to 1} (x^2 - 2x + 1)/(x - 1)$.

In Problems 7 through 14 find the limits and justify your procedure.

7. $\lim_{x\to 1} 1/x$.

8. $\lim_{x\to 2} \dfrac{x^2 - 3x + 2}{x - 2}$.

9. $\lim_{x\to 0} \dfrac{Ax^n + 2}{Bx^m + 3}$.

10. $\lim_{x\to 0} \dfrac{x + \epsilon}{x - \epsilon}$.

11. $\lim\limits_{x \to 1} (3 + x)^{25}$.

12. $\lim\limits_{h \to 0} \dfrac{(x + h)^3 - x^3}{h}$.

★13. $\lim\limits_{h \to 0} (1/h)[\sqrt{x + h} - \sqrt{x}]$.

14. $\lim\limits_{h \to 0} \dfrac{(x + h)^4 - x^4}{h}$.

15. Argue that all polynomials are continuous everywhere. Argue that all rational functions are continuous everywhere except at the zeros of their denominators.

16. Show by induction that if f is continuous at a point, so is f^n for n a natural number.

★17. Show that if there is a neighborhood of a in which $f(x)$ is bounded, and if $\lim_{x \to a} g(x) = 0$, then $\lim_{x \to a} f(x)g(x) = 0$. Express this in symbols.

18. Show that $[\lim_{x \to a} f(x) = L] \leftrightarrow [\lim_{x \to a} (f(x) - L) = 0]$.

★19. Give a geometric argument for $\lim_{x \to 0} \sin x = 0$.

20. Use the result in Problem 19 and (4–12–29) to show that sin is everywhere continuous by showing that $\lim_{x \to a} (\sin x - \sin a) = 0$.

★21. Show that cos is continuous everywhere.

★22. Discuss the continuity of tan and cot.

23. Is $|I|$ continuous at the origin?

★24. Show that if $\lim_{x \to a} f(x) = L$, then there is a deleted neighborhood of a in which $f(x)$ is bounded; that is, $\exists \delta, M \; 0 < |x - a| < \delta \to |f(x)| < M$.

★25. Show that if $\lim f(x)$ exists and is positive, then there is a positive number A and a deleted neighborhood of a in which $f(x) > A$.

★26. State and justify a result similar to Problem 25 when the limit is nonzero.

★27. Show that if $\lim_{x \to a} f(x) = L$ and $\lim_{x \to a} f(x) = M$, then $M = L$. This means that the limit, if it exists, is unique.

★28. Consider $(\sin \theta)/\theta$. Is it defined for $\theta = 0$? Does it have a limit as $\theta \to 0$? Use Fig. 7–16 to guess the limit and justify your results by a geometric argument. (Note that θ is in radians!)

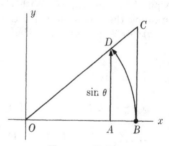

FIGURE 7–16

★29. Show that $\lim\limits_{x \to 0} \dfrac{\sin 2x}{x} = 2$.

★30. Find $\lim\limits_{x \to 0} \dfrac{1 - \cos x}{x}$.

★31. Prove (5) by noting that $|f(x)g(x) - AB| = |f(x)g(x) - Ag(x) + Ag(x) - AB| = |g(x)(f(x) - A) + A(g(x) - B)| \leq |g(x)| \, |f(x) - A| + |A| \, |g(x) - B|$, and using the result of Problem 24.

★32. Prove (6) by noting that $|1/f(x) - 1/A| = |(1/Af(x))(f(x) - A)|$ and using the result of Problem 26.

★33. If f is continuous at a and g is continuous at $f(a)$, then $g(f)$ is continuous at a, that is, the composite of continuous functions is continuous. Prove this.

★34. Construct a function that is continuous at the origin but discontinuous everywhere else.

★35. Show that if $f(x)$ is bounded from above (below) in some neighborhood of a, then $\lim_{x \to a} f(x)$ is not greater (less) than this upper (lower) bound.

★36. Show that if $\lim g(x) = \lim h(x)$ as $x \to a$ and in some neighborhood of a $g(x) \leq f(x) \leq h(x)$, then $\lim f(x) = \lim g(x)$ as $x \to a$.

ANSWERS TO EXERCISES

(a) 2.   (b) $2/(4n + 3)$ where n is integral.   (c) If δ_1 is a value of δ for which the implication holds, then so is any smaller value of δ.   (d) When x is in an $(\epsilon/2)$-neighborhood of a, $2x$ is in the ϵ-neighborhood of $2a$.   (e) $\epsilon/4$, $\epsilon/10$.   (f) $|x| < \epsilon/10 \to |x| < \epsilon$.   (g) If $x < \sqrt{\epsilon}$, then $x^2 < \epsilon$.   (h) ϵ. (i) For (2), any $\delta \leq \epsilon$. For (3), $\underline{c}(x) - c = 0$, so any number will do for δ. (j) Otherwise the right member is undefined.   (k) The sum and product of the limits is the limit of the sum and product. The limit of the reciprocal is the reciprocal of the limit, provided the limit is not zero.   (l) (2–10–5) and the following discussion; (2–10–9).   (m) (3–9–11).   (n) For example, for (7) apply $(5)(g{:}\underline{c}, B{:}c)$.   (o) Step functions are discontinuous at the steps. $1/I$ is discontinuous as $x = 0$.   (p) Yes, since it is undefined there. With the additional point it is continuous.   (q) Since they are sums of constant functions and constant multiples of the identity function.   (r) Since I is continuous everywhere, so is $I \cdot I$ or I^2 by (10).   (s) For $n = 1$, we have the identity function. If I^k is continuous, so is $I \cdot I^k$ or I^{k+1} by (10).   (t) All quadratics are sums of constant functions and constant multiples of I and I^2, all of which are continuous everywhere.   (u) $2ax + h$ is linear in h.

ANSWERS TO PROBLEMS

1. No. No. 2. $g(x) \to 0$ as $x \to 0$, but $g(0)$ is undefined. With $(0, 0)$ the function is continuous there. No. 3. b/d by use of (2), (7), (4), and (9). 5. If two functions coincide in a deleted neighborhood of a, then their limits (if any) as $x \to a$ must be the same. 7. 1. 9. $2/3$. 11. 2^{50}. 13. Multiply and divide by $\sqrt{x + h} + \sqrt{x}$ and assume that $I^{1/2}$ is continuous. The continuity is not hard to prove by noting that $\sqrt{x} - \sqrt{a} = (x - a)/(\sqrt{x} + \sqrt{a})$. If $|x - a| < \delta_1$, $x > a - \delta_1$, $\sqrt{x} > \sqrt{a - \delta_1}$, and $\sqrt{x} + \sqrt{a} > \sqrt{a - \delta_1} + \sqrt{a} = M$. Hence $1/(\sqrt{x} + \sqrt{a}) < 1/M$ for $|x - a| < \delta_1$. Now let δ be the smaller of δ_1 and ϵM. Then for $0 < |x - a| < \delta$, $|\sqrt{x} - \sqrt{a}| < |(\epsilon M)(1/M)| = \epsilon$ and hence $\sqrt{x} \to \sqrt{a}$ as $x \to a$. 15. By Exercise (s) and theorem (10). 17. Apply (1). 23. Yes. 27. Assume $M \neq L$ and use the definition (1) to find a contradiction.

7–4 The derivative. We define the *derivative* Df of the function f by

(1) **Def.** $$Df(x) = \lim_{h \to 0} \frac{f(x + h) - f(x)}{h}.$$

Note that Df is a function whose domain is the domain of f less those points (if any) where the limit does not exist. Thus D is a function that transforms a function f into its derivative Df.

Now we can make the following precise definitions. *The rate of change* of $f(x)$ with respect to x at a is $Df(a)$. The *slope* of f at $(a, f(a))$ is $Df(a)$. The *tangent line* to f at $(a, f(a))$ is the line through $(a, f(a))$ having slope $Df(a)$. Note that "slope" up to now has applied only to straight lines, but this definition means that we can speak of the slope of any curve at any point where the derivative exists.

(a) Complete: The slope of a curve at a point is the slope of the _____.
(b) Justify (2).

(2)
The equation of the line tangent to f at $(x_0, f(x_0))$ is

$$y - f(x_0) = [Df(x_0)](x - x_0).$$

(c) Apply (2) to find the equation of the tangent line to $3I^2$ at $(-1, 3)$.
(d) What is the slope of $\underline{3}$ at $(0, 3)$?   (e) What is the slope of $I^2/5$ at -5?
(f) Review the discussion of rates of change and derivatives in Sections 7–1 and 7–2.

The number of situations in which the derivative is a useful tool is enormous. Let x be the output of a firm and y the total cost of producing x. We have $y = f(x)$, where f is the cost function. A possible graph is sketched in Fig. 7–17. In this case the slope of the curve at any point (the

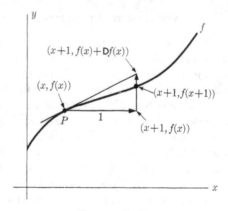

FIGURE 7–17

derivative, the rate of change of cost with respect to output) is called the *marginal cost*. It represents the change in the cost per unit change in output if cost increases at a constant rate equal to the slope of the tangent line. Thus, in the figure at P, if output were increased 1 unit and the rate of increase remained constant (which it may not), then cost would increase by $\mathbf{D}f(x)$. Of course, we have distorted the scale in Fig. 7–17. On an actual cost curve a change of one unit in x would be very small, and $f(x)$ and $\mathbf{D}f(x)$ would be almost the same as $f(x+1)$ and $f(x+1) - f(x)$.

(g) What vector represents $\mathbf{D}f(x)$ in Fig. 7–17?   (h) Describe the behavior of $\mathbf{D}f(x)$ in Fig. 7–17 as x increases.

Copy Fig. 7–17 and sketch $\mathbf{D}f$ on the same graph. Do not worry about scale, but make $\mathbf{D}f(x)$ increase or decrease as indicated by the figure. What do we call $\mathbf{D}f(x)$ where $y = f(x)$ and y and x are: (i) distance, time,   (j) velocity, time, (k) water in a tank, time,   (l) altitude, horizontal distance?

Definition (1) and the laws concerning limits can be used to find derivatives. However, the work is much simplified by using the following theorems.

(3) $$\mathbf{D}\underline{c} = \underline{0}.$$

(4) $$\mathbf{D}I = \underline{1}.$$

(5) $$\mathbf{D}(f + g) = \mathbf{D}f + \mathbf{D}g.$$

(6) $$\mathbf{D}(f \cdot g) = f \cdot \mathbf{D}g + g \cdot \mathbf{D}f.$$

(7) $$\mathbf{D}(\underline{c} \cdot f) = \underline{c} \cdot \mathbf{D}f.$$

(8) $$\mathbf{D}(f - g) = \mathbf{D}f - \mathbf{D}g.$$

(9) $$\mathbf{D}(1/f) = -\mathbf{D}f/f^2.$$

(10) $$\mathbf{D}(f/g) = \frac{g \cdot \mathbf{D}f - f \cdot \mathbf{D}g}{g^2}.$$

(11) $$\mathbf{D}I^n = nI^{n-1}.$$

(m) What is the geometric interpretation of (3)?   (n) Complete the proof of (3) suggested by $\mathbf{D}\underline{c}(x) = \lim (1/h)[(\underline{c})(x+h) - \underline{c}(x)] = \lim 0/h = 0$. (o) Similarly prove (4) by noting that $\mathbf{D}x = \lim [(x + h - x)/h]$.

To illustrate the usefulness of these theorems, let us find the derivative of $3I^2 + 2I - \underline{10}$ without and with their use. We have

(12) $\mathbf{D}(3I^2 + 2I - \underline{10})(x)$
$$= \lim_{h \to 0} [3(x + h)^2 + 2(x + h) - 10 - 3x^2 - 2x + 10]/h$$

$$(13) \qquad\qquad = \lim_{h \to 0} (6hx + 3h^2 + 2h)/h$$

$$(14) \qquad\qquad = \lim_{h \to 0} (6x + 2 + 3h)$$

$$(15) \qquad\qquad = 6x + 2$$

$$(16) \qquad\qquad = (6I + 2)(x).$$

The justification of (14) is that $(6hx + 3h^2 + 2h)/h = 6x + 2 + 3h$ for all $h \neq 0$, and the limit does not depend on the situation at $h = 0$. The same idea is embodied in (7-3-11). Step (15) is justified by the fact that $6x + 2 + 3h$ is linear in h and hence continuous, so that we can merely substitute $h = 0$ in it. By using the theorems of this section the same result is obtained much more easily.

(17) $\mathsf{D}(3I^2 + 2I - 10) = \mathsf{D}(3I^2) + \mathsf{D}(2I) - \mathsf{D}(10)$ $\qquad$ (5)

(18) $\qquad\qquad = 3\mathsf{D}(I^2) + 2\mathsf{D}I - \underline{0}$ $\qquad$ (7), (3)

(19) $\qquad\qquad = 6I + 2$ $\qquad\qquad$ (11)(n:2), (n:1).

(p) What is the slope of $3I^2 + 2I - 10$ at -3? Find the equation of the tangent line there. (q) Find $\mathsf{D}(5I^2 + 3I)$ without and with the theorems of this section.

To suggest the way in which laws (5) through (11) are proved, we give a proof of (5), leaving the others to the problems.

(20) $[\mathsf{D}(f + g)](x) = \lim\limits_{h \to 0} [f(x + h) + g(x + h) - f(x) - g(x)]/h$ $\quad$ (1)

$$(21) \qquad\qquad = \lim_{h \to 0} \left[\frac{f(x + h) - f(x)}{h} + \frac{g(x + h) - g(x)}{h} \right]$$

$$(22) \qquad\qquad = \lim_{h \to 0} \frac{f(x + h) - f(x)}{h} + \lim_{h \to 0} \frac{g(x + h) - g(x)}{h}$$

$$(7\text{-}3\text{-}4)$$

(23) $\qquad\qquad = \mathsf{D}f(x) + \mathsf{D}g(x)$ $\qquad\qquad\qquad\qquad$ (1).

The variables "f," "g," "$\underline{c}$" in the laws stand for functions. Indeed, their ranges are sets of functions having derivatives. The laws are identities relating functions. Of course, if we are interested in particular points we must insert an independent variable to get relations between images under functions and their derivatives. For example, (3) becomes $\mathsf{D}\underline{c}(x) = \underline{0}(x)$. But since $\underline{c}(x) = c$ and $\underline{0}(x) = 0$, we have $\mathsf{D}c = 0$. Similarly (4) becomes

$DI(x) = \underline{1}(x)$ or $Dx = 1$. Similarly (5) becomes $D(f + g)(x) = Df(x) + Dg(x)$. The left member means the image of x under $D(f + g)$. It is customary to write this $D(f(x) + g(x))$, so that (5) becomes $D(f(x) + g(x)) = Df(x) + Dg(x)$. Thus $D(x^2 + 3x) = D(I^2 + 3I)(x)$.

(r) Rewrite (6) through (11) similarly in terms of the independent variable. (s) A body is moving so that its distance y is given by $y = x^2 + 3x$, where x is the time. Find its velocity when $x = 3$.   (t) Find the slope of $I^2 + 3I + 3$ at $x = 3$.

There are many different notations for the derivative. Common synonyms for "$Df(x)$" are "$f'(x)$," "$D_x f(x)$," "$df(x)/dx$," and "$f_x(x)$." The last three are useful when an expression contains several variables. For example, if a person is asked to find the derivative of $2ax + 3y$, he may not know whether the function is $2aI + \underline{3y}$, $2xI + \underline{3y}$, or $\underline{2ax} + 3I$. These three possibilities may be indicated by writing $D_x(2ax + 3y)$, $D_a(2ax + 3y)$, $D_y(2ax + 3y)$. The first synonym, "$f'(x)$," has the advantage of brevity. Usually derivatives are treated in terms of images so that one sees $D(x^2 + x - 2) = 2x + 1$ rather than $D(I^2 + I - 2) = 2I + 1$. Actually $[D(I^2 + I - 2) = 2I + 1] \leftrightarrow [D_x(x^2 + x - 2) = 2x + 1]$. The reader should use whichever notation is convenient.

Still other notations are used in various contexts. Often Δx is written for h and Δy for $f(x + h) - f(x)$. Then the average rate of change is $\Delta y/\Delta x$, and $Df(x)$ is the limit of this (if any) as $\Delta x \to 0$. We call Δx and Δy the *increments* of x and y respectively. Then $Df(x)$ is expressed by dy/dx, y', Dy, or $D_x y$.

Find: (u) $D(x^3 - 5x^2 + 3)$.   (v) $D(I^3 - 5I^2 + \underline{3})$.   (w) $D(1/I)$.   (x) $D(1/I^2)$.   (y) $D(1/x^2)$.   (z) $D(x^2(x - 3x^3))$.

<h2 style="text-align:center">Problems</h2>

In Problems 1 through 8 find the derivative by the definition and the laws of limits without using the laws of this section. Check by finding the derivative *with* the aid of the laws in this section. You may assume that (11) holds for all real values of n.

1. DI^4.
2. $D(1/I)$.
3. $D(I^{1/2})$.
4. $D(1 - 1/I^2)$.
5. $D(I^{-1/2})$.
6. $D(aI^2 + bI + c)$.
7. $D(5x - x^{-3})$.
8. $D(x + 1)/(x - 1)$.

9. Find $D|I|$. *Suggestion:* Consider $x > 0$, $x = 0$, $x < 0$ separately and keep the geometric interpretation in mind. What is its domain?

10. Describe the derivative of a step function.

11. Do the same for a piecewise linear function.

★12. If the limit of $(f(a + h) - f(a))/h$ as $h \to 0$ exists, that is, if $Df(a)$ exists, we say that f is *differentiable* at a. Show that

(24) $(f$ is differentiable at $a) \to (f$ is continuous at $a)$.

13. Show that the converse of (24) is false.

14. Show that a function cannot be differentiable at a point where it is discontinuous.

15. Prove (6) by using the fact that $f(x + h)g(x + h) - f(x)g(x) = f(x + h)g(x + h) - f(x + h)g(x) + g(x)f(x + h) - f(x)g(x) = f(x + h)[g(x + h) - g(x)] + g(x)[f(x + h) - f(x)]$.

16. Prove (7) from (3) and (6).

17. Prove (8) from (5) and (7).

18. Prove (9) by noting that $1/f(x + h) - 1/f(x) = [f(x) - f(x + h)]/f(x)f(x + h) = -[f(x + h) - f(x)]/f(x)f(x + h)$.

19. Prove (10) from (6) and (9).

20. Prove $D(f + g + h) = Df + Dg + Dh$.

21. Prove (11) for positive integral n by induction. (We define I^0 as $\underline{1}$.)

22. Show that (11) holds for $n = 0$.

23. Prove $D(f \cdot g \cdot h) = f \cdot g \cdot Dh + f \cdot h \cdot Dg + g \cdot h \cdot Df$.

★24. Prove (11) for negative integral n by using (10).

★25. Prove (25) and (26) by applying (1), (4–12–9), (4–12–10), and Problems 28 and 30 in Section 7–3.

(25) $D \sin = \cos$.

(26) $D \cos = -\sin$.

26. What is the slope of sin at the origin? at $\pi/2$? at π? at $3\pi/2$? at $\pi/4$?

Answers to Exercises

(a) tangent line at that point.   (b) The point-slope form derived in Exercise (q) of Section 4–5.   (c) $D(3I^2) = 6I$, $6I(-1) = -6$, $y - 3 = -6(x + 1)$.   (d) 0.   (e) $(2I/5)(-5)$ or -2.   (g) The vector from $(x + 1, f(x))$ to $(x + 1, f(x) + Df(x))$.   (h) $Df(x)$ is initially positive and decreases for a time but remains positive. About halfway across the graph it begins and continues to increase.   (i) Velocity.   (j) Acceleration.   (k) Rate of flow.   (l) Gradient.   (m) That a horizontal line has zero slope.   (n) $\underline{c}(x) = c$ for all x. Insert "$h \to 0$" to complete the limit notation.   (o) $(x + h - x)/h = 1$ for all $h \neq 0$.   (p) $D(3I^2 + 2I - 10) = 6I + 2$. $(6I + 2)(-3) = -16$. $y - 11 = -16(x + 3)$.   (q) $D(5I^2 + 3I)(x) = \lim (5(x + h)^2 + 3(x + h) - 5x^2 - 3x)/h = \lim (10hx + 5h^2 + 3h)/h = \lim (10x + 3 + 5h) = 10x + 3$. $D(5I^2 + 3I) = D(5I^2) + D(3I) = 5DI^2 + 3DI = 10I + 3$.   (r) $D(f(x)g(x)) = f(x)Dg(x) + g(x)Df(x)$. $Dcf(x) = cDf(x)$. $D(f(x) - g(x)) = Df(x) - Dg(x)$. $D(1/f(x)) = -Df(x)/f^2(x)$. $D(f(x)/g(x)) = [g(x)Df(x) - f(x)Dg(x)]/g^2(x)$. $Dx^n = nx^{n-1}$.   (s) $Df(x) = 2x + 3$. $Df(3) = 9$.   (t) 9.   (u) $3x^2 - 10x$.   (v) $3I^2 - 10I$.   (w) $-1/I^2$.   (x) $-2/I^3$.   (y) $-2x^{-3}$.   (z) $x^2(1 - 9x^2) + 2x(x - 3x^3)$ or $3x^2 - 15x^4$.

Answers to Problems

1. $4I^3$. 3. $(1/2)I^{-1/2}$. 5. $(-1/2)I^{-3/2}$. 7. $5 + 3x^{-4}$. 9. The domain does not contain the origin. For $x > 0$, $D|x| = 1$ and for $x < 0$, $D|x| = -1$. Hence $D|x| = |x|/x$, $D|I| = |I|/I$. 10. Derivative is zero throughout its domain, which is the domain of the function less the endpoints of segments. 13. $|I|$ is a counterexample. 15. This trick of adding and subtracting the same thing is very frequently useful. We have, with the aid of (7–3–4) and (7–3–5), $D[f(x) \cdot g(x)] = [\lim f(x + h)] \cdot [\lim [g(x + h) - g(x)]/h] + g(x) \lim [f(x + h) - f(x)]/h = f(x)Dg(x) + g(x)Df(x)$. Here we assume that as $h \to 0$, $f(x + h) \to f(x)$, that is, that f is continuous, which must be true anyway if Df is to exist, as indicated in (24). 16. $D(\underline{c} \cdot f) = \underline{c} \cdot Df + f \cdot D\underline{c} = \underline{c} \cdot Df + 0$. 17. $D(f - g) = D[f + (-1) \cdot g] = Df + (-1)Dg$.
18. $D[1/f(x)] = -\lim [f(x + h) - f(x)]/h \cdot \lim [1/f(x) \cdot f(x + h)] = -Df(x)/[f(x)]^2$. Note necessity of continuity of f again. 19. $D[f/g] = D[f \cdot (1/g)] = f \cdot D(1/g) + (1/g) \cdot Df = -fDg/g^2 + Df/g = (gDf - fDg)/g^2$. 21. Let $p(n) = (11)$. Then $p(1) = [DI = 1]$, $p(n + 1) = [DI^{n+1} = (n + 1)I^n]$. Assume $p(n)$ by Hyp. Then, $DI^{n+1} = D(I \cdot I^n) = I \cdot DI^n + I^n \cdot DI = I \cdot nI^{n-1} + I^n \cdot \underline{1} = nI^n + I^n = (n + 1)I^n$. 24. For $n = -m$ with $m > 0$, $DI^{-m} = D(1/I^m) = (I^m D\underline{1} - 1 \cdot DI^m)/I^{2m} = -mI^{m-1}/I^{2m} = -mI^{-m-1} = nI^{n-1}$.
25. $D \sin x = \lim_{h \to 0}[\sin (x + h) - \sin x]/h = \lim_{h \to 0}[\sin x \cos h + \cos x \sin h - \sin x]/h = \lim_{h \to 0}[\sin x(\cos h - 1)/h + \cos x \sin h/h] = [\sin x \lim_{h \to 0}(\cos h - 1)/h] + [\cos x \lim_{h \to 0}(\sin h)/h]$. Then see Problem 28 in Section 7–3.

7–5 Polynomials. *Polynomials* are functions that can be constructed from constant functions and the identity function by a finite number of operations of addition and multiplication. (See Section 5–12.) A polynomial has a defining formula of the form $\sum_{i=0}^{i=n} c_i x^i$, with $c_n \neq 0$, and hence a name of the form $\sum_{i=0}^{i=n} c_i I^i$. We call n the *degree* of the polynomial for $n \geq 0$. We consider $\underline{0}$ a polynomial without a degree. Other constant functions are polynomials of degree zero, linear functions are polynomials of degree one, and quadratic functions are polynomials of degree two. Third, fourth, and fifth degree polynomials are sometimes called *cubics*, *quartics*, and *quintics*. The c_i are called *coefficients*.

Since polynomials are continuous, their graphs have no breaks. Also, since

(1) $$D \sum_{i=0}^{i=n} c_i I^i = \sum_{i=1}^{i=n} ic_i I^{i-1},$$

we see that the derivative of a polynomial is a polynomial. It follows that the derivative is continuous. This means that as the tangent line moves along the graph of the original polynomial it turns gradually without any sudden changes. In other words, the graph of a polynomial is *smooth*, without any sharp corners or other points at which there is no tangent. The fact that polynomials are both continuous and smooth justifies us completely in graphing one by plotting selected points and joining them by a smooth curve.

(a) Cite a continuous function whose derivative is not continuous and which has a sharp point at the place where its derivative does not exist. (b) Why are polynomials continuous everywhere? (c) Why does the summation in the right member of (1) run from 1 to n instead of from 0 to n? (d) Write (1) for the special case of a quadratic. (e) Justify (1).

An important subset of the polynomials is the *power functions* which have names of the form cI^n with n a non-negative integer. We have already considered I and I^2. Figure 7–18 shows I^3. Since $DI^3(0) = 3I^2(0) = 0$, we know that the graph does level off at the origin as shown. This fact, together with our knowledge of continuity and smoothness, permits us to draw the curve with only a few points. We know that the curve always rises from left to right since $Dx^3 = 3x^2 \geq 0$.

Graph in $(\underline{-2\quad 2})$: (f) $I^3/2$, (g) $-I^3$, (h) $2I^3$. (i) Why does $Dx^3 \geq 0$ imply that the curve is rising from left to right?

Power functions of the form I^{2n}, that is, even power functions, are similar to I^2. We show I^2 and I^4 in Fig. 7–19. Since $Dx^{2n} = 2nx^{2n-1} < 0$ for $x < 0$, $= 0$ for $x = 0$, and > 0 for $x > 0$, I^{2n} is decreasing at the left of the origin, horizontal at the origin, and increasing to the right of the origin. Since $x^{2n} \geq 0$, the graph never goes below the x-axis. I^{2n} is symmetric with respect to the y-axis, since $x^{2n} = (-x)^{2n}$.

Functions of the form I^{2n+1}, that is, odd power functions, have graphs similar to I^3. Since $Dx^{2n+1} = (2n+1)x^{2n} \geq 0$, the curve never decreases and has a horizontal tangent at the origin. Since $(-x)^{2n+1} = -x^{2n+1}$, the curve is symmetric with respect to the origin, i.e., for each point on the curve there is another such that the segment joining the two passes through the origin and is bisected by it.

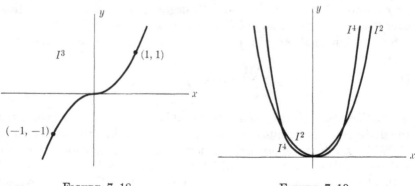

FIGURE 7–18   FIGURE 7–19

(j) Sketch I^3 and I^5 on the same axes.　(k) For what positive values of x is $x^{n+1} < x^n$?　(l) Review the discussion of symmetry in Sections 5–14 and 5–15 and justify the above statements about symmetry.

We have indicated that in order to graph a polynomial it is sufficient to plot key points and then join them by a smooth curve. Among the key points are those where the graph crosses the x-axis. If P is the polynomial, we seek the solutions of the equation $P(x) = 0$, that is, the zeros of P. We have simple methods for finding zeros of linear and quadratic functions. But for polynomials of higher degree, it is usually convenient and sometimes necessary to resort to approximation. All methods of approximation are based on finding a first approximation to a root, i.e., an x for which $f(x) \doteq 0$, and then improving the approximation by some systematic method. One of the most effective is called *Newton's method*.

To illustrate Newton's method we consider f defined by $f(x) = x^3 - 5x^2 + 6x - 3$, which is similar to functions arising in genetics, economics, and many other fields. By trial we find that $f(3) = -3$ and $f(4) = 5$. Since f is continuous, it is evident that f must have a zero between 3 and 4, as suggested in Fig. 7–20. It seems reasonable to take 3.5 as a first guess for the zero. We find $f(3.5) = -0.375$, so this guess is not very far off. But how can we get a better approximation? In Fig. 7–21 we show the first approximation x_1 and the corresponding point $(x_1, f(x_1))$ on the graph of f. In our problem $(x_1, f(x_1)) = (3.5, -0.375)$. We show also the tangent line to f at this point. Since the tangent is near the curve in a small neighborhood of the point of tangency, we take as our second approximation the point $(x_2, 0)$ where this tangent meets the x-axis.

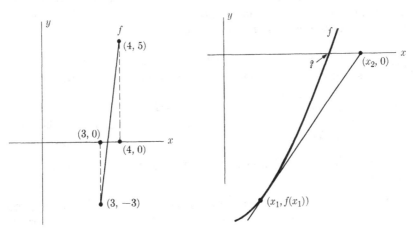

FIGURE 7–20　　　　　　　　FIGURE 7–21

It is easy to find this point, for we know by (7–4–2) that the equation of the tangent is $y - f(x_1) = Df(x_1)(x - x_1)$. To find x_2, we substitute $(y:0, x:x_2)$ and get

$$(2) \qquad\qquad x_2 = x_1 - f(x_1)/Df(x_1).$$

Since $Df(x) = 3x^2 - 10x + 6$ and $Df(3.5) = 7.75$, we find $x_2 = 3.5 - (-0.375)/7.75 \doteq 3.5 + 0.0484 \doteq 3.55$. The calculations were done on a slide rule, hence the approximate equalities. We find $f(3.55) = 0.025$ and $f(3.54) = -0.054$. Hence 3.55 seems a satisfactory approximation to two decimal places.

However, there is nothing to stop our continuing the process by using (2) with $(x_1:x_2, x_2:x_3)$ to find a third approximation. Indeed, we can find a sequence of approximations defined by the recursion formula

$$(3) \qquad\qquad x_{n+1} = x_n - f(x_n)/Df(x_n).$$

If we wish to carry the calculation beyond one or two decimal places, it is convenient to have a desk calculator. With these or other computing aids, roots may be found to many decimal places very rapidly.

(m) Let $x_1 = 3.6$ and find x_2 by (2) for the above problem.   (n) Do the same for $x_1 = 4$.   (o) Do the same for $x_1 = 3$.   (p) Use this method to find the larger one of the roots of $x^2 + 2x - 4 = 0$ to two decimal places and check your result by using the quadratic formula.

★ *Theory of equations.* The branch of mathematics that deals with finding the zeros of polynomial functions is called the *theory of equations.* The interested reader should consult treatises with this title. Two of the best known are by C. C. MacDuffee and by J. V. Uspensky. We give here in brief some of the most useful elementary theorems.

$$(4) \qquad \begin{array}{l} \textit{If } P \textit{ is a polynomial and } P(x) = (x - r)Q(x) + R, \textit{ where} \\ Q \textit{ is a polynomial and } R \textit{ is a number, then } P(r) = R. \end{array}$$

This is the famous *remainder theorem.* It tells us that if we divide a polynomial by $x - r$, then the constant remainder is the same as the image of r under the polynomial function.

(q) Rewrite the equation in theorem (4) so as to indicate the way in which division is involved.

The remainder theorem has an obvious application. If we wish to find $P(r)$, we may do so by dividing by $x - r$ and noting the remainder. Division can be systematized so that it is easier to carry through than substitution and evaluation.

Another application is to derive the *factor theorem*.

(5)   *A number r is a zero of the polynomial P if and only if x — r is a factor of P(x).*

A most useful application of this theorem is the following. Suppose that we have found one zero r_1 of P. Then we know that $x - r_1$ is a factor, that is, $P(x) = (x - r_1)Q(x)$, where Q is a polynomial of degree one less than P. Now, in searching for the other roots of P, we need consider only the equation $Q(x) = 0$, called the *reduced equation*. As an example, let P be the polynomial used to illustrate Newton's method and let r_1 be the zero we found approximately. We find by long division

(6) $x^3 - 5x^2 + 6x - 3 \doteq (x - 3.55)(x^2 - 1.45x + 0.85) + 0.02.$

If the root had been found exactly, the remainder term would have been zero instead of 0.02. Now, to find the other zeros we need consider only $x^2 - 1.45x + 0.85$. However, the discriminant of this quadratic is less than zero and hence the function has no real zeros. Hence P has only the one real root we have approximated.

Another useful conclusion that can be drawn from the factor theorem is

(7)   *A polynomial of degree n has at most n roots.*

This is easily seen as follows. By the fundamental theorem of algebra (see Problem 25 in Section 6–13), a polynomial P of degree n has at least one zero. Call it r_1. Then $P(x) = (x - r_1)Q(x)$, where Q is of degree $n - 1$. But by the same theorem, Q has at least one zero. Calling it r_2, we have $P(x) = (x - r_1)(x - r_2)R(x)$. Continuing in this way we find $P(x) = (x - r_1)(x - r_2) \ldots (x - r_n)A$, where A, r_1, $r_2, \ldots, r_n$ are complex numbers. Since some of the r_i may be the same, we know only that there are at least one, and at most n, roots.

Most equations that appear in applications do not have integral or rational roots. However, such simple equations often appear in textbooks and occasionally in real life. The following theorem is helpful in finding rational roots if they exist.

(8)   *If a rational p/q (where p and q are integers without a common prime factor) is a zero of the polynomial $\sum_{i=0}^{i=n} m_i I^i$, where the m_i are integers, then p must be a factor of m_0 and q must be a factor of m_n.*

For example, we easily see that $I^3 - 5I^2 + 5I - 3$ has no rational zeros because the only possibilities are 1, -1, 3, -3 and none of these satisfy the equation $x^3 - 5x^2 + 5x - 3 = 0$.

(r) Use this to give a quick proof that $\sqrt{2}$ is irrational by showing that $I^2 - 2$ has no rational zeros.   (s) List the possible roots of $30x^4 - 7x^3 + 199x^2 + 21 = 0$.   (t) Find all roots of $2x^3 - 5x^2 - 9 = 0$, and sketch the graph by plotting a few additional points.   (u) Factor $x^4 - 1 = 0$ into linear factors.

PROBLEMS

In Problems 1 through 8, sketch the polynomial, showing its zeros. Find its derivative and sketch it on the same graph. Discuss the relation between the graphs.

1. $y = (x - 1)^3$.
2. $y = (x - 1)^2 x$.
3. $y = (x - 1)x(x + 1)$.
4. $y = (x^2 + 2x - 1)(x^2 - 4)$.
5. $y = (x^2 + 2x - 1)(x^2 + 4)$.
6. $y = x(x - 2)(x + 1)(x - 3)$.
7. $y = -x(x - 1)^4$.
8. $y = (x + 2)^2(x - 3)^2$.

9. Try finding the zero of the function in Problem 1 by Newton's method, beginning with $x_1 = 2$.

10. Do the same for Problem 7 starting with $x_1 = 2$.

11. Do the same for Problem 7 starting with $x_1 = 0.5$.

12. What happens in Newton's method if we guess the root exactly at any stage?

13. Show that the recurrence relation (3) becomes $x_{n+1} = (x_n + a/x_n)/2$ for solving $x^2 - a = 0$. Use this to calculate $\sqrt{2}$ to four decimal places.

★14. Work out a formula for finding principal cube roots, and generalize for nth roots.

In Problems 15 through 20 find all zeros with the aid of the theorems of this section. Begin by testing for rational roots. When one root is found, use the reduced equation. Sketch the polynomials.

★15. $I^3 - 6I^2 + 6I - 1$.

★16. $I^3 + 5I^2 + 6I - 3$.

★17. $I^4 - 4I^3 - 7I^2 + 34I - 24 = 0$.

★18. $x^3 - 5 = y$.

★19. $x^4 + 4x^3 - x - 4 = y$.

★20. $I^4 - 2I^3 - 3I^2 + 4I + 4$.

★21. A function f for which $f(x) = f(-x)$ is called an *even function*. What is the corresponding geometric property? Suggest an explanation of the term. Cite examples other than polynomials.

★22. A function for which $f(x) = -f(-x)$ is called an *odd function*. Discuss as in Problem 21.

★23. Prove (4) on the assumption that for any polynomial P there exists a polynomial Q such that $P(x) = (x - r)Q(x) + R$ where R is a number.

★24. Prove (5) from (4).

★25. Prove (8).

★26. By making use of (4) written in the form $P(x)/(x - r) = Q(x) + P(r)/(x - r)$, show that $DP(r) = Q(r)$. Illustrate. Restate with quantifiers to make clear that the last equation is not an identity holding for a fixed Q and any r.

★27. Newton's method is applicable to functions other than polynomials. Use it to find approximately the roots of $x - 2 \sin x = 0$. Begin by graphing I and $2 \sin$ to locate a first approximation.

★28. Show that if r is a zero of a polynomial with real coefficients, then so is its conjugate $\bar{r}$.

ANSWERS TO EXERCISES

(a) $|I|$.   (b) (7–3–10).   (c) The derivative of the constant term in the left member is 0. The right member would be unchanged in value if the lower limit of summation were changed from 1 to 0.   (d) $D(AI^2 + BI + C) = 2AI + B$. (e) From (7–4–11) and (7–4–7) we have $D(AI^n) = nAI^{n-1}$, which we apply to each term of the left member of (1).   (f) Like Fig. 7–18, but each y half as large.   (g) The mirror image in the x-axis of Fig. 7–18.   (h) Each y twice that of Fig. 7–18.   (i) Since Df is the slope of the tangent line, $Df(x) > 0$ means that the tangent line is sloping up from left to right. See Fig. 4–27. (j) I^5 should be below in $(-\infty \underline{\quad} -1)$ and $(0 \underline{\quad} 1)$ and above elsewhere. (k) $(0 \underline{\quad} 1)$.   (m) 3.55.   (n) 3.6.   (o) 4.   (q) $P(x)/(x - r) = Q(x) + R/(x - r)$.

(r) The only possibilities are ± 2, ± 1.   (s) ± 1, ± 3, ± 7, ± 21, $\pm 1/2$, $\pm 1/3$, $\pm 1/5$, $\pm 1/6$, $\pm 1/10$; $\pm 1/15$, $\pm 1/30$, $\pm 3/2$, $\pm 3/5$, $\pm 3/10$, $\pm 7/2$, $\pm 7/3$, $\pm 7/5$, $\pm 7/6$, $\pm 7/10$, $\pm 7/15$, $\pm 7/30$, $\pm 21/2$, $\pm 21/5$, $\pm 21/10$.   (t) 3, $(-1 \pm i\sqrt{23})/4$.   (u) $(x + 1)(x - 1)(x + i)(x - i)$.

ANSWERS TO PROBLEMS

1 through 8. Check zeros by substitution.   1. Fig. 7–18 translated one unit to the right.   2. Tangent to x-axis at $(1, 0)$.   7. Tangent to x-axis at $(1, 0)$. 8. Tangent to x-axis at $(-2, 0)$ and $(3, 0)$.   13. Here $f = I^2 - \underline{a}$, $Df = 2I$. 15. 1, $(5 \pm \sqrt{21})/2$.   17. 1, 2, -3, 4.   19. Real roots are -4, 1; others are imaginary.

7–6 Maxima and minima.   Consider again the polynomial $f = I^3 - 5I^2 + 6I - 3$, one of whose zeros we found approximately by Newton's method in Section 7–5. To graph f we may begin by finding a few points corresponding to simple values of x. In this way we get $(-1, -15)$, $(0, -3)$, $(1, -1)$, $(2, -3)$, $(3, -3)$, and $(4, 5)$.

(a) Make a large sketch showing these points and $(3.55, 0)$ found in Section 7–5, use it to follow the discussion below, and complete it as you get additional information.

Apparently the graph rises from $(-1, -15)$ to $(0, -3)$ and to $(1, -1)$, but how much farther? Does it cross the x-axis? If so it must turn down again and cross the axis a second time to get to $(2, -3)$. Then, presumably, it drops from $(2, -3)$, but how far before turning up again? Does it continue to rise to the right of $(3.55, 0)$? To answer these questions we use the derivative. Where the curve turns from going up to going down or from going down to going up it appears that it should have a horizontal tangent, that is, $Df(x) = 0$. [See (3) below.] But $Df(x) = 3x^2 - 10x + 6 = 0$ at $x = (5 \pm \sqrt{7})/3 \doteq 2.55$ or 0.78. Since $f(2.55) \doteq -3.6$ and $f(0.78) \doteq -0.89$, we see that the highest point between $(0, -3)$ and $(2, -3)$ is $(0.78, -0.89)$ and the lowest point between $(2, -3)$ and $(3, -3)$ is $(2.55, -3.6)$. Since these are the only points where the curve is horizontal, we know that it cannot cross the x-axis except at the point found by Newton's method above. This can be further checked by showing that $Df(x) > 0$ for $x < 0.78$ and for $x > 2.55$ and that $Df(x) < 0$ for $0.78 < x < 2.55$.

(b) Complete your graph of $y = x^3 - 5x^2 + 6x - 3$. On the same axes sketch its derivative. (c) For what value of x is $Df(x)$ minimum? How does this show up on the graph?

We call a point where a function changes from increasing to decreasing or from decreasing to increasing an *extremum*. At an extremum of f, $f(x)$ is either larger than at any nearby points, in which case it has what is called a *relative maximum*, or else $f(x)$ is smaller than at any nearby points, in which case it has a *relative minimum*. We have seen that if a function is differentiable at an extremum, then it appears plausible that its derivative is zero there. This means that a good way to find extrema is to set the derivative equal to zero and solve. However, two cautions should be noted: 1. If the derivative fails to exist at a point, a function may have a high or low point there without having a zero derivative. An example is $|I|$, which has a minimum at $(0, 0)$. 2. Even if the derivative exists and is zero, it does not necessarily follow that the point is an extremum. It may be simply a point at which f levels out momentarily, as in Fig. 7-18. In other words, at a point where the derivative exists, a zero derivative is a necessary but not a sufficient condition for an extremum.

(d) Can the first case occur for polynomials?

We call a point $(x, f(x))$ where $Df(x) = 0$ a *critical point* of f. At such points the tangent line is horizontal. To graph a function by finding its zeros and critical points, we begin by plotting easily found points, find the derivative and its zeros, plot the critical points, and determine the zeros of the original function, by approximation if necessary. These methods apply to other functions as well as to polynomials.

(e) Use the above method to find the vertex of the quadratic function $2I^2 - 3I - 4$ considered at the beginning of Section 5–8. (f) Derive (5–8–4) and (5–8–5) by calculus. (g) What is the greatest possible number of extrema for a polynomial of degree n? (h) Find the critical points of the function of Problem 15 in Section 7–5.

The above methods are satisfactory for graphing simple polynomials. However, we need a more general and precise theory because problems of maxima and minima arise very frequently in mathematics and its applications. For example, very often we are interested in doing something in the cheapest, quickest, or "best" way, and many laws of physics, chemistry, and biology are formulated in terms of least time, shortest distance, or minimum energy.

We begin by distinguishing between absolute and relative maxima. The *absolute maximum* of a function f is the maximum of its range, that is, max $[\text{Rge}\,(f)]$. It is the largest value of $f(x)$ for all values of x in the domain of f. Of course, some functions have absolute maxima and others do not. For example, the absolute maximum of the signum function is 1, since its range is just $\{-1, 0, 1\}$. But I, I^2, and I^{-1} have no maximum. The *absolute minimum* is defined similarly.

(i) Define the absolute minimum and give examples. Review Section 3–6.
(j) Find, or indicate the nonexistence of, min $|I|$,   (k) of min $(2 \cdot I - 1)$,
(l) of min $\{(x, y)|y = 2x + 1 \wedge -3 \leq x \leq 1\}$.

As suggested in Exercise (l), a function that has no maximum or minimum may acquire one if the domain is restricted. Indeed the following theorem can be proved from the properties of real numbers.

(1)   *If f is continuous on a closed interval $(a\ \ b)$, then f has an*
absolute maximum and an absolute minimum in that interval;
that is, $\exists x\, x \in (a\ \ b) \wedge f(x) = \max \text{Rge}\,[f \cap (a\ \ b) \times R]$
and $\exists x\, x \in (a\ \ b) \wedge f(x) = \min \text{Rge}\,[f \cap (a\ \ b) \times R]$.

(m) Show that it is necessary to require the interval in (1) to be closed by citing a function that is continuous on the open interval (0_1) but has no maximum there.   (n) What are the max and min of I^3 on the closed interval $(-1\ \ 3)$?   (o) Why does I^3 have no max on $(-1\ \ 3)$?

We say that f has a *relative maximum* at a when there is neighborhood of a in which $f(a)$ is the absolute maximum of f. A relative maximum is also called a *local maximum* because it is a maximum with respect to the other nearby values of $f(x)$. Formally,

(2) **Def.**   (*f has a relative maximum at a*)
$= \exists \epsilon[\epsilon > 0 \wedge \forall x|x - a| < \epsilon \to f(x) \leq f(a)]$.

(p) Formulate a similar definition of a relative minimum at a.   (q) Define "extremum" in terms of "maximum" and "minimum."

The above concepts are illustrated well by the function you sketched in Exercise (a). It has a local maximum at 0.78, a local minimum at 2.55, but no absolute maximum or minimum. However, if we restrict the domain to $(-1 \quad 4)$, the function has an absolute minimum of -15 and an absolute maximum of 5, both occurring at the endpoints of the interval.

Indicate intervals in which the function of Exercise (a) has: (r) an absolute maximum at 0.78 and an absolute minimum at 2.55,   (s) an absolute maximum of -3,   (t) an absolute minimum of -3.

Finding extrema is very much facilitated by the following theorem.

(3)   *If f is differentiable at a and has an extremum at a,*
        then $Df(a) = 0$.

(u) Give a plausibility argument for (3).   (v) Show that the converse is not true.   (w) Show that a function may have a relative minimum without having a zero derivative at the point.   (x) Show that an absolute maximum (minimum) is always a relative maximum (minimum).   (y) Argue that (3) means that if a function is continuous on a closed interval, its extrema are at points where $Df(a) = 0$, at points where the function is not differentiable, or at the endpoints of the interval.

Use of the second derivative. The above discussion tells us that $Df(a) = 0$ is neither a necessary nor a sufficient condition for an extremum. When the function f is differentiable at a, $Df(a) = 0$ is a necessary condition, but it does not tell us whether f has a local minimum, local maximum, or neither. In many cases one can tell from inspection of the graph, but it would be good to have sufficient conditions. We now introduce the theory necessary for these.

The operation D maps each differentiable function f into a function Df. If Df is also differentiable, we may apply D again to get $D(Df)$, which we call the *second derivative* and write D^2f. The reader should note immediately that the exponent on D indicates composition of the operation D with itself. Thus $D^2f = (D[D])(f)$ and $D^2f \neq (Df)^2$. This is inconsistent with the previous use of exponents to indicate numerical squaring of functions. However, the usage is universal. The convention is that exponents on functions whose range consists of numbers indicate numerical squaring, while exponents on operations (functions) whose range consists of functions indicate composition. To find D^2f, we simply differentiate f and then differentiate the result. For example, $D^2(I^2) = D(DI^2) = D(2I) = 2$ and $D^2(I^{-1}) = D(-I^{-2}) = 2I^{-3}$.

What is the geometrical interpretation of $D^2f(x)$? Since $Df(x)$ is the rate of change of $f(x)$ or the slope of the curve f at $(x, f(x))$, $D^2f(x)$ is the rate of change of $Df(x)$ or the rate of change of the slope. If y is distance

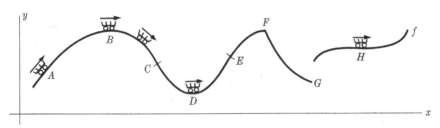

FIGURE 7–22

and x time, Dy is the speed and D^2y is the rate of change of speed, i.e., the acceleration.

(z) Show that if a body falls so that its distance is given in terms of time by $y = 16x^2$, then its acceleration is 32.

Let us imagine the graph of f to be a roller coaster, as suggested in Fig. 7–22. A short segment of the tangent line becomes a car (with a short wheel base) moving along the curve. It carries a vector parallel to the tangent line, as indicated in the figure. For brevity we let $Df(x) = m$. At A, $m > 0$. Then m decreases as the car approaches B. At B, $m = 0$, and m continues to decrease to the right of B. But at C, m stops decreasing and starts increasing, that is, m has a minimum of C. This suggests that $D^2f(x) < 0$ in the interval A to C and $D^2f(x) = 0$ at C.

In an interval where $D^2f(x) < 0$, the slope of the curve is decreasing and we say that the curve is *concave downward*. That is, the tangent line is rotating clockwise as it moves along the curve. It appears evident that at a point such as B, where $Df(a) = 0$ and $D^2f(a) < 0$, f has a local maximum. $Df(x) > 0$ immediately to the left, and $Df(x) < 0$ immediately to the right. This means that $f(x)$ increases as x approaches a from the left and decreases as it moves past.

(4)   *If $Df(a) = 0$ and $D^2f(a) < 0$, then f has a relative maximum at a.*

(aa) State a similar theorem for $D^2f(a) > 0$, and argue for it by considering the interval C to E in the figure. (At a point where $D^2f(x) > 0$, we say the curve is *concave upward*.)   (bb) Show that the converse of (4) is false.   (cc) Illustrate the above results for the functions of Fig. 7–23 by finding the derivatives at the critical points.   (dd) Find the local maximum and minimum of $I^3 - 3I^2 - 6I + 8$ without graphing.   (ee) Prove (5–8–4) by (4).

Theorems such as (4) are useful in finding extrema. However, note that absolute maxima or minima may occur at the endpoints of an interval in which the function is defined, at points where the function does not have a derivative, and at points where the derivative exists and is zero.

Setting the derivative equal to zero detects only points of the third kind. The other two cases must be considered separately.

(ff) At what points in Fig. 7–22 is there a relative maximum or minimum but no derivative?   (gg) At what point does $Df(x) = 0$ without there being an extremum?   (hh) Where is the absolute maximum of f?

Problems

In Problems 1 through 8, graph the function and its derivative. Find the critical points.

1. $3I^2 - 2I + 2$.
2. $5I^3 + 2I^2 - 1$.
3. $I^4 - I^2$.
4. I^{-1}.
5. $1/(1 + I^2)$.
6. $1/(1 - I)^2$.
7. $I^3/(I^2 - 1)$.
8. $I^{1/2} + I$.

9. In the graph of I^{-1}, as x increases does y ever increase? Does the curve approach the horizontal as x gets very large?

10. Show that for any real value of n such that $n > 1$, I^n has a horizontal tangent at the point $(0, 0)$. What if $n \leq 1$?

11. What is the minimum of $\sqrt{x^3 - 2x^2 + x}$?

★12. Let $R(x)$ and $C(x)$ be the revenue and cost respectively of producing and selling x units of a commodity. We call $DR(x)$ and $DC(x)$ the *marginal revenue* and marginal cost. Show that the equality of marginal revenue and marginal cost is a necessary condition for maximum profit if we assume that R and C are differentiable. Write a sufficient condition.

★13. If $u(x)$ is the utility of x, then economists call $Du(x)$ the *marginal utility*. Why would you expect $Du(x) > 0$? What would $Du(x) < 0$ mean? $Du(x) = 0$?

★14. Usually economists assume decreasing marginal utility. What does this mean in terms of derivatives? Why do they make the assumption?

★15. Looking upon D as a function, what is Dom (D)? Rge (D)? The image of a function under D?

★16. In statistics it is frequently of interest to have a measure of the differences between a set of numbers and some number that is representative of the whole set of numbers. Such representative numbers are often called *measures of central tendency*. A very commonly used measure of central tendency is the *mean $\bar{x}$* defined by

(5) $$\bar{x} = \frac{1}{n} \sum_{i=1}^{i=n} x_i,$$

where $x_1, x_2, \ldots, x_n$ are the numbers involved. Now let x be some number. We call $x_i - x$ the *deviation* of x_i from x. We call

(6) $$\sum_{i=1}^{i=n} (x_i - x)^2$$

the *sum of the squares of the deviations* from x. Now for fixed x_i, (6) depends only upon x. Moreover, (6) defines a quadratic function of x, for

(7)
$$\sum_{i=1}^{i=n} (x_i - x)^2 = \sum_{i=1}^{i=n} (x_i^2 - 2xx_i + x^2)$$

(8)
$$= nx^2 - 2\left(\sum_{i=1}^{i=n} x_i\right) x + \sum_{i=1}^{i=n} x_i^2.$$

The reason for the term nx^2 is that x^2 is a constant relative to the summation, and we apply (6–7–19). Show that the minimum of (6) occurs when $x = \bar{x}$. This result means that the sum of the squares of the deviations from the mean is smaller than from any other number.

★17. A point at which a curve changes its concavity, from concave downward to upward as at C in Fig. 7–22 or from upward to downward as at E, is called a *point of inflection*. Inflection points of f are the extrema of Df. Argue that if $(a, f(x))$ is a point of inflection and $D^2f(a)$ exists, then $D^2f(a) = 0$.

★18. Is $D^2f(a) = 0$ sufficient for an inflection?

★19. State sufficient conditions for an inflection.

★20. Find the points of inflection of the graph in Fig. 7–23.

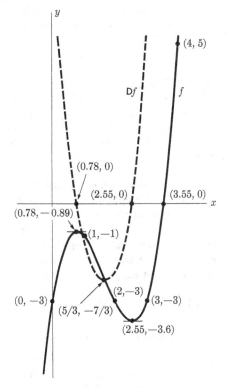

FIGURE 7–23

★21. Assuming (7-4-25) and (7-4-26), use (7-4-10) to show

(9) $$D \tan = \sec^2.$$

★22. What is the slope of the graph of sin at the origin? Where are its maxima and minima? its inflections?

★23. What is the slope of the graph of tan at the origin?

★24. Show that $D \sin (2I) = 2 \cos (2I)$.

★25. Show that a square is the rectangle that encloses the maximum area for a given perimeter.

★26. Suppose that a rancher wishes to enclose the maximum area by having a fence on three sides of a rectangle and a river on the fourth. If he has 1 mile of fence, what dimensions should he select?

★27. Find the shortest distance from the point $(1, 0)$ to the curve $y^2 = Cx$. (*Suggestions:* The distance will be minimum when its square is minimum. Do not forget endpoints, which are different depending on C. Draw Sketches!)

★28. The height above the ground at a time t of a projectile fired at an angle θ with the horizontal and at an initial speed of v_0 is given by $y = (v_0 \sin \theta)t - (\frac{1}{2})gt^2$. Find the time to reach the highest point and the height there.

★29. The range of a projectile is given by $R = (2v_0^2 \sin \theta \cos \theta)/g$. Show that the maximum range corresponds to $\theta = 45°$.

★30. Economists usually assume that marginal utility decreases with increasing consumption. Express this in terms of derivatives.

★31. A psychologist wrote: "The temporal course of adaptation is negatively accelerated, proceeding rapidly at first and gradually slowing in rate." Express in terms of derivatives.

★32. A sociologist wrote: "The proportion of Republican votes goes up with the socio-economic status of the voters." State this in terms of derivatives.

★33. J. M. Keynes wrote in his famous *General Theory of Employment, Interest, and Money:* "The fundamental psychological law...is that men are disposed, as a rule and on the average, to increase their consumption as their income increases, but not by as much as the increase in their income." Express this in terms of derivatives.

★34. Let x and y be the amounts of two commodities owned by an individual. An indifference curve is a set of points which all have the same utility to the individual. It is usually assumed that such curves are "negatively sloped." What does this mean? It is also assumed that such curves are convex. What does this mean in terms of derivatives? Complete the following: Convexity of the indifference curves means that the more a consumer has of one commodity the _____ is the increment in it which is required to compensate for a decrease in the other.

★35. Show that $\sqrt{f(x)}$ has extrema at the same values of x as does $f(x)$.

Answers to Exercises

(b) See Fig. 7-23. (c) Where the derivative of the derivative $6x - 10$ is zero, that is, at $(5/3, -7/3)$. It is the value of x where the original curve is dropping most rapidly, that is, has the smallest slope. (d) No. (g) $n - 1$, since derivative is of degree $n - 1$.   (h) $(3.41, -10.7)$, $(0.59, 0.66)$.   (i) The absolute minimum of a function is the minimum of its range. (j) 0. (k) None. (l) -5.   (m) I^{-1}.   (n) -1 and 27.   (o) There is no greatest number less than 27, but no number less than 27 is an upper bound of the values of x^3 for $x < 3$.   (p) In (2) change "maximum" to "minimum" and $\leq$ to $\geq$.   (q) A relative maximum or minimum.

(r) $(0\ \underline{\quad}\ 3)$.   (s) $(\underline{-5}\ \underline{\quad}\ 0)$.   (t) $(0\ \underline{\quad}\ 2)$.   (u) If the curve has a tangent, this tangent must be horizontal, otherwise the curve would be rising or falling and the point would be neither a maximum nor a minimum. (Note use of proof by contradiction.)   (v) I^3 at origin.   (w) $|I|$ at origin.   (x) Immediate from definitions.   (y) $(p \wedge q) \rightarrow r = [\sim r \rightarrow \sim p \vee \sim q]$.   (z) $D^2(16I^2) = D(D16I^2) = D(32I) = 32$.   (aa) If $Df(a) = 0$ and $D^2f(a) > 0$, then f has a relative minimum at a.   (bb) $-I^4$ at origin.   (dd) $D(x^3 - 3x^2 - 6x + 8) = 3x^2 - 6x - 6 = 0 \leftrightarrow x^2 - 2x - 2 = 0 \leftrightarrow x = 1 \pm \sqrt{3}$.   (ff) G, F. (gg) H.   (hh) At the right-hand endpoint.

Answers to Problems

4. No endpoints, maxima, or mimima.   5. Maximum at $(0, 1)$.   8. Minimum at $(0, 0)$.   9. Study behavior of $D(I^{-1})$ as x increases.   11. Zero, since the principal square root can never be negative and the radical is zero for $x = 1$! No calculus needed here!

7-7 Derivative of the composite. Suppose we wish to find the derivative of $(I + 2)^{10}$. To use $DI^n = nI^{n-1}$ directly we would first have to expand $(I + 2)^{10}$. However, we see that $(I + 2)^{10} = I^{10}[I + 2] = (I + 2) \circ I^{10}$. That is, $(I + 2)^{10}$ is the composite of $I + 2$ and I^{10}.

(a) Review Section 5–9 if necessary.

Now, if we had a formula for finding the derivative of the composite of two functions, we could use it to find $D(I + 2)^{10}$ directly. Such a formula exists. Indeed, we have the following law.

(1) $$D(g[f]) = Dg[f] \cdot Df.$$

In words, the derivative of g-of-f is Dg-of-f times Df. In the above example, we have $D(I^{10}[I + 2]) = (DI^{10}[I + 2]) \cdot D(I + 2) = 10I^9[I + 2] \cdot \underline{1} = 10(I + 2)^9$. It is the purpose of this section to explain this law, which is called the *chain rule*.

First we remind the reader of some facts about the composite. We recall that $g[f]$ is defined by $(g[f])(x) = g(f(x))$, that is, the image of x under $g[f]$ is the image under g of the image under f of x. To evaluate $g(f(x))$ we evaluate $f(x)$, then g-of-$f(x)$. The composite $g[f]$ is not to be confused with the product $g \cdot f$ under which the image of x is $g(x) \cdot f(x)$. We use brackets to indicate composition and insert the dot for multiplication. But some writers use the ordinary parentheses or omit the dot; in such cases it is necessary to tell from the context which is meant. We have $I^2[I^3] = (I^3)^2 = I^3 \cdot I^3 = I^6$, but $I^2 \cdot I^3 = I^5$. From (5-9-11) we have

$$(2) \qquad\qquad f[I] = I[f] = f.$$

(b) Justify (2) by inserting the variable x.   (c) Justify (3).   (d) Justify (4).

$$(3) \qquad\qquad g[f] = \{(x, y) \mid y = g(f(x))\}.$$

$$(4) \qquad\qquad (f[I])(x) = f(x).$$

The law (4) means that we may always get a name of a function by putting "I" in place of "x" in an appropriate defining formula. This is consistent with our notation for polynomials. For example, $(I^3 - I)(x) = x^3 - x$. Also,

$$(5) \qquad\qquad g[f] = (g[I])[f].$$

Note that we get a name for $g[f]$ by putting a name for f in place of "I" in a name for g. For example, above we had $I^{10}[I + 2] = (I + 2)^{10}$ and $10I^9[I + 2] = 10(I + 2)^9$.

(e) Justify (5). Review Section 5–12.   (f) $I^3[I^3] = ?$ (g) $I^3 \cdot I^3 = ?$

To see the plausibility of (1), let us write it in terms of images.

$$(6) \qquad\qquad \mathsf{D}(g(f(x))) = \mathsf{D}g(f(x))\mathsf{D}f(x).$$

Now, by definition, the left member is the limit as $h \to 0$ of $[g(f(x + h)) - g(f(x))]/h$. Let $k = f(x + h) - f(x)$. Then the derivative we seek is the limit of $[g(f(x) + k) - g(f(x))]/h$. If $k \neq 0$, this is the same as

$$(7) \qquad \frac{g(f(x) + k) - g(f(x))}{k} \cdot \frac{f(x + h) - f(x)}{h}.$$

Since the limit of the product is the product of the limits, the limit of (7) as $h \to 0$ is the product of the limits of the two fractions. As $h \to 0$, $k \to 0$, and the limit of (7) is $\mathsf{D}g(f(x)) \cdot \mathsf{D}f(x)$.

(h) Why must we have $k \neq 0$?   (i) Why is this not a proof of (6)? (j) Where did we assume that f is continuous?   (k) If we use the notation df/dx for the derivative, (6) becomes

$$\frac{dg}{dx} = \frac{dg}{df}\frac{df}{dx}$$

Would it be a proof of this to just "cancel" "df" in the right member?
To illustrate the use of the chain rule, consider $\mathsf{D}(I^2 + 2)^2$. We have

$$\mathsf{D}(I^2 + 2)^2 = \mathsf{D}(I^2[I^2 + 2])$$

$$= \mathsf{D}I^2[I^2 + 2] \cdot \mathsf{D}(I^2 + 2) = 2I[I^2 + 2] \cdot 2I$$

$$= 2(I^2 + 2) \cdot 2I = 4I(I^2 + 2) = 4I^3 + 8I.$$

Without using the chain rule, we have $\mathsf{D}(I^2 + 2)^2 = \mathsf{D}(I^4 + 4I^2 + 4) = 4I^3 + 8I$. The chain rule does not seem to be shorter here, but we have chosen a very simple example and our procedure for using the chain rule was very detailed. In practice we write $\mathsf{D}(I^2 + 2)^2 = 2(I^2 + 2) \cdot 2I$. We do this by thinking of the $I^2 + 2$ as though it were I, and finding the derivative of $(I^2 + 2)^2$ "with respect to $I^2 + 2$," then multiplying by the derivative of $I^2 + 2$. For example, $\mathsf{D}(I^2 + 2)^3 = 3(I^2 + 2)^2 2I$.

(l) Verify this result by expanding $(I^2 + 2)^3$ and then differentiating it.
(m) Justify (8).   (n) Find $\mathsf{D}(1 - I)^4$ two ways.   (o) Find $\mathsf{D} \sin [3I]$.
(p) Find $\mathsf{D} \sin 3x$.   (q) Find $\mathsf{D}(I^2 - 1)^{1/2}$.

(8) $\mathsf{D}(f^n) = nf^{n-1}\mathsf{D}f.$

★ *The binomial theorem.* The following law is called the *binomial theorem.*

(9) $(a + x)^n = \displaystyle\sum_{i=0}^{i=n} \frac{n!a^{n-i}x^i}{i!(n - i)!}.$

(r) Write out (9) for $n = 1, 2, 3$ and expand the right member in each case. Note that for $i = 0$ and $i = n$ we get in the denominator 0!, which was defined as 1 in (6-6-9). (s) Let $\binom{n}{i} = n!/i!(n - i)!$. Show that $\binom{n}{0} = 1, \binom{n}{1} = n,$ $\binom{n}{2} = n(n - 1)/2,$ and $\binom{n}{3} = n(n - 1)(n - 2)/3!$. (t) Show that $\binom{n}{i} = \binom{n}{n - i}$. (u) Expand $(x + 2y)^5$.

To prove (9) we first note that $(a + x)^n$ is a polynomial of degree n; that is,

(10) $(a + x)^n = c_0 + \displaystyle\sum_{i=1}^{i=n} c_i x^i$

for *some* c_i. This is fairly obvious, but it can be proved by induction. Now, to prove (9) we must prove that $c_i = n!a^{n-i}/i!(n - i)!$ Since (10) is an identity for properly chosen c_i, we may apply $(x:0)$, which yields $a^n = c_0$. But $n!a^{n-0}/0!(n-0)! = a^n$, so this result is as we wish. Now we differentiate both members of (10) to get

$$(11) \qquad n(a+x)^{n-1} = c_1 + \sum_{i=2}^{i=n} ic_i x^{i-1}.$$

We omit the term with $i = 0$, since it vanishes. Now in (11) we apply $(x:0)$ to get $na^{n-1} = c_1$. Since $n!a^{n-1}/1!(n-1)! = na^{n-1}$, the desired result (9) holds for $i = 1$. Now suppose we differentiate both members of (10) j times. We claim that the result is

$$(12) \qquad (n!/(n-j)!) \cdot (a+x)^{n-j} = c_j + \sum_{i=j+1}^{i=n} (i!/(i-j)!)c_i x^{i-j}.$$

This certainly holds for $j = 1$, since that is just (11). Assuming that it holds for j, it is easy to show that it holds for $j + 1$ by finding the derivative of both members. Since (12) does not have meaning for $j > n$, this induction argument shows that (12) holds for $j \le n$. Setting $(x:0)$ in (12) we find $n!a^{n-j}/(n-j)! = j!c_j/(j-j)!$, or $c_j = n!a^{n-j}/j!(n-j)!$, as we wished to prove. Hence (9) is proved by induction.

(v) Write out (9), (10), and (11) for $n = 3$ without the summation notation. Determine c_0, c_1, c_2, c_3, for the expansion of $(a+x)^3$ by differentiating your (10) successively and setting x equal to zero in the results. (w) Formulate the induction argument to prove that $(a+x)^n$ is a polynomial of the nth degree.

PROBLEMS

1. Show that the graph of $(I - a)^{10}$ is tangent to the x-axis at $(a, 0)$.

In Problems 2 through 9 find each derivative by using the chain rule and verify by another method.

2. $D(2I - 1)^2$.   3. $D(x^2 + 2)^2$.   4. $D(x + a)^2$.
5. $D(a - x)^2$.   ★6. $D \cos^2$.   ★7. $D \cos 2I$.
★8. $D \sin^3$.   9. $D(x^2 + 2ax + b)^2$.

In Problems 10 through 15 find the derivative by the aid of the chain rule.

10. $D(2x + a)^{1/2}$.   11. $D(a^2 + x^2)^{1/2}$.   ★12. $D \sin (x^2 - 1)$.
★13. $D \sin^3 x$.   ★14. $D \sin^2 x^2$.   ★15. $D (\sin^2 \cdot I^2)$.

16. Prove that $(a + x)^2 = a^2 + 2ax + x^2$ by setting $x = 0$, differentiating and setting $x = 0$ in the result, then differentiating again and setting $x = 0$ in the result.

★17. Give plausibility arguments for (13) and (14).

(13)   *If f is continuous at a, and g is continuous at f(a), then g(f) is continuous at a, and* $\lim_{x \to a} g(f(x)) = g(f(a))$.

(14)   *If g is continuous at* $\lim_{x \to a} f(x)$, *then* $\lim_{x \to a} g(f(x)) = g(\lim_{x \to a} f(x))$.

★18. Find $\lim_{x \to 2} (3x^2 + 2x)^5$ and justify your procedure.
★19. Complete the following: If $g(f)$ is discontinuous at a, then _____.
★20. Find the slope of the tangent line to the graph of $y = (x^2 - 3)^3$ where $x = 1$.

21. A projectile is moving in the xy-plane so that its x-coordinate is increasing at 10 ft/sec and its y-coordinate is increasing at 2 ft/sec. At what rate is its distance from the origin changing when $x = y = 1$?

★22. A searchlight 10 miles from shore is rotating at the rate of one revolution per minute. At what rate is the light moving along the shore when the beam is perpendicular to the shore line?

23. If a searchlight on the earth were rotating at the rate of 1 degree per minute, how fast would the light be moving on a perpendicular surface as far away as the sun, that is, at a distance of 93,000,000 miles?

24. A cylindrical tank of radius 3 ft is being filled by water flowing in at the rate of 15 cubic feet per minute. How fast is the depth of the fluid increasing?

25. Answer the same question for a tank in the shape of an inverted cone with height 10 ft and radius of base 2 ft.

Expand the binomials by (9) in Problems 26 through 31.

★26. $(a + x)^4$ ★27. $(a + 2x)^3$ ★28. $(x - 2)^5$
★29. $(2x + 3y)^3$ ★30. $(a + b + c)^3$ ★31. $(1 - x)^6$

★32. Carry through the proof of (9) for $n = 4$.
★33. What is the term involving x^{10} in the expansion of $(1 + x)^{28}$?
★34. Answer the same question for the term involving b^8 in the expansion of $(b - x)^{15}$.
★35. Use the binomial expansion to approximate $(1.01)^{50}$.
★36. If one dollar is placed at compound interest of 2% for 1000 years, to what will it amount?
★37. How long does it take money to double at 6%?
★38. Show that $\binom{n}{i-1} + \binom{n}{i} = \binom{n+1}{i}$.

Answers to Exercises

(b) $f(I(x)) = f(x)$; $I(f(x)) = f(x)$.   (c) Since $(g[f])(x) = g(f(x))$ by definition.   (d) $f[I] = f$.   (e) $g[I] = g$.   (f) I^9.   (g) I^6.   (h) If $k = 0$, the first fraction is undefined.   (i) Because k may be zero for some value of $h \neq 0$.   (j) When we assumed $k \to 0$ as $h \to 0$.   (k) No, in this alternative notation for the derivative we have not given any meaning to the symbols in the numerator and denominator separately. (l) $D(I^2 + 2)^3 = D(I^6 + 6I^4 + 12I^2 + 8) = 6I^5 + 24I^3 + 24I = 6I(I^4 + 4I^2 + 4) = 6I(I^2 + 2)^2$. (m) $(1)(g{:}I^n)$. (n) $4(1 - I)^3(-1)$. (o) $3 \cos[3I]$. (p) $3 \cos 3x$. (q) $(1/2)(I^2 - 1)^{-1/2}(2I) = I(I^2 - 1)^{-1/2}$.

(r) Verify by multiplication. (s) For example, $\binom{n}{0} = n!/0!(n - 0)! = 1$.

(t) $n!/i!(n - i)! = n!/(n - i)!(n - (n - i))!$ (u) $x^5 + 10x^4y + 40x^3y^2 + 80x^2y^3 + 80xy^4 + 32y^5$. (v) $(a + x)^3 = \dfrac{3!}{0!3!} a^3x^0 + \dfrac{3!}{1!2!} a^2x + \dfrac{3!}{2!1!} ax^2 + \dfrac{3!}{3!0!} a^0x^3 = a^3 + 3a^2x + 3ax^2 + x^3$. $(a + x)^3 = c_0 + c_1x + c_2x^2 + c_3x^3$.

$3(a + x)^2 = c_1 + 2c_2x + 3c_3x^2$. $c_0 = a^3$; $c_1 = 3a^2$; $c_2 = 3a$; $c_3 = 1$. (w) If $(a + x)^n$ is a polomial of degree n, $(a + x)(a + x)^n = a(a + x)^n + x(a + x)^n$ is of degree $n + 1$.

Answers to Problems

1. Letting $f = (I - a)^{10}$, $Df(a) = 10(a - a)^9 = 0$. 3. $4x(x^2 + 2)$. 5. $-2(a - x)$. 7. $-2 \sin 2I$. 9. $4(x + a)(x^2 + 2ax + b)$. 11. $x(a^2 + x^2)^{-1/2}$. 13. $3 \sin^2 x \cos x$. 15. $2 \sin \cdot \cos \cdot I^2 + \sin^2 \cdot 2I = 2I \sin (I \cos + \sin)$. 21. $d = (x^2 + y^2)^{1/2}$; $D_t d = (1/2)(x^2 + y^2)^{-1/2}(2xD_tx + 2yD_ty) = (xD_tx + yD_ty) (x^2 + y^2)^{-1/2} = (10x + 2y)(x^2 + y^2)^{-1/2}$. When $x = y = 1$, $D_t d = 12/\sqrt{2}$.
23. $s = (9.3 \times 10^7)\theta$; $D_t s = (9.3 \times 10^7)D_t\theta = (9.3 \times 10^7) \left(\dfrac{\pi}{180}\right)$ mi/min.
25. The volume of water is given by $V = \pi(x/5)^2x/3 = (\pi/75)x^3$, where x is the depth of liquid. $D_t V = (\pi/25)x^2 D_t x$. $D_t V = 15$. Hence $D_t x = 375/\pi x^2$. See Fig. 7-24.

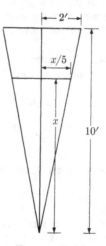

FIGURE 7-24

7–8 Functions defined implicitly. By using the chain rule we can find the derivative of a function that is defined implicitly, provided we are able to find the derivatives of the functions involved in its definition. For example, suppose we wish to find the derivative of $I^{1/2}$. This function is defined implicitly by $y^2 = x \wedge y \geq 0$. Letting $f = I^{1/2}$, we have $f^2 = I$. Now we find the derivative of each member of this identity to obtain $2f\mathsf{D}f = 1$, from which we have $\mathsf{D}f = 1/2f = 1/2I^{1/2} = (1/2)I^{-1/2}$. If we prefer to work in terms of images, we have $y^2 = x$, $\mathsf{D}_x y^2 = \mathsf{D}_x x$, $2y\mathsf{D}_x y = 1$, $\mathsf{D}_x y = 1/2y = 1/2\sqrt{x} = (1/2)x^{-1/2}$.

(a) Review Section 5–4 as necessary. (b) Obtain the same result by finding the derivative of each member of the identity $(I^{1/2})^2 = I$. (c) Prove that $\mathsf{D}I^{1/3} = (1/3)I^{-2/3}$ by finding the derivative of both members of $(I^{1/3})^3 = I$. (d) Show that these results conform to (7–4–11).

We can now prove that (7–4–11) holds for all positive rational powers. Let p and q be positive integers. Then

(1) $$(I^{p/q})^q = I^p.$$

by (6–8–9). Differentiating both members, we find

(2) $$q(I^{p/q})^{q-1}\mathsf{D}I^{p/q} = pI^{p-1}.$$

Note that we use (7–4–11) for n a positive integer here. Dividing both members of (2) by the coefficient of $\mathsf{D}I^{p/q}$, we find

(3) $$\mathsf{D}I^{p/q} = (p/q)I^{p/q-1},$$

which is just (7–4–11)$(n{:}p/q)$, as desired.

(e) Carry through the algebra to verify that (3) follows from (2). (f) (3) shows that (7–4–11) holds for all positive rationals. Show now that it holds for all rationals by letting r be a positive rational and completing $\mathsf{D}I^{-r} = \mathsf{D}(1/I^r) = \ldots$. Find the derivatives of: (g) $I^{-1/2}$, (h) $I^{3/5}$, (i) $\sqrt{x-1}$, (j) $\sqrt{x^2+2}$.

Sometimes it is hard to find the derivative of a function directly, but fairly easy to find the derivative of its inverse. In such cases the following law is helpful.

(4) $$\mathsf{D}f* = 1/\mathsf{D}f[f*],$$

or, in terms of images,

(5) $$\mathsf{D}f*(x) = 1/\mathsf{D}f(f*(x)).$$

(k) Prove from (4) that $Df = 1/Df*[f]$.   (l) Use (4) to show that $DI^{1/2} = (1/2)I^{-1/2}$.

To prove (4) we write $f(f*(x)) = x$ by (5–10–9). Finding the derivative of both members, we have

(6) $D(f[f*])(x) = (Df[f*])(x) \cdot Df*(x) = 1.$

Hence $Df*(x) = 1/(Df[f*])(x) = 1/Df(f*(x))$, which is just (5).

To see the plausibility of (4) and (5), let $y =$ distance and $x =$ time, so that $y = f(x)$ gives distance in terms of time and $x = f*(y)$ gives time in terms of distance. Then $Df(x)$ is the speed, i.e., the distance per unit time, and $Df*(y)$ is the rate of change of time with distance, i.e., the time per unit distance. Evidently one is the reciprocal of the other.

Or, let $k = f(x + h) - f(x)$, so that $f(x + h) = f(x) + k = y + k$ and $f*(y + k) = x + h$. Then $Df(x) = \lim_{h\to 0} (k/h)$ and $Df*(y)$ $\lim_{k\to 0} (h/k) = \lim_{h\to 0} (h/k)$, since $k \to 0$ and $h \to 0$ together. But the limit of the reciprocal is the reciprocal of the limit, that is, $\lim (h/k) = 1/\lim (k/h)$.   Hence $Df*(y) = 1/Df(x) = 1/Df(f*(y))$.   This is just $(5)(x{:}y)$.

The plausibility of (4) is still more evident if we adopt the frequently used Δ-notation.   (See Section 7–4.)   Then $Df(x) = \lim (\Delta y/\Delta x)$, and $Df*(y) = \lim (\Delta x/\Delta y)$.   Again we see the reciprocal relation between the two.

To illustrate the applications of the ideas of this section, consider a point moving around the circumference of the circle $x^2 + y^2 = r^2$. Suppose that x and y depend on time in such a way that this equation is always satisfied.   Then we have $D_t x^2 + D_t y^2 = D_t r^2$, which becomes $2xD_t x + 2yD_t y = 0$ or $D_t y/D_t x = -x/y$. In Fig. 7–25 we show the circle with the

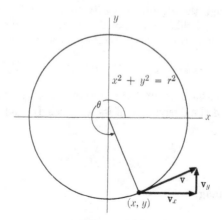

FIGURE 7–25

vectors $\mathbf{v}_x = (\mathsf{D}_t x, 0)$ and $\mathbf{v}_y = (0, \mathsf{D}_t y)$, which are the time rates of change of x and of y respectively, or the velocities in the x- and y-directions respectively. Their sum, $\mathbf{v}$, is the velocity vector. The slope of the vector $\mathbf{v}$ is $\mathsf{D}_t y / \mathsf{D}_t x = -x/y = -1/(y/x)$. Since y/x is the slope of the radius vector (x, y), this shows that the vector $\mathbf{v}$ is perpendicular to the radius vector. Hence it is tangent to the circle at (x, y). This proves that the velocity of a point moving in a circle is in the direction of the tangent to its path.

(m) Justify the assertion about perpendicularity.   (n) Letting $|\mathbf{v}| = v$, $|\mathbf{v}_x| = v_x$, $|\mathbf{v}_y| = v_y$, find v in terms of v_x and v_y.   (o) Assuming that v is the speed measured along the circumference, express it in terms of θ, r, and t. (p) Letting $\mathsf{D}_t \theta = \omega$, express v_x, v_y, and v in terms of ω (ω is called the *angular velocity*).

The chain rule is very useful when functions are defined implicitly and it is not convenient to solve explicitly for a defining formula. For example, the equation $y^3 - 8x^2 y^2 + 3x^5 y + x + 100 = 0$, if we limit x and y to real values, defines a function whose domain is the real numbers. To get an explicit formula would be awkward. (It *could* be done in this case, but no simple formula could be obtained for, say, a fifth degree equation in y.) However, we can easily find $\mathsf{D}_x y$. Finding the derivative of both members of the equation, and keeping in mind that y depends on x by some functional relation, we find

(7) $3y^2 y' - 16xy^2 - 16x^2 yy' + 15x^4 y + 3x^5 y' + 1 = 0.$

(8) $y' = (16xy^2 - 15x^4 y - 1)/(3y^2 - 16x^2 y + 3x^5).$

Thus we can find a formula for the rate of change of y with respect to x without finding an explicit formula for y in terms of x!

(q) Find y' where $2y^5 + xy^4 + 5x^5 y + y/x - 5 = 0$.   (r) Find y' where $x = \mathrm{Sin}\, y$.

We handled these problems in terms of the variables x and y. We could have dealt with them in terms of the functions directly by putting f in place of y and I in place of x, since the function f defined by a sentence $s(x, y)$ must satisfy the sentence $s(I, f)$ because of the conventions we are using for naming functions in terms of I. It is more usual to handle problems in terms of the variables when a convenient name for the function is not available, i.e., when the function is defined only implicitly.

PROBLEMS

Find the derivatives in Problems 1 through 6 by applying (7–4–11) directly and by using the defining equation.

1. I^{-1}; $xy = 1$.   2. I^{-2}; $x^2 y = 1$.   3. $I^{1/5}$; $y^5 = x$.

4. $(1 - I^2)^{1/2}; x^2 + y^2 = 1 \wedge y \geq 0.$
5. $(1 - I)^{1/2}; x + y^2 = 1 \wedge y \geq 0.$
6. $-(1 - I)^{1/2}; x + y^2 = 1 \wedge y \leq 0.$

7. Show that as x gets very large in absolute value the slope of the hyperbola $b^2x^2 - a^2y^2 = a^2b^2$ approaches b/a or $-b/a$, that is, the slope of the asymptote.

8. Show that there are only two points where the slope of the graph of $b^2x^2 + a^2y^2 = a^2b^2$ is zero. Where does it have a vertical tangent?

9. Find y' where $ay^2 + by + c = x$. Check by solving for y and finding the derivative of the result.

★10. Show that there are at most two points where the graph of any conic has a given slope, unless the conic is degenerate.

★11. Find D Arcsin from the defining equation $x = \text{Sin } y$.

★12. Find it by making use of (4).

★13. Treat Arccos similarly.

14. Assuming that a function is defined by $x^2y^3 + 2xy^2 + 18 = 0$, find D_xy.

★15. Show that the velocity of a point moving along the curve given by $y = f(x)$ is tangent to the curve.

16. Find the derivative of $I^{1/5}$ by using (4).

In Problems 17 through 22 find the derivative of the function, assuming that f is an unknown function.

17. $I^2f + 2f^2.$

18. $f^{1/2} - I^3f^3.$

19. $1/f^2 \sin.$

20. $3f^4 - f^3I^{-1}.$

21. $2I^2 - If + f^3.$

22. $\sin [f^2].$

23. Prove that the curves $2x + 3y + x^5 - xy^3 = 0$ and $2y + x^3y^4 - 3x = 0$ are perpendicular at the origin.

24. Find a formula giving the equation of the tangent line to the ellipse $b^2x^2 + a^2y^2 = a^2b^2$ at the point (x_0, y_0).

25. Solve a similar problem for the hyperbola.

26. We call cI^{-1} an *inverse variation*. When $y = cx^{-1}$ we say that y *varies inversely* with x. We say also that y is *inversely proportional* to x with c the *constant of proportionality*. Since $c(-x)^{-1} = -cx^{-1}$, an inverse variation is an odd function and its graph is symmetric with respect to the origin. The special case I^{-1} is sketched in Fig. 7–26. Since $DI^{-1} = -I^{-2}$, we see that the slope of I^{-1} is always negative, that is, $f(x)$ is always decreasing.

Suppose that a company has an overhead of c. Then if $y =$ the overhead per unit output, and $x =$ output, $y = cx^{-1}$. The function is graphed in Fig. 7–27. The horizontal line is the constant function giving overhead. We see that as production increases, y decreases toward zero. On the other hand, as x goes toward zero, overhead per unit increases without bound. Find the marginal unit overhead, $D_x(cx^{-1})$. Sketch it. Is it increasing or decreasing?

27. Show that if (x_1, y_1) and (x_2, y_2) belong to cI^{-1}, then $y_2/y_1 = x_1/x_2$ and $x_1y_1 = x_2y_2$.

28. Why is $D(1/x) < 0$?

29. Is I^{-1} everywhere continuous?

30. Find the slope of I^{-1} at $(1, 1)$.

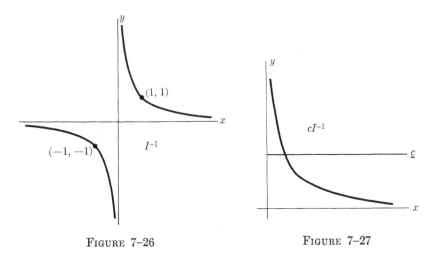

FIGURE 7-26   FIGURE 7-27

★31. If the cost is given by $y = mx + b$, then unit cost q is given by $q = y/x = m + b/x$. Graph the unit cost function for $m = 20$, $b = 100$.

★32. Let C be the total cost and $C = x^2 + 100x + 100$, where x is output. Graph $y = C/x$, the *unit cost function*. Also graph its derivative, the *marginal unit cost function*.

★33. Let f be the function of Fig. 7-6. Graph f/I and its derivative. What is f/I if f is the cost function of a firm? What is f/I if f is the function giving the distance traveled by a car as in Section 7-2?

★34. Let f be the function graphed in Fig. 7-7. Sketch f/I and its derivative. Interpret your results in terms of cost and unit cost.

ANSWERS TO EXERCISES

(b) $2I^{1/2}DI^{1/2} = 1$, therefore $DI^{1/2} = 1/2I^{1/2} = (1/2)I^{-1/2}$.   (c) First step is $3(I^{1/3})^2DI^{1/3} = 1$.   (d) By appropriate substitutions in (7-4-11). (e) $(I^{p/q})^{q-1} = I^{p-p/q}$. Then $I^{p-1}I^{p/q-p} = I^{p/q-1}$.   (f) $= (I^rD\underline{1} - 1DI^r)/I^{2r} = -rI^{r-1}/I^{2r} = -rI^{-r-1}$.   (g) $-(1/2)I^{-3/2}$.   (h) $(3/5)I^{-2/5}$.

(i) $(1/2)(x - 1)^{-1/2}$.   (j) $x(x^2 + 2)^{-1/2}$.   (k) $(f{:}f*)$.   Then use law $f** = f$.   (l) $DI^{1/2} = 1/DI^2[I^{1/2}] = 1/2I[I^{1/2}] = 1/2I^{1/2} = (1/2)I^{-1/2}$. Here $f* = I^{1/2}$, $f = I^2$.   (m) (4-7-8).   (n) $v^2 = v_x^2 + v_y^2$.   (o) $v = rD_t\theta$.   (p) $v = r\omega$; $v_x = -v\sin\theta$, $v_y = v\cos\theta$.   (q) $y' = y(1 - x^2y^3 - 25x^6)/x(10xy^4 + 4x^2y^3 + 5x^6 + 1)$. (r) $1 = (\text{Cos } y)y'$; $y' = 1/\text{Cos } y = 1/\sqrt{1 - x^2}$.

ANSWERS TO PROBLEMS

7. $y' = b^2x/a^2y = \pm b^2x/a^2(b/a)(x^2 - a^2)^{1/2} = \pm(b/a)(x/(x^2 - a^2)^{1/2})$. As x gets very large, $x^2 - a^2 \doteq x^2$.   17. $2If + I^2f' + 4ff'$.   19. $-f^{-3}\sin^{-2}(2f'\sin + f\cos)$.   21. $4I - f - If' + 3f^2f'$.   23. From the first

428 CALCULUS [CHAP. 7

equation, $2 + 3y' + 5x^4 - y^3 - 3xy^2y' = 0$ and $y' = (y^3 - 5x^4 - 2)/(3 - 3xy^2)$. From the second equation, $2y' + 3x^2y^4 + 4x^3y^3y' - 3 = 0$ and $y' = (3 - 3x^2y^4)/(2 + 4x^3y^3)$. At the origin, the first has a slope of $-2/3$ and the second of $3/2$.

7–9 The antiderivative. According to Newton's laws of motion, a force F acting on a body of mass m gives it an acceleration a according to the relation $F = ma$. Now $a = \mathsf{D}_t v$, where v is the velocity measured along an axis in the line of action F, and t is the time. Hence $F = m\mathsf{D}_t v$. Let us use this to find a description of the action of a body falling freely from rest. For such a body the force of gravity is proportional to the mass. Letting the constant of proportionality be g, we have $F = mg$. Substituting in the previously derived equation, we have $mg = m\mathsf{D}_t v$, or

(1) $$\mathsf{D}_t v = g.$$

It is because of (1) that g is called the *acceleration of gravity.*

To describe the motion we should like to find the relation between v and t, that is, some function f such that $v = f(t)$. From (1) $\mathsf{D}f = g$. A function with this property is gI, so that $f(t) = gt$ is a *possibility.* Of course $f(t) = gt + C$ for any C would do. However, we wish the body to be falling from rest, so that $f(0) = 0$. Hence we choose $f(t) = gt$ and write

(2) $$v = gt.$$

But $v = \mathsf{D}_t s$, where s is the distance measured along some axis in the line of action of F. Hence we now have

(3) $$\mathsf{D}_t s = gt.$$

Here we wish to find s in terms of t. What function F has the property that $F'(t) = gt$ or $\mathsf{D}F = gI$? One possibility is $(1/2)gI^2$. Also, for any $\underline{C}$, $(1/2)gI^2 + \underline{C}$ has the same derivative. However, if we measure s from the point where $t = 0$, we have

(4) $$s = (1/2)gt^2.$$

This is the famous law of motion of a freely falling body, which the student may have seen in physics books, and which may have seemed mysterious. It follows quite inevitably from Newton's axioms. Of course it is confirmed by experiments, which is one of the facts that makes Newton's axioms acceptable. If the pound-foot-second system is used, measurements give $g = 32$ ft/sec/sec; if the gram-centimeter-second system is used, $g = 980$ cm/sec/sec.

(a) Suppose that the acceleration of a body varies directly with time with a constant of proportionality k. Find the velocity and distance at time t, assuming that it starts from rest at $t = 0$.

In (1) we have an equation which we wish to solve for an unknown function f. Indeed (1) is equivalent to the equation $\mathsf{D}f = \underline{g}$. To solve it we have to answer the question: What function has a derivative equal to a constant function $\underline{g}$? Equations which involve derivatives of unknown functions are called *differential equations*. Equation (1) is an example. A slightly more complicated one is

$$(5) \qquad \mathsf{D}^2f(x) + 2\mathsf{D}f(x) + 3f(x) = x^2.$$

Differential equations of the form

$$(6) \qquad \mathsf{D}F(x) = f(x)$$

occur frequently in mathematics. Eq. (1) is a special case with $(F{:}f, f{:}\underline{g})$. In previous sections we have been solving problems in which F is given and f is to be found. Here we are interested in the inverse problem, given the derivative f to find the function F. We call any solution of (6) for F *an antiderivative of f*.

Find an antiderivative of: (b) $2I$,   (c) $mI + b$,   (d) $3I^2$.   Find a solution of each of the following: (e) $\mathsf{D}f(x) = x^2 - 3x$,   (f) $\mathsf{D}f(x) = -1/x^2$, (g) $\mathsf{D}f(x) = 1$.

Suppose that $\mathsf{D}F = f$, so that F is an antiderivative of f. Then

$$\mathsf{D}(F + \underline{C}) = \mathsf{D}F + \mathsf{D}\underline{C} = \mathsf{D}F + \underline{0} = f.$$

Hence $F + \underline{C}$ is also an antiderivative. Hence, if a function has an antiderivative, it has infinitely many formed by adding arbitrary constant functions to any one of them. But are there still other antiderivatives not obtainable by adding constants? The answer is no, according to the following theorem.

$$(7) \qquad (\mathsf{D}F = \mathsf{D}G) \rightarrow (\exists \underline{c}\, F = G + \underline{c}).$$

In words, if two functions have the same derivative, i.e., are antiderivatives of the same function, then they differ by a constant function. To prove this we need to develop some further theory, which is of great importance for other purposes as well.

Consider the function pictured in Fig. 7–28. The slope of the secant line is $[f(b) - f(a)]/(b - a)$. It appears obvious that there is some point in $(\underline{a}\ \ \underline{b})$ at which the tangent line is parallel to this secant; i.e., there is

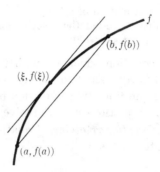

FIGURE 7–28

some ξ (pronounced kseye) for which $a < \xi < b$ and $Df(\xi) = [f(b) - f(a)]/(b - a)$. More generally, we state the *law of the mean*.

(8)   *If f is continuous in $(a\_\_b)$ and is differentiable in $(a\_\_b)$, then $\exists \xi\ (a < \xi < b) \wedge [f(b) = f(a) + (b - a)Df(\xi)]$.*

This can be proved, but we give only some exercises to indicate its plausibility.

(h) Prove (8) for quadratic functions by showing that the required point is actually the midpoint of the interval in this case. (*Suggestion:* Let the quadratic function be given by $f(x) = Ax^2 + Bx + C$ rather than by $ax^2 + bx + c$ to avoid confusion with the letters giving the interval.)   (i) What is ξ for a linear function?   (j) Give an example to show that continuity in the closed interval is necessary.   (k) Give an example to show that differentiability in the closed interval $(a\_\_b)$ is not necessary.   (l) Give an example to show that differentiability in the open interval $(a\_\_b)$ is necessary.   (m) Show that the velocity of a freely falling body at the middle of any time interval is equal to its average velocity during the whole time interval.

We can now prove the following.

(9) $$(Df = \underline{0}) \to (\exists \underline{c}\ f = \underline{c}).$$

In words, if the derivative of a function is everywhere zero, then the function is a constant function. Equivalently,

$$[\forall x\ Df(x) = 0] \to [\exists c\ \forall x\ f(x) = c].$$

This is easily proved from (8). For suppose, by hypothesis, that $Df = \underline{0}$, and let $f(a) = c$ where a is any number. Then by (8),

$$f(b) = c + (b - a)Df(\xi)$$

for some ξ in (a_b) or in (b_a). But $Df(x)$ is zero for *any* x. Hence $f(b) = c$. Since this argument holds for any b, $f = \underline{c}$.

(n) In the above proof we have said nothing about the continuity or differentiability of f, as required by (8). Fill in these omissions.

Now we can prove (7), for consider the function $F - G$. We have $D(F - G) = DF - DG = 0$ by hypothesis. Hence by (8), $F - G = \underline{c}$ or $F = G + \underline{c}$!

(o) Find a formula that gives all the solutions of $Df = 3I + 1$.

We looked at D as an operation (function) in Section 7–6. What, then, is $D*$, the inverse of D? We have $D = \{(f, g)|g = Df\}$, where we use f as the independent variable and g as the dependent variable. D maps each function f into its derivative. Then $D* = \{(g, f)|g = Df\}$. Here $D*$ maps each function g into its antiderivatives. *Note the plural*, since corresponding to each g there is an infinity of f's.

We see that $D*$ is not a function according to our definition of this term. However, it is customary to view it as an operation that maps a function into the set of all its antiderivatives. More common symbols for $D*g$ are $D^{-1}g$ and $\int g$, and for $D*g(x)$, $D^{-1}g(x)$ and $\int g(x)\,dx$. The "dx" serves merely to indicate the independent variable, like the "x" in "$D_x f(x)$." The origin of the symbol $\int$ will be indicated in Section 7–12. $D^{-1}g$ is used as an abbreviation for a formula that stands for any one of the antiderivatives of g. Since $D(D^{-1}g) = g$ and $D^2(D^{-1}g) = Dg$, the negative exponent on D appears to behave like an exponent on numbers. However, it represents the inverse with respect to composition, *not* numerical reciprocation. Thus $D^{-1}g \neq 1/Dg$. Sometimes the antiderivative is called the *indefinite integral*.

The relation between D^{-1} and D is expressed formally as follows

(10) **Def.** $[D^{-1}g = f] = [Df = g]$.

When we speak of finding *the* antiderivative of a function g, we mean finding the solutions of $Df = g$, that is, finding a formula that gives *all* the antiderivatives of g. For example, $D^{-1}I = I^2/2 + \underline{C}$, where $\underline{C}$ is an arbitrary constant function. Checking by (10), we find $D(I^2/2 + \underline{C}) = 2I/2 + \underline{0} = I$.

In terms of images,

(11) $[D^{-1}g(x) = f(x)] = [Df(x) = g(x)]$.

Hence $D^{-1}(x) = x^2/2 + \underline{C}$, where $\underline{C}$ is an arbitrary constant, since $D(x^2/2 + \underline{C}) = 2x/2 + 0 = x$. It is essential to include the arbitrary constant, because $Df = g$ has many solutions. Thus, although $D^{-1}(x) = x^2/2$ according to (11), the right member gives only one of the possible values.

Find: (p) $D^{-1}(2I)$, (q) $D^{-1}(2)$, (r) $D^{-1}(mI + b)$, (s) $D^{-1}(2 + x + 3x^2 + x^3)$, (t) $D^{-1}(1 - x)^2$.

If we wish to find $Df(x)$, we may apply (7–4–1), or rules developed from it. But to find $D^{-1}g(x)$ no such direct procedure is available. Just as we divide by using our knowledge of multiplication, so we look for anti-derivatives by using our knowledge of derivatives. Since this kind of indirect searching can be tedious, mathematicians have prepared tables of antiderivatives, usually called tables of integrals. Among the most widely used are *Table of Integrals and Other Mathematical Data*, by H. B. Dwight; *C.R.C. Standard Mathematical Tables*, edited by C. D. Hodgman; *A Short Table of Integrals*, by B. O. Peirce, *The Handbook of Mathematical Tables and Formulas*, by R. S. Burington, and *Handbook of Calculus, Difference and Differential Equations*, by E. J. Cogan and R. Z. Norman.

Suppose, for example, that we wish to solve the differential equation $Df(x) = x(ax + b)^{-3}$. Offhand (or even not offhand!) it is not easy to think of an expression whose derivative is $x/(ax + b)^3$. However, we find on page 60 of Burington,

$$(12) \qquad \int \frac{x\,dx}{(ax + b)^3} = \frac{b}{2a^2(ax + b)^2} - \frac{1}{a^2(ax + b)}.$$

This means that the expression on the right is an antiderivative of $x(ax + b)^{-3}$. By (11), this is to say that the derivative of the right member equals $x(ax + b)^{-3}$. Let us check this.

$$D[(b/2a^2)(ax + b)^{-2} - a^{-2}(ax + b)^{-1}]$$

$$= b2^{-1}a^{-2}(-2)(ax + b)^{-3}a - a^{-2}(-1)(ax + b)^{-2}a$$

$$= -ba^{-1}(ax + b)^{-3} + a^{-1}(ax + b)^{-3}(ax + b)$$

$$= (ax + b)^{-3}(-ba^{-1} + x + a^{-1}b) = x(ax + b)^{-3},$$

as desired. Hence all the antiderivatives are given by the right member of (12) plus an arbitrary constant.

Integral tables contain thousands of formulas such as (12). Because of (11) they may be read also as formulas for derivatives. There is also an extensive theory of how to find antiderivatives with or without such tables, which the reader may locate in calculus books under the heading "methods of integration." (See also Section 7-18.)

(u) Prove that

$$\int \frac{x\,dx}{\sqrt{ax^2 + c}} = \frac{1}{a}\sqrt{ax^2 + c}.$$

PROBLEMS

1. Show that constant functions are the only ones having zero derivative at every point.

2. Show that straight lines are the only curves having the same slope at every point.

In Problems 3 through 10 a function is given. Find the family of antiderivatives in each case.

3. $I - 3I^2$.   4. $\sqrt{I + 1}$.

5. $I^{1/2} + 1$.   ★6. sin.

7. $I^4 - 3 + 2I$.   8. $3 - 2I^{-3}$.

9. $\underline{0}$.   10. $1 + I$.

11. Show that $(I^2 + 1)^{3/2}$ is not an antiderivative of $(I^2 + 1)^{1/2}$.

12. Show that $\sim\exists n\, \mathsf{D}I^n = 1/I$.

In Problems 13 through 16 $f(x)$ is given. In each case find $\mathsf{D}^{-1}f(x)$.

13. $x\sqrt{x}$.   14. $x - 3x^2$.   15. 1.   16. $\cos x$.

17. The slope of a certain curve at any one of its points (x, y) is $x - 3$. Find the family of possible curves. Find the curve if we know that it passes through the origin. Sketch.

18. The slope of a curve is $x^2 - 1$ and it passes through $(2, -5)$. Find the curve.

★19. A body is moving in a straight line with an acceleration given by $3t - 2$, where t is the time. Find its velocity if it is motionless at $t = 0$. Find its distance from the position at $t = 0$.

★20. Work Problem 19 with the initial velocity 20.

★21. A boy throws a ball downward with an initial velocity of 20 ft/sec from the top of a building 100 ft high. When will it strike the ground? How much later would it strike the ground if he just dropped it?

★22. Show that a body falling freely with a velocity of v_0 and a position given by s_0 measured from some origin in its line of motion at time t_0 will move according to the equation $s = s_0 + v_0(t - t_0) + (\frac{1}{2})g(t - t_0)^2$.

★23. Show that $v^2 = v_0^2 + 2g(s - s_0)$ in the situation described in Problem 22.

★24. Show that $\mathsf{D}_t^2 s = v\mathsf{D}_s v$ in the same context. (*Hint:* Chain rule.)

★25. Practice differentiation by verifying formulas in a table of integrals.

ANSWERS TO EXERCISES

(a) $v = kt^2/2$. (b) I^2. (c) $mI^2/2 + bI$. (d) I^3. (e) $x^3/3 - 3x^2/2$. (f) $1/x$. (g) x. (h) $Ab^2 + Bb + C = Aa^2 + Ba + C + (b - a)(2A\xi + B) \leftrightarrow \xi = (a + b)/2$. (i) Any number in (a_b). (j) sg in $(\underline{0}\quad 1)$. (k) $|I|$ in $(\underline{0}\quad 1)$. (l) $|I|$ in (-1_1). (m) Since the distance is given by a quadratic, you may apply Exercise (h). (n) Since the derivative is zero everywhere, the function must be continuous and have a derivative. (o) $3I^2/2 + I + C$. (p) $I^2 + C$. (q) $2I + C$. (r) $mI^2/2 + bI + C$. (s) $2x + x^2/2 + x^3 + x^4/4 + C$. (t) $-(1 - x)^3/3$. (u) Differentiate the right member.

ANSWERS TO PROBLEMS

Always check antiderivatives by finding their derivatives! 11. The derivative is $(3/2)(I^2 + 1)^{1/2}(2I)$. We are not yet in a position to find a function whose derivative is $(I^2 + 1)^{1/2}$. 17. Given by $y = x^2/2 - 3x + C$. If it passes through origin, $0 = C$.

7-10 Measure and area. Often we are interested in the length, area, volume, mass, or some other measure of an object or set. Such measures tell us in some sense "how much" of something is present. For example, when confronted with a finite set of objects we might count them to get a measure of the size of the set. When confronted with a physical body we might be interested in its mass, its dimensions, or its volume. We might be interested in the length of a straight line or curve and in the area of a region enclosed by a curve, as suggested in Fig. 7–29.

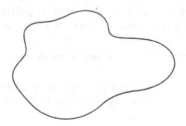

FIGURE 7–29

As suggested in Section 7–1, the problem of length and area is by no means simple for curves other than broken lines. We shall find it helpful to consider some properties common to all measures. To get at these, we consider the particular case of measuring finite sets by counting them. Imagine, then, a certain finite set together with all its subsets. Let $M(A)$ be the measure of a subset A obtained by counting. Among the subsets is the null set $\emptyset$, and

(1) $$M(\emptyset) = 0.$$

For any other subset A, $M(A)$ is a natural number. Hence

(2) $$M(A) \geq 0.$$

From (6-3-2) we have

(3) $$(A \cap B = \emptyset) \rightarrow [M(A \cup B) = M(A) + M(B)].$$

Also,

(4) $$(A \subseteq B) \rightarrow [M(A) \leq M(B)].$$

(a) Argue for (4).   (b) When would the equality hold in the conclusion of this implication?

The above properties are not surprising, but it is interesting that (1) through (4) hold also when we consider some other kinds of measure. For example, the mass of an empty set is zero (1), the mass of any body is a non-negative real number (2), the mass of a body consisting of two disjoint bodies together is the sum of the masses (3), and the mass of a part of a body is not greater than the mass of the body (4).

Or let us consider the set of all finite segments on a straight line. The measure consisting of the length of a segment satisfies (1) through (4) if we consider $\emptyset$ to be a degenerate interval. There are some additional properties of interest. For example, the length of a segment consisting of a single point is zero, so that $\emptyset$ is not the only set of points on the line with zero length. We may have set A a proper subset of set B and yet $M(A) = M(B)$. For example, the lengths of $(a\ \ b)$ and $(a\ \ b)$ are the same since the intervals differ by but a single point. Also we note that two congruent segments have the same measure.

Suppose we wish to extend the idea of measure to segments on curves other than straight lines or to sets other than intervals on a straight line. It would be natural to try to do this in such a way that (1) through (4) still would hold. In the following questions, assume that you are trying to define the measure of sets of points on a straight line so that the measure of an interval is its length and so that (1) through (4) hold.   (c) How would you define the measure of the segment $(a\ \ b)$?   (d) Could you then *prove* that $M(a\ \ a) = 0$?   (e) How would you define the measure of a set consisting of the union of two disjoint intervals?   (f) What would be the measure of any finite set of points?

Now let us consider in the same way a plane region and all rectangular regions in it. We define the area of a rectangle to be the product of its two dimensions. Then (1) through (4) hold if we consider the null set as a kind of degenerate rectangle. We should like to extend the concept of area by appropriate definitions to cover regions other than those bounded by

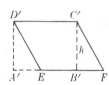 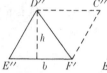

FIGURE 7-30

rectangles. More precisely, we wish to define the area of any region enclosed by a curve in such a way that (1) through (4) hold, the area of a rectangle is the product of its dimensions, and the areas of two congruent figures are equal. Figure 7-30 suggests how we can derive formulas for the area of regions enclosed by parallelograms, triangles, and any curve consisting of straight lines.

(g) Show how we get the formula for the area of a region bounded by a parallelogram by using (3) and the congruence assumption. (h) Explain how the area for a triangular region is derived.

Now we consider more general regions, bounded by curves other than straight lines. For definiteness, consider the region bounded by the interval $(0 \quad 2)$ on the x-axis, the segment from $(2, 0)$ to $(2, 4)$, and the arc of the graph of I^2 between $(0, 0)$ and $(2, 4)$. The region is shaded in Fig. 7–31. To calculate its area is quite beyond our present methods, since it clearly cannot be divided up into triangles. However, consider the subset of our region shaded in Fig. 7–32. It is that portion of the region lying below the step function whose horizontal segments, each of length $1/2$, have their left ends on the curve at $(0, 0)$, $(1/2, 1/4)$, $(1, 1)$, and $(3/2, 9/4)$. We can easily calculate the area of this region by (3). It is $(1/2)(1/4 + 1 + 9/4)$. But since this region is a subset of the region in which we are interested, its area is no greater by (4). Letting A be the area sought and $\underline{A}_4$ be the area of the union of rectangular regions, we have $\underline{A}_4 \leq A$. Similarly, the area under the step function whose horizontal segments have their right endpoints on the curve (the upper boundary is given by a dotted line in Fig. 7–32) is $(1/2)(1/4 + 1 + 9/4 + 4)$. Calling this area $\overline{A}_4$, we have $\overline{A}_4 \geq A$.

(i) Justify the formulas for $\underline{A}_4$ and $\overline{A}_4$. (j) Why are the notations $\underline{A}_4$ and $\overline{A}_4$ (read "A sub 4 lower" and "A sub 4 upper") appropriate? (k) How is (3) used?

The significant thing about the above maneuver is that we have "boxed" A between two areas that can easily be found, that is, $\underline{A}_4 \leq A \leq \overline{A}_4$. We have done this by dividing the interval $(0 \quad 2)$ into subintervals and constructing on these subintervals rectangles, the area of whose union approximates the area sought. By dividing the interval into a larger number

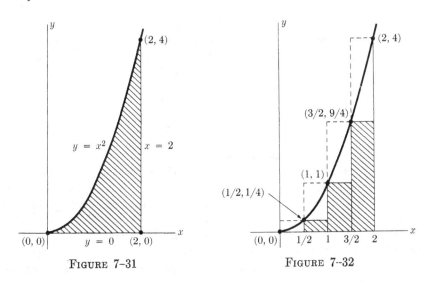

FIGURE 7–31   FIGURE 7–32

of subintervals we might hope to enclose A as tightly as we wish between two areas. This follows because the tops of the rectangles would be closer to the curve if we made their bases smaller.

(l) Draw a large figure similar to Fig. 7–32 but with eight subintervals of equal length. Calculate $\underline{A}_8$ and $\overline{A}_8$.   (m) Find $\underline{A}_4$ and $\overline{A}_4$ as single fractions and find the difference $\overline{A}_4 - \underline{A}_4$.   (n) If we approximate A by $(\underline{A}_4 + \overline{A}_4)/2$, what is the largest possible error we might make?   (o) Do the last two exercises for the 8 subintervals.

Now, let us imagine that we have divided the interval from the origin to $(2, 0)$ into n congruent subintervals, each of length $2/n$. The values of x at the points of division are then $0, 2/n, 4/n, 6/n, 8/n, \ldots, (2n - 2)/n$, 2. The heights of the rectangles are $0, 4/n^2, 16/n^2, 36/n^2, \ldots, 4$. We consider first the rectangles below the curve. The height of the ith rectangle is $(2i/n)^2$ or $4i^2/n^2$. Since its base is $2/n$, its area is $8i^2/n^3$. Since i runs here from 1 to $n - 1$, the area of the union of these rectangular regions is

(5) $$\sum_{i=1}^{i=n-1} 8i^2/n^3 = (8/n^3) \sum_{i=1}^{i=n-1} i^2.$$

Applying $(6–7–23)(n{:}n - 1)$ to the summation, and letting $\underline{A}_n$ stand for area given by (5), we have

(6) $$\underline{A}_n = (8/n^3)(n - 1)n(2n - 1)/6$$

(7) $$= (8/3)(1 - 1/n)(1 - 1/2n).$$

(p) Verify that the right members of (6) and (7) are synonymous.

If we work with the rectangles whose tops are above the curve, the ith rectangle again has height $4i^2/n^2$ and area $8i^2/n^3$, but now i runs from 1 to n. Calling the area of the union of these rectangular regions $\overline{A}_n$, we have

$$(8) \qquad \overline{A}_n = (8/n^3) \sum_{i=1}^{i=n} i^2 = (\,8/n^3)(n)(n+1)(2n+1)/6$$

$$(9) \qquad\qquad\qquad\quad = (8/3)(1 + 1/n)(1 + 1/2n).$$

It appears plausible that as we let n get very large the difference between $\underline{A}_n$ and $\overline{A}_n$ gets quite small, so that they become very good approximations to A. We wish now to *define* the area A in some way consistent with these ideas. To do this, consider the set of all $\underline{A}_n$ for n a natural number. Clearly this set of real numbers is bounded from above, since they are all less than $\overline{A}_4$, for example. Hence this set of numbers has a least upper bound by (6–9–6). We *define* the area A to be this least upper bound. An alternative definition might come from considering the set of all $\overline{A}_n$. It is bounded from below, so has a greatest lower bound. We could define A as this greatest lower bound.

It is not hard to see that the least upper bound of the $\underline{A}_n$'s and the greatest lower bound of the $\overline{A}_n$'s ought to be the same. For as we let n get larger we see from (7) that $\underline{A}_n$ gets larger. Indeed, for large enough n we can make $1/n$ and $1/2n$ so small that $\underline{A}_n$ is as close as we like to $8/3$. Similarly, as n gets very large we see from (9) that $\overline{A}_n$ decreases and gets as close as desired to $8/3$. Hence $8/3$ is the lub, glb, and the desired area.

(q) Use this method to find the exact area of the region bounded by the arc of I^2 from $(1, 1)$ to $(3, 9)$, the interval $(\underline{1 \quad 3})$ on the x-axis, the segment from $(1, 0)$ to $(1, 1,)$, and the segment from $(3, 0)$ to $(3, 9)$.

PROBLEMS

In Problems 1 through 8 find $\underline{A}_4$, $\overline{A}_4$, $\underline{A}_n$, $\overline{A}_n$, and A by seeing what happens when n gets very large. In each case we are concerned with the area of the region bounded by the graph of f between $(a, f(a))$ and $(b, f(b))$ and the segments joining $(a, f(a))$, $(a, 0)$, $(b, 0)$, and $(b, f(b))$, in that order. (See Fig. 7–33). Draw figures carefully.

    1. $f = I^2$, $a = 0$, $b = 3$.   2. $f = I^2$, $a = 2$, $b = 3$.
    3. $f = I^2$, $a = -1$, $b = 1$.   4. $f = I$, $a = 0$, $b = 2$.
    5. $f = I$, $a = 1$, $b = 3$.   6. $f = 2I^2$, $a = 0$, $b = 2$.
    7. $f = I^2$, $a = 0$, $b = c$.   8. $f = \underline{c}$, $a = 0$, $b = c$.

★9. Use Problem 13 in Section 6–7 to find the area defined by $f = I^3$, $a = 0$, $b = 1$.

    10. Generalize for $f = I^3$, $a = 0$, $b = c$.

    11. Generalize Problem 7 to $f = I^2$, a and b unspecified.

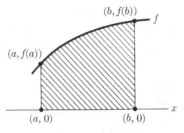

FIGURE 7–33

12. Find by the methods of this section the area of the region between the graph of $mI + b$ and the x-axis between the lines $x = a$ and $x = c$. (Assume that m, b, a, and c are such that this region lies entirely above the x-axis.) Check your result against the area found by formulas of plane geometry.

13. How could the methods of this section be used to find the area of a figure such as that of Fig. 7–29?

Answers to Exercises

(a) If A is a proper subset of B, we would have members of B left over if we attempted to pair them.   (b) When $A = B$.   (c) $|b - a|$.   (d) $|a - a| = 0$.
(e) As the sum of the measures.   (f) Zero.   (g) In Fig. 7–30, $A'D'E \cong B'C'F$.   (h) In Fig. 7–30, $E''D''F' \cong C''F'D''$. Hence their areas are equal and equal to half that of the parallelogram.   (i) The base of each rectangle has length 1/2. The heights are 0, 1/4, 1, 9/4, and 4.   (j) They suggest the number of subdivisions.   (k) We are adding the areas of disjoint regions.
(l) 35/16, 51/16.   (m) 7/4, 15/4, 2.   (n) 1.   (o) Difference is 1, largest error 1/2.   (p) Recall that $(1/a^3)bcd = (b/a)(c/a)(d/a)$.   (q) 26/3.

Answers to Problems

1. 9.   3. 2/3.   5. 4.   7. $c^3/3$.   9. 1/4.   11. $b^3/3 - a^3/3$.   13. Split the region by straight lines so that it appears as the union of disjoint subregions, each of which is similar to Fig. 7–33. Of course, we would need to know the functions giving the curve bounding each region on top.

7–11 The definite integral. We wish to generalize the ideas of the last section in order to find general methods of calculating areas quickly without the tedious devices used in the last section. We wish to be able to deal with curves both rising and falling and crossing the x-axis. Moreover, it will become apparent that the generalization has many uses besides that of finding areas.

We consider a function f continuous in the closed interval $(a \quad b)$. Partition the interval into n subintervals, by the points of division $a = x_0 < x_1 < x_2 < \ldots < x_{i-1} < x_i < x_{i+1} < \ldots < x_n = b$.   These subintervals are not necessarily of equal length. Now f is continuous on each subinterval, since it is continuous on the entire interval. In the ith subinterval, $(x_{i-1} \quad x_i)$, f has an absolute maximum and an absolute minimum according to (7-6-1). Let these be M_i and m_i respectively.

(a) Review the discussion of continuity in Sections 7-3 and 7-6. (b) Complete: M_i is the largest value of _____ for values of x in _____. (c) Complete: m_i is _____. (d) To get a physical picture of the situation, let us imagine $f(x)$ to be the rate of flow of water into a tank (say in gallons per minute) and x the time. Then what is $(x_{i-1} \quad x_i)$? What is $(x_i - x_{i-1})$? What is M_i? What is m_i?

We call this partition of the interval $(a \quad b)$ into subintervals p. Then we form the sums $\underline{I}_p$ and $\overline{I}_p$ defined by

$$(1) \qquad \underline{I}_p = \sum_{i=1}^{i=n} m_i(x_i - x_{i-1})$$

$$(2) \qquad \overline{I}_p = \sum_{i=1}^{i=n} M_i(x_i - x_{i-1})$$

We call $\underline{I}_p$ and $\overline{I}_p$ the *lower* and *upper sums* corresponding to p.

(e) Referring back to the physical interpretation in Exercise (d), what are $\underline{I}_p$ and $\overline{I}_p$?

Now the function f has an absolute maximum and minimum on the whole interval $(a \quad b)$. Let these be M and m respectively. It is easy to show that

$$(3) \qquad m(b - a) \le \underline{I}_p \le \overline{I}_p \le M(b - a).$$

(f) Show the first inequality. (g) The third. (h) The middle one. (i) Referring to Exercise (d), what are $m(b - a)$ and $M(b - a)$?

Consider the set of all values of $\underline{I}_p$ for all possible partitions. Clearly this set of real numbers is bounded from above by $M(b - a)$. Hence it has a least upper bound $\underline{I}$. Similarly, the set of all upper sums is bounded from below by $m(b - a)$. Hence it has a greatest lower bound $\overline{I}$.

(j) Show that $\underline{I}_p \le \underline{I}$ and $\overline{I} \le \overline{I}_p$. ★(k) Show that $\underline{I} \le \overline{I}$. (l) Referring to Exercise (d), what are $\underline{I}$ and $\overline{I}$?

Now we are going to show that for f a continuous function (which is all we are considering here), $\underline{I} = \overline{I}$. To do this we shall make use of the following theorem about continuous functions, which we state here without proof.

(4)   *If f is continuous in the closed interval $(\underline{a} \quad b)$, then for any positive ϵ there is a δ such that in any subinterval of length less than δ, the maximum and minimum of $f(x)$ differ by less than ϵ.*

Suppose now that we are given any arbitrarily small ϵ. We choose a partition of the interval $(\underline{a} \quad b)$ in which all the subintervals are of length less than the corresponding δ, whose existence is guaranteed by (4). Then for this partition,

$$(5) \qquad \overline{I}_p - \underline{I}_p = \sum_{i=1}^{i=n} (M_i - m_i)(x_i - x_{i-1})$$

$$(6) \qquad < \sum_{i=1}^{i=n} \epsilon(x_i - x_{i-1})$$

$$(7) \qquad < (b - a)\epsilon.$$

Now $b - a$ is fixed and ϵ may be chosen arbitrarily small. Hence $\overline{I}_p - \underline{I}_p$ may be made arbitrarily small. But by the results of Exercises (j) and (k), $\underline{I}_p \leq \underline{I} \leq \overline{I} \leq \overline{I}_p$. Hence the difference between $\overline{I}$ and $\underline{I}$ can be shown to be arbitrarily small. It follows that $\overline{I} - \underline{I} = 0$, since only 0 is non-negative and smaller than any positive number.

We now define the *definite integral of f from a to b* as $\underline{I}$ (or $\overline{I}$) when $\underline{I} = \overline{I}$. We designate this definite integral by

$$(8) \qquad \int_a^b f \qquad \text{or} \qquad \int_a^b f(x)\, dx.$$

We prefer the first form, but the second is most commonly used. Its origin is as follows. In the above discussion let $\Delta x_i = x_i - x_{i-1}$. Then the lower and upper sums (1) and (2) are sums of terms of the form $f(x_i^*)\, \Delta x_i$, where x_i^* is the value of x in the ith interval that yields the minimum or maximum. The $\int$ is an elongated S suggesting "sum," and the dx suggests the Δx_i. The reason that the $\int$ is used also to designate the antiderivative will be explained in Section 7-12.

(m) Suppose that $f(x)$ is the velocity of a moving body and x is the time. Interpret $(x_{i-1} \quad x_i)$, $x_i - x_{i-1}$, m_i, M_i, each term in (3), $\underline{I}$, and $\overline{I}$. (n) Do the same letting $f(x)$ be the rate of growth of national income and x the time. (o) Do the same if we let $f(x)$ be the y-coordinate of a point on the graph of f where $f(x) > 0$ in $(\underline{a} \quad b)$. Relate the symbols of this section to those of Section 7-10.

In the next section we shall present a relatively easy way of evaluating definite integrals, but in this one we are interested in meaning rather than calculation. For this purpose we now derive another expression for the definite integral. As above, let f be a continuous function in $(a\ \ b)$. We choose the special partition into n intervals of equal length, $(b - a)/n$. Then consider the sum

(9) $$S_n = \sum_{i=1}^{i=n} f(a + i(b - a)/n)(b - a)/n.$$

Here each term is the product of the length of the intervals, $(b - a)/n$, and the function evaluated at the right-hand end of the interval. Clearly S_n lies between the lower and upper sums corresponding to this partition.

(p) Why?

Now suppose we let n get larger. It appears that the corresponding upper and lower sums will come closer together and approach the integral I. Hence as n gets larger, $S_n \to$ the definite integral. We write

(10) $$\int_a^b f = \lim_{n \to \infty} \sum_{i=1}^{i=n} f(a + i(b - a)/n)(b - a)/n.$$

We have not yet defined "$\lim_{x \to \infty}$." This is done as follows.

(11) **Def.** $[\lim_{x \to \infty} f(x) = L]$

$$= \forall \epsilon\ \epsilon > 0 \to \exists N\ x > N \to |f(x) - L| < \epsilon.$$

Roughly speaking, this says that "the limit of $f(x)$ as x approaches infinity is L" means, by definition, that we can make $f(x)$ arbitrarily close to L by taking x large enough. Our proof that $\underline{I} = \overline{I}$ showed also that

(11′) $$\lim_{n \to \infty} \underline{I}_n = \lim_{n \to \infty} \overline{I}_n = I,$$

where $\underline{I}_n$ and $\overline{I}_n$ are the sums corresponding to the partition of (9). Since $\underline{I}_n \leq S_n \leq \overline{I}_n$, by taking n large enough we can make the summation in (10) differ from the integral by an arbitrarily small amount. Hence (10) holds.

It is not necessary that we take the intervals to be congruent nor is it necessary to evaluate $f(x)$ at the right endpoint of each interval. Suppose that p is an arbitrary partition such as we introduced at the beginning of this section, and let x_i' be any point in the ith interval, that is, $x_{i-1} \leq x_i' \leq x_i$. Then we have

(12) $$\int_a^b f = \lim_{n \to \infty} \sum_{i=1}^{i=n} f(x_i')(x_i - x_{i-1}).$$

Here it is understood that as $n \to \infty$, we form the corresponding partitions so that the maximum length of the intervals in each partition approaches zero.

(q) Argue for (12).   (r) Why is it not surprising that when we let the number of intervals get large and their lengths get small, we get the same result whether we evaluate $f(x)$ at the left end of each interval, the right end, the place where $f(x)$ is maximum, the place where it is minimum, or anywhere else in the interval?

PROBLEMS

1. In Fig. 7–34(a) divide the interval $(a \quad b)$ into four congruent intervals and draw verticals at the points of division. Draw the rectangles corresponding to (1).

2. Copy Fig. 7–34(a) and draw the rectangles corresponding to (2).

3. On your drawings for Problems 1 and 2 show rectangles whose areas are $m(b - a)$ and $M(b - a)$ and explain how your drawing makes (3) appear plausible.

4. Make another copy of Fig. 7–34(a) and sketch the rectangles corresponding to (9).

5. Make another similar drawing illustrating (12).

6. Carry through Problems 1 through 5 for Fig. 7–34(b).

7. Having defined the definite integral precisely, we can now use it to *define* the area of a region bounded by the graph of a function, the x-axis, and vertical lines. In fact we define the *area* of the region bounded by the graph of f between

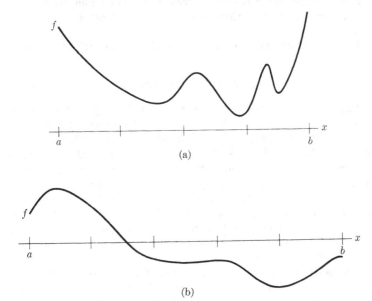

(a)

(b)

FIGURE 7–34

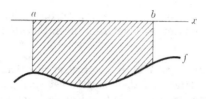

FIGURE 7-35

$(a, f(a))$, $(b, f(b))$, and the segments joining $(a, f(a))$, $(a, 0)$, $(b, 0)$, and $(b, f(b))$, in that order, where $a < b$ and $f(x) \geq 0$ for all x in $(a \quad b)$, as $\int_a^b f$. Suppose we found the integral in the case illustrated in Fig. 7-34(b). Why would we fail to get the area between the graph of f, the x-axis, and the lines $x = a$ and $x = b$? What would we get?

8. Argue that $\int_a^b |f|$ gives the area of the region bounded by the graph of f, the x-axis, and the lines $x = a$ and $x = b$ regardless of the sign of $f(x)$ in the interval.

9. Suppose we wish the area of the region graphed in Fig. 7-35? How is this area related to $\int_a^b f$?

10. Suppose $f(t)$ is the velocity at time t along a straight line. Does $\int_a^b f$ represent the distance traveled in the time from a to b, or does it represent the net distance traveled, i.e., the distance traveled in the positive direction less the distance traveled in the negative direction? Why?

11. In each part of Fig. 7-36 evaluate $\int_{-1}^{1} f$, where the broken line is the graph of f.

12. Suppose that $f(t)$ is the rate of growth of population, where population may be increasing at some times and decreasing at others. What does $\int_a^b f$ represent?

13. Discuss (10) and (12) in terms of the physical image of Exercise (d).

ANSWERS TO EXERCISES

(b) $f(x)$; the ith interval.   (c) the smallest value of $f(x)$ in the ith interval. (d) The ith time interval. The length of the ith time interval. The largest rate of flow during the ith time interval. The smallest rate of flow during that interval.   (e) The total flow calculated on the assumption that in each interval the flow was constant and equal to the smallest actual rate of flow during that interval. Same, with (smallest:largest).   (f) We make each term in (1) no larger if we put m in place of m_i. Then m is a common term and the expression becomes m times the sum of the lengths of the intervals, that is, $m(b - a)$. (g) Similarly.   (h) Since each term in (1) is not larger than the corresponding term in (2).   (i) The total flow if the rate of flow had remained constant at the value m. Same but $(m:M)$.

(j) Any member of a set is no greater than the least upper bound of the set. (k) Since we can find an element in a set arbitrarily close to the least upper bound (greatest lower bound) of the set, $\underline{I} \geq \overline{I}$ would require some upper sum to be smaller than some lower sum. But this cannot be, for consider any $\underline{I}_p$

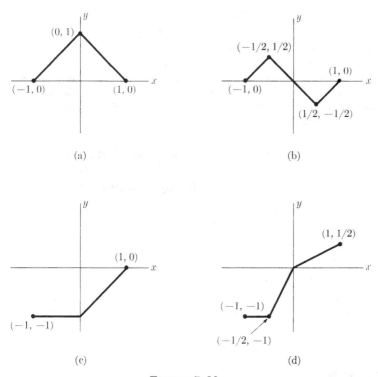

FIGURE 7-36

and $\bar{I}_q$. Let r be the partition formed by using the points of division of both p and q. Then $\underline{I}_r \leq \bar{I}_r$ by (3). But $\underline{I}_p \leq \underline{I}_r$, because the minima in the new intervals cannot be smaller than those in the old. Similarly $\bar{I}_q \geq \bar{I}_r$. Hence $\underline{I}_p \leq \bar{I}_q$.   (l) The total flow.   (m) Interval of time, elapsed time in interval, smallest velocity in ith interval, largest, distance if velocity constantly equal to smallest, distance if in each interval velocity had been equal to smallest velocity in that interval, same but largest velocity in each interval, same but largest velocity over entire period, actual net distance moved.

(n) Time interval, length of time interval, smallest and largest rate of growth in ith interval, net growth calculated at minimum rate over entire period, minimum in each interval, maximum in each interval, maximum over entire period, and actual net growth.   (o) Interval on x-axis, length of same, area of rectangle of base the total interval and height equal to minimum of f in interval, area of rectangles below curve, area of rectangles above, area of rectangle with base the interval and altitude the maximum of f in the interval, the actual area.   (p) Because the value of the function at any point in the ith interval must lie between the maximum and minimum there.   (q) The argument for (10) holds.   (r) As interval gets small enough it makes little difference where we evaluate $f(x)$, *provided it is continuous!*

ANSWERS TO PROBLEMS

1. The top of each rectangle should not cross the curve but should touch it at least once, and no part of the curve should be below it in that interval. 3. These rectangles are those we could get with just one subinterval! 4. Each rectangle touches the curve at the right endpoint of the interval regardless of the nature of the curve in the interval. 7. We would get the area of the part above the x-axis less the area of the part below. 9. It is the negative of the integral. 11. (a) 1, (b) 0, (c) $-3/2$, (d) $-1/2$.

7–12 The fundamental theorem of integral calculus. The theorem after which this section is named will enable us to evaluate many definite integrals with considerable ease. It may be stated for the case of continuous functions as follows.

If f is continuous in $(a \quad b)$, then

$$(1) \qquad \int_a^b f = \mathsf{D}^{-1}f(b) - \mathsf{D}^{-1}f(a).$$

This says that the definite integral of f from a to b is equal to an antiderivative of f evaluated at b less the same antiderivative evaluated at a. It is sometimes stated in the following equivalent form.

$$(2) \qquad \int_a^b f = F(b) - F(a), \text{ where } \mathsf{D}F = f.$$

(a) Review Section 7–9. (b) Why would it make no difference which antiderivative were used in (1) or (2)?

To illustrate the use of the theorem, we consider again the area of Fig. 7–31. Here $f = I^2$. By the fundamental theorem,

$$(3) \qquad \int_0^2 I^2 = (I^3/3)(2) - (I^3/3)(0) = 8/3.$$

(c) Similarly use the fundamental theorem to do exercise (q) in Section 7–10. (d) Verify some other results in the problems of Section 7–10. (e) Do Problem 11 in 7–10.

To prove the fundamental theorem we shall need some properties of the definite integral. In

$$(4) \qquad \int_a^b f \qquad \text{or} \qquad \int_a^b f(x) \, dx$$

we call a the *lower limit of integration*, b the *upper limit of integration*, and f or $f(x)$ the *integrand*. (Note that this use of "limit" has nothing to do with the meaning in Section 7–3.) It is evident that the value of (4) depends only on a, b, and f. The letter x in the second form is a dummy, since the value of (1) is unchanged if we replace it by any other letter.

This is evident from the definitions in Section 7–11, since the upper and lower sums depend only on the images of the x_i under f. The dummy "x" serves merely as a suggestive indication of the independent variable. Hence it is natural, and increasingly common, to omit "(x)" and "dx," as we have done in the first form. Often "x" in (4) is called the *variable of integration*.

(5) $$\int_a^b f(x)\, dx = \int_a^b f(t)\, dt.$$

In Section 7–11 we assumed $a < b$. The meaning of (1) is extended by the following definitions.

(6) **Def.** $$\int_b^a f = -\int_a^b f.$$

(7) **Def.** $$\int_a^a f = 0.$$

One way of looking at (6) is to note that if we integrate from b to a with $b > a$, the Δx_i in the lower and upper sums are negative.

(f) Show that (7) is essential if (6) is to hold when $a = b$.   (g) Verbalize (6) in terms of upper and lower limits.   (h) Interpret (7) in terms of areas.

We assume as before that f is continuous. Let m and M be the glb and lub respectively of the range of f in $(a \quad b)$. Then

(8) $$m(b - a) \le \int_a^b f \le M(b - a).$$

Geometrically this means that the area under f is less than that of the rectangle whose height is M and greater than the rectangle whose height is m, as suggested in Fig. 7–37. Analytically, it follows from (7–11–3) and the definition of the integral as $\underline{I}$ (or $\overline{I}$).

It is intuitively evident from Fig. 7–37 that there is some rectangle on the base $(a \quad b)$ whose area is exactly equal to the integral; that is,

(9) $$\exists H\; m \le H \le M \wedge \int_a^b f = H(b - a).$$

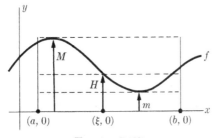

FIGURE 7–37

The top of this rectangle is a line parallel to the x-axis between the lines $y = m$ and $y = M$, as suggested in the figure. Of course (9) is true independently of geometrical argument since $H = (\int_a^b f)/(b - a)$ is a number satisfying (9).

The line $y = H$ meets the graph of f at some point in the interval since the graph must pass continuously from its highest point on $y = M$ to its lowest point on $y = m$. Hence there is some point ξ in the interval such that $f(\xi) = H$. That is, if f is continuous on $(a \quad b)$, then

(10)
$$\exists \xi \; a \leq \xi \leq b \wedge \int_a^b f = (b - a)f(\xi).$$

This theorem is called the *law of the mean for integrals*. Its proof need not depend on geometric imagery. It follows from a theorem about continuous functions that we state here without proof.

(11)
If f is continuous in $(a \quad b)$ and m and M are its minimum and maximum there, then for any y in the interval $(m \quad M)$ there is an x in $(a \quad b)$ such that $f(x) = y$.

Now, if we divide the members of (8) by $(b - a)$, we have

(12)
$$m \leq \frac{\int_a^b f}{b - a} \leq M.$$

Hence the middle expression represents a number in $(m \quad M)$, and by (11) there must be an x (call it ξ) such that (10) holds.

(i) Interpret (8) through (10) in terms of flow of water into a tank.

Since definite integrals are limits of summations by (7–11–10) and (7–11–12), we would expect them to have some properties similar to those of summations. Indeed

(13)
$$\int_a^b c \cdot f = c \int_a^b f.$$

(14)
$$\int_a^b (f + g) = \int_a^b f + \int_a^b g.$$

(15)
$$\int_a^b f = \int_a^c f + \int_c^b f.$$

(j) Argue for (13) through (15) in terms of areas.   (k) Argue for them in terms of flow of water into a tank.   (l) From them prove (16).

(16)
$$\int_a^b (f - g) = \int_a^b f - \int_a^b g.$$

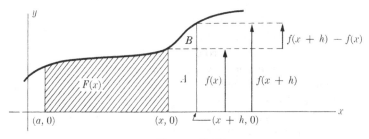

FIGURE 7–38

If we accept theorems (10) and (15), we are in a position to prove the following theorem, from which the fundamental theorem can easily be derived.

(17) $$\mathsf{D}_x \int_a^x f = f(x).$$

This seems quite plausible in terms of our various interpretations, for suppose that $f(x)$ is the rate of flow of water into a tank. Then $\int_a^x f$ is the total flow into the tank in the interval of time from a to x, and the rate of change of total flow is the rate of flow. For simplicity of symbols, let $F(x)$ be the integral from a to x of f. Let $F(x)$ be interpreted as the area under the graph of f between the vertical lines through $(a, 0)$ and $(x, 0)$, as suggested in Fig. 7–38. Now $F(x + h) = F(x) + A + B$, as indicated in the figure. But $A = hf(x)$. Hence, by definition, $\mathsf{D}F(x) = \lim [hf(x) + B]/h = \lim [f(x) + B/h] = f(x) + \lim (B/h)$, where the limits are as $h \to 0$. But $B \doteq (1/2)h[f(x + h) - f(x)]$, since it is practically a triangle of base h and altitude $f(x + h) - f(x)$. Hence $B/h \doteq [f(x + h) - f(x)]/2 \to 0$ as $h \to 0$ if f is continuous, and $\mathsf{D}F(x) = f(x)$.

(m) Why is this *not* a proof of (17)?

To prove (17), we let $F(x) = \int_a^x f$ by definition and prove $\mathsf{D}F = f$ as follows.

(18) $\qquad \mathsf{D}F(x) = \lim\limits_{h \to 0} \left[\int_a^{x+h} f - \int_a^x f \right]/h$ $\qquad\qquad$ (7–4–1)

(19) $\qquad\qquad = \lim\limits_{h \to 0} \left[\int_a^x f + \int_x^{x+h} f - \int_a^x f \right]/h$ $\qquad\qquad$ (15)

(20) $\qquad\qquad = \lim\limits_{h \to 0} \left[\int_x^{x+h} f \right]/h$

(21) $\qquad\qquad = \lim\limits_{h \to 0} [hf(\xi)/h]$ for some ξ in $(x \quad x + h)$ $\qquad$ (10)

(22) $\qquad\qquad = \lim\limits_{h \to 0} f(\xi) = f(x).$

The final equality holds because $x \leq \xi \leq x + h$, so that as $h \to 0$, $\xi \to x$. Then $f(\xi) \to f(x)$ because f is continuous.

Now we can complete the proof of the fundamental theorem itself. Since $DF = f$, we have for any antiderivative $D^{-1}f$, $F = D^{-1}f + \underline{c}$, where $\underline{c}$ is some constant function. [See 7–9–7.] Hence for every x in $(\underline{a \quad b})$, $F(x) = D^{-1}f(x) + c$. In particular, let $x = a$. Then $F(a) = D^{-1}f(a) + c$. But by (7), $F(a) = 0$. Hence $c = -D^{-1}f(a)$, and for all x we have $F(x) = D^{-1}f(x) - D^{-1}f(a)$. Now we let $x = b$ to get $F(b) = D^{-1}f(b) - D^{-1}f(a)$, which is just (1)!

Problems

The fundamental theorem opens up vast possibilities for finding areas and other quantities given by definite integrals. If we can find an antiderivative of f, we can immediately find any definite integral of which it is the integrand. In doing the calculations it is convenient to adopt the notation $[F]_a^b$ for $F(b) - F(a)$. Then the fundamental theorem can be written briefly; i.e.,

$$(23) \qquad \int_a^b f = [D^{-1}f]_a^b.$$

Thus we write

$$(24) \qquad \int_0^2 I^2 = [I^3/3]_0^2 = 8/3 - 0/3 = 8/3.$$

Sometimes, in fact most often, one sees such calculations written in terms of images using a dummy variable. For example, (24) becomes

$$(25) \qquad \int_0^2 x^2 \, dx = [x^3/3]_0^2 = 8/3 - 0/3 = 8/3.$$

The calculation in terms of the function itself is easier and briefer, but the form in terms of images may remind the reader that the integral is the limit of a sum of terms of the form $x^2 \, \Delta x$. It may therefore be preferable in writing about applications.

In Problems 1 through 20 evaluate the definite integrals by using the fundamental theorem. Present your work in one of the forms (24) or (25).

1. $\int_1^3 I^3.$

2. $\int_1^3 x^3 \, dx.$

3. $\int_{-1}^1 I^5.$

4. $\int_1^2 (I - I^2 + 2I^3).$

5. $\int_{-10}^{-8} dx/x^2.$

6. $\int_2^4 (x^{1/2} + x^{1/3} + x^{1/4}) \, dx.$

7. $\int_0^1 \sqrt{x^2 + 1}\,(2x) \, dx.$

8. $\int_a^b (I + 3)^{10}.$

9. $\int_1^0 (2I - I^{1/2})$.

10. $\int_0^0 x^2\sqrt{1 - x^3}\, dx$.

11. $\int_0^\pi \sin$.

12. $\int_0^\pi \cos$.

13. $\int_0^1 \sqrt{1 - t}\, dt$.

14. $\int_0^{\pi/2} \cos [2I]$.

15. $\int_1^0 \cos \cdot \sin^2$.

16. $\int_1^2 (I^{1/2} + I^{1/3} + I^{1/4})$.

17. $\int_0^1 I(I^2 + 1)^{1/2}$.

18. $\int_0^1 (1 - x)^{1.3}\, dx$.

19. $\int_1^2 I \cos [I^2]$.

20. $\int_a^b I^2(1 - I^3)^2$.

21. The rate of production of a firm from time t_1 to time t_2 is given by $mt + b$. Find total production during the period.

22. Find the area under $-I^2 + I + 6$ between the two points where it crosses the x-axis.

23. Find $\int_{-a}^a I^3$ and interpret your result.

24. Show that (6), (7), and (13) through (16) follow from (2). (Note that this is the opposite of the logical development we used.)

★25. Find the H and ξ for which (9) and (10) hold for the integrals in Problems 1 and 13.

★26. Argue from Fig. 7-39 that if the graphs of f and g meet at $(a, f(a))$ and $(b, f(b))$, and $f(x) - g(x) \geq 0$ in $(a\quad b)$, then the area between the curves is given by the integral of $f - g$ from a to b.

★27. Find the area bounded by $y = x^2$ and $y = x$.

★28. Find the area bounded by $y = x^2$ and $y = x^{1/2}$.

★29. Find the area bounded by $y = x^2$ and $y = -x^4 + 5$.

★30. Argue from a figure that the area bounded by the graphs of f and g and the lines $x = a$ and $x = b$ is given by the integral over the interval of $|f - g|$.

★31. Find the area bounded by the graphs of $y = x^3$, $y = -x^2 + 10$, and $x = \pm 1$.

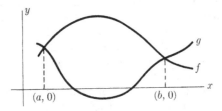

FIGURE 7-39

★32. Water is flowing into a tank at the rate of $5t$, where t is the time, and is flowing out at the rate of $-3t^2$. What is the change in the water in the tank from $t = 0$ to $t = 1$? from $t = 0$ to $t = 2$? How much water is in the tank after 3 units of time if the original amount in the tank is 100? Graph the amount of water in the tank as a function of time from $t = 0$ to $t = 3$.

★33. Prove (13), (14), and (15) by using the definitions and theorems of Section 7–11.

★34. Argue for (26) through (29). [Law (27) is called the formula for *integration by parts.*]

(26)
$$\left| \int_a^b f \right| \le \int_a^b |f|.$$

(27)
$$\int_a^b f \cdot Dg = [f \cdot g]_a^b - \int_a^b g \cdot Df.$$

(28)
$$\exists \xi \int_a^b f \cdot |g| = f(\xi) \cdot \int_a^b |g| \land (a \le \xi \le b).$$

(29)
$$[x\epsilon(\underline{a \quad b}) \to f(x) \le g(x)] \to \int_a^b f \le \int_a^b g.$$

★35. When a force of magnitude F moves through a distance s in the direction of the force, the *work* done is defined as Fs. If the magnitude of the force depends on x (the distance measured from an origin in the line of action of the force), then the work done in going from $x = a$ to $x = b$ is defined as the integral from a to b of $F(x)$; that is, $W = \int_a^b F(x)\,dx$. Argue for the reasonableness of this definition.

★36. Suppose the force is given by $\sin x$. Find the work done in moving from $x = \pi$ to $x = 0$.

★37. In a physics textbook the following integral appears: $H = \int_0^T i^2 R\,dt$, where $i^2 R = I_m^2 R \sin^2 wt$ and T is the period of $\sin 2wt$. Show that $H = (1/2) I_m^2 R T$. [Here R, I_m, and w are independent of t. *Suggestion:* To find an indefinite integral of $\sin^2 wt$, use the fact that $\sin^2 wt = (1/2)(1 - \cos 2wt)$.]

ANSWERS TO EXERCISES

(b) Using another antiderivative would add the same constant to $F(b)$ and $F(a)$.   (f) $(\int_a^a f = -\int_a^a f) \to (\int_a^a f = 0)$.   (g) Interchanging the limits multiplies the integral by -1.   (h) The area of a line segment is zero.   (i) (8): Total flow is between what it would be if rate of flow had been at the minimum and at the maximum.   (9): There is some constant rate of flow at which total flow would have been the same; this rate is the average rate. [*Note:* The number H defined by (9) is called the *average value of* $f(x)$ *in* $(\underline{a \quad b})$.]   (10): At some time the rate of flow was equal to the average rate of flow.   (j) (13): Multiplying each ordinate by c multiples the area of each rectangle by c.   (14): The area under $f + g$ is the sum of the areas under f and g.   (15): The area below f from a to b equals the sum of the areas obtained by breaking up the region at some intermediate point.   (The law holds, however, whether or not $a \le c \le b$.)

(k) Multiplying the rate of flow by a number multiplies the total flow by that number. The total flow due to two flows is the sum of the total flows due to them separately. The total flow over a period of time is the sum of the flows over any two intervals into which the time interval is divided. (l) $(13)(c:-1)$ and (14). (m) It depended on a figure in which $f(x) > 0$ and on intuitively justified approximations.

Answers to Problems

1. 20. 3. 0. 5. 1/40. 7. Note that $D_x(x^2 + 1)^{3/2} = (3/2)(x^2 + 1)^{1/2}2x$ by the chain rule. Hence an antiderivative is $(2/3)(x^2 + 1)^{3/2}$. 9. $-1/3$. 11. Since $D \cos = -\sin$, an antiderivative is $-\cos$; 2. 13. 2/3. 15. $D \sin^3 = 3 \sin^2\cos$. Hence an antiderivative is $(1/3) \sin^3$. 17. Note that $D(I^2 + 1)^{3/2} = (3/2)(I^2 + 1)^{1/2}2I$. 19. $D \sin [I^2] = \cos [I^2]2I$. 21. $(m/2)(t_2^2 - t_1^2) + b(t_2 - t_1)$. 22. 125/6. 23. 0. The area below just balances the area above because the function is odd.

7–13 Exponential and logarithm. The function defined by $y = b^x$ is called the *exponential function to the base b*. We call it $\exp_b$ or b^I. We have defined b^x for all rational x (Section 6–8). To avoid imaginaries, we assume $b > 0$. The range of x in b^x was extended to all reals by (6–9–7), which we repeat here.

(1) **Def.** $b^x = \mathrm{lub} \{y \,|\, \exists r\, r \in Ra \wedge r < x \wedge y = b^r\}$ $(b \geq 1)$.

This defines b^x for irrational x in terms of b^r for rational r. It means, in effect, that we approximate b^x for irrational x by means of b^r for r a rational approximation of x. Now, the laws of exponents can be proved to hold for any real exponents, though we omit the proof. The domain of $\exp_b$ is Re. In Fig. 7–40 we show 2^I.

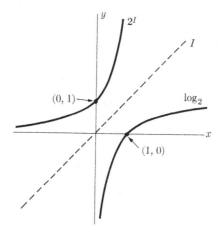

Figure 7–40

(a) What is 1^I? (b) What is the range of b^I? (c) Graph 10^I and 2^I together. (d) Why does b^I pass through $(0, 1)$ for any b?

It appears plausible from the graph that b^x increases with x when $b > 1$, and that it is continuous. This means that any horizontal line above the x-axis crosses it once and only once, i.e., that $y = b^x$ has just one solution for x for each positive y. From this it follows that the converse of b^I, defined by $x = b^y$, is also a function. We call this inverse the *logarithm to the base b*. It is defined by

(2) **Def.** $$\log_b = (b^I)*.$$

The graph of $\log_2$ is sketched in Fig. 7–40. Sometimes b^I is called the *antilogarithm to the base b*.

(e) Graph $\exp_3$ and $\log_3$ together. (f) What about $\exp_b$ for $b < 1$? (g) What are $\mathrm{Rge}\ (\log_b)$ and $\mathrm{Dom}\ (\log_b)$?

From (2) we have immediately

(3) $$(y = \log_b x) \leftrightarrow (x = b^y = \exp_b y).$$

Then from (5–10–8) and (5–10–9),

(4) $$\exp_b(\log_b y) = b^{\log_b y} = y,$$

(5) $$\log_b(b^x) = \log_b(\exp_b x) = x.$$

From the laws of exponents,

(6) $$b^u \cdot b^v = b^{u+v},$$

(7) $$(b^u)^v = b^{uv}.$$

From these we have

(8) $$\log_b(u \cdot v) = \log_b u + \log_b v,$$

(9) $$\log_b(u^v) = v \cdot \log_b u.$$

To prove (8) we write

(10) $$\log_b(u \cdot v) = \log_b(b^{\log_b u} \cdot b^{\log_b v}) \quad (4)$$

(11) $$= \log_b(b^{\log_b u + \log_b v}) \quad (6)$$

(12) $$= \log_b u + \log_b v \quad (5).$$

(h) Prove (9) similarly, and derive (13) through (16).

(13) $$\log_b(1/v) = -\log_b v,$$

(14) $$\log_b(u/v) = \log_b u - \log_b v,$$

(15) $$\log_b a = 1/\log_a b,$$

(16) $$\log_a x = \log_b x \cdot \log_a b.$$

The laws (8) through (16) make logarithms very valuable computational aids. If we have a way to find logarithms and antilogarithms of numbers, we can reduce multiplications to additions and finding powers to multiplications. The most convenient base for such purposes is 10. We write $\log x$ for $\log_{10} x$. Below is a table of $\log x$.

(17)

x 10^y	0.001	0.01	0.1	1	10	100	1000	10,000
$\log_x$ y	-3	-2	-1	0	1	2	3	4

(i) Justify the entries in (17).   (j) Argue that $2 < \log_{10} 157.3 < 3$, $-1 < \log_{10} 5 < 1$.   (k) $\log_2 4 = ?$ (l) $\log_4 2 = ?$ (m) $\log_2 16 = ?$ (n) $\log\sqrt{10} = ?$ (o) $\log_{0.5} 0.25 = ?$ (p) $\log_4 x = -1/2$, $x = ?$ (q) Verify that (8) through (16) hold in (17).

Table (17) serves to place the logarithm of any number to the base ten in an interval of unit length. For example, $\log_{10} 150 = 2 + d$, where $0 < d < 1$. The decimal part d can be found in tables and is called the *mantissa*. The integral part is called the *characteristic*. Logarithms to the base 10 are called *common*. Since

(18) $$\log (10 \cdot x) = \log 10 + \log x = 1 + \log x,$$

moving the decimal point in a decimal changes the characteristic of its common logarithm but has no effect on the mantissa. Hence, in looking up mantissas we ignore decimal points. Then the characteristic is determined by a knowledge of table (17), by noting that it is the exponent on 10 when the number is written in scientific notation (see Section 6–11), or else by the rule of Problem 4 below.

Table III in the Appendix gives three-digit approximations of the mantissas of common logarithms of two-digit numbers, the first digit of the number appearing at the left and the second at the top. Thus $\log 2 \doteq 0.301$, $\log 5 \doteq 0.699$, $\log 8.6 \doteq 0.934$. The characteristics of all these logarithms are 0, since the numbers 2, 5, and 8.6 all lie between 1 and 10. However, $\log 86 \doteq 1.934$, $\log 860 \doteq 2.934$, and $\log 8600 \doteq 3.934$, as can be seen from (17). On the other hand, $\log 0.86 \doteq -1 + 0.934$, $\log 0.086 \doteq -2 + 0.934$, $\log 0.0086 \doteq -3 + 0.934$. We could not

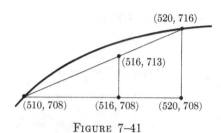

FIGURE 7–41

write −1.934 for −1 + 0.934. To avoid negative mantissas, which are not tabulated, we write log 0.86 = 9.934 − 10, log 0.086 = 8.934 − 10, and so on. This amounts to adding 10 to the characteristic, then subtracting 10 from the logarithm.

Find the common logarithm of: (r) 1,   (s) 1.2,   (t) 530,   (u) 5300, (v) 0.0053,   (w) 0.00053.   (x) Make a large-scale graph of log in the interval (1   10) by using Table III.

As is evident from the graph of log and from Table III, the curve is very closely approximated by straight line segments. Thus in Table III the difference between successive values of $\log x$ remains practically the same over a number of values of x, i.e., the slope of the curve is almost constant. We can use this fact to estimate logarithms by linear interpolation from Table III. For example, to find log 5.16 we note that log 5.1 = 0.708 and log 5.2 = 0.716. In Fig. 7–41 we show the points (5.1, 0.708) and (5.2, 0.716) on the curve, and the secant line. The point (5.16, y) on the secant can easily be found, because the equality of the slopes of the segments on the secant implies that the corresponding differences of x and $\log x$ are proportional. Hence $(5.16 − 5.1)/(5.2 − 5.1) = (y − 0.708)/(0.716 − 0.708)$ and $y = 0.708 + 0.005 = 0.713$. The calculation is conveniently arranged as follows.

$$x \qquad\qquad 10gx$$

(19)

$$\begin{array}{c|c} & 510 \\ & \downarrow\, 6 \\ 10 & 516 \\ & \downarrow\, 520 \end{array} \qquad \begin{array}{c|c} & 708 \\ h\,\downarrow & \\ & y \quad\big|\; 8 \\ & 716 \end{array}$$

Then $h/8 = 6/10$, $h = 48/10 \doteq 5$, $y = 708 + 5 = 713$. Note that decimal points are ignored and all numbers are written with three digits. Calculations are rounded off to the nearest third digit. (Compare the discussion in Section 4–11.)

Three-digit logarithms give roughly three-digit accuracy in computing. For more accurate work, tables of 4, 5, or even 15 or 20 digits are available.

Table III(a) in the Appendix is an expansion of Table III for $1.0 \leq x \leq$ 2.0, where the interpolation from Table III is less accurate. In deciding how many digits to keep in calculations we follow the rules given in Sections 6–11 and 6–12.

Find mantissas for the following numbers by interpolation in Table III: (y) 517,   (z) 515. Find the mantissas for the following numbers by interpolating in Table III and compare with the results given by Table III(a): (aa) 191,   (bb) 155.

To do computations with logarithms one must first decide on the procedure and make a form. For example, to find $A = (51.6) \cdot (0.238)$ we decide to use $(8)(u{:}51.6, v{:}0.238)$, then write down the form

$$\log 51.6 \quad = \ldots$$
$$+ \log 0.238 = \ldots$$

(20)
$$\overline{}$$
$$\log A \quad = \ldots$$
$$A \quad = \ldots$$

Then we find the logs, insert them, perform the addition to find $\log A$, then go into Table III "backward" (using it as a table of antilog x) to find A. From (19) and (17) $\log 51.6 = 1.713$. Similarly, $\log 0.238 = 9.376 - 10$. Adding, we find $\log A = 1.089$. Ignoring the characteristic, we look in Table III for 089. The table for interpolating in reverse is

(21)

	$\log x$			x	
		079			120
35		089 $\downarrow$ 10		$h\downarrow$	x 10
		114 $\downarrow$		130 $\downarrow$	

Hence $h/10 = 10/35$, $h = 100/35 \doteq 3$, and $x = 123$. Now we note that the characteristic of $\log A$ is 1. Hence $A = 12.3$. Interpolation in Table III(a) yields practically the same result.

Find the antilogarithms of the following: (cc) 0.883,   (dd) 1.155. Carry out the following computations, checking your work by arithmetic: (ee) 2(4.18), (ff) 4/3,   (gg) 2^3. Find the following, being sure to make a good form before beginning the calculation: (hh) (3.14)(78).

PROBLEMS

1. Graph $(1/2)^I$.

★2. Prove as follows that b^x increases with x when $b > 1$. Note that we wish to prove that $b^{x+h} > b^x$ for $h > 0$, that is, $b^x(b^h - 1) > 0$. Hence it is sufficient to prove $b^h > 1$ for $h > 0$. Why? To prove $b^{h.} > 1$, prove it first for $h \in N$ by induction. Then prove that $0 \le c \le 1 \to c^n \le 1$ for $n \in N$. Now prove that for $n \in N$ and $b > 1$, $b^{1/n} > 1$, by assuming the contrary, that is, $b^{1/n} \le 1$, and showing that it contradicts $b > 1$. Combine these results to show that $b^h > 1$ for $h \in Ra$. Argue for the plausibility of the same result for $h \in Re$.

3. Show that $\log_b x = [y \ni b^y = x]$.

4. Justify the following rules: Count the number of decimal places from the space at the right of the first significant digit to the decimal point, counting positive to the right and negative to the left; the characteristic is this integer. To place the decimal point in an antilogarithm begin at the place to the right of the first significant digit and count a number of places equal to the characteristic.

In Problems 5 through 14 find the logarithm of the given number.

5. 218.	6. 3.14.	7. 0.031.
8. 215.	9. 0.00891.	10. 51.3
11. 8.27	12. 0.0000194.	13. 0.101.
14. 0.0777.		

15. Show that a negative number does not have a logarithm that is a real number.

16. Why is interpolation less accurate when the number is near 1 in Table III?

In Problems 17 through 24 find the antilogarithm of each number.

17. 0.114.	18. 2.708.	19. 9.814–10.
20. 8.211–10.	21. 0.329.	22. 5.128.
23. 3.889–10.	24. 3.777.	

Use logarithms to compute the numbers in Problems 25 through 34. Make a form before starting computations, and approximate the answer by drastic rounding off.

25. $(3.14)(8.26)^2$.	26. $\sqrt{26}$.
27. $3\sqrt{153}$.	28. $(856)(35.2)^{-1}(0.0154)$.
29. $1/256$.	30. $(28.1)(252)/(27.35)$.
31. $(2/3)^5$.	32. $(1.06)^{15}$.
33. $(0.99)^{100}$.	34. $(0.999)^{100}$.

ANSWERS TO EXERCISES

(a) 1.   (b) Positive reals.   (c) They cross at $(0, 1)$. The first is below at the left and above at the right.   (d) Since $b^0 = 1$.   (e) They should be mirror images in the graph of I since they are inverses.   (f) The graph of $\exp_b$ is the mirror image in the y-axis of the graph of $\exp_{(1/b)}$, since $b^x = (1/b)^{-x}$.   (g) Range is all reals, domain is positive reals.   (h) For (9): $\log_b(u^v) = \log_b((b^{\log_b u})^v) = \log_b(b^{v \log_b u}) = v \log_b v$.   (13) is (9)$(u{:}v, v{:}{-}1)$.   (14) is (8)$(v{:}1/v)$.   (15) $\leftrightarrow$ $(a = b^{1/\log_a b}) \leftrightarrow (a^{\log_a b} = b) \leftrightarrow (b = b)$.   (16) Similar to proof of (15).

(i) Refer to (3).   (j) Since $a < c < b \rightarrow \log a < \log c < \log b$.   (k) 2.
(l) 1/2.   (m) 4.   (n) 1/2.   (o) 2.   (p) 1/2.   (q) Substitute for u and v.
(r) 0.   (s) 0.079.   (t) 2.724.   (u) 3.724.   (v) 7.724–10.   (w) 6.724–10.
(x) Similar to arc of $\log_2$ in Fig. 7–40.   (y) 0.714.   (z) 0.712.   (aa) 0.281.
(bb) 0.190.   (cc) 7.64.   (dd) 14.3.   (ee) Should agree when rounded to
three significant digits. Same for (ff), (gg), (hh).

ANSWERS TO PROBLEMS

1. Mirror image in y axis of 2^I. 3. (3). 5. 2.338. 7. 8.491–10. 9. 7.950–10.
11. 0.918. 13. 9.004–10. 15. Since $b^x > 0$ for $b > 0$. 17. 1.3. 19. 0.651.
21. 2.14. 23. 0.000000775, 25. Check by arithmetical calculation or use of
tables.

★7–14 Growth and decay. From the definition of the derivative and
the properties of log,

(1) $$D \log_b x = \lim_{h \to 0} [\log (x + h) - \log x]/h$$

(2) $$= \lim_{h \to 0} \left[(1/h) \log \left(\frac{x + h}{x} \right) \right]$$

(3) $$= \lim_{h \to 0} [(1/x)(x/h) \log (1 + h/x)]$$

(4) $$= (1/x) \lim_{h \to 0} [\log (1 + h/x)^{x/h}].$$

(a) Justify each step.

Evidently in order to find $D \log_b x$ we must evaluate the limit as $h \to 0$
of $\log_b(1 + h/x)^{x/h}$. Since $\log_b$ is everywhere continuous, we can do this
provided we can find the limit of $(1 + h/x)^{x/h}$. Note that as $h \to 0$,
$h/x \to 0$, $1 + h/x \to 1$, but $x/h \to \infty$. Hence as $h \to 0$, we have larger
and larger powers of numbers that get nearer and nearer one. The limit
is a certain irrational number, approximately equal to 2.72, which is
designated by the letter e. Formally,

(5) Def. $$e = \lim_{a \to \infty} (1 + 1/a)^a.$$

We omit the proof that this limit exists. However, the student may approximate it by expanding $(1 + 1/a)^a$ for large integral values of a by
the binomial theorem and observing that only a few of the terms in the
expansion contribute significantly to the result.
Letting $a = x/h$, $a \to \infty$ as $h \to 0$, and we have from (4)

(6) $$D \log_b x = (1/x) \lim_{a \to \infty} \log_b[(1 + 1/a)^a]$$

(7) $$= (1/x) \log_b[\lim_{a \to \infty} (1 + 1/a)^a]$$

(8) $$= (1/x) \log_b e.$$

(b) Justify (7) and (8).

Now suppose we choose e as the base. Since $\log_e e = 1$,

(9) $$D \log_e x = 1/x.$$

It is this result that makes e the most convenient base for scientific work. Logarithms to the base e are called *natural logarithms*. The logarithm to the base e is usually designated by ln, defined by

(10) **Def.** $\ln = \log_e$.

Table IV in the Appendix gives ln x in terms of x. By reading it backward we have a table of e^x.

From (9),

(11) $D \ln = I^{-1}$.

It is interesting to observe that prior to this we had no function whose derivative was $1/I$. Also, we were unable to find the antiderivative of $1/I$. Now we can find an antiderivative of any power function I^n.

Now, by applying (7–8–4) and recalling (7–13–2), we find

(12) $D e^I = e^I$ (that is, $D e^x = e^x$),

(13) $D \exp_b = \exp_b \cdot (\log_e b)$ (that is, $D b^x = b^x \cdot \log_e b$).

We have found a function that is its own derivative! This means that we can solve an equation of the form $Df(x) = f(x)$; that is,

(14) $(Df = f) \leftrightarrow (\exists c\, f = \underline{c} \cdot e^I)$.

More generally,

(15) $(Df = f \cdot g) \leftrightarrow (\exists c\, f = \underline{c} \cdot e^{D^{-1}g})$.

This reduces to (14) by $(g:1)$. To find (15) we might write the hypothesis in the form $(Df)/f = g$. But $D[\ln [f]] = (Df)/f$ by the chain rule. Hence this is equivalent to $D[\ln [f]] = D[D^{-1}g]$. Then by (7–9–7), $\ln (f) = D^{-1}g + \underline{C}$ where $\underline{C}$ is some constant. By (7–13–3) this is equivalent to $f = \exp (D^{-1}g + \underline{C}) = [\exp (\underline{C})][\exp (D^{-1}g)] = \underline{c} \cdot \exp (D^{-1}g)$. This argument holds only for f such that $f(x) > 0$, since the domain of ln is the positive reals. (See Problems 20 and 21.)

Imagine a population (human, animal, or botanical). It is natural to assume that the rate of growth is proportional to the size of the population y; that is, if x is the time and $y = f(x)$ for some f, $\mathsf{D}f = \underline{k} \cdot f$. This is on the assumption that the rate of reproduction per person and the death rate per person are fixed, so that k is the net rate of growth. Then from $(15)(g{:}k)$, $f = c \cdot \exp{(\mathsf{D}^{-1}\underline{k})} = \underline{c} \cdot \exp{(kI)}$, that is, $f(x) = ce^{kx}$ for some constant c. Letting the population be y_0 when $x = 0$, we have $y_0 = ce^0$, $c = y_0$, and $y = y_0 e^{kx}$.

(16) $(\mathsf{D}f = \underline{k} \cdot f) \to [f(x) = f(0)\ e^{kx}]$.

This is the law of growth (or decay if $k < 0$) of a population with a constant rate of growth k per unit.

(c) Suppose that the population rate of growth is 3%, that is, that $k = 3/100$. What will be the population after 3 years if it starts at 100,000,000?   (d) Suppose that the population decreases at the rate of 3%. Answer the same question as that in Exercise (c).   (e) Suppose that a body decays so as to lose 1/10 of its weight every 10 years. Find its weight in terms of time if its original weight was 5 pounds.   (f) What is the solution of $\mathsf{D}f(x) = kf(x)$ if $f(x_0) = y_0$, that is, if we are given a value of $f(x)$ corresponding to x_0 instead of to 0?

PROBLEMS

1. If interest is assumed to be compounded "instantaneously" at a rate i, the total deposit y "grows" at a rate given by $\mathsf{D}y = iy$. Find the instantaneous compound interest law; i.e., find f so that $y = f(t)$.

2. At instantaneous rate of 5% what is the value 5 years from the present of \$100?

3. What amount must have been deposited 5 years ago to amount to \$100 now at 5% compounded continuously?

4. Suppose that the rate of decay of a radioactive substance is $-k$, where $k > 0$, so that $\mathsf{D}y = -ky$, where y is the amount of the substance. Find the law of decay.

5. The *half-life* of radioactive material is the time in which one half of the material radiates away, i.e., the x for which $f(x) = f(0)/2$. Determine this in terms of the k in Problem 4. Graph f for $k = 1$.

6. The German psychologist Fechner (1801–87) assumed that for small increments ΔR and ΔS in the stimulus R and the response S (yes, that is the way the letters are used by psychologists!), $\Delta S = c \, \Delta R/R$, i.e., the change in response is proportional to the ratio of the change in the stimulus to the stimulus. Writing this $\Delta S/\Delta R = c/R$, letting $\Delta R \to 0$, and thinking of S as depending on R by some function given by $S = f(R)$, we have $\mathsf{D}S = c/R$. From this derive *Fechner's law*, $S = c \ln{(R/R_0)}$, where R_0, called the absolute threshold, is given by $f(R_0) = 0$.

7. When the electromotive force is cut off in a wire in which a current i is flowing, the rate of change of the current with time, $\mathsf{D}_t i$, is related to the re-

sistance R and the inductance L by $LD_t i + Ri = 0$. Find the current in terms of time if $i = i_0$ when $t = 0$.

8. When the motor of a boat is cut, its acceleration $D_t^2 s$ is related to its velocity $D_t s$ by $D_t^2 s = -kD_t s$. Let $D_t s = v$, so that the equation becomes $D_t v = -kv$, and solve to get v and finally s in terms of t if $s = s_0$ and $v = v_0$ when $t = 0$.

9. If a body is falling in a resisting fluid with velocity v, then $D_t v = A - Bv$, where A and B are fixed. Show that $v = (A/B)(1 - e^{-Bt})$ satisfies the equation.

★10. A more realistic assumption about rate of population growth in a limited environment is that it tends to decrease as the population approaches a maximum Y determined by the environment. (Imagine the growth of a fly colony in a bottle!) A possible hypothesis is $Dy = (1 - y/Y)ky$ or $Df(x)/(1 - f(x)/Y)f(x) = k$. Show that the left member is the derivative of $-\ln(1 - Y/f(x))$ and so solve this equation in order to find the *logistic law* of population growth, $y = Y(1 + ce^{-kx})^{-1}$, where $c = (Y - y_0)/y_0$ and $y_0 =$ the population at the time $x = 0$. Sketch the curve for $Y = 100{,}000{,}000$, $y_0 = 25{,}000{,}000$, and find the points of inflection and asymptotes. Show that $Dy \to 0$ as $x \to \infty$ and as $x \to -\infty$.

★11. Computations can be done with natural logarithms. However, (7–13–18) does not hold, and moving the decimal point changes the logarithm by the irrational number $\log_e 10$. Table IV in the Appendix gives the natural logarithms of some numbers. By reading it backward we have a table of e^I. The exponential function to the base e appears so frequently in scientific work that the "exponential function" and "exp" without indication of base mean $\exp_e$. Note, in contrast, that "log" without base means $\log_{10}$. We have $\ln = \exp^*$ and $\log = (\exp_{10})^*$. Make a large drawing of $\ln$ and $\exp$.

★12. Prove that the slope of $\ln$ at $(1, 0)$ is 1.

★13. From Table IV find $\ln 2$, $\ln 3$, and $\ln 6$.

★14. From Table IV verify that the slope of the secant joining two nearby points on $\ln$ is approximately $1/x$.

15. Find $D^2 \ln$ and explain why linear interpolation for finding $\log x$ is so accurate for large x but is less accurate for x near 1.

★16. Find the area under $1/I$ and between 1 and 3.

★17. What is $[D^{-1} I^{-1}]_0^x$?

★18. Show how (7–13–16) may be used to construct a table of natural logarithms if we have a table of common logarithms, or vice versa.

★19. Show that $i = (E/R)(1 - e^{-Rt/L})$ is a solution of the differential equation $LD_t i + Ri = E$.

★20. Prove

$$(17) \qquad\qquad D \ln |I| = I^{-1},$$

that is, $D_x \ln |x| = 1/x$. From this show that $D \ln |f| = Df/f$.

ANSWERS TO EXERCISES

(a) (1) by definition of derivative; (2) by properties of logarithms; (3) since $1/h = (1/x)(x/h)$; (4) by (7–3–7).   (b) Since $\log_b$ is a continuous function;

by definition (5).   (c) $y = 100{,}000{,}000e^{0.03t}$; $100{,}000{,}000e^{0.09}$, which can be evaluated by using Table IV backward.   (d) $100{,}000{,}000e^{-0.09}$.   (e) $5e^{0.01t}$ approximately.   (f) $y_0e^{k(x-x_0)}$.

ANSWERS TO PROBLEMS

1. $y = y_0e^{it}$.   3. $100e^{-0.25}$.   5. The solution of $0.5 = e^{-kx}$, that is, $x = 0.693/k$.   7. $i_0e^{-(R/L)t}$.   9. Differentiate and substitute.

★7-15 Parametric representation. We have occasionally defined relations by sentences of the form

(1) $\exists t \; x = g(t) \;\wedge\; y = h(t) \;\wedge\; t \in A.$

Usually the quantifier is omitted and the range A of the parameter t is left implicit, but the pair of parametric equations $x = g(t)$, $y = h(t)$, where A is the range of t, defines the relation

(2) $\{(x, y) \,|\, \exists t \; x = g(t) \;\wedge\; y = h(t) \;\wedge\; t \in A\}.$

(a) Review (4–5–5), Problem 20 in Section 4–5, Problem 14 in Section 5–8, Exercise (e) and Problem 18 in Section 5–14, and Exercise (e) in Section 5–15.

Usually the range of the parameter is some interval of the real numbers. For example, consider $x = t^2$, $y = t^3$, $t \in Re$. Solving the first equation for t and substituting in the second, we find

$$(y = x^{3/2} \;\vee\; y = -x^{3/2}) \;\wedge\; x \geq 0.$$

The curve is sketched in Fig. 7–42. The equations resulting from eliminating the parameter are helpful, but the parametric equations are more convenient in finding points, since the square and cube are easier to calculate than the three-halves power. We can also get considerable information about the graph from the parametric equations directly. For example, we see immediately that the minimum value of x is zero, that the curve is symmetric with respect to the x-axis, and that y takes all real values. Note that the points on the curve are in the same order as the corresponding values of t, so that the parametric equations map the real line on to the curve in a way that preserves order.

(b) Justify these statements.

Under what conditions does (1) define a function? If $g*$ is a function, then we can solve for $t = g*(x)$ in the first parametric equation and substitute in the second to find $y = h(g*(x))$, which defines a function. If we restrict the range of x to $\mathrm{Rge}(g)$, this function is the same as that defined by (1), that is, in this case

(3) $[\exists t \; x = g(t) \;\wedge\; y = h(t) \;\wedge\; t \in A] \leftrightarrow [y = h(g*(x)) \;\wedge\; x \in \mathrm{Rge}(g)]$

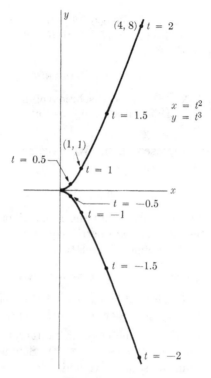

FIGURE 7–42

However, it is not necessary that $g*$ be a function in order that (1) define a function. For example, consider $x = t^2$, $y = t^2$, $t \in Re$. Eliminating t, we find $y = x \wedge x \geq 0$ even though $g*$ is not a function here.

(c) Show that $\exists t \; x = t^2 \wedge y = t^2 \wedge t \in Re \leftrightarrow y = x \wedge x \geq 0$. (d) Find and sketch the function defined by $x = \cos^2 t$, $y = \sin^2 t$, $t \in Re$. (e) Sketch the relation defined by $x = \cos^2 t$, $y = 2 \sin t$, $t \in Re$. Is it a function? (f) Sketch $x = \cos t$, $y = 3 \sin^2 t$, $t \in Re$. Is it a function? (g) Sketch $x = 2|t|$, $y = 1 - t$. Is it a function?

Often when the relation defined by parametric equations is not a function it can conveniently be expressed as the union of two functions. Indeed if $g*$ is the union of two or more functions, then (1) defines the union of the functions corresponding to each of these. For example, the relation of Fig. 7–42 is evidently $I^{3/2} \cup -I^{3/2}$. $I^{3/2}$ corresponds to the nonnegative values of the parameter, and $-I^{3/2}$ to the nonpositive values, since $(x = t^2 \wedge y = t^3 \wedge t \in Re) \leftrightarrow [(x = t^2 \wedge y = t^3 \wedge t \geq 0) \vee (x = t^2 \wedge y = t^3 \wedge t \leq 0)]$. The parametric representation is now expressed as the disjunction of two sentences in each of which the converse of the function giving x is also a function.

(h) Justify the above manipulations and statements.   (i) Exhibit the relation of Exercise (e) as the union of two functions.   (j) Exhibit the relation defined by $x = \sin t$, $y = \cos t$, $\wedge\ t \in (\underline{0\ \ \ } 2\pi)$ as the union of two functions. Sketch.

Whenever parametric equations define a function (or the union of functions) it may be of interest to find the derivative, if it exists. Since in this case we have $y = h(g*(x))$ and $x = g(t)$, the chain rule gives immediately $D_t y = D_x y D_t x$ or

(4) $$D_x y = D_t y / D_t x.$$

From (4) we can find the slope of a curve defined by parametric equations without eliminating the parameter. For example, applying (4) to $x = t^2$, $y = t^3$, we find $y' = 3t^2/2t = (3/2)t$. This formula gives us the slope of the graph of Fig. 7–42 at the point corresponding to t. When $t > 0$ it gives the values of the derivative of $I^{3/2}$, and when $t < 0$ it gives those of the derivative of $-I^{3/2}$. The point $(0, 0)$ lies on both functions. Strictly speaking there is no derivative there, since neither function is defined for $x < 0$. However, if we define a *right-hand derivative* by (7–4–1) with h restricted to positive values, $(3/2)t$ gives the right-hand derivative for both functions.

★(k) Formulate a definition of *right-hand limit* so that the right-hand derivative can be defined as the right-hand limit of the fraction in (7–4–1). (The notation used is $\lim_{x \to a+}$.)

In each case find $D_x y$ in terms of the parameter: (l) $x = 3t$, $y = t^3$; (m) $x = \sqrt{1 - t}$, $y = \sqrt{1 + t}$;   (n) $x = e^{-t}$, $y = t^2 e^t$;   (o) $x = 2 \sin t$, $y = 3 \cos t$.

We have seen above how to deal with given parametric equations. We now wish to suggest how it is convenient to introduce parameters in order to find defining sentences of relations. Suppose we wish to discover the path that will be followed by an object near the earth's surface (a rocket, for example) released with an initial velocity $\mathbf{v}_0$ and subject only to the force of gravity. (We ignore such other factors as air resistance.) We place a coordinate system with origin at the point of release and y-axis through the earth's center, as suggested in Fig. 7–43. Let $\mathbf{v}_0 = (v_0, \alpha_0)_p = (v_0 \cos \alpha, v_0 \sin \alpha)$ and $\mathbf{v} = (v_x, v_y)$, as indicated on the figure. It is a fact of physics that the components v_x and v_y of the velocity vector $\mathbf{v}$ at any time are independent of each other. Since there are by assumption no forces in the x-direction, we have $v_x = D_t x = v_0 \cos \alpha_0$ from which $x = (v_0 \cos \alpha_0)t$. In the y-direction, we have a freely falling body with initial velocity $v_0 \sin \alpha_0$ and acceleration $-g$. Solving this differential equation by the methods of Section 7–9, we find

$$y = (v_0 \sin \alpha_0)t - (1/2)gt^2.$$

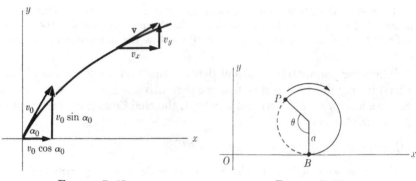

FIGURE 7–43   FIGURE 7–44

These are the equations of Problem 14 in Section 5–8, from which we can find out everything about the path.

(p) An airplane is flying on a course due west at an air speed of 275 mi/hr. The wind is blowing due south at a speed of 55 mi/hr. Find parametric equations of the path in terms of time.

PROBLEMS

In Problems 1 through 10 graph the relation defined by the parametric equations. Exhibit nonfunctional relations as unions of functions where convenient. Identify the parameter geometrically if possible. In graphing you may find it helpful to eliminate the parameter, but try to find out also as much as is convenient from the parametric equations. For example, the intercepts may be found by solving $g(t) = 0$ and $h(t) = 0$, symmetry may be investigated, vertical and horizontal tangents may be found by solving $g'(t) = 0$ and $h'(t) = 0$, $\mathbf{D}_{x}y$ may be determined from (4), and so on. Note that points found should be linked in order of the corresponding values of t.

1. $x = 2 \cos t,\ y = 3 \sin t,\ 0 \leq t < 2\pi$.
2. $x = 2 \sin t,\ y = 3 \cos t,\ t \in Re$.
3. $x = 2t + 1,\ y = 1,\ x \in (-\infty\_\_2)$.
4. $x = 1 - t,\ y = 3t + 5,\ x \in (-1\_\_1)$.
5. $x = 2t^3,\ y = t^2$.
6. $x = 1/t,\ y = 1/t^2$.
★7. $x = t \sin t,\ y = t \cos t$.
8. $x = 1 + t^3,\ y = 1 - t^3$.
9. $x = (1 + t)^{-1},\ y = (1 - t)^{-1}$.
★10. $x = 10 + e^{-t},\ y = \sin t$.

11. A bomb is dropped from an airplane flying 585 mi/hr. Find the path, neglecting air resistance.

12. Solve the same problem if the plane is diving at 700 mi/hr at an angle of 45° below the horizontal.

13. A circle rolls along the x-axis. What is the path described by a fixed point on its circumference? (*Suggestion:* Let the parameter be θ in Fig. 7–44,

where the point P starts at the origin so that OB has the same length as the arc PB.) The resulting curve is called a *cycloid*. Sketch it.

14. Find formulas giving the velocity vector of the point in Problem 13, assuming that the circle rolls with constant angular velocity. Prove that when the point is at the top of the circle it is moving twice as fast as the center. What happens when it is on the x-axis?

★15. Prove that the normal to the cycloid of Problem 13 always passes through the point of tangency B, and that the tangent passes through the other end of the diameter through B.

★16. A circle rolls on the inside of a circle of four times the radius. Show that the path of a point on the circumference is given by $x^{2/3} + y^{2/3} = a^{2/3}$, where the circle is given by $x^2 + y^2 = a^2$, and the path goes through the point $(a, 0)$. The curve is called a *hypocycloid*.

★17. Find parametric equations of the path of a point on a circle of radius a that rolls on the outside of a circle of radius b. It is called an *epicycloid*.

★18. Derive parametric equations for Problem 16 with radii a and b. What happens when $a = b/2$?

★19. Find parametric equations of the square whose vertices are $(1, 1)$, $(1, -1)$, $(-1, -1)$, $(-1, 1)$.

★20. Find parametric equations of the triangle with vertices (x_1, y_1), (x_2, y_2), and (x_3, y_3).

★21. Show that (2) is synonymous with $\{(x, y) \mid \exists t \, (x, y) = (g(t), h(t)) \wedge t \in A\}$ and $\{z \mid \exists t \, z = (g(t), h(t)) \wedge t \in A\}$. How would you interpret $\{(g(t), h(t)) \mid t \in A\}$? Define it.

22. Find the equation of the tangent line to $x = at/(1 + t^3)$, $y = at^2/(1 + t^3)$ at $t = 1$.

★23. Graph $x^3 + y^3 - 3xy = 0$ by introducing a parameter through the substitution $y = tx$ and so determining parametric equations of the locus. What is the interpretation of t? This curve is called the *folium of Descartes*.

★24. Graph $x = a(3 - t^2)/(1 + t^2)$, $y = at(3 - t^2)/(1 + t^2)$, with $a > 0$. Eliminate t and find a single defining equation. The curve is called the *trisectrix of Maclaurin*.

Answers to Exericses

(b) Since $t^2 \geq 0$ and $0^2 = 0$; for every point (t^2, t^3) we have a point $((-t)^2, (-t)^3) = (t^2, -t^3)$; and t^3 takes all real values.   (c) If there is a t for which $x = t^2$ and $y = t^2$, then certainly $y = x$ and $x \geq 0$. Conversely, if $y = x$ and $x \geq 0$, we can find a t, namely $\sqrt{x}$ for which $y = t^2$ and $x = t^2$.

(d) $\{(x, y) \mid y = 1 - x \wedge 0 \leq x \leq 1\}$.   (e) $\{(x, y) \mid 4x + y^2 = 4 \wedge 0 \leq x \leq 1\}$. No.   (f) $\{(x, y) \mid 3x^2 + y = 3 \wedge -1 \leq x \leq 1\}$. Yes.   (g) $\{(x, y) \mid x = 2|1 - y|\}$. No.   (h) $t \in Re \leftrightarrow t \geq 0 \vee t \leq 0$ and (2-5-4).

(i) $\{(x, y) \mid y = 2(1 - x)^{1/2} \wedge 0 \leq x \leq 1\} \cup \{(x, y) \; y = -2(1 - x)^{1/2} \wedge 0 \leq x \leq 1\}$.   (j) $(1 - I^2)^{1/2} \cup -(1 - I^2)^{1/2}$.   (k) In (7–3–1) omit the absolute value signs around $x - a$.   (l) t^2.   (m) $-(1 - t)^{1/2}(1 + t)^{-1/2}$.   (n) $-e^{2t}(t^2 + 2t)$.   (o) $-(3/2) \tan t$.   (p) $x = -275t + x_0$, $y = -55t + y_0$.

ANSWERS TO PROBLEMS

1. Ellipse. 3. Half-line. 5. $y = (x/2)^{2/3}$. 7. Spiral. 9. Hyperbola.
11. Rocket problem with $\alpha_0 = 0$, $v_0 = 585$. 13. $x = a(\theta - \sin \theta)$,
$y = a(1 - \cos \theta)$.

★7–16 **Arc length.** We now return to the problem of the length of arc discussed briefly in Section 7–1. Suppose that the curve is given by $x = g(t)$, $y = h(t)$, $a \leq t \leq b$. We partition the interval $(a \;\; b)$ into subintervals given by $a = t_0 < t_1 < t_2 < \ldots < t_{n-1} < t_n = b$. The points on the curve corresponding to the points of division are (x_0, y_0), $(x_1, y_1), \ldots, (x_i, y_i), \ldots, (x_n, y_n)$, where $x_i = g(t_i)$, $y_i = h(t_i)$. Imagine these successive points of division joined by straight line segments as in Fig. 7–2. The length of the ith segment is $\sqrt{(x_i - x_{i-1})^2 + (y_i - y_{i-1})^2}$, so that the length of the broken line approximating the curve is

$$(1) \qquad \sum_{i=1}^{i=n} \sqrt{(x_i - x_{i-1})^2 + (y_i - y_{i-1})^2}.$$

It seems natural to define the length of the curve as the limit of this sum as the maximum of the $(t_i - t_{i-1})$ approaches zero.

(a) Argue that if the maximum of the $(t_i - t_{i-1})$ approaches zero, then $n \to \infty$. (b) Is it certain that the limit of (1) always exists?

To put (1) in a form so that its limit may be calculated by evaluating a definite integral, we use the law of the mean (7–9–8) to write $x_i - x_{i-1} = g(t_i) - g(t_{i-1}) = g'(t_i')(t_i - t_{i-1})$ and $y_i - y_{i-1} = h'(t_i'')(t_i - t_{i-1})$. Substituting in (1) and simplifying, we have

$$(2) \qquad \sum_{i=1}^{i=n} \sqrt{[g'(t_i')]^2 + [h'(t_i'')]^2}\,(t_i - t_{i-1}).$$

This is *almost* in the form of (7–11–12) with $f(x): \sqrt{[g'(t)]^2 + [h'(t)]^2}$. It would be exactly in this form if $t_i' = t_i''$.

(c) Why does it make no difference that we use t instead of x here? (d) Why $t_i'' = t_i'$? (e) Why can we not assume $t_i'' = t_i'$?

If $t_i' = t_i''$ in (2), its limit as the maximum of the $t_i - t_{i-1}$ approaches zero would be just

$$(3) \qquad \int_a^b \sqrt{[\mathsf{D}_t x]^2 + [\mathsf{D}_t y]^2}\; dt.$$

It appears that as the lengths of the subintervals approach zero the difference of t_i' and t_i'' become negligible. Accordingly it seems reasonable to take (3) as the definition of the length of the arc of the curve given by the parametric equations.

(f) What is one advantage of using parametric equations in this discussion?

As an illustration of the use of (3) we find the familiar formula for the circumference of a circle. Letting the circle be defined by $x = a \cos t$, $y = a \sin t$, $t \in (0 \quad 2\pi)$, we have

(4) $$C = \int_0^{2\pi} a\sqrt{\sin^2 t + \cos^2 t} \, dt$$

(5) $$= a \int_0^{2\pi} dt = [at]_0^{2\pi} = 2\pi a.$$

(g) Find similarly a formula for the length of arc of a circle with central angle θ.

If the curve happens to be the graph of a function, $y = h(g*(x)) = f(x)$, then we have $y_i - y_{i-1} = f'(x_i')(x_i - x_{i-1})$, and (1) becomes

(6) $$\sum_{i=1}^{i=n} \sqrt{1 + [f'(x_i')]^2} \, (x_i - x_{i-1}),$$

from which we have

(7) $$\int_\alpha^\beta \sqrt{1 + (D_x y)^2} \, dx,$$

where $\alpha = g(a)$ and $\beta = g(b)$. Note that the limits are the values of x corresponding to the endpoints of the arc.

(h) Suppose the curve is the graph of $x = g(h*(y)) = F(y)$. Write an integral for the arc length in terms of y.   (i) Show that if s is the arc length measured along the curve given by $x = g(t)$, $y = h(t)$ from $(g(a), h(a))$ to $(g(t), h(t))$, then

(8) $$D_t s = [(D_t x)^2 + (D_t y)^2]^{1/2}.$$

(j) Show that if s is the arc length measured along the graph of $y = f(x)$ from $(x_1, f(x_1))$ to $(x, f(x))$, then

(9) $$D_x s = [1 + (D_x y)^2]^{1/2}.$$

(k) Find the length of the arc of the curve of Fig. 7–42 from $(1, 1)$ to $(4, 8)$ by using the parametric form. Check by using the equation $y = x^{3/2}$.

PROBLEMS

In Problems 1 through 10 find an integral giving the arc length on the curve between the indicated points. Sketch. Evaluate the integral if you can.

1. $y = x$, $x = 0$ to $x = 1$.   2. $y = x^2$, $x = 0$ to $x = 1$.

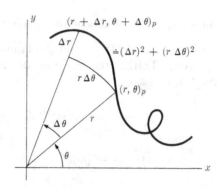

FIGURE 7-45

3. $y = |x|$, $x = -1$ to $x = 2$.

4. $y = (e^x + e^{-x})/2$, $x = 0$ to $x = 1$.

5. $y = \cos x$, $x = 0$ to $x = 2\pi$.

6. $y = \ln x$, $x = 1$ to $x = 10$.

7. $x = 3 \sin t$, $y = 3 \cos t$, $t = 0$ to $t = 5$.

8. $x = 3 \sin t$, $y = 2 \cos t$, $t = 0$ to $t = \pi/2$.

9. $y = x^2 + 4x - 5$, $x = -5$ to $x = 1$.

10. $x = t \sin t$, $y = t \cos t$, $t = 0$ to $t = 1$.

★11. Suppose a curve is given in polar coordinates parametrically by $r = g(t)$, $\theta = h(t)$, $a \leq t \leq b$. With the aid of Fig. 7–45 argue for the plausibility of the following formula for the arc length.

$$(10) \qquad \int_a^b \sqrt{(D_t r)^2 + (r D_t \theta)^2} \, dt.$$

★12. Similarly show that if the curve is given by $r = f(\theta)$, then the arc length from $(\theta_1, f(\theta_1))$ to $(\theta_2, f(\theta_2))$ is

$$(11) \qquad \int_{\theta_1}^{\theta_2} \sqrt{r^2 + (D_\theta r)^2} \, d\theta.$$

★13. Find a formula for the arc length if the curve is given by $\theta = f(r)$.

★14. Use polar coordinates to find the circumference of a circle.

★15. Find an integral giving the arc length of $r = 2\theta$ between $\theta = 0$ and $\theta = 1$.

★16. In terms of polar coordinates find the length of the straight line joining $(0, 0)$ and $(1, 1)$.

ANSWERS TO EXERCISES

(a) Since the number of intervals must be no less than $(b - a)$ divided by the maximum. (b) No. Not every curve has a length. (c) These variables are dummies. (d) In (7–11–12) $f(x_1')$ means the result of substituting for x in

$f(x)$, and we cannot substitute two different things. (e) The values asserted to exist by the law of the mean may not be the same. (f) The curve may not represent a function. (g) (4-9-1). (h) The integral from $h(a)$ to $h(b)$ of the square root of $1 + (D_y x)^2$. (i) (3) and (7-12-17). (j) (7) and (7-12-17).

ANSWERS TO PROBLEMS

1. $\sqrt{2}$. 3. $3\sqrt{2}$. 5. $\int_0^{2\pi}\sqrt{1 + \sin^2 x}\,dx$. 7. 15. 9. Integral of $(4x^2 + 16x + 17)^{1/2}$ from -5 to 1.

★7-17 The differential. The definite integral has a geometrical interpretation that is easily visualized—the net area between the graph and the x-axis. The derivative, on the other hand, is a little elusive. True, it may be conceived as the rate of change of the dependent variable with respect to the independent variable or as the slope of the line tangent to the graph. But there is no immediately evident geometric entity whose measure is the derivative.

As pointed out in Section 7-4, the derivative is equal to the change that would occur in the dependent variable if the rate of change were constant and the independent variable changed by one unit. This is illustrated in Fig. 7-46. Since the slope of the tangent line T is equal to the derivative, we have $\overline{AB}/1 = Df(x)$ and $\overline{AB} = Df(x)$. If f were a linear function we would have $Df(x) = f(x + 1) - f(x) = \Delta f(x)$. But ordinarily $\Delta f(x) - Df(x) = \overline{BC} \neq 0$. The derivative is an approximation to the actual change in the function when the independent variable increases by one unit. Economists find this a convenient way to think of the derivative, because in economics one unit is often "relatively small" and $\overline{BC} \doteq 0$. However, 1 may often be a "large" change, i.e., the error in thinking of $Df(x)$ as the change in $f(x)$ may be "large."

In Fig. 7-47 we indicate an unspecified increment Δx in the independent variable. Again the slope of the tangent line is equal to the derivative, that is, $\overline{AB}/\overline{PA} = Df(x)$, or $\overline{AB} = Df(x)\,\Delta x$, as indicated in the figure. Here $Df(x)\,\Delta x$ is an approximation to the change in $f(x)$, $\Delta f(x)$. The error,

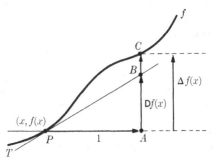

FIGURE 7-46

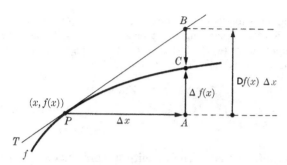

FIGURE 7-47

$\overline{BC} = \Delta f(x) - \mathsf{D}f(x) \, \Delta x$ depends on Δx, but evidently approaches zero as $\Delta x \to 0$. More than that, the error approaches zero faster than Δx in the sense that $\overline{BC}/\Delta x = [\Delta f(x)/\Delta x - \mathsf{D}f(x)] \to [\mathsf{D}f(x) - \mathsf{D}f(x)] = 0$ as $\Delta x \to 0$, This means that as $\Delta x \to 0$, the error is small compared with Δx.

We adopt the symbol $df(x)$ for $\mathsf{D}f(x) \, \Delta x$. In Fig. 7–48 we show the increment Δx in x, the increment $\Delta f(x)$ in $f(x)$ and $df(x)$. Note that $df(x)$ is the directed distance to the tangent line and is an approximation to $\Delta f(x)$.

(a) Draw a figure in which $\Delta x < 0$, $0 < \Delta f(x) < df(x)$. Label it carefully to correspond with the figures in the book.   (b) Do the same for $\Delta x > 0$, $\Delta f(x) < df(x) < 0$.   (c) Do the same for $\Delta x > 0$, $df(x) < \Delta f(x) < 0$. (d) Do the same for $\Delta x > 0$, $\Delta f(x) < 0$, $df(x) > 0$.

It is obvious from the above discussion that "$df(x)$" is not a constant. Its value depends on x and Δx as well as on f. It gives an approximation to $\Delta f(x)$ that depends on the function, the point on it, and the change in the independent variable. We shall sometimes use the notation "dy" in place of "$df(x)$" to suggest an approximation to the increment Δy, as indicated in Fig. 7–49. Then we write $dy = \mathsf{D}f(x) \, \Delta x$. Evidently, for a given f, the variables here are "x," "dy," and "Δx." For fixed x and hence fixed $\mathsf{D}f(x)$, we have a direct variation in which the constant of proportionality is $\mathsf{D}f(x)$, that is, the direct variation $\mathsf{D}f(x)I$ in which "dy" is the dependent variable and "Δx" is the independent variable.

(e) Find dy and Δy for $y = x^2$, $x = 2$, $\Delta x = 1$, 0.1, and 0.01.   (f) Do the same for $y = \ln x$, $x = 1$, $\Delta x = 2, 1, 0.5, 0.1, 0.01$.   (g) Estimate $(2.000001)^2$ by using the fact that $dy \doteq \Delta y$.   (h) Similarly estimate $\ln (2.0035)$.

A very interesting result appears if we apply the chain rule to find $df(g(t))$. Suppose $y = f(x)$ and $x = g(t)$, so that $y = f(g(t))$. By the chain rule, $\mathsf{D}_x y = \mathsf{D}f(g(t))\mathsf{D}g(t)$. Hence $dy = \mathsf{D}f(g(t))\mathsf{D}g(t) \, \Delta t$. But we

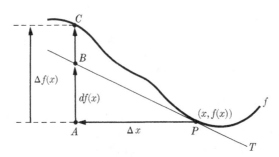

FIGURE 7–48

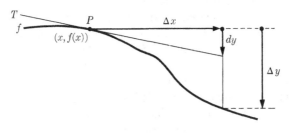

FIGURE 7–49

have also $dy = Df(x)\,\Delta x$ and $dx = Dg(t)\,\Delta t$. Now for given values of t and Δt, we have corresponding values of x and Δx for which we should like the two formulas for dy to give the same result. But the first one becomes $dy = Df(x)\,dx$ if we replace $Dg(t)\,\Delta t$ by dx and $g(t)$ by x. Now we have $f'(x)\,dx = f'(x)\,\Delta x$, which implies $dx = \Delta x$. This is a contradiction, since there is no reason to expect $dx = g'(t)\,\Delta t$ to be the same as $\Delta x = g(t + \Delta t) - g(t)$.

The above contradiction can be eliminated if we use "dx" in place of "Δx." Accordingly we write $dy = Df(x)\,dx$ according to the definition

(1) **Def.** $[dy = Df(x)\,dx] = [D_x y = Df(x)]$

We call dy and dx the *differentials* of y and x. By *definition*, to say that the differential of y is given by a certain expression times the differential of x is the same as to say that the derivative of y with respect to x is given by this expression. For example, $dy = 2x\,dx$ means the same as $D_x y = 2x$. Definition (1) does not define "dy" or "dx." These are simply variables.

(i) $dy = e^x dx$, $D_x y = \,?$ (j) $y = (1 - x)^2$, $dy = \,?$

Now, if we *interpret dx* as the increment Δx, then dy is *approximately* the corresponding increment $\Delta y = f(x + \Delta x) - f(x)$. This is the convenient interpretation if we think of x as the independent variable. On the other hand, if we think of x in its turn as depending on a third variable t, according to $x = g(t)$, then we have by (1) $dx = Dg(t) \, dt$ and $dy = Df(x) \, dx$. We get $dy = Df(g(t))Dg(t) \, dt$ either by replacing dx by $Dg(t) \, dt$ in the second equation or by applying (1) directly to $y = f(g(t))$ and using the chain rule.

Clearly if we know the differential of a function we also know its derivative, and conversely. It appears that nothing new has been added. But the notation has the advantage that it suggests the relation between the change in the independent variable and the change in the dependent variable. More precisely we may think of the values of the differential as approximating the changes in the dependent variable corresponding to given increments in the independent variable. This advantage is important to scientists who find it more convenient to think in terms of small changes than in terms of rate of change. If $dx \neq 0$, $[dy = Df(x) \, dx] \leftrightarrow [dy/dx = Df(x)]$. Suppose y is distance, x is time. Then the left member of the last equation suggests the change in distance (actually, dy is only approximately the change in distance) divided by the change in time, i.e., the velocity. Also $dy = Df(x) \, dx$ says that the distance traveled (approximately dy) equals the speed ($Df(x)$) times the time elapsed (dx). Moreover, as we have seen the error is small relative to dx, so that scientists can safely *think* in terms of small changes in order to get results that are precisely correct in terms of rates of change.

Another reason for the use of the differential notation is that many results become "obvious" by mechanical rules of manipulation if we use differentials. For example, suppose we have the parametric equations $x = g(t)$, $y = h(t)$. We write $dx = g'(t) \, dt$, $dy = h'(t) \, dt$, and by division $dy/dx = h'(t)/g'(t)$, which is just (7-15-4). Of course, this is not proof of (7-15-4)! We have simply defined differentials so that they give the same result. Now that we have defined them, we may treat the "dy" and "dx" in "dy/dx" as separate variables to be manipulated like numbers. The way in which we have formulated the definition makes this possible without contradiction.

(k) Clearly, in terms of differentials $dy/dx = (dy/du)(du/dx)$. Why? Is this consistent with our previous results? (l) In Fig. 7–38 show that the area A equals the differential $dF(x)$ of the total area $F(x)$.

As suggested by the last exercise, the differential notation has conceptual advantage in relation to the definite integral. If we compare the two expressions

(2) $$\sum f(x_i) \, \Delta x_i \quad \text{and} \quad \int_a^b f(x) \, dx,$$

we see that the second is reminiscent of the first. Indeed, in setting up a definite integral we may think in terms of small increments but write in terms of differentials. For example, work W is defined as force times distance Fs when the force is constant. If the force varies, we may think as follows: the force is almost constant for a short distance ds, so the increment dW of work in that short distance is about Fds, and the total work done is the sum of such increments. But $dW = Fds$ implies $dW/ds = F$, $W = \int Fds$, and $W_a^b = \int_a^b Fds$, where W_a^b is the work done in going from $s = a$ to $s = b$. Thus we *think* in terms of small changes and sums, but we *write* differentials and integrals. The notation is so designed that the results are correct, *provided* our intuition leads us to use differentials properly in a given problem.

(m) Compare the discussion of (7–11–8). (n) Discuss and show $F'(x) = \int dF(x)$. (o) Suppose an organism's weight W depends on time t. Interpret $W = \int_0^t dW$ in terms of small changes in W. Interpret $W = \int_0^t W'dt$ in terms of small changes in time.

For reasons indicated above and for others, one of which we indicate in the next section, the differential notation is very widely used. We did not introduce it sooner because it is a notational device that distracts from the basic ideas of calculus and may lead to serious misunderstanding. Indeed, it is largely a device for doing calculus manipulations without thinking in terms of the essential underlying idea of limits. It was invented by Leibniz (1646–1716) at a time when the limit concept and the fundamental ideas of calculus were not understood and one had to depend upon intuition to avoid errors. Accordingly, the differential notation was a very helpful device and still is useful. The person who continues to use calculus should use the derivative or differential notation according to convenience. Thus he may write $d \cos x = -\sin x \, dx$ or $\mathsf{D} \cos x = -\sin x$ (or $\mathsf{D} \cos = -\sin$) according to convenience.

(p) What is wrong with $d \cos x = -\sin x$ or $d \cos = -\sin$?

On the other hand, the student should be aware that the manipulative advantages of the differential are very special. It does not follow that this notation can be extended with equal advantage. For example, the differential notation d^2y/dx^2 is often used for $\mathsf{D}_x^2 y$. Suppose we define a second differential by $d^2y = \mathsf{D}_x^2 y (dx)^2$. Now if $y = f(x)$, $x = g(t)$, and $y = f(g(t))$, then by the chain rule $\mathsf{D}_t y = f'(g(t))g'(t)$, $\mathsf{D}_t^2 y = f'(g(t))g''(t) + f''(g(t))(g'(t))^2$, and $d^2y = f'(g(t))g''(t)(dt)^2 + f''(g(t))(g'(t))^2(dt)^2$. But directly from the definition, $d^2y = f''(x)(dx)^2 = f''(x)(g'(t)\,dt)^2 = f''(x)(g'(t))^2(dt)^2$. This contradiction suggests that the differential notation does not apply to higher derivatives with any advantage. Though we may safely think of dy/dx as the ratio of two differentials, we cannot safely think of d^2y/dx^2 as the ratio of d^2y and dx^2, since with this latter interpretation our results depend on what variables we use! Instead,

d^2y/dx^2 must be thought of as $\dfrac{d}{dx}\dfrac{dy}{dx}$, that is, as the result of applying

$\dfrac{d}{dx}$ twice. The $\mathsf{D}_x^2 y$ (or y'') notation is much more convenient.

(q) Did we solve the problem posed in the first paragraph of the section by finding a visualization for the derivative? (r) What would be the objection to defining the second differential as the differential of the first differential? ★(s) How would it work to define the second differential as the differential of the first derivative? (t) Carry through the differentiations in the previous paragraph.

PROBLEMS

1. When does $dy = \mathsf{D}_x y$?
2. Show that if f is a constant function, $df = 0$.
3. For what functions does $dy = dx$?
4. Show that $d(uv) = u\,dv + v\,du$.
5. Use the result in Problem 4 to show that the relative error in a product is approximately the sum of the relative errors in the factors. (Note that if we think of u and v as the true values and du and dv as the errors, du/u is the relative error.)
6. Show that $d(1/y) = -dy/y^2$.
7. Use differentials to estimate $(8.993)^2$.
8. Estimate $(5.0003)^3$.
9. Estimate $\sin 0.45$ by differentials and compare with the estimate obtained by interpolation in Table II.
10. Why is it reasonable from an intuitive point of view that we should have $dy = 0$ at a maximum or minimum if we think of dy as the change in y when x changes slightly?

In Problems 11 through 16 find dy.

11. $y = \sin x^2$.   12. $y = e^{-x^2/2}$.
13. $y = \ln(\sin x)$.   14. $x^2 + y^2 = 4$.
15. $x = \cos y + e^{g(x)}$.   16. $y = (a - x^2)^5 - x^2 \ln h(x)$.

17. In terms of differentials it appears obvious that $\mathsf{D}_y x = 1/\mathsf{D}_x y$ since $dy/dx = 1/(dx/dy)$. Why is this justified by (7-8-5)?

18. Often the simplest differential equations are written in terms of differentials instead of derivatives. Indeed, this is the origin of the name. Solve the differential equation $dy/y = dx/x$.

★19. The derivative was at one time called the "differential coefficient." Explain.

★20. Before the development of a logically satisfactory theory of calculus, the derivative was conceived of as the ratio of two very small changes (differentials) that were smaller than any finite quantities yet not actually zero. Criticize this formulation and explain why it does nevertheless serve for practical purposes in elementary calculus.

21. Show that the relative error in a quotient is approximately the difference of the relative errors.

22. Show that the relative error of an approximation is cut approximately in half when we take the square root.

★23. Show that the relative error in the nth power is approximately n times the relative error in the base.

★24. Show that the relative error in $\ln x$ equals approximately $1/\ln x$ times the relative error in the number x.

25. Rework some of the Problems in Section 7–15, using differentials.

★26. Prove that

$$(3) \qquad (ds)^2 = (dx)^2 + (dy)^2,$$

where ds is the differential of arc length. Draw a sketch showing the geometrical interpretation of ds.

★27. Show how (7–16–7) can easily be found from (7–16–3) by using differentials.

★28. Show that in polar coordinates $(ds)^2 = (dr)^2 + r^2(d\theta)^2$. Draw a sketch showing ds in this case.

★29. When a string is stretched an amount x, the force required is kx, where k is the force constant. Find dW, the differential of the work done in moving a distance dx, and so set up and evaluate an integral for the work done in stretching a spring from $x = 0$ to $x = b$.

★30. Show, using differentials, that $a = vD_x v$, where $v = D_t x$ and $a = D_t^2 x$.

★31. In what sense may the equation $dy = y' dx$ be interpreted as the equation of the tangent line?

★32. By definition the work done by a constant force F exerted through a distance x in the direction of the force is Fx. The average power $\overline{P}$ is defined as the work divided by the time, Fx/t. In a textbook on physics we find, "If the time interval is made extremely short" then this last formula becomes $F dx/dt$, so that the instantaneous power P is Fv, where v is the velocity. Explain this. Derive $P = Fv$, by using derivatives or differentials, from the definition $P = dW/dt$. (Incidentally, in the same book, this definition is justified by saying, "If the rate of doing work is not uniform, the power at any instant is the ratio of the work done to the time interval, when both are extremely small.")

★33. Pressure on a surface is defined as the force on the surface divided by the area. However if the force is not uniformly distributed, this merely gives average pressure. In Sears and Zemansky's *University Physics* we find, "We ... define the pressure at any point as the ratio of the normal force dF exerted on a small area dA including the point, to the area dA." Explain.

★34. For further examples of reasoning using differentials, see in Sears and Zemansky's *University Physics* the discussions of Pascal's Law, forces against a dam, specific heat, thermal conductivity, the differential form of the first law of thermodynamics, electric intensity, electrical potential energy, etc. Note that in every case the reasoning is in terms of "small" or "infinitesimal" changes but that the result is a derivative or integral expressed in terms of differentials.

ANSWERS TO EXERCISES

(a) through (d) There are still other possibilities! (e) $dy = 4$, 0.4, 0.04; $\Delta y = 5$, 0.41, 0.0401. (f) $dy = 2$, 1, 0.5, 0.1, 0.01; $\Delta y = 1.099$, 0.693, 0.405, 0.095, interpolation is not accurate for ln (1.01) in Table IV, but a more accurate table yields 0.00995. (g) $y + \Delta y \doteq y + dy = f(x) + df(x)$, with $f(x) = x^2, x = 2, dx = 0.000001; 4.000004$. (Check error by squaring 2.000001!) (h) $f(x) = \ln x, dx = 0.0035, x = 2; y + dy \doteq 0.693 + 0.00175 \doteq 0.695$. (i) e^x. (j) $-2(1 - x) dx$. (k) Rule for multiplying fractions. Yes, it is just the chain rule! (l) $dF(x) = f(x) dx$, since $F' = f$, and we interpret dx as Δx here. (n) It is the same as $F(x) = \int F'(x) dx$, which is true by definition. It suggests that $F(x)$ is the sum of the changes in $F(x)$.

(o) W at any time is the sum of the increments to the weight from the beginning to the time. An increment is the rate of change of W times the time interval. This amounts to thinking of a continuous process in terms of small finite changes. (p) It is a mixture of the two notations. A differential always ends with the differential of some variable; $\mathsf{D} \neq d!$ (q) We indicated two, of which the first was the special case of the second for $dx = 1$. (r) The first differential is a function whose domain consists of pairs of real numbers, so this would not make sense.

ANSWERS TO PROBLEMS

1. When $dx = 1$. 3. $I + c$. 5. Divide through by uv. 7. $dx = -0.007; y + dy = 80.874$. 9. 0.43, the same for both methods to two significant figures. 11. $2x \cos x^2 \, dx$. 13. $\cot x \, dx$. 15. $dx = -\sin y \, dy + e^{g(x)}g'(x) \, dx$ and solve for dy. 17. Here we are not interchanging variables, so (7–8–5) becomes $\mathsf{D}f*(y) = \mathsf{D}_y x = 1/\mathsf{D}f(f*(y)) = 1/\mathsf{D}f(x)$, since $f*(y) = x$ and $y = f(x)$. 21. Find differential of quotient and divide by the quotient.

★**7–18 Manipulation of integrals.** In Section 7–9 we indicated that in practice integrations are performed by using tables of integrals. In this section we discuss some of the manipulations that are involved.

In any table of integrals we find the formula

(1) $$\int \frac{dx}{\sqrt{1 - x^2}} = \text{Arcsin } x \qquad |x| < 1.$$

Of course, this expression is by definition synonymous with (2) and (3).

(2) $$\mathsf{D}_x \text{ Arcsin } x = (1 - x^2)^{-1/2},$$

(3) $$d \text{ Arcsin } x = (1 - x^2)^{-1/2} dx.$$

Forms (2) and (3) have the advantage of being identities to which we are sure that we may apply the Rule of Substitution, whereas the left member of (1) stands for any one of many formulas and the equality means that any antiderivative of $(1 - x^2)^{-1/2}$ is Arcsin x plus some constant.

(a) Why $|x| < 1$ in (1)?   (b) Write (2) using the identity function I, without using the independent variable.   ★(c) If you did not do Problem 11 in Section 7–8, derive (2) by writing $x = \sin y$, $\mathbf{D}_x x = \mathbf{D}_x \sin y$, $1 = (\cos y)\mathbf{D}_x y$, and observing that $\cos y = (1 - \sin^2 y)^{1/2}$.   (d) Why do we take the positive square root in this last sentence?

Suppose we wish to find the antiderivative of $(1 - 4I^2)^{-1/2}$, that is,

(4) $$\int \frac{dx}{\sqrt{1 - 4x^2}} = ?$$

Comparing it with (1) we might be tempted to complete (4) by substituting $(x{:}2x)$ in the integrand in (1) to get the alleged indefinite integral Arcsin $2x$. But $\mathbf{D}$ Arcsin $2x = 2(1 - 4x^2)^{-1/2}$ by (2) and the chain rule, so that this result is wrong. Of course, we see how to correct it by a factor of 2 to get the correct integral $(1/2)$ Arcsin $2x$. However, the fact that our substitution gave a wrong result indicates that either it or our notation is faulty.

The error will become clear if we substitute $(x{:}2x)$ in (2) and (3) to get

(2′) $\mathbf{D}_{2x}$ Arcsin $2x = (1 - 4x^2)^{-1/2}$,

(3′) d Arcsin $2x = (1 - 4x^2)^{-1/2} d(2x) = 2(1 - 4x^2)^{-1/2}$.

The correct result can be found from (2′) by noting that $\mathbf{D}_x$Arcsin $2x = \mathbf{D}_{2x}$Arcsin $2x\mathbf{D}_x 2x$, or from (3′) by dividing both members by 2. This suggests that we would have obtained the correct answer had we substituted $(x{:}2x)$ throughout in (1), including in the differential. In short, we would not have gone astray here if we had followed the Rule of Substitution mechanically.

This may not seem surprising, since we expect that a correct application of the Rule of Substitution would not lead to error. However, the expectation is justified *only* if our notation is well conceived. Suppose, for example, that we write (2) $\mathbf{D}$ Arcsin $x = (1 - x^2)^{-1/2}$ without explicit indication of the variable of differentiation. Then substitution throughout here would lead to $\mathbf{D}$ Arcsin $2x = (1 - 4x^2)^{-1/2}$, which is incorrect if the differentiation is supposed to be with respect to x. We see that the Rule of Substitution in integrals may lead to error unless the differential notation is used or the variable of integration is indicated explicitly and subjected to substitution throughout.

(e) Is the same caution applicable to derivative formulas?

The following theorem indicates the way in which substitution can be used safely in integrals.

(5) $\left[\int f(x)\, dx = F(x) \right] \leftrightarrow \left[\int f(g(x))g'(x)\, dx = F(g(x)) \right].$

The proof is immediate if we note that the two members are synonymous with $F'(x) = f(x)$ and $D_x F(g(x)) = f(g(x))g'(x)$. Since by the chain rule $D_x F(g(x)) = F'(g(x))g'(x)$, the first member implies the second by replacement. Also the second can be true only if $F' = f$. But since $dg(x) = g'(x)\,dx$, (5) tells us that *we can substitute in an indefinite integral formula for the variable of integration, provided we do it throughout, including the differential.*

Of course, we are not obliged to use the differential notation. We can write (5) in the form

$$(5')\qquad\qquad \left(\int f = F\right) \leftrightarrow \left(\int f[g]g' = F[g]\right).$$

Then we can get new integration formulas by applying (5') with appropriate choice of g. Since $f[I] = f$ and $F[I] = F$, the symbolic manipulations will be similar to those using the differential notation, though somewhat more concise. However, since the differential notation is almost universal, we shall use it below. The reader may find it instructive to repeat the work, using the names of functions instead of names of images and differentials.

In applying (5) it is customary to use a new variable in the substitution (which amounts to expressing the integrand in parametric form), carry through the integration, and then resubstitute to express in terms of the original variable. Thus we would compute (4) as follows.

$$(6)\qquad \int \frac{dx}{\sqrt{1 - 4x^2}} = \int \frac{d(t/2)}{\sqrt{1 - 4(t/2)^2}} \qquad (x = t/2)$$

$$(7)\qquad\qquad\qquad = (1/2)\int \frac{dt}{\sqrt{1 - t^2}} \qquad \text{since } d(t/2) = dt/2$$

$$(8)\qquad\qquad\qquad = (1/2)\,\mathrm{Arcsin}\,t \qquad (1)(x\!:\!t)$$

$$(9)\qquad\qquad\qquad = (1/2)\,\mathrm{Arcsin}\,2x \qquad (t = 2x).$$

We chose the transformation $x = t/2$ so that we would get an integral identical with (1) except for the variable used. We thus transformed the integral to make it like the known formula, rather than the reverse. This is perfectly justified by the logical equivalence in (5). Also, (5) justifies the equation (6) because by (5) the left member equals some $F(x)$ if and only if the right member equals $F(t/2)$, and these are identical for all x and t if $x = t/2$.

As a more complicated example, consider the following manipulations, where we choose $x = \sin t$ because it promises the elimination of a square root.

(10) $\int (1 - x^2)^{1/2}\, dx = \int (1 - \sin^2 t)^{1/2} \cos t\, dt$ $(x = \sin t)$

(11) $= \int \cos^2 t\, dt$

(12) $= \int \dfrac{1 + \cos 2t}{2}\, dt$ $(4\text{-}12\text{-}25)$

(13) $= \left(\dfrac{t}{2}\right) + \left(\dfrac{1}{4}\right) \sin 2t$ (check by differentiating)

(14) $= \left(\dfrac{t}{2}\right) + \left(\dfrac{1}{2}\right) \sin t \cos t$

(15) $= \left(\dfrac{1}{2}\right) \text{Arcsin } x + \left(\dfrac{1}{2}\right) x\sqrt{1 - x^2}$

(since $x = \sin t$).

(f) Justify (14).

In these manipulations there are certainly some dubious steps. For example, if $x = \sin t$, it does not follow that $(1 - x^2)^{1/2} = \cos t$. The most we can say is that $\cos t = \pm(1 - x^2)^{1/2}$. Nor does it follow that $t = \text{Arcsin } x$. However, such matters are not important *if* we check the final result by differentiation. The manipulations are to *find* an integral; only the differentiation proves that the discovery is correct. In the above case we have

(16) $\mathsf{D}[(1/2) \text{Arcsin } x + (1/2)x(1 - x^2)^{1/2}]$

$= (1/2)(1 - x^2)^{-1/2}$

$+ (1/2)(1 - x^2)^{1/2}$

$+ (1/4)x(1 - x^2)^{-1/2}(-2x)$

(17) $= (1 - x^2)^{1/2}.$

(g) Simplify the expression in (16) to find (17).   (h) What would happen if we had chosen the negative square root?   (i) Does this show that our final result holds only for certain values of x?

Evaluate the following integrals, using the form (6) through (9) and (10) through (15):   (j) $\int(1 - 9x^2)^{-1/2}dx$,   (k) $\int(1 - 4x^2)^{1/2}dx$.

(l) Repeat (10) through (15), but using the substitution $x = \cos t$.

PROBLEMS

1. To the question, "What is an indefinite integral of $\cos x$?" a student gave $\int_0^x \cos x\, dx$ as an answer. Is this correct? Do you think this is what the professor

wanted? Can you formulate the question more precisely so that the student could not dodge the issue so easily?

2. Show that

(18) $$\int cf = c\int f,$$

(19) $$\int (f + g) = \int f + \int g,$$

and explain precisely what the laws mean.

In Problems 3 through 10, make the indicated substitution, find an antiderivative in terms of the new variable, and then express in terms of the original variable according to the pattern in the section. Make a list of integral formulas by utilizing previously found formulas for derivatives, and add to it as you go along. In this way you will make a small integral table of your own.

3. $\int (a^2 - x^2)^{-1/2}\, dx,\ x = a \sin t.$

4. $\int x(ax^2 + c)^{-1/2}\, dx, t = ax^2 + c.$

5. $\int \dfrac{x\, dx}{a^2 + x^2}, t = a^2 + x^2.$

6. $\int x(1 + x)^{1/2}\, dx, t = (1 + x)^{1/2}.$

7. $\int \dfrac{dx}{1 - x}, t = 1 - x.$

8. $\int \tan x\, dx, t = \cos x.$

9. $\int \cot x\, dx, t = \sin x.$

10. $\int \dfrac{dx}{1 + x}, t = 1 + x.$

★11. Justify (20) by differentiating the right member.

(20) $$\int \frac{dx}{a^2 - x^2} = \frac{1}{2a} \ln \left| \frac{a + x}{a - x} \right|.$$

★12. Show how to find (20) by using the fact that $(a^2 - x^2)^{-1} = (1/2a)$ $((a + x)^{-1} + (a - x)^{-1})$.

★13. Find a formula for $\int (x^2 - a^2)\, dx$.

★14. Justify (21).

(21) $$\int \sec x\, dx = \ln |\sec x + \tan x|.$$

★15. Find a formula for $\int \csc x\, dx$.

★16. Show that

(22) $$D \text{ Arctan} = (1 + I^2)^{-1}.$$

★17. Justify (23). Show how it might be found by using (22).

(23) $$\int \frac{dx}{a^2 + x^2} = (1/a) \text{ Arctan} (x/a).$$

18. Justify (24).

(24) $$\int (x^2 + a^2)^{-1/2} dx = \ln |x + \sqrt{x^2 + a^2}|.$$

19. Show how (24) might be found by the substitution $x = a \tan t$ and the use of (21).

20. Explain and justify (25)

(25) $$\int u \, dv = uv - \int v \, du \quad \text{(Integration by parts)}.$$

21. Use (25)(u:x, v:sin x) to find $\int x \cos x \, dx$.

22. Find $\int \ln x \, dx$ by applying (25)(u:ln x, v:x).

23. Find $\int xe^x \, dx$.

24. Argue for (26) from the definition of the two members as limits of sums.

(26) $$\int_a^b f(x) \, dx = \int_{g^*(a)}^{g^*(b)} f(g(t))g'(t) \, dt.$$

25. Argue for (26) by using the fundamental theorem of integral calculus.

26. Is the right member of (26) what we would obtain by the substitution $(x$:$g(t))$ in the left member?

ANSWERS TO EXERCISES

(a) Otherwise the integrand is imaginary or undefined.   (b) D Arcsin $= (1 - I^2)^{-1/2}$.   (d) Since $y = \text{Arcsin } x$, $y \in (-\pi/2 \quad \pi/2)$ and $\cos y \geq 0$. (e) Yes. No trouble arises if we use identities in terms of the functions themselves, but if we write formulas in terms of images, the variable needs to be indicated explicitly.   (f) (4-12-22).   (g) Do it!   (h) If we had done so in both (11) and (15) we should have found the same final result.   (i) No. Our check shows the contrary.   (j) (1/3) Arcsin $3x$.   (k) (1/4) (Arcsin $2x + 2x\sqrt{1 - 4x^2}$).

ANSWERS TO PROBLEMS

1. Yes, it is correct. No, he wanted an expression for this antiderivative in terms of familiar functions. Possibly "Express the following in terms of polynomials, trigonometric functions, exponentials, or logarithms, without any indicated integrations and in as simple a form as possible." 2. If $DF = f$, then $DcF = cf$, and conversely, and similarly for (19). 3. Arcsin (x/a). 5. (1/2) ln $(a^2 + x^2)$. 7. $-\ln |1 - x|$. 9. ln $|\sin x|$.

FURTHER READING

This chapter was designed to give the reader an appreciation of the basic ideas of calculus and a minimum skill in its use. However, we could only touch the edges of the vast amount of theory and technique that has developed in this area during the last several hundred years. The student who continues mathematics will undoubtedly take a course that concentrates on calculus. Those who are unable to do so may find the following references helpful.

Introductory Calculus with Analytic Geometry, by Edward G. Begle, Henry Holt and Co., 1954. In spirit and notation very similar to this book.

Calculus and Analytic Geometry, by George B. Thomas, Addison-Wesley, 1951. Contains a large and interesting collection of problems.

Differential and Integral Calculus, by R. Courant, (two volumes), Interscience Publishers, 1937. The great classic textbook on elementary calculus. Difficult but worth the effort.

★CHAPTER 8*

PROBABILITY

8-1 The nature of probability. The word "probability" is used with many different meanings and connotations. For example, in "He is probably happy" it suggests a degree of belief or confidence. On the other hand, in "The probability of throwing a seven with two dice is one-sixth" it refers to an expectation that if we throw two dice many times, the total of seven will occur about one-sixth of the time.

The mathematical theory of probability has been developed to deal with problems of the type suggested by the second example above. The typical problem dealt with by the theory takes the following form: We know (or assume) the probability of certain events; that is, we know how often these events occur. We wish to calculate the probabilities of certain related events in order to be able to predict how often they will occur. Such problems arise, of course, only when we are dealing with events whose occurrence we cannot predict with absolute certainty.

Because of the various connotations of "probability," there exist differences of opinion among philosophers and mathematicians as to the meaning of the mathematical theory of probability. However, there is virtually unanimous agreement about the fundamental laws and methods of calculation. It is the purpose of this chapter (1) to familiarize the reader with the fundamental ideas of probability, (2) to show how the theory of probability may be derived from a few simple axioms, and (3) to indicate how the theory of probability may be applied to a wide variety of situations. Difficult proofs are omitted.

Let us imagine a situation in which any one of a set of events may occur. For example, if two dice are thrown, the set of events consists of all the different ways that they may come to rest. Letting a = the number that comes up on the first die and b = the number that comes up on the second die, an outcome is an ordered pair (a, b). The 36 possible events are sketched in Fig. 8-1. Long experience by countless experimenters indicates that with "fair" dice each of these 36 outcomes occurs about equally often. Accordingly, it seems reasonable to say that the probability of any one of them is 1/36, i.e., each may be expected to occur about 1/36 of the time.

---

* Chapter 8 is prerequisite for Chapter 9 but not for Chapter 10.

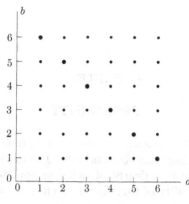

FIGURE 8–1

As a rule we are interested not so much in the probability of just a single one of the possible outcomes as we are in the probability that any one of a set of outcomes occurs. For example, we may be interested in the probability that the sum of the faces of the dice is 7. In the figure we note that this may happen in 6 ways, i.e., the subset defined by $a + b = 7$ has 6 members. Since each has a probability of $1/36$, it seems reasonable that the probability that one of them occurs is $6/36$ or $1/6$. This seems plausible because if each occurs about $1/36$ of the time, we may expect to observe one or another of them about 6 times as often.

Find the probability that the total of the dice is: (a) 2,   (b) 3,   (c) 4, (d) 5,   (e) 6,   (f) 7,   (g) 8,   (h) 9,   (i) 10,   (j) 11,   (k) 12,   (l) 13.

Along the lines suggested by the previous example, we always think of a probability situation in terms of sets. We visualize a situation (experiment, observation) as a set of conceivable outcomes. We refer to this set of conceivable outcomes as a *sample space* and to the outcomes as *points*. In the experiment of throwing dice the sample space consists of 36 points. We refer to each set of points of the sample space as an *event*. An event may be a singleton consisting of just one point, or it may consist of more than one point. For example, the event that the sum of the dice is 2 has just one member, $(1, 1)$, whereas the event that the sum is 3 has two members, $(1, 2)$ and $(2, 1)$.

Since we think of probability situations in terms of a sample space, and of each event as a subset of this sample space, we may apply all the ideas of set theory. Letting our sample space be the universe of discourse U, the null set $\emptyset$ represents an impossible event (one that contains no members of the sample space); the complement A' represents the event that consists of A not occurring; $A \cup B$ is the event that consists of either A or B or both occurring; and $A \cap B$ is the event that consists of both A and B occurring.

Referring again to the dice and Fig. 8–1, let $A = \{(a, b) \mid a + b = 7\} =$ the occurrence of a total of 7, and $B = \{(a, b) \mid a + b = 11\} =$ the occurrence of a total of 11. Then $A' =$ the nonoccurrence of 7, $A \cup B =$ the occurrence of either 7 or 11, and $A \cap B =$ the occurrence of both 7 and 11. Evidently in this case $A \cap B = \emptyset$, which corresponds to the fact that 7 and 11 cannot both turn up at the same time.

Usually probability problems (in real life as much as in books) are not stated in terms of sample space and its subsets. For example, what is the probability that two dice come up with a total of 7 or 11? Letting $P[X]$ stand for the probability of X, this may be translated into set terms as a request for $P[A \cup B_1]$. By counting the points in the figure, we conclude that $P[A \cup B] = 8/36 = 2/9$. It seems reasonable to write $P[A \cap B] = 0$. Similarly, the probability that one die or the other shows 6 is $P\{(x, y) \mid x = 6 \lor y = 6\} = 11/36$, as the reader can find by counting the points in the indicated set. Note that in this case one cannot simply count the points where $x = 6$ and the points where $y = 6$ and then add, since these two sets overlap.

If an event (subset of the sample space) is defined by a simple sentence, it is customary to use the sentence in place of the full name of the set. Thus we have $P[$the sum of the dice is $7] = P[x + y = 7] = P[A] = P[\{(x, y) \mid x + y = 7\}]$.

Find the following by reference to Fig. 8–1:   (m) The probability that the sum of the dice is 8.   (n) The probability that the sum is either 2 or 3. (o) $P[x + y = 4]$.   (p) $P[x - y = 2]$.   (q) $P[x = 2 \lor x = 4]$. (r) $P[x = 2 \lor y = 3]$.   (s) $P[x \neq y]$.   (t) $P[x = y]$.   (u) The probability that one of the dice shows twice the other.   (v) The probability that both dice show an even number.   (w) The probability that both show an odd number.   (x) The probability that neither shows an even number.   (y) $P[2 \leq x + y \leq 12]$.   (z) $P[x + y < 2]$.

ANSWERS TO EXERCISES

(a) 1/36.   (b) 1/18.   (c) 1/12.   (d) 1/9.   (e) 5/36.   (f) 1/6.   (g) 5/36. (h) 1/9.   (i) 1/12.   (j) 1/18.   (k) 1/36.   (l) 0.   (m) $P[x + y = 8] = 5/36$. (n) $P[x + y = 2 \lor x + y = 3] = 1/12$.   (o) 1/12.   (p) 1/9.   (q) 1/3. (r) 11/36.   (s) 5/6.   (t) 1/6.   (u) 1/6.   (v) 1/4.   (w) 1/4.   (x) 1/4.   (y) 1. (z) 0.

8–2 Fundamental laws of probability.   To construct an axiomatic theory of probability we imagine a universe of discourse U which we refer to as sample space and view as the set of all conceivable outcomes of an experiment. Each of the subsets of U is called an event. We wish to associate with each event X a probability $P[X]$ so that the following axioms are satisfied.

(1) $\qquad P[X] \in Re,$

(2) $\qquad P[X] \geq 0,$

(3) $\qquad (A \cap B = \emptyset) \rightarrow (P[A \cup B] = P[A] + P[B]),$

(4) $\qquad P[U] = 1.$

These axioms assert that the probability of any event is a non-negative (2) real number (1), that the probability of either of two disjoint events is found by adding their probabilities (3), and that the probability of the sample space is 1 (4).

The reasonableness of (3) is apparent from the examples relating to two dice considered in Section 8–1. Letting $P[A] = P[x = 6]$ and $P[B] = P[y = 6]$, we have $P[A] = P[B] = 1/6$, but $P[A \cup B] = P[x = 6 \lor y = 6] = 11/36 \neq 1/6 + 1/6$. Here A is not disjoint with B, and we see that the right member of (3) fails. However, letting $C = \{(x, y) \,|\, x = 5\}$, $P[C] = 1/6$ and $P[A \cup C] = P[x = 6 \lor x = 5] = 1/3 = P[A] + P[C]$. Here $A//C$ and the right member of (3) holds with $(B{:}C)$.

The probability associated with a set appears to be a kind of measure of the set. Accordingly, we would expect that the properties (7–10–1) through (7–10–4) would hold. We already have (7–10–2) and (7–10–3) included in the above axioms. We shall derive (7–10–1) and (7–10–4) below.

To derive

(5) $\qquad\qquad\qquad P[\emptyset] = 0$

we note that $1 = P[U] = P[U \cup \emptyset] = P[U] + P[\emptyset] = 1 + P[\emptyset]$ since U and $\emptyset$ are disjoint. From $1 = 1 + P[\emptyset]$, we have (5) immediately. We interpret $\emptyset$ as the impossible event, the event that consists in the occurrence of no one of the set of all possible outcomes of the experiment. Hence (5) says that an impossible event has probability zero, that is, $A = \emptyset \rightarrow P(A) = 0$. The converse of this, however, cannot be proved from (1) through (4). Actually there are situations in which an event that is possible has probability zero. For example, the probability that a point chosen at random on an axis is at a rational point is zero. The proof of this involves considerably more mathematics than is contained in this book, but it may be made plausible by considering how the irrationals are much more tightly packed on the line than the rationals.

Since A and its complement A' are disjoint, we have $P[A] + P[A'] = P[A \cup A'] = P[U] = 1$. Hence

(6) $\qquad\qquad\qquad P[A'] = 1 - P[A].$

Since A' occurs if and only if A does not, (6) says that the probability that an event does not happen is 1 minus the probability that it does happen. For example, referring to the two dice in Section 8–1,

$$P[3 \leq x + y \leq 11] = 1 - P[x + y = 2 \vee x + y = 12].$$

The calculations suggested by the right member are easier than those suggested by the left.

(a) Find $P[3 \leq x + y \leq 11]$ by adding the seven probabilities.   (b) Get the same result by using (6).

Use (6) to find the probability that: (c) 5 does not come up when one die is thrown,   (d) 5 is not the total when two dice are thrown,   (e) the sum of the faces is greater than 3 when two dice are thrown.

We now derive

(7) $$A \subseteq B \rightarrow P[A] \leq P[B],$$

which is just (7–10–4). To do so we note that $A \subseteq B \leftrightarrow B = A \cup (A' \cap B)$, where $A' \cap B$ and A are disjoint. Hence $P[B] = P[A] + P[A' \cap B] \geq P[A]$, since $P[A' \cap B] \geq 0$.

(f) Justify each of the statements in the proof of (7)   (g) Prove (8) by using (4) and (3–8–4).

(8) $$P[A] \leq 1.$$

Axiom (3) covers only the case of disjoint sets. A more general law is

(9) $$P[A \cup B] = P[A] + P[B] - P[A \cap B].$$

To prove it we note that $A \cup B = (A \cap B') \cup B$, where $A \cap B'$ and B are disjoint. Hence $P[A \cup B] = P[A \cap B'] + P[B]$. Now we note that $(A \cap B) \cup (A \cap B') = A$, where $(A \cap B)$ and $(A \cap B')$ are disjoint. Hence $P[A \cap B] + P[A \cap B'] = P[A]$. Combining these equations we arrive at (9). This law will be very useful as soon as we have a procedure for finding $P[A \cap B]$. The necessary theory is included in Section 8–5.

(h) Justify the set identities and do the final manipulation.

We may summarize the laws of this section as follows. (1), (2), (8): The probability of an event is a real number in the closed interval from zero to one. (5), (4): An impossible event has zero probability and a certain event has probability of one. (3): Probabilities of mutually exclusive events are additive. (7): The probability of an event cannot be greater

than that of an event that always occurs whenever the first one does.
(9): The probability that either one of two events occurs is the sum of
the probabilities that either one occurs less the probability that both
occur.

We have already pointed out that there is nothing in the axioms to prevent a possible event (a non-null set) from having zero probability. Similarly there is nothing to prevent a noncertain event (one not equal to the
universe of discourse) from having a probability of one. We may write
(4) as $A = U \to P[A] = 1$. The converse of this is not true. We shall
give examples in later sections.

PROBLEMS

1. Show that if $A//B$, $B//C$, and $A//C$, then $P[A \cup B \cup C] = P[A] + P[B] + P[C]$.

2. By induction prove a similar theorem for any number of mutually disjoint sets.

3. Prove that $A \cap B \cap C = 0$ is not a sufficient condition for $P[A \cup B \cup C] = P[A] + P[B] + P[C]$.

4. State and prove a theorem that generalizes (9).

5. Prove (10).

(10) $$P[A \cup B] \leq P[A] + P[B].$$

6. Illustrate the laws of this section by reference to the example of Section 8–1.

Problems 7 through 12 refer to the situation of two dice described in 8–1.

7. What is the probability that the two dice come up the same?

8. What is the probability that one die is even and the other odd?

9. That both dice are even?

10. That one die is even or that the dice come up the same?

11. That the sum of the faces is 5 and that one face is even?

12. That the sum of the faces is 5 or that one face is even?

13. Under what conditions does the equality hold in the right member of (7)?

14. Generalize (10).

15. Statistics indicate that there are 13 persons injured by tornados for every
one killed by them. The figures for Arkansas show that a person living 68 years
there has one chance in 1,490 of being killed, one in 115 of being injured. What
is the chance of being involved through injury or death in a tornado in Arkansas?

Prove and interpret 16 through 19 verbally.

16. $P[A \cap B] \leq P[A]$.

17. $P[A \cup A'] = 1$.

18. $P[A \cap A'] = 0$.

19. $P[A \cap B] = P[A] - P[A \cap B']$.

★20. How many members are there in the domain of P?

ANSWERS TO EXERCISES

(a) 17/18. (c) 5/6. (d) 8/9. (e) 11/12. (f) $A \cup (A' \cap B) = (A \cup A') \cap (A \cup B) = U \cap (A \cup B) = A \cup B = B$ by (3-8-12) when $A \subseteq B$. The disjointness follows from $A \cap (A' \cap B) = (A \cap A') \cap B = \emptyset \cap B = \emptyset$. (g) $A \subseteq U$ and $P[U] = 1$. Then (7)(B:U). (h) Use laws of Section 3-7 as above. Compare with the proof of (6-3-3).

ANSWERS TO PROBLEMS

1. $P[A \cup (B \cup C)] = P[A] + P[B \cup C]$ since $A//B$ and $A//C$ imply $A//(B \cup C)$. 3. For two dice let A, B, and C be respectively the events that the first die shows 1, the sum of the dice is 7, and the points on the two dice are the same. Then the left member is 4/9 and the right member is 1/2. 5. From (9), since $P[A \cap B] \geq 0$. 7. 1/6. 9. 1/4. 11. 1/9. 13. When $P[A' \cap B] = 0$. (It is not necessary that $A = B$.) 15. About 0.0093. 17. (3-7-19). 19. (3-7-24), the terms in whose right member are disjoint.

8-3 Relative frequency. In the previous section we assumed that P is a function whose domain is some collection of subsets of some universe and which satisfies the stated axioms. We did not further define P, and indeed we are free to specify P in any way we wish as long as the axioms are satisfied.

In particular, if the sample space is a finite set, it suffices to assign a non-negative probability to each point in such a way that the sum of all probabilities is 1, and then let the probability of any event be the sum of the probabilities associated with its points. For example, suppose that a die is loaded so that the probabilities of points from 1 to 6 are 1/6, 7/36, 1/9, 5/36, 2/9, 1/6, respectively. Then the probability of getting an even point is 7/36 + 5/36 + 1/6.

In this situation what is the probability that the point is: (a) less than 3? (b) odd? (c) even or divisible by 3? (d) greater than 1?

(e) Let the sample space consist of n points $a_1, \ldots, a_n$ with which are associated probabilities $p_1, \ldots, p_n$, such that $p_i > 0$ and $\sum p_i = 1$. Let $P[X] = \sum p_j$, where the summation is carried over all j for which $a_j \in X$. Show that the axioms are satisfied.

In a very large number of situations the sample space is chosen so that the probabilities associated with the sample points are equal, i.e., the sample points are equally likely. Letting $\mathfrak{N}(X)$ be the number of members of X, which we call the *frequency* of X (see Section 6-2), we see that the probability associated with each point is $1/\mathfrak{N}(U)$, i.e., the reciprocal of the number of points in the sample space. Then the schema of Exercise (e) suggests the following definition of probability.

(1) $P[X] =_d \mathfrak{N}(X)/\mathfrak{N}(U)$.

With this definition, the probability of an event is simply the number of favorable cases, $\mathfrak{N}(X)$, divided by the total number of cases, $\mathfrak{N}(U)$. It is called the *relative frequency* of the event.

(f) How do we know immediately from Exercise (e) that (1) yields a probability satisfying the axioms?   (g) Without relying on Exercise (e), show that (1) satisfies the axioms.   (h) Why is it not necessary to prove (8–2–5) through (8–2–10) directly from (1)?   (i) Nevertheless, prove (8–2–5) through (8–2–10) directly from (1).

(j) According to the American Experience Table of Mortality, of 90,471 people living at age 23, 45,291 live to age 67. State this in probability terms. What is the sample space? What is the set corresponding to the probability? What does this result mean for a particular individual age 23?   (k) In 1952, 46.8% of the auto accidents resulting in deaths and injuries involved exceeding the speed limit. State this in probability terms after indicating the universe of discourse.

Most probability statements that are made in daily affairs and in science can be reformulated precisely in relative frequency terms, but some care may be required in order to choose the sample space properly. In some cases probability statements are merely ways of reporting an actual relative frequency that has been observed. Exercise (k) is an example of this. In other cases the probability is thought to indicate the relative frequency with which the event will occur "in the long run." Thus Exercise (j) reports an observed relative frequency, but it also suggests the proportion of persons 23 years old who may be expected to live to 68.

(l) "In Iowa...there is one chance in 1,203 that any given square mile will be struck by a tornado in any one year." What does this mean?

To calculate probabilities from (1) we need to be able to find the frequencies involved. Usually the problem takes the form: how many ways can such and such an event take place? For example, to calculate the probability of getting a one-suit hand in the game of bridge we need to know the total number of different possible hands and the number of these in which one gets 13 cards of one suit. The branch of mathematics that deals with such questions is called *combinatorial analysis*. Its fundamental law is:

(2)   *If a certain event can take place in n_1 ways and then another in n_2 ways, then the ordered pair of events can occur in $n_1 n_2$ ways.*

For example, one die can come up in 6 ways, a second in 6 ways, and the two together in 36 ways. This law is really only a different way of stating (6–3–6), since we can think of each different way of doing the two things as an ordered pair of a way of doing the first and a way of doing the second.

(m) In how many ways can two houses be painted if the first house can be painted three ways and the second four?   (n) In how many ways can three names be assigned to three houses, so that no two have the same name?   (o) In how many ways can three dice come up?

In statistics a set of objects under study is often called a *population*. An ordered set $(a_1, a_2, \ldots, a_r)$ whose components are drawn from a population is called a *sample*. (Often the sample space is a set of samples drawn from a population, and this is the origin of the name.) We may visualize a sample formed by choosing members of the population one after the other. If the method of choice is such that all possible samples are equally likely, we call the result a *random sample*.

One method of drawing a random sample consists of choosing an object (so that each object has equal probability) and then replacing the object before making the next choice. This is called *sampling with replacement*. Suppose the population has n members and we wish to draw a sample of r members. Then the first element in the sample can be chosen in n ways. So can the second and all succeeding elements, since the original situation is restored after each choice. Hence, from (2),

(3)   *In sampling with replacement, the number of different samples of r members chosen from a population of n members is n^r.*

(p) How is (2) used in (3)?

For example, how many different sequences of 10 symbols can be formed with the plus and minus signs? Here $n = 2$, since the population is $\{+, -\}$, and $r = 10$. Hence the number of sequences is 2^{10}. Of these only one consists of all $+$'s and only one of all $-$'s. Hence the probability that all symbols are the same is $2/2^{10}$ or $1/2^9$.

(q) How many different sequences of four symbols can be formed from the digits, where a digit may be used more than once?

A second method of forming a sample is to choose a first element, then choose the second element from the remaining elements, and so on. This is called *sampling without replacement*. Since the first element can be chosen in n ways, the second in $n - 1$, and so on, it is easy to see that the number of samples without replacement is $n(n - 1)(n - 2) \ldots (n - r + 1)$ or $n!/(n - r)!$. This number is often called the number of *permutations* of n things taken r at a time. It is the number of different ways we can choose and arrange r things from among n things. A common symbol for it is P_r^n.

(4) $$P_r^n = n!/(n - r)!.$$

For example, in how many ways can we choose a president, vice-president, and secretary from among ten people? Here $r = 3$, $n = 10$ and $P^{10}_3 = 10 \cdot 9 \cdot 8 = 720$. When $n = r$ we have the number of arrangements of n things, that is, $P^n_n = n!$ For example, a squad of eight soldiers may be lined up in 8! ways.

(r) In how many ways can a sergeant and a corporal be chosen from a squad of 12 men?   (s) How many ways can the squad be arranged in a line?

To calculate probabilities from (1) we find the frequencies of the event and of the universe and divide. For example, to find the probability that eight soldiers of different heights will line up in order of their heights from right to left if they are arranged at random, we note that there is just one way to arrange them correctly, i.e., the frequency of our event is 1. The frequency of the universe is 8!. Hence the required probability is 1/8!.

(t) What is the probability that in six throws of a die the numbers 1 through 6 turn up in that order?   (u) If a student is to match three items on a test with three items from a list of five items, what is the probability that he will succeed by chance alone?

PROBLEMS

1. Extend (2) to s events and prove your result.

2. The A auto company offers 16 body styles, three different engines, 50 items of optional equipment, and 286 exterior and interior color variations. The B company offers 16 basic body styles, 22 optional items of equipment, and 20 colors offered in up to three possible combinations on each auto. For each such color or combination, there are two interior colors. (Data from *Wall Street Journal*, March 21, 1955) Which company makes the greatest number of different kinds of cars?

3. A psychologist wishes to present the subjects of an experiment with three different stimuli simultaneously. If the first stimulus has three possible values, the second 15, and the third 50, how many different combinations can he present?

4. In how many ways can a football team of 11 be assigned to the positions?

5. In how many ways could a first-string football team be organized from among 12 men? 13 men? 22 men?

6. In how many ways can the ten digits be arranged in a line? How many of these give integers written in the usual way?

7. Studies of chess play have shown that in typical positions there are about 30 legal moves. Hence for two moves there are about 10^3 possibilities. Why?

8. If a typical chess game lasts about 40 moves, about how many different games are possible?

9. Suppose that an electronic brain were to try out all possible chess games in order to see the final result of each, and suppose that it can go through a million different games in one-millionth of a second. How long would it take the machine to try out all games?

10. Suppose a rat is run through a T-maze in which it has the choice of turning right (R) or left (L), and suppose that R brings reward and L a punishment.

Suppose that we consider the rat to have "learned" if it makes the correct choice five times in succession. We let the rat run 20 times to see whether he is able to learn in that number of trials. Let U be the set of all different ways in which the rat can run these 20 trials, i.e., all different sequences of 20 R's and L's. What is $\mathfrak{N}(U)$? Let X be the event of the rat's having learned. Describe X as a subset of U. (*Stochastic Models for Learning*, by R. R. Bush and F. Mosteller)

11. Suppose that in a batch of 100 light bulbs there is one defective bulb. What is the probability of picking it at random?

12. Suppose in this situation we pick a random sample without replacement of five light bulbs. How many such samples might we pick? How many of them contain the defective bulb? What is the probability that a random sample of five bulbs contains the defective one?

13. Suppose that there are 20 male rats and 20 female rats, some of which are cancerous. The 40 rats are divided into four disjoint subsets, the male cancerous, female cancerous, male noncancerous, and female noncancerous. The number of members in these subsets is known as soon as we know an ordered pair of numbers (x, y) giving the numbers of male and female cancerous rats. We consider a sample space consisting of all the points (x, y). How many members does it have?

★14. For examples in which the points in a sample space do not have equal probabilities, see the article by L. Guttman in *Psychometrika*, Vol. 11, 1946, pp. 81–95, and the article by A. P. Horst in the *Journal of Educational Psychology*, Vol. 24, 1933, pp. 229–232.

15. Suppose four objects are ranked by an observer, i.e., given ranking numbers from 1 through 4. We are interested in a particular pair of objects and we record the ranks given to them as an ordered pair of numbers. For example, (2, 3) would indicate that the objects were assigned the orders 2 and 3 respectively. How many ways can these two objects be assigned ranks? Assuming that each assignment is equally likely, what is the probability that the sum of the ranks is 4?

16. Solve Problem 15 if there are six objects in which we are interested in a subset of three objects.

17. Prove (4) by induction.

18. Probabilities are often stated in terms of odds or chances, according to the following definitions.

(5) $\qquad$ The odds are r to s for $X = [P[X] = r/(r + s)]$

(6) $\qquad$ The odds are s to r against $X = [P[X] = r/(r + s)]$

(7) $\qquad$ The chances are r to s that $X = [P[X] = r/s]$

Suppose that the probability of a coin coming up "heads" is 0.51. State this in terms of odds and chances.

19. When two pairs of identical twins with the same first names appeared in the same graduating class, a professor said that the chances were 1,140,000,000 to 1 against the coincidence. What did he mean?

20. What probability corresponds to odds of two-to-one in favor?

21. According to an article in the bulletin of the Commonwealth Club of San Francisco of January 21, 1946, "When a section reaches a conclusion after long and careful study, chances are 13 to 1 that the club, if it takes a ballot, will vote the same way." What does this mean?

★22. Referring to Exercise (u), what is the probability that the student will match just two out of three?

★23. Just one out of three?

★24. None at all?

Answers to Exercises

(a) 13/36.   (b) 1/2.   (c) 11/18.   (d) 5/6.   (e) Immediate by considering each axiom.   (f) This is a special case of (e) with $p_i = 1/\mathfrak{N}(U)$ for all i.   (g) Consider each axiom. (3) follows from (6–3–2).   (h) Since the others follow from the axioms, and we have shown the axioms apply.   (i) Use (6–3–3) and other properties of cardinals of sets.   (j) The probability that a person alive at 23 will live to be 67 or older is 45,291/90,471, or about 1/2. Sample space is a set of people alive at the age of 23, probably a set of insured Americans. Nothing; a particular person may be more or less likely to live that long, depending on health, etc. The probability does not refer to an individual but to our choice of individuals. If we choose individuals at random, about half of them will live to 67 or more.

(k) Sample space (universe of discourse) is set of auto accidents resulting in deaths or injuries in 1952. Probability of the subset consisting of those in which speed limit was exceeded is 0.468.   (l) About 1/1203 of the square-mile regions in Iowa are struck in any one year. Also, we would expect a given square mile to be struck about once in every 1,203 years.   (m) 12.   (n) 6.   (o) 6^3. (p) Repeatedly, r times.   (q) 10^4.   (r) 132.   (s) 12!   (t) $1/6^6$.   (u) $1/5^3$ if repetitions are allowed; otherwise 1/60.

Answers to Problems

1. Use induction. 3. 2,250. 5. 12!, 13!/2, 22!/11!. 7. $30^2 \doteq 10^3$. 9. 10^{40} years approximately. 11. 1/100. 13. 400. 15. 12; 1/6. 19. Probability is 1/1,140,000,001, or approximately $1/10^9$. 21. The rest of the context indicates that this summarizes the fact that during the past 10 years the committees and the club membership differed on about 1/14 of the questions on which both voted. However, in addition to reporting this observed relative frequency, there is the suggestion that one may expect it to continue, i.e., from the experience of the past 10 years one may estimate a relative frequency that will be likely to appear in the future. If we think of the past experience, the sample space is the set of all questions on which both club and a committee voted. If we think of the future, the sample space is an imagined set of future questions.

8–4 Partitions. We have seen in the previous section that knowing how to find the number of samples (with or without replacement) of a given size that can be chosen from a given set is often useful in finding relative frequencies. Of course, not all problems in combinatory analysis can be treated in this way, as is evident from the problems in Section 8-3. Indeed, one cannot expect to find formulas to fit every situation, and the student should be prepared to attack problems directly in terms of the fundamental law (8-3-2). However, many combinatorial problems can be handled in terms of the concept of a partition.

Often we wish to break up a set into subsets in such a way that every object belongs to one and only one of the subsets, i.e., so that the union of the subsets is the whole set and every pair of the subsets is disjoint. More precisely, a class of sets $\{A_1, A_2, \ldots, A_n\}$ such that $A_i // A_j$ for all $i \neq j$ and $A_1 \cup A_2 \cup \ldots \cup A_n = A$ is said to be a *partition* of A.

(a) The process of assigning members of a set to subsets is often called *classification*. When all members of the set are classified, the classification is called *exhaustive*. When no member is assigned to more than one subset, the classification is said to be *mutually exclusive*. Give an example of an exhaustive and mutually exclusive classification.   (b) Give an example of a classification that is not a partition.   (c) Did we use partitions in Chapter 7?

A partition whose sets are ordered is called an *ordered partition*. Note that the sets are ordered, but not the members of the sets. Thus an ordered partition is simply an ordered set $(A_1, A_2, \ldots, A_n)$ of subsets of some set where $\{A_1, A_2, \ldots, A_n\}$ is a partition of A. For example $\{\{1, 2\}, \{3, 4\}\}$ is a partition of $\{1, 2, 3, 4\}$ and $\{\{3, 4\}, \{1, 2\}\}$ is the same partition. But $(\{1, 2\}, \{3, 4\})$ and $(\{3, 4\}, \{1, 2\})$ are different ordered partitions of $\{1, 2, 3, 4\}$.

The simplest partition of a set A is $\{A_1, A_1'\}$, where the complement is taken with respect to A. We simply divided A into a set A_1 and all other elements of A. The corresponding classification is called a *dichotomy*. Numerous problems in probability are expressible in terms of ordered dichotomies. For example, what is the probability that a person playing bridge gets all cards of one suit specified in advance? Here our sample space is the set of all different 13-card hands that can be picked from a 52-card deck. How many members does this space have? Now choosing a 13-card hand amounts to deciding on an ordered dichotomy of the 52 cards into two subsets of 13 and 39 cards respectively. Accordingly, to solve the problem we need to know how many different ways we can perform such a dichotomy.

More generally, we are interested in knowing how many different ordered dichotomies there are in which a set of n members is partitioned into a set of r members and a set of $n - r$ members. Often a set of r objects chosen (without regard to order) from a set of n objects is called a *combination* of n things taken r at a time. The symbol for the number of combinations is C_r^n.

To find a formula we make use of the fact that we know how to find the number of permutations of n things taken r at a time, i.e., the number of sets of r objects (ordered) that can be chosen from n objects. This is given by $P_r^n = n!/(n - r)!$. Now in order to form a permutation we can first form a combination and then arrange this combination in some order. The first can be done in C_r^n ways, the second in $r!$ ways by (8–3–4). Hence $P_r^n = r!C_r^n$ and

(1) $C_r^n = n!/r!(n - r)!.$

Returning to the problem of bridge hands, the number of points in the sample space is C_{13}^{52}. Since there is just one point in the set whose probability we seek, the probability is $1/C_{13}^{52}$ or $13!39!/52!$, if we assume all points equally likely.

★(d) Estimate this to order of magnitude.   (e) Show that

(2) $C_r^n = C_{n-r}^n.$

(f) What is the probability of getting *any* particular hand in bridge specified in advance?   (g) What is the probability of getting four aces in a five-card hand dealt from a 52-card pack?   (h) Calculate C_2^5 and verify your result by writing down the two member subsets of $\{a, b, c, d, e\}$.

The results for partitioning into two subsets can be generalized to partitioning into k subsets. Indeed,

(3)   *The number of different ordered partitions of a set of n members into k subsets with $r_1, r_2, \ldots, r_k$ members is*

$$n!/r_1!r_2!r_3! \ldots r_k!.$$

We prove (3) by induction. For $k = 2$, $r_2 = n - r_1$ and (3) reduces to (1). Hence (3) holds for $k = 2$. Suppose (3) holds for k subsets. Then a partitioning into $k + 1$ subsets can be obtained by first partitioning the n objects into k subsets with $r_1, r_2, \ldots, r_{k-1}, r_k + r_{k+1}$ members, and then partitioning the last set into two sets of r_k and r_{k+1} members. The first operation can be performed in $n!/r_1!r_2! \ldots (r_k + r_{k+1})!$ ways, by the induction assumption, the second in $(r_k + r_{k+1})!/r_k!r_{k+1}!$ ways. Multiplying we get (3)($k{:}k + 1$).

(i) Why does $r_1 + r_2 + \ldots + r_k = n$?

To illustrate the use of (3), consider the number of different ways that four hands can be dealt in bridge, that is, the number of different ordered partitions of 52 cards into four subsets of 13 cards each. Note that it is an ordered partition that applies, since the same hands redistributed among North, South, East, and West would constitute a different situation. According to (3), we have $52!/(13!)^4$.

(j) How many different ways can five-card hands be dealt to six poker players? (Note that we must partition the 52 cards into seven subsets, including the set left over after the deal.)

We adopt the notation $C^n_{r_1, r_2, \ldots, r_k}$ for the formula in (3).

Calculate: (k) $C^4_{2,1,1}$,   (l) $C^5_{2,2,1}$,   (m) $C^3_{2,0,1}$.

(n) What is the probability that one of six players is dealt four aces in a five-card hand?

An experiment with two possible outcomes a and a' is performed 10 times in order.   (o) In how many different ways can it happen that a occurs three times and a' seven times?   (p) a five times and a' five times?   (q) a seven times and a' three times?   (r) a ten times?   (s) Generalize to an experiment performed n times and give the number of ways in which a could occur x times and a' $n - x$ times.   (t) How many different ways can x 0's and $n - x$ 1's be arranged in a straight line?

★*Stirling's formula.* As the reader has seen, the evaluation of probabilities often involved calculating $n!$ for large n. This is facilitated by the following approximation, known as *Stirling's formula.*

(4) $$n! \doteq (2\pi)^{1/2} n^{n+1/2} e^{-n}.$$

Here e is the base of the system of natural logarithms. The formula is a good approximation in the sense that

(5) $$\lim_{n \to \infty} (2\pi)^{1/2} n^{n+1/2} e^{-n} / n! = 1.$$

Actually the approximation is always too small, but the percent error approaches zero, and for $n = 10$ it is already less than 1%.

★(u) Compare the two members of (4) for $n = 1, 2, 5, 10$, and 100.   ★(v) Use (4) to evaluate some of the frequencies and probabilities in problems of this section.

PROBLEMS

1. What is the relation between C^n_r and $\binom{n}{r}$? (See Section 7–7.)

2. Show that the number of combinations of $n + 1$ things i at a time is equal to the sum of the number of combinations of n things i at a time and the number of combinations of n things $i - 1$ at a time.

3. Show that $1 + C^n_1 + C^n_2 + C^n_3 + \ldots + C^n_n = 2^n$.

4. Interpret Problem 3 in terms of combinations and sampling.

5. The reliability of a test is often estimated by comparing the results on half the questions with those for the other half. In how many ways can a test of $2n$ items be split in half? What conclusions do you draw about this method of estimating reliability?

6. In how many ways can the five letters a, a, b, b, c be arranged if we count as different arrangements only those that appear to be different, i.e., if we do not distinguish different occurrences of the same letter?

7. How many different signals can be made with dots, dashes, and pauses by using a sequence of five such elements?

8. What is the probability of getting a full house (three of one kind and two of another) in a five-card poker hand?

9. If a set is classified by several different dichotomies, (A_1, A_1'), (A_2, A_2'), $\ldots$, (A_n, A_n'), it is partitioned into sets each one of which is the intersection of n sets, one chosen from each dichotomy. For example, one of the sets is $A_1 \cap A_2' \cap A_3 \cap A_4' \cap \ldots$. These sets are called *cross classifications*. How many cross classifications are determined by n dichotomies?

10. In a bag are three black balls, b_1, b_2, b_3, and two white balls, w_1, w_2. How many different pairs can be drawn? (What is the sample space?)

11. Assuming that each pair in Problem 10 is equally likely, what is the probability of any one pair? What is the probability of drawing two black balls? two white balls? a black and a white?

★12. What is the probability that in a sample of r people all have different birthdays? What is the probability that at least two have the same birthday?

★13. Find the answer to Problem 12 for $r = 10$, 20, and 30.

14. What is the probability that in six throws of two dice, seven does not turn up?

15. Suppose that each of n sticks is broken into one long and one short part. The $2n$ parts are arranged into n pairs from which new sticks are formed. Find the probability that the parts will be rejoined as before (i.e., each original stick reformed), and find the probability that all new sticks will be formed from one short and one long part. (This problem is related to the genetic effects of radiation. It is taken from *An Introduction to Probability Theory and Its Applications*, by William Feller. In the first four chapters of that book the reader will find numerous other interesting problems illustrating combinatorial analysis.)

16. Suppose in a certain population there are n_1 blue elements and n_2 green elements. A random sample of r elements is chosen without replacement. Show that the probability that exactly x elements are blue is $C_x^{n_1} C_{r-x}^{n-n_1}/C_r^n$, where $n = n_1 + n_2$.

★17. We have seen in Problem 1 that C_r^n is the coefficient of x^r in the expansion of $(a + x)^n$. For this reason the C_r^n are called the *binomial coefficients*. Show that $C_{r_1, r_2, \ldots, r_k}^n$ is the coefficient of $x_1^{r_1} x_2^{r_2} \ldots x_k^{r_k}$ in the expansion of $(x_1 + x_2 + \ldots + x_k)^n$. For this reason these numbers are called the *multinomial coefficients*.

★18. Prove the binomial theorem by a combinatorial argument. (*Suggestion:* How is each term formed in the expansion of $(a + x)^n = (a + x)(a + x) \ldots (a + x)$ before collecting terms?)

★19. Generalize by stating and proving a multinomial theorem.

★20. A device of very general utility for attacking combinatorial problems is the use of "trees." This method is presented in *An Introduction to Finite Mathematics*, by J. G. Kemeny, J. L. Snell, and G. L. Thompson. Study this method as there presented and apply it to problems of your choice.

★21. Many interesting examples in combinatorial analysis may be found in *Choice and Chance*, by William Whitworth.

★22. Show that $C_x^r C_{n_1-x}^{n-r}/C_{n_1}^n = C_x^{n_1} C_{r-x}^{n-n_1}/C_r^n$.

★23. Show that $\sum_{i=0}^{i=n}(-1)^i C_i^n = 0$.

★24. What is the probability that when five-card poker hands are dealt to three players each one has just one ace?

★25. What is the probability that one of three poker players has three aces, another three kings, and a third three queens, when five cards have been dealt to each?

ANSWERS TO EXERCISES

(a) People into males and females. Molecules into substances. Any partition.
(b) People into adults, males, and females.   (c) In definition of definite integral we partitioned the interval into subintervals, but permitted the subintervals to have endpoints in common.   (d) $1/10^{12}$.   (e) Immediate from (1).   (f) $1/C_{13}^{52}$ as above! No hand is any more improbable than any other. However, there are more mediocre than good hands, and therefore the probability of getting one of the set of "good" hands is less than that of getting one of the "bad" ones.   (g) $1 \cdot 48/C_5^{52}$.   (h) 10.   (i) Definition of partition.

(j) $C_{5,5,5,5,5,22}^{52}$.   (k) 12.   (l) 30.   (m) 3.   (n) The total number of possible ways of dealing is given by Exercise (j). Then to get a deal of the kind desired we may first choose the player to have the aces (in six ways), decide on the fifth card (in 48 ways), and then distribute the remaining 47 cards to the other players (in $C_{5,5,5,5,22}^{47}$ ways). Hence the probability is $288C_{5,5,5,5,22}^{47}/C_{5,5,5,5,5,22}^{52}$. (*Note:* The problem is here understood to mean that any player may get the four aces. If the player is specified in advance, 288 would be replaced by 48.)   (o) C_3^{10}, since it is a matter of assigning the 10 outcomes to two boxes, the a box and the a' box.   (p) C_5^{10}.   (q) C_3^{10}.   (r) 1.   (s) C_x^n.   (t) C_x^n.

ANSWERS TO PROBLEMS

1. The same. 3. Consider $(1 + 1)^n$. 5. $C_n^{2n} = \dfrac{(2n)!}{(n!)^2}$. One wonders whether the choice of dichotomy might not make a difference. 7. 3^5 (assuming that signals such as five pauses are allowed). 9. 2^n. 11. $1/C_2^5$, C_2^3/C_2^5, C_2^2/C_2^5, $3 \cdot 2/C_2^5$. 13. For $r = 23$, the probability is approximately $1/2$. 15. $2^n/[(2n)!/n!]$, $2^n/C_n^{2n}$.

8–5 Conditional probability. Suppose that the sum of the faces of two dice is seven. What is the probability that the faces differ by five? The desired event can happen in two ways, symbolized by (1, 6) and (6, 1). The sample space here is not the 36 ways in which two dice may come up, but merely the six ways in which a total of seven may occur. Hence the desired probability is 1/3. On the other hand, without the indication that the sum of the dice is seven, the probability that the faces differ by five is 2/36 or 1/18. We see that the probability of an event depends upon the set of possible events we are considering.

Let $P[A\,|\,B]$ be the probability that A occurs given that B occurs, or the *conditional probability* of A on the hypothesis B. Then $P[A\,|\,U]$ is the usual probability $P[A]$ calculated with respect to the original sample space. Sometimes it is called *absolute probability*. Suppose that U is a finite set of equally likely points. Then $P[A \mid B]$ is the relative frequency of A in B, i.e., the probability calculated as if B were the sample space.

(1) $$P[A \mid B] = \mathfrak{N}(A \cap B)/\mathfrak{N}(B).$$

Also $P[B] = P[B \mid U] = \mathfrak{N}(B)/\mathfrak{N}(U)$, and $P[A \cap B] = \mathfrak{N}(A \cap B)/\mathfrak{N}(U)$. We note that the following holds for the relative frequencies.

(2) **Def.** $$P[A \mid B] = P[A \cap B]/P[B].$$

We take it as a *definition* of $P[A \mid B]$ in all cases where $P[B] \neq 0$. When $P[B] = 0$ we do not define $P[A \mid B]$.

For example, letting $U =$ the cases when two dice are thrown, $A =$ the cases where the faces differ by five, and $B =$ the cases where the sum of the faces is seven, $P[A \cap B] = 1/18$, $P[B] = 1/6$, and $P[A \mid B] = (1/18)/(1/6) = 1/3$, as before.

(a) Let $A =$ the faces differ by four, $B =$ the sum of the faces is six. Find $P[A]$, $P[B]$, $P[A \cap B]$, and $P[A \mid B]$ by enumeration in Fig. 8–1, and verify that (2) holds. (b) Do the same, but with $B =$ the sum of the faces is even.

From (2) we have immediately

(3) $$P[A \cap B] = P[B]P[A \mid B] = P[A]P[B \mid A].$$

The first equation is found by multiplying both members of (2) by $P[B]$, the second by $(B:A, A:B)$. This law says that the probability that both A and B occur is the probability that B occurs (the absolute probability) times the conditional probability of A given B, or the probability of A times the conditional probability of B given A. This formula is often useful because it frequently happens that it is easier to find the probabilities in the right member than to find $P[A \cap B]$ directly by enumeration. Sometimes events that consist in the intersections of events (i.e., the simultaneous occurrence of events) are called *compound events*.

For example, consider the following experiment. We have an urn containing seven blue and three gold chips, and a second containing five blue and five gold. We draw a chip at random from the first urn and place it in the second, *then* draw a chip at random from the second. What is the probability that we draw gold chips both times? Here $A =$ gold chip on first draw, $B =$ gold chip on second. We have $P[A] = 3/10$, $P[B \mid A] = 6/11$, and $P[A \cap B] = 9/55$. An enumeration of all possible pairs of drawings is more tedious, though not difficult in this case.

(c) What is the probability of drawing both blue chips? (d) first gold, then blue? (e) first blue, then gold? (f) What is the probability of drawing gold on the second draw? (g) of drawing blue on the second draw? (h) Can we use (3) to find $P[A \cap B]$ if either $P[A]$ or $P[B]$ is zero?

From the previous examples it is clear that the absolute probability $P[A]$ may not be equal to the conditional probability $P[A \mid B]$. Usually the probability of an event changes if the hypotheses (i.e., the given conditions or sample space) change. However, it may happen that $P[A \mid B] = P[A \mid U] = P[A]$. In this case (3) becomes $P[A \cap B] = P[A]P[B]$. When this equation holds we say that A and B are *statistically independent*. Statistical independence does not mean that there is no relation between A and B! On the contrary, there is a *very special relation*, namely that $P[A \cap B]/P[B] = P[A]$. In terms of relative frequencies, this becomes $\mathfrak{N}(A \cap B)/\mathfrak{N}(B) = \mathfrak{N}(A)/\mathfrak{N}(U)$. In words, the relative frequency of A in B is the same as the relative frequency of A in the sample space. The situation may be visualized by a Venn diagram in which areas are proportional to frequencies. In Fig. 8-2, area $(A \cap B)$/area (B) = area (A)/area (U). This is a very special situation. The proportionality of the frequencies means that the occurrence or nonoccurrence of B has no effect on the probability of the occurrence of A.

Statistical independence must not be confused with mutual exclusiveness. If A and B are mutually exclusive, the occurrence of one precludes the occurrence of the other, and they are certainly *not* independent.

In practice we assume that two events are independent if the occurrence of one has no effect on the conditions influencing the occurrence of the other. For example, if we throw a die and then throw it again after thorough shaking, experience shows that the relative frequencies of the faces on the second throw are the same regardless of the outcome of the first throw. We say that the throws are independent. The probability of

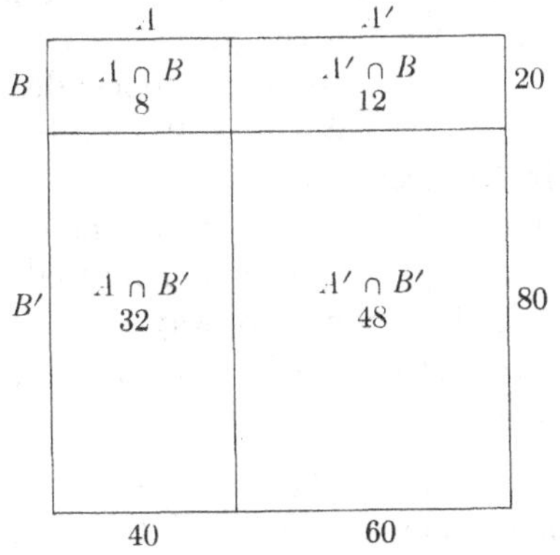

FIGURE 8-2

throwing a seven twice in succession with two dice is $(1/6)(1/6)$ or $(1/36)$.

Find the probability that: (i) a coin comes up heads twice in succession, (j) a pair of dice totals seven on the first throw and six on the second, (k) both of two cards drawn with replacement from a deck of 52 cards are aces, (l) both of two cards drawn from a deck without replacement are aces.

We should like to extend the above ideas to more than two events. For n events, $A_1, A_2, \ldots, A_n$, (3) becomes

$$(4) \quad P[A_1 \cap A_2 \cap \ldots \cap A_n] = P[A_1]P[A_2 \mid A_1]P[A_3 \mid A_1 \cap A_2] \ldots$$
$$P[A_n \mid A_1 \cap A_2 \cap \ldots \cap A_{n-1}].$$

Under what conditions can we replace the conditional probabilities in the right member by the corresponding absolute probabilities? At first sight we might guess that it would be sufficient for each pair of events to be independent. Actually, this pairwise independence is not enough. We shall not go into details, but the interested student may consult Chapter 5 of the book cited in Problem 15 of Section 8–4. We call a set of events $\{A_1, \ldots, A_n\}$ *mutually independent* if the probability of the intersection of any subset is the product of the corresponding absolute probabilities.

$$(5) \quad \begin{array}{c} \textit{If } A_1, A_2, \ldots, A_n \textit{ are mutually independent,} \\ P[A_1 \cap A_2 \cap \ldots \cap A_n] = P[A_1]P[A_2] \ldots P[A_n]. \end{array}$$

In practice, we use this formula when we have reason to believe that no combination of occurrences or nonoccurrences of any set of the events has any effect on the chances of any other event. Of course, this belief may be tested by experimental determination of relative frequencies.

Formula (5) is one of the most frequently used laws in probability and statistics. For example, what is the probability that a coin will come up heads 10 times in a row? Assuming independence of the throws, the answer is 2^{-10} or $1/1024$. What is the probability of the sequence HTHTHTHTHT in 10 throws of a coin? The answer is the same, 2^{-10}. What is the probability that in 10 throws there will be five heads and five tails? This result can occur in many mutually exclusive ways, each with probability 2^{-10}. The number of ways is the number of partitions of 10 objects into two sets of five each, that is, $10!/(5!)^2$. Hence the probability is $2^{-10}10!/(5!)^2$.

(m) What law from Section 8–2 did we use to get the last result? (n) From a box containing nine red and one white ball, ten balls are drawn with replacement. What is the probability that they are all red? (o) If the probability of a plane's being shot down during a mission is $1/10$, is it certain that it will get shot down if it undertakes ten missions?   (p) From a box containing five red and five white balls, three balls are drawn, each one being replaced if and only if it is red. What is the probability that the third ball is red?

PROBLEMS

1. What is the probability that a five-card poker hand has no pairs? Solve this problem by direct enumeration and by using theorems of this section.

2. Suppose that the probability that life appears on a planet similar to the earth is 10^{-10}. Suppose that the probability of a sun similar to ours having a planet similar to the earth is 10^{-5}. What is the probability that a sun similar to ours has a planet on which life exists?

3. Suppose that of the students who cheat on examinations, one out of 100 is caught. What is the probability that a person who cheats ten times gets caught? one hundred times?

4. Suppose the chromosomes of one parent contain one gene for brown eyes and one for blue, while those of the other contain two genes for blue eyes. The chromosome of the child contains one gene from each parent. The eyes will be brown if at least one gene for brown is present. What is the probability that this will happen, assuming that either gene from each parent is equally likely?

5. What is the probability that a child of the parents in Problem 4 has blue eyes? that two children in a row have blue eyes? that of four children, all have blue eyes?

6. Answer the questions of Problems 4 and 5 assuming that both parents have one gene for blue and one for brown.

7. Suppose that one-tenth of the pennies in circulation have heads on both sides. Suppose that a penny is picked at random and tossed. What is the probability that it shows heads?

8. In the situation in Problem 7, if a penny is tossed and shows heads, what is the probability that it is a two-headed penny?

9. A die is to be thrown n times. What is the probability that six comes up at least once in the n throws?

10. That four comes up exactly three times? exactly once?

11. Suppose that $D = A \cup B \cup C$, where A, B, and C are independent. Show that $P[D] = 1 - P[A']P[B']P[C']$.

12. If a head has come up six times in a row on the toss of an evenly balanced coin, what is the probability of its coming up again?

13. The following quotation is from *The Criminality of Women*, by Otto Pollak: "Another type of fraud in which female accomplices are used is the disposition of fake jewelry by the professional pawner. The faker usually buys a quantity of rolled trinkets and works a small piece of genuine gold into each. The woman helper takes the product to a pawnbroker and offers it for what it may be worth, making no statement as to its being real gold. The pawnbroker tests it by acid application and, if he happens to hit the piece of real gold, buys the whole trinket as such. If he hits another spot, he refuses to buy. According to the law of probability, the woman tries other pawnbrokers until she has disposed of the whole stock." Explain the last sentence.

14. Suppose that the probability of the pawnbroker hitting the gold is 1/3. What is the probability that the woman would be successful with the third pawnbroker? What is the probability that she would have to visit no more than three pawnbrokers?

15. Answer the questions in Problem 14 if the probability of a pawnbroker hitting gold is only 1/10.

16. This problem is adapted from *First Course in Probability and Statistics*, by J. Neyman. A procedure for diagnosing cancer of the lung consists of three steps: 1. Collect a sample of sputum from the patient. 2. Put a small subsample of this sample on a slide. 3. Examine the slide for cancerous cells. Let p_1 = probability that sputum from a cancerous patient contains cancerous cells, p_2 = probability that the subsample contains cancerous cells given that the sample does, and p_3 = probability that cancerous cells will be observed on a slide containing some such cells. If this procedure is followed once, what is the probability that a cancer will be discovered in a cancerous patient?

17. What is the probability that it will not be discovered?

18. What is the probability of correct diagnosis if the procedure is followed twice?

19. Ten times?

20. Suppose k samples of sputum are taken, m slides made from each sample, and each slide examined n times. Assuming that all kmn operations are statistically independent, find a formula for the probability of a correct diagnosis.

21. If $p_1 = 0.80$, $p_2 = 0.25$, $p_3 = 0.90$, how large must k, m, and n be so that the probability of correct diagnosis is at least 95%?

★22. A coin is tossed until heads appears twice in succession. What is the probability that this happens on an even throw? odd throw?

★23. Answer the same questions for tails.

★24. Answer the same questions when either heads or tails comes up twice in succession.

★25. Suppose that p is the probability that an individual in a community favors a certain verdict in a certain jury-trial case. Find the probability that the majority of a 12-man jury picked at random favors this verdict.

ANSWERS TO EXERCISES

(a) 1/9, 5/36, 1/18, 2/5. (b) 1/9, 1/2, 1/9, 2/9. (c) 21/55. (d) 3/22. (e) 7/22. (f) 53/110. (g) 57/110. (h) The conditional probability is not defined, but obviously $P[A \cap B] = 0$. (i) 1/4. (j) $(1/6)(5/36)$. (k) $(1/13)^2$. (l) Events not independent; $(1/13)(3/51)$. (m) $(8 \cdot 2 \cdot 3)$. (n) $(0.9)^{10}$. (o) No, the probability of its survival is the same as the probability in Exercise (n). (p) Events not independent. There are four mutually exclusive ways this can happen, RRR, RWR, WRR, and WWR. The probability is $(5/10)(5/10)(5/10) + (5/10)(5/10)(5/9) + (5/10)(5/9)(5/9) + (5/10)(4/9)(5/8)$.

ANSWERS TO PROBLEMS

1. $4^5\binom{13}{5}\bigg/\binom{52}{5} \doteq 0.507$. 3. $1 - (99/100)^{10} \doteq 1/10$; $1 - (0.99)^{100} \doteq 0.73$.

5. 1/2; 1/4; 1/16. 7. $(9/10)(1/2) + 1/10$. 9. $1 - (5/6)^n$. 11. $P[D] = 1 -$

$P[D'] = 1 - P[A' \cap B' \cap C']$. 13. By chance, *some* pawnbroker is very likely to hit the gold. 15. $(9/10)^2(1/10)$; $1/10 + (9/10)(1/10) + (9/10)^2(1/10)$. 17. $1 - p_1p_2p_3$.   19. $1 - (1 - p_1p_2p_3)^{10}$.   20. $1 - (1 - p_1p_2p_3)^{kmn}$. If independence is not assumed the answer is $1 - (q_1 + p_1(q_2 + p_2q_3^n)^m)^k$. 21. $kmn \geq 16$.

8–6 Discrete distributions. Let $v =$ the sum of the faces when two dice are thrown, let $f(x) = P[v = x]$ and $F(x) = P[v \leq x]$. The graphs of f and F are sketched in Figs. 8–3 and 8–4. Each of these functions tells us everything about the probabilities associated with different values of v, for f tells us directly the probability that $v = x$, since $f(x)$ is this probability, and F tells us indirectly, for the probability that $v = x$ is given by $F(x) - F(x - 1)$, the height of the step in F at $x = v$. More generally $P[x_1 < v \leq x_2] = F(x_2) - F(x_1)$. If we think of zero probabilities associated with values of v other than the 11 possibilities, the domains of f and F are the real numbers and their ranges are subsets of the interval $(0 \quad 1)$.

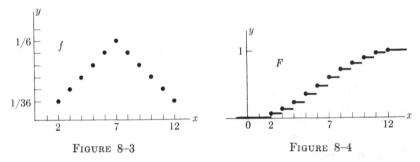

FIGURE 8–3   FIGURE 8–4

Let S be a sample space, let r be a function whose domain is S and whose range is a subset of the real numbers, and let $v = r(u)$. Then we call v a *random variable*. Of course, v is just an ordinary variable in every way. The term "random" merely indicates that we are using it as the dependent variable in a function that maps a sample space into real numbers. In the dice example, S is the set of points in Fig. 8–1, $u = (a, b)$, where a is the face on the first die, b is the face on the second, and $v = a + b$ gives the mapping.

Now let $f(x) = P[v = x] =$ the probability that the random variable has the value x. We call f the *frequency function* of v. It gives the probability (relative frequency) of each value of v. Let $F(x) = P[v \leq x] =$ the probability that v has a value less than or equal to x. We call F the *distribution function* of v. Sometimes F is called the cumulative distribution function. The word "distribution" comes from thinking of a total probability of 1 as distributed over the range of v. In these definitions there is no limitation on the nature of S nor any requirement that the range of r

be finite. In this section we consider only random variables whose ranges are finite. The corresponding distributions are naturally called *discrete*. (See Section 6–8, Problem 15.)

(a) Let a coin be tossed twice. Let heads be represented by 1, tails by 0. Letting x = result of first toss, y = result of second, graph the sample space of couples of tosses. Letting v = the number of heads in two tosses, find and graph the frequency function and distribution function of v. (b) Do the same letting v = the number that turns up when one die is thrown.

We note that for a discrete frequency function

(1) $$f(x) \geq 0,$$

(2) $$\sum f(x) = 1,$$

where the summation extends over the domain of f.

(3) $$F(x) = \sum_{t \leq x} f(t),$$

(4) $$x_2 \geq x_1 \to F(x_2) \geq F(x_1),$$

(5) $$\lim_{x \to -\infty} F(x) = 0, \qquad \lim_{x \to \infty} F(x) = 1.$$

Any function that satisfies (1) and (2), i.e., is non-negative and whose values add to one, is the frequency function of some random variable. Also, any function that satisfies (4) and (5), i.e., is monotonic increasing from 0 to 1, is the distribution function of some random variable.

Let g be a function that maps the range of a random variable v into a set of real numbers. Then we define the *expectation* of $g(x)$ by

(6) $$E[g(x)] = \sum g(x)f(x),$$

where f is the discrete frequency function of v and the summation extends over the domain of f. Suppose, for example, that an individual is gambling on the value of v and that $g(x)$ is the payoff he receives corresponding to $v = x$, where $g(x) < 0$ means that he has to pay out $|g(x)|$. Then if he gambles many times under the same conditions, he may expect to receive $g(x_1)$ a proportion $f(x_1)$ of the time, $g(x_2)$ a proportion $f(x_2)$ of the time, and so on for all values of x. On the average, then, he may expect to receive $E[g(x)]$. If he plays 100 times he may expect to receive $100E[g(x)]$, since $g(x_i)[100f(x_i)]$ is the amount he might expect to win from the occasions when $v = x_i$.

For example, if two people bet even money on the throws of two dice with the understanding that the thrower wins \$1 if he throws a seven and pays out the same if he does not, we have $g(x) = $ the payoff to the first player when the outcome is x. Then $g(7$ comes up$) = 1$ and $g(7$ does not come up$) = -1$. The corresponding probabilities are $1/6$ and $5/6$, so that the player's expection is $(1/6)(1) + (5/6)(-1) = -2/3$. That is, he may expect to lose $2/3$ of a dollar on the average each time he plays this game.

(c) What should the bets be in this game in order for it to be fair, i.e., so that the expectation of each player is zero?   (d) Suppose A agrees to pay B an amount equal to the sum of the faces of two dice that he throws. What is B's expectation?   (e) What should A require B to pay him each time he plays this game in order to break even in the long run?

Imagine an experiment in which there are only two possible outcomes, success with probability p, and failure with probability $q = 1 - p$. Suppose that this experiment is repeated n times under identical conditions, so that each experiment is independent of the others. Let $v = $ the number of successes in the n trials. What is the frequency function of v? We have $f(x) = $ the probability of exactly x successes out of n trials. Letting 1 be a success and 0 a failure, a possible outcome of the trials may be represented by a sequence such as $011010111 \ldots 10$. By (8–5–5) the probability of any such outcome is $p^x q^{n-x}$. But there are many different sequences in which x 1's and $(n - x)$ 0's appear. Indeed, from Exercise (s) in Section 8–4, there are just C_x^n. Hence we have C_x^n mutually exclusive ways, each with probability $p^x q^{n-x}$, and

$$(7) \qquad\qquad f(x) = C_x^n p^x q^{n-x}.$$

This distribution is known as the *binomial distribution*.

(f) Explain the name.

Sketch the frequency function of a binomial variable for $n = 4$ and p as follows: (g) 0.1,   (h) 0.2,   (i) 0.3,   (j) 0.4,   (k) 0.5,   (l) 0.6,   (m) 0.7,   (n) 0.8,   (o) 0.9. (Keep your results for future reference in Chapter 9.)

The frequency function or the distribution function of a random variable (given by a table, graph, or defining formula) tells us everything about the situation. However, it is sometimes convenient to find certain numbers that are characteristic of a distribution and that suggest its important features. Among these the most common is the *first moment* or *mean* μ (pronounced "mew"), defined as the expectation of x, that is,

$$(8) \qquad\qquad \mu = E[x] = \sum xf(x).$$

The term "moment" arises in the following way. Let us imagine the probabilities $f(x)$ as weights located at the points $(x, 0)$ along the x-axis.

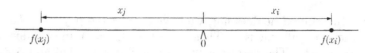

FIGURE 8–5

If we place a fulcrum at the origin, the moment of $f(x)$ about it is $xf(x)$, the weight times the lever arm. (See Fig. 8–5.) The sum (8) is the total moment of all the weights. The total weight is one by (2). If we place a weight of one at a distance $(E[x], 0)$, the system will be in balance. Hence $E[x]$ is both the first moment and the average lever arm.

(p) Let x = the number of heads that appear when a coin is tossed once. Find the frequency function and first moment of x.   (q) Find the first moments of the distributions in Exercises (g) through (o).

Another frequently used number associated with a distribution is its *variance* defined as the expectation of $(x - E[x])^2$. It is symbolized by σ^2 (sigma squared), and σ is called the *standard deviation*. We have

$$(9) \qquad \sigma^2 = E[(x - \mu)^2] = \sum (x - \mu)^2 f(x).$$

The variance is simply the weighted average of the squares of the distances of each value of x from the mean. It is a measure of the extent to which the points are spread out, the greater the variance the greater the average distance. Squaring the differences eliminates negatives, which is desirable since we are interested in how far away each point is and not in the direction. The same result could be accomplished by using $E[|x - \mu|]$, called the *mean deviation*, but the variance is more widely used.

(r) Find the variance of the distribution of Fig. 8–3.   (s) Find the mean and variance of the distribution in Exercise (a).   (t) Do the same for Exercise (b). (u) For Exercise (g).   (v) For Exercise (k).

PROBLEMS

1. If out of 100 people age 5, 1 dies at age 10, 1 at age 20, 2 at age 30, 2 at age 40, 6 at age 50, 11 at age 60, 19 at age 70, 27 at age 80, 23 at age 90, and the rest at age 100, what is the expectation of the age at death (life expectancy) of a person in the group? What is meant when it is said that the life expectancy of a person in the United States is 65 years?

2. A coin is tossed 10 times. What is the probability that it comes up heads five times and tails five times?

3. If the probability of success of an experiment is $1/100$, how many trials are necessary to have a probability of $1/2$ of getting at least one success?

4. Show that for the binomial distribution, the first moment $\mu = np$. Why is this reasonable?

★5. Show that the variance of the binomial distribution is npq.

6. A distribution is called *uniform* if the frequency function is a constant function. Give an example. What is the appearance of the distribution function?

7. Suppose that a "wheel of fortune" has 12 numerals, 1 through 12, and is equally likely to stop at any one. Let v = the number at which it stops. Sketch the frequency function and the distribution function.

8. In the situation in Problem 7, if the sponsor gives away $1000 if the wheel stops at the number chosen by the contestant, what is the expected cost?

9. Suppose in the above situation, the sponsor merely gives the contestant a chance to win $1000 by correctly answering a question on which three out of four people fail. What then is the expected cost?

10. Let v = the number of aces in a five-card hand. Find the frequency function and sketch it. Find the first moment and the variance.

★11. Since the values of $f(x)$, where f is a discrete frequency function, are bounded from above, there must be a maximum. If $f(x)$ is this maximum, then x is called a *most probable value* of x. What is (or are) the most probable value (or values) of a binomial random variable?

★12. The formula in Problem 16 of Section 8–4, where n_1, n, and r are fixed, is a defining formula for a frequency function called the *hypergeometric distribution*. Sketch it for $n = 5$, $n_1 = 2$, $r = 3$. Find its first moment and variance.

13. Let v be the number of hearts in a 13-card hand from a full 52-card deck. Write the frequency function of v and sketch it. What is the first moment? What is the most probable value or values?

★*Poisson distribution.* Imagine a sequence of trials to which the binomial distribution applies, and suppose that the number n of trials is large but the probability p is small, so that their product np is of moderate size. The combinatorial coefficient in (7) is difficult to compute. Under these conditions it can be proved (see Feller's *Probability Theory and Its Applications* or Neyman's *First Course in Probability and Statistics*) that the following approximation is a good one.

(10) $$C_x^n p^x q^{n-x} \doteq (np)^x e^{-np}/x!.$$

More precisely, it is true that if $n \to \infty$ and $p \to 0$ in such a way that np remains fixed, then the left member of (10) approaches $e^{-np}(np)^x/x!$ Letting $\lambda = np$, we have

(11) $$p(x) = \frac{e^{-\lambda}\lambda^x}{x!},$$

which is called the *Poisson distribution*. Its first moment is λ.

As an example of the use of this approximation, consider the probability of the appearance of five cases of a disease in a population of 100,000,000 where the probability of getting this disease is 1/50,000,000. The left member of (10) would be unmanageable! But here $\lambda = 2$, and the approximation gives us $e^{-2}2^5/5!$, or about 0.037.

14. In this example, find the approximate probabilities of the appearance of 0, 1, 2, 3, 4, 5, 6 cases and graph your results.

In practical problems when we are faced with an event that occurs rather rarely (p small) in many trials (n large), we take the observed average number of occurrences to be an estimate of λ. Then we use $p(x)$ as an estimate of the probability of x occurrences. Typically we have an event that occurs λ times on the average during a time interval, and we wish the probabilities of different numbers of occurrences during the time interval.

15. Suppose a radioactive substance emits an average of four particles per unit time. What is the probability that it will emit none, 1, 2, 3, 4, and 5 particles in a particular unit time interval? (Here we can imagine each of the n particles it *could* emit as a trial.) Graph your results.

16. Suppose that the average number of bacteria per unit area on a plate is three. What is the probability that a particular unit area contains bacteria? What does this mean in terms of the number of unit areas that may be expected to contain no bacteria?

Answers to Exercises

(a) $\{(1, 0), (0, 1), (0, 0), (1, 1)\}$; $f(0) = f(2) = 1/4, f(1) = 1/2$. (b) $\{1, 2, 3, 4, 5, 6\}$; $f = 1/6$. (c) The thrower should pay \$1 when he loses and receive \$5 when he wins. (d) 7. (e) \$7. (f) The distribution is given by $(p + q)^n$. (p) $f(0) = f(1) = 1/2, \mu = 0(1/2) + 1(1/2) = 1/2$. (q) 0.4, 0.8, 1.2, 1.6, 2, 2.4, 2.8, 3.2, 3.6. (r) 35/6. (s) 1, 1/2. (t) 7/2, 35/12. (u) 0.36. (v) 1.

Answers to Problems

1. 74.9. 2. 63/256. 3. Probability of at least one success is $1 - (0.99)^n$. For this $\geq 1/2$, $(0.99)^n \leq 0.5$, $n \log 0.99 \leq \log 0.5$, or $n \geq \log 0.5 / \log 0.99$. (Note change of direction because $\log 0.99 < 0$.) Then $n > 69$. 7. Uniform, $f = 1/12$. 9. About \$21.

13. $f(x) = C_x^{13} C_{13-x}^{39}/C_{13}^{52}$. 15. $e^{-4} 4^x/x!$. 16. $1 - e^{-3}$. We may expect about 5% to contain no bacteria.

8-7 Continuous distributions. Consider, as suggested in Fig. 8-6, a pointer free to rotate about its center and balanced so that it is as likely to come to rest at one point as at another. It is easy to see that the probability that it comes to rest at any particular point is zero. For, suppose that the probability of a point is $\epsilon > 0$. There are more than $1/\epsilon$ points, indeed there are an infinite number of points. Hence the total probability would be more than $(1/\epsilon)\epsilon$, i.e., more than 1, which contradicts (8-2-8).

To deal with the situation we must consider the probability that the pointer comes to rest in some interval. Since we naturally feel that the pointer is twice as likely to come to rest in an interval twice as long, we are led to define the probability of an interval as the length of the in-

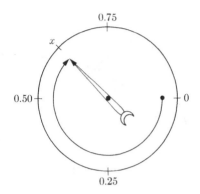

FIGURE 8-6

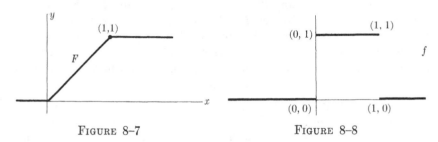

FIGURE 8-7   FIGURE 8-8

terval divided by the total length of the circumference, that is, $P[I] = L(I)/L(U)$, where U is the sample space. Note the analogy of this with (8–3–1). In each case we take the probability of a set as a measure of the set divided by the same measure of the sample space.

In the case of the rotating pointer, $L(U)$ is the length of the circumference. If we choose units so that this length is 1, designate a point on the circumference as origin, and let $v =$ the shortest clockwise distance on the circumference from the origin to the point of rest, then $P[v \leq x] = x$. The cumulative distribution function of v is as sketched in Fig. 8–7.

Let $x_2 > x_1$, then $F(x_2) - F(x_1)$ is the probability of the interval $(x_1 \quad x_2)$; and $F(x_2) - F(x_1)/(x_2 - x_1)$, the slope of F, is the probability per unit length. If we think of the probability as a unit mass distributed over the interval $(0 \quad 1)$, this slope is the mass per unit length, i.e., the density. Let $f(x) =$ the slope of F at $(x, F(x))$. In this case $f(x)$ is 1 in $(0 \quad 1)$, undefined at 0 and 1, and 0 elsewhere. It is sketched in Fig. 8–8.

(a) Suppose that the circumference of the dial is 2. Sketch F and f.   (b) Do the same for circumference 1/3.   (c) Do the same for circumference c.

To generalize the above example let us imagine a random variable v with a distribution function F that is continuous and that has a continuous

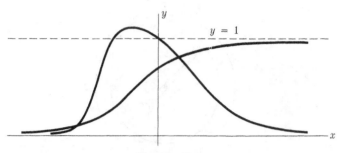

FIGURE 8–9

derivative everywhere except possibly at a finite number of points. Since
it must satisfy (8–6–4) and (8–6–5) it will have an appearance something
like Fig. 8–9. We call DF the *density function*, designate it by f, and call
$f(x)$ the *probability density*. Where probability is more concentrated, F
is steep and $f(x)$ is large. Where F is horizontal, $f(x)$ is zero. When $f(x)$
is constant over an interval and zero elsewhere, we call the distribution
uniform. Figure 8–8 is typical of a uniform distribution.

(d) Is a distribution function F necessarily steepest at $x = 0$? If it is, what
can one say of the density function f at $x = 0$? (e) What is the appearance
of f if F is symmetric about the point where it crosses the y-axis? (f) Show
that a continuous distribution function F must satisfy (8–6–4) and (8–6–5).
(g) Show that a density function f satisfies (8–6–1).

To get theorems corresponding to (8–6–2) and (8–6–3), we must replace the summations by integrals. For, since $DF = f$,

(1) $$P[x_1 < v \le x_2] = F(x_2) - F(x_1) = \int_{x_1}^{x_2} f.$$

That is, the probability over any interval is the integral of the density
function over that interval. Letting $F(-\infty) = \lim_{x \to -\infty} F(x)$, $F(\infty) = \lim_{x \to \infty} F(x)$, and defining

(2) **Def.** $$\int_{-\infty}^{x} f = \lim_{a \to -\infty} \int_{a}^{x} f,$$

(3) **Def.** $$\int_{-\infty}^{\infty} f = \int_{-\infty}^{x} f + \lim_{b \to \infty} \int_{x}^{b} f,$$

we have

(4) $$F(x) = \int_{-\infty}^{x} f,$$

which corresponds to (8–6–3), and

$$(5) \qquad \int_{-\infty}^{\infty} f = 1,$$

which corresponds to (8–6–2). We use $-\infty$ and ∞ as limits of integration because we wish to leave open the possibility that the density may not be zero no matter how far out we go. Of course, if $f(x) = 0$ outside some interval $(a \quad b)$, then we may change $-\infty$ to a and ∞ to b in the above formulas.

(h) Explain carefully the justification of (1).   (i) Derive (4) from (1), (2), and (8–6–5).   (j) Do the same for (5).   (k) Verify that (1), (4), and (5) hold for the distribution of Fig. 8–7.

Sketch each of the following density functions, find and sketch the distribution function, and verify that (5) holds: (l) $f(x) = 2x$ in $(0 \quad 1)$, zero elsewhere; (m) $f(x) = 1/x$ in $(1 \quad e)$, zero elsewhere;   ★(n) $f(x) = e^{-x}$ for $x \geq 0$, zero elsewhere.

In strict analogy to (8–6–8) and (8–6–9), we define the mean and variance of a continuous distribution by

$$(6) \qquad \mu = \int_{-\infty}^{\infty} xf(x)\, dx,$$

$$(7) \qquad \sigma^2 = \int_{-\infty}^{\infty} (x - \mu)^2 f(x)\, dx.$$

(o) Find the mean and variance of the distribution of Fig. 8–7.   (p) Do the same for the distribution of Exercise (a).   (q) Do the same for Exercise (c).

PROBLEMS

1. We may regard Fig. 5–19 as giving the distribution function for which $F(x) = $ the probability that the pitch x will be discriminated $= $ the probability that the minimum pitch discriminated is less than or equal to x. Graph the density function f. What is $f(x)$? Find the mean and variance.

2. Argue that if f is the probability density, then $f(x)\, \Delta x$ gives approximately the probability of the interval $(x \quad x + \Delta x)$.

3. Suppose $F(x) = $ the probability that an individual dies at or before the age of x. What would you expect the graph of F to look like? Sketch a plausible graph. Sketch the corresponding f. What is $f(x)$?

4. The expectation of $g(x)$, where x has the continuous probability density $f(x)$, is defined by

$$(8) \qquad E[g(x)] = \int_{-\infty}^{\infty} gf.$$

Argue for the reasonableness of this and its consistency with (8–6–6).

5. Show that (6) follows from the first equation in (8-6-8) and from (8) above.

6. Derive (7) similarly.

★7. Show that if ce^{-I} is to be a probability density function in the interval $(0\_\_\infty)$, then $c = 1$.

8. Suppose $f(x) = c$ for $x \in (a\_\_b)$ and $f(x) = 0$ for $x \notin (a\_\_b)$. Find the value of c so that f is a density function.

9. Find the first moment and variance for the distribution in Problem 8.

10. Sketch the distribution function corresponding to f of Problem 7.

11. Determine c so that f, defined by $f(x) = c(mx + b)$ for $x \in (0\_\_\_10)$ and $f(x) = 0$ otherwise, is a density function. Sketch it and the corresponding distribution function for $m = 1$, $b = 2$.

12. What is the expected value of x in the pointer problem?

13. Suppose a point is chosen at random inside a circle of radius 2. What is the probability that it lies inside the concentric circle of radius 1?

14. To answer the question in Problem 13 we need to extend the definition of probability to cover areas, since the sample space is a plane area and the events are subsets of it. Suggest a definition for probability of a subset of a plane area on the assumption that probability is uniformly distributed.

15. With the definition chosen in Problem 14, what is the probability that a point chosen at random in a unit circle lies at the center?

16. Lies on a specified diameter?

17. Lies in a region of unit area?

★18. A board is ruled with a series of equidistant parallel lines. A needle whose length is less than the distance between the lines is thrown at random on the board. Letting d be the length of the needle and h the distance between lines, show that the probability that the needle intersects a line is $2d/\pi h$. (Since the probability should be approximated by the relative frequency in many trials, this formula can be used to make an experimental determination of π. The idea is due to the Comte de Buffon (1707–1788), and the problem is called Buffon's needle problem. See *Introduction to Mathematical Probability*, by J. V. Uspensky, p. 112, and elsewhere.)

★19. Determine a so that $F(x) = 1 - ax^{-b}$ is a distribution function. Find f. Graph F and f for $b = 1.5$. (This is the Pareto distribution, in which $F(x)$ is the proportion of the population having an income of x or less. See *The Theory of Econometrics*, by H. T. Davis, p. 23.)

★20. Show that if a measure M is defined for subsets of a sample space U and probability is defined by $P(.1) = M(A)/M(U)$, then (8-2-1) through (8-2-4) are satisfied.

Answers to Exercises

(a) $F(x) = x/2$ for $0 \le x \le 2$, $F(x) = 0$ for $x < 1$, $F(x) = 1$ for $x > 2$. $f(x) = 1/2$ in the same interval, 0 elsewhere.   (b) $F(x) = 3x$ in $0 \le x \le 1/3$, $f(x) = 3$.   (c) $F(x) = x/c$ in $0 \le x \le c$, $f(x) = 1/c$ in $0 < x < c$.   (d) No. Has its maximum.   (e) f is symmetric about the line $x = 0$.   (f) Since $F(x_2) - F(x_1) = P[x_1 < v \le x_2]$, $F(x_2) - F(x_1) \ge 0$ by (8-2-2). This implies (8-6-4). Also, since $F(x) =$ the probability that $v \le x$, and since the probability is 1 that v takes some value in $(-\infty\_\infty)$, (8–6–5) follows. Note that these arguments are quite independent of whether the distribution is discrete or continuous. (g) Since $F(x)$ is nowhere decreasing by (8–6–4), its derivative $f(x)$ is nowhere negative. (h) The first equation comes from the definition of $F(x)$. We have $P[x_1 < v \leq x_2] = P[v \leq x_2] - P[v \leq x_1] = F(x_2) - F(x_1)$. The first equation here follows from $(-\infty\_x_1) \cup (x_1\_x_2) = (-\infty\_x_2)$. (i) In (1) let x_1 approach $-\infty$. (k) Note that in this case the integrals have finite limits. (l) through (n) f is undefined at endpoints. (o) 1/2, 1/12. (p) 1, 1/3. (q) $c/2$, $c^2/12$.

Answers to Problems

1. $f(x) \Delta x$ is approximately the probability that a pitch in the interval $(x\_x + \Delta x)$ will first be discriminated. 3.5, 1/3. 3. Starts at origin, rises steeply for very short interval (infant mortality), then rises rather slowly and steadily until a more rapid rise after age 50, approaches 1 after age 80. Same general shape as first graph in Fig. 8–9. 7. Integrate ce^{-I} in the interval from 0 to b and then let $b \to \infty$. 9. $a + (b - a)/2$, $(b - a)^2/12$. 11. $(50m + 10b)^{-1}$. 13. 1/4. 14. Area of set over area of space. 15. Zero. 16. Zero. 17. $1/\pi$.

8–8 The normal distribution. The most useful of all continuous distributions is the *normal distribution* whose density function g is given by

$$(1) \qquad g(x) = \frac{1}{\sigma\sqrt{2\pi}} e^{-\frac{1}{2}\left(\frac{x-\mu}{\sigma}\right)^2}$$

over the interval $(-\infty\_\infty)$. This function obviously satisfies (8–6–1), and it also satisfies (8–7–5), though the proof is too advanced to be given here. The parameters μ and σ are respectively the mean and standard deviation of the distribution.

Since we can always make the mean zero by choosing it as the origin for our measurements, and since we can make $\sigma = 1$ by choosing an appropriate unit of measurement, we work with the *standard normal distribution* defined by

$$(2) \qquad f(z) = (2\pi)^{-1/2} e^{-z^2/2}.$$

The relation between x and z is ·

$$(3) \qquad z = (x - \mu)/\sigma \quad \text{or} \quad x = \sigma z + \mu.$$

Table VI in the Appendix gives $f(z)$ for selected values of z. Since the function is even, we do not need a separate tabulation for $z < 0$.

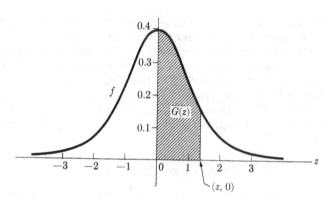

FIGURE 8–10

Figure 8–10 shows the standard normal density function. Since

$$(4) \qquad\qquad Df(z) = f(z) \cdot (-z),$$

by (7–14–12) and (7–7–1), we see that $Df(z) = 0$ only at $z = 0$. Hence the curve has no other maxima or minima. Since

$$(5) \qquad D^2 f(z) = f(z) \cdot (-1) + f(z) \cdot (-z)^2 = f(z)(z^2 - 1),$$

the curve has inflections at $(\pm 1, (2\pi e)^{-1/2})$. It evidently approaches the z-axis asymptotically.

The more general normal distribution (1) is similar, with maximum at $x = \mu$ and inflections at $x = \mu \pm \sigma$.

To find the probability that a normally distributed random variable lies in an interval $(a \quad b)$ we need to evaluate the integral

$$(6) \qquad\qquad F(z) = \int_{-\infty}^{z} (2\pi)^{-1/2} e^{-I^2/2}.$$

We cannot do this by finding a simple formula for the antiderivative, since there is no elementary function whose derivative is the integrand. (An *elementary function* is one constructible from the constant functions, I, e^I, $\log_e$, and the trigonometric functions, and their inverses by the operations of addition, subtraction, multiplication, division, root extraction, and composition.) However, the integral (6) can be approximated by various means. It turns out to be most convenient to work with the integral

$$(7) \qquad\qquad H(z) = F(z) - F(0) = \int_{0}^{z} (2\pi)^{-1/2} e^{-I^2/2}.$$

Table VI gives $H(z)$ for selected values of z. Since $F(0) = 0.5$, we have

$F(z) = H(z) + 0.5$ for $z > 0$. For $z < 0$, we have $F(z) = 1 - F(-z) = 0.5 - H(-z)$.

(a) Why does $F(0) = 0.5$?   (b) Why does $F(z) = 1 - F(-z)$ for $z < 0$?
(c) Justify (7).

By making use of the Table VI and (8-7-1) we can find the probability that any normally distributed variable lies in a certain interval. For example, what is the probability that a variable following the standard normal distribution has a value less than 1? From Table VI, $H(1) = 0.341$; hence $F(1) = 0.841$. The probability that it is greater than 1 is therefore $1 - 0.841 = 0.159$. What is the probability that it deviates from the origin by less than 1, i.e., that $-1 < v < 1$? Clearly the probability is twice that for the interval $(0\ \ 1)$, that is, $2H(1)$, or 0.682. The probability that it deviates more than 1 is therefore 0.318. In general, $P[-a < v < a] = 2H(a)$. It is because probabilities of this kind are often needed that $H(z)$ is tabulated instead of $F(z)$.

What is the probability that a standard normal random variable is: (d) greater than 2?   (e) less than 2?   (f) greater than -2?   (g) less than -2?   (h) deviates from the origin by less than 2?   (i) deviates from the origin by more than 2?   (j) lies in the interval $(1\ \ 2)$?

By using (3) we can answer similar questions for a normal random variable with mean μ and variance σ^2. If x is given, we find the corresponding z from (3) and use Table VI as before. For example, what is the probability that a normal random variable deviates from the mean by less than the standard deviation? This deviation means that $x - \mu = \sigma$, hence $z = 1$, and we have from above the probability 0.682. This means that if we sample from a population following the normal distribution, we expect deviations from the mean of less than the standard deviation about 2/3 of the time.

What is the probability that a normal random variable differs from the mean by: (k) less than twice σ?   (l) less than 3σ?   (m) more than σ?   (n) more than 2σ?   (o) more than 3σ?

Assume that the weights of people are normally distributed, so that the actual weights of a set of individuals may be viewed as a random sample from an infinite normal population. Suppose that the average weight is 175 lb and the standard deviation is 25. Out of 1000 people, how many would you expect to have weights: (p) between 150 and 200?   (q) between 125 and 225?   (r) more than 225?   (s) less than 125?

(t) Suppose that an instructor of a large class grades "on the curve" by finding the mean μ and standard deviation σ of all numerical grades, dividing the interval $(-\infty\ \ \infty)$ into five intervals at the points $\mu - 1.5\sigma$, $\mu - 0.5\sigma$, $\mu + 0.5\sigma$, and $\mu + 1.5\sigma$, then assigning E, D, C, B, or A according to the interval in which the numerical grade lies. About what percentages of students get each grade?

(u) Complete the following: In a population that is normally distributed, about __ percent of the members lie within one standard deviation of the mean, __ percent within two standard deviations, and __ percent within three standard deviations.

PROBLEMS

1. Show that $g(x)$ and $f(z)$ satisfy (8-6-1).

2. Find $f(0)$.

3. Sketch (1) for $\mu = 2$, $\sigma = 3$. Show the points of inflection.

4. Suppose v is distributed normally. Let E be a number such that $P(\mu - E \leq v \leq \mu + E) = 1/2$. Find E in terms of μ and σ.

5. When measurements of a physical entity are made, it often happens that they are distributed approximately normally with μ equal to the true value. Supposing this to be the case, what percentage of the measurements would be expected to fall within 0.67σ of the true value?

★ 6. Evaluate $\int_1^{1.1} f$ from the table for $H(z)$. Evaluate it, using (7-12-10), and compare.

7. Make a table showing the probabilities that a normally distributed variable lies more than 1, 2, 3, 4 standard deviations from the mean.

8. Suppose 2500 students are tested on an examination in which the mean is 500 and the standard deviation 100. Assuming normality, how many might be expected to score 750 or over?

Problems 9 through 14 refer to the examination described in Problem 8. Find the expected percentage who score.

9. Between 400 and 600.   10. Less than 400.   11. Less than 300.

12. Over 600.   13. Over 790.   14. Over 799.

★15. A manufacturer guarantees batteries to last 24 months. He wishes less than 10% of his batteries to last more than 30 months, but he wants to cut to a minimum the number that fail before 24 months. Assuming the mean life to be 24, what σ should be the engineer's aim in design?

★16. In the same situation, suppose the manufacturer wishes to have less than 1% of the batteries fail before 18 months and wishes to minimize the number that last more than 25 months. If $\mu = 24$, for what σ should the engineer aim?

★17. If the engineer may choose both μ and σ, what should be his choice in Problem 16?

ANSWERS TO EXERCISES

(a) The graph of f is symmetric with respect to $x = 0$, since $f(-z) = f(z)$. Since the whole area is 1, $F(0) = 0.5$.   (b) $F(z) = P[v \leq z] = 1 - P[v > z] = 1 - P[v \leq -z]$.   (c) (8-7-1).   (d) 0.023.   (e) 0.977.   (f) 0.977.   (g) 0.023.   (h) 0.954.   (i) 0.046.   (j) $0.477 - 0.341 = 0.136$.   (k) 0.954.   (l) 0.998.   (m) 0.318.   (n) 0.046.   (o) 0.002.   (p) 682.   (q) 954.   (r) 23.   (s) 23.   (t) 7; 24; 38; 24; 7.   (u) 68, 95, 99.8.

ANSWERS TO PROBLEMS

1. Since $\forall x\ e^x > 0$.   3. Max at $(2, 1/3\sqrt{2\pi})$. Inflections at $x = -1$ and $x = 5$.   5. About 50%.   7. 31.8%, 4.6%, 0.2%.   9. 68%.   11. 2.3%.   13. 0.2%.

8-9 Laws of large numbers. From the beginning (Section 8-1) we have viewed probability as a number that gives an expected relative frequency in the long run. If the probability is p, we expect the event to occur about np times in n independent trials. Indeed, if this does not happen, we usually conclude that the probability is not correctly calculated. The kind of behavior we expect is illustrated in Fig. 8-11, which gives the observed proportion x/n of successes x in n independent trials in each of which the probability is 1/2. We note that as n gets larger, x/n seems to tend toward 1/2, though there is considerable oscillation. The graph was constructed from Table V in the Appendix, which gives a record of successes (1) and failures (0) in 2500 trials. If we think of 1 as heads and 0 as tails, reading along successive lines in Table V yields the sort of experimental results that would be obtained by tossing a perfectly balanced coin 2500 times.

(a) Make a chart similar to Fig. 8-11 for line 2 in Table V.

It is because of results such as those indicated in Fig. 8-11 that we find probability theory very practical. It enables us to predict the relative frequency of events with a margin of error that tends to become smaller

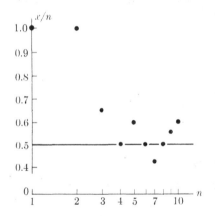

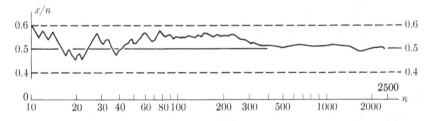

Fig. 8-11. Proportion x/n of successes in n independent trials with $p = 0.5$, $n = 1, 2, \ldots, 2500$. [Logarithmic scale for n used to contract graph horizontally; i.e., the point corresponding to $(n, x/n)$ is plotted at $(\log_{10} n, x/n)$.]

as the number of observations gets larger. Of course, we cannot go so far as to say that the relative frequency *always* approaches the probability as the number of trials gets large. If x is the number of successes in n independent trials with probability p, we cannot assert that $x/n \to p$ as $n \to \infty$. It is quite possible, *though very unlikely*, that in some particular experiment x/n will not approach p at all. However, from the theory of probability we can prove that the probability that x/n differs from p by *less* than any quantity, however small, approaches one as n approaches infinity. Such theorems, which give a precise form to our intuitive notion of the relation between probability and the results of repeated experiments in the long run, are called *laws of large numbers*.

The following theorem, due to the nineteenth century Russian mathematician P. L. Chebyshev, gives insight into the meaning of the variance of a distribution and is useful in proving laws of large numbers.

(1)   *Let v be a random variable with mean μ and variance σ^2.*
        Then for any $t > 0$, $P[|v - \mu| \geq t] \leq \sigma^2/t^2$.

The proof is easy. In (8–6–9), σ^2 is given as a sum over *all* values of x. Since all terms are positive, we have

(2) $$\sigma^2 \geq \sum (x - \mu)^2 f(x),$$

where the sum is over just those values of x such that $|x - \mu| \geq t$. Since each squared term in the right member of (2) is greater than or equal to t, we have $\sigma^2 \geq \sum t^2 f(x) = t^2 \sum f(x) = t^2 P[|x - \mu| \geq t]$. Dividing both members of the last result by t^2 gives (1).

(b) Explain each step in this proof.   (c) Our proof was for discrete distributions. Give a similar proof for continuous distributions.

Chebyshev's inequality tells us that the probability that a random variable differs from its mean by more than any number is less than the variance divided by the square of this number. The larger we make the number, the smaller is the probability. Roughly speaking, the variable is likely to be near its mean and very unlikely to be very far from it. More specifically, substituting $(t:n\sigma)$, we have

(3) $$P[|v - \mu| \geq n\sigma] \leq 1/n^2.$$

For example, the probability that a random variable differs from its mean by more than twice the standard deviation is less than $1/4$.

(d) What can we conclude from (3) about the probability that a random variable differs from its mean by more than three times the standard deviation?

(e) How does this compare with the probability that a normally distributed variable differs from its mean by more than 3σ? Why the difference?

From Chebyshev's inequality we can derive the following theorem, which is sometimes called *the law of large numbers*.

(4)   *Let x be the number of successes in n independent trials in each of which the probability of success is p. Then*

$$\lim_{n \to \infty} P[|x/n - p| > \epsilon] = 0.$$

In words, the probability that the relative frequency differs from the probability of success by more than any amount, however small, approaches zero as n approaches infinity. An equivalent statement is that the probability that x/n differs from p by not more than any quantity, however small, approaches one as n approaches infinity.

(f) Symbolize the last statement.   (g) Derive it from (4).

To prove (4) we note that x is a binomial variable and hence has mean np and variance npq (see Problems 4 and 5 in Section 8–6). Hence, from (1),

(5) $P[|x - np| \geq t] \leq npq/t^2$

holds for any $t > 0$. The substitution $(t:n\epsilon)$ yields $P[|x - np| \geq n\epsilon] \leq npq/n^2\epsilon^2$, which is equivalent to $P[|x/n - p| \geq \epsilon] \leq pq/n\epsilon^2$. Since the right member approaches zero as $n \to \infty$, the proof is complete.

(h) Can we say from (4) that $|x - np|$ is likely to be small for large n? (i) Explain in detail the steps in this proof.   (j) Relate (4) to Fig. 8–11.

The law of large numbers (4) applies only to the binomial distribution, but it can be generalized. Suppose that v is a random variable with mean μ. Suppose that we make n independent observations of v, $x_1, x_2, \ldots, x_n$. Let $\bar{x} = (x_1 + x_2 + \ldots + x_n)/n$. Then the probability that $\bar{x}$ differs from μ by more than any ϵ, however small, approaches zero as n approaches infinity, i.e.,

(6) $\lim_{n \to \infty} P[|\bar{x} - \mu| > \epsilon] = 0.$

This law is sometimes called the *weak law of large numbers*. We may view the n independent observations of v as the result of sampling with replacement from a population consisting of the values of v, the probability of choosing the values being given by the frequency function. Then (6) means that when we sample from a population, the mean of the sample may be expected to approximate the mean of the population, with the probability of a deviation of any given size approaching zero as the sample size increases. It is interesting to note that for (6) to hold all that is required is the existence of the mean. Granted this, the law holds regardless of the nature of the distribution. The proof of (6) and the remaining theorems of this section are beyond the scope of this book.

There are many laws of large numbers. One of them, known as the *strong law of large numbers*, asserts that

(7) $$P[\lim_{n \to \infty} |\bar{x} - \mu| = 0] = 1.$$

(k) Explain the difference between (7) and (6). (l) Relate (7) to Fig. 8–11.

In a very large number of cases we can be much more specific about the nature of the distribution of the mean of a sample. Suppose that x_1, x_2, ..., x_n are n independent observations of a random variable with mean μ and variance σ^2. Then as $n \to \infty$,

(8) $$P\left[a \leq \frac{\bar{x} - \mu}{\sigma/\sqrt{n}} \leq b\right] \to F(b) - F(a),$$

where $\bar{x}$ is defined as above in (6) and F is the standard normal distribution function (8-8-6).

It is this theorem and its generalizations that make the normal distribution so valuable. No matter what the original distribution (provided it has a mean and variance), the mean of a sample tends to follow the normal distribution as the sample size is increased. Since probabilities associated with normal distributions are rather easily calculated, as we have seen in Section 8–8, it is much easier to deal with normal distributions than with the many different, and often unknown, distributions that may be encountered. But (8) tells us that if we are dealing with the mean of a large sample, we may use the normal distribution almost without regard to the original distribution.

To illustrate (8) we consider a random variable with two values 0 and 1 each with probability 1/2. Here $\mu = 1 \cdot (1/2) + 0 \cdot (1/2) = 1/2$, and $\sigma^2 = (1/2)^2(1/2) + (-1/2)^2(1/2) = 1/4$. Now $\bar{x} = x/n$, where x is the number of successes in n trials. Hence $(\bar{x} - \mu)/(\sigma/\sqrt{n}) = (x/n - 1/2)(2\sqrt{n})$. This expression should, according to (8), be approximately normally distributed with mean 0 and variance 1 for large n.

Now we may view each half-line in Table V as the record of 25 independent trials, and the whole table as the record of 100 such experiments. Here $n = 25$ and x is the number of 1's in each half-line. Corresponding to each x we have a value of z given by $z = (x/25 - 1/2)(2\sqrt{25})$ or $(x - 12.5)(0.4)$. In the hundred half-lines we count the frequency of

x	z	Frequency	Relative frequency	Cumulative relative frequency	$F(z + 0.2)$
5	−3.0	0	0	0	0
6	−2.6	0	0	0	0.01
7	−2.2	0	0	0	0.02
8	−1.8	2	0.02	0.02	0.06
9	−1.4	4	0.04	0.06	0.12
10	−1.0	13	0.13	0.19	0.21
11	−0.6	18	0.18	0.37	0.34
12	−0.2	14	0.14	0.51	0.50
13	0.2	21	0.21	0.72	0.66
14	0.6	12	0.12	0.84	0.79
15	1.0	5	0.05	0.89	0.88
16	1.4	6	0.06	0.95	0.94
17	1.8	3	0.03	0.98	0.98
18	2.2	1	0.01	0.99	0.99
19	2.6	1	0.01	1.00	1.00
20	3.0	0	0	1.00	1.00

each value of z in order to obtain the above table, in which we have omitted the values for $x < 5$, since all entries are zero. The last column gives the cumulative probability for the standard normal distribution obtained from Table VI. These cumulative probabilities should correspond to the cumulative relative frequencies observed, if the distribution of x is to be approximately normal. We use $F(z + 0.2)$ in order to get the cumulative probabilities corresponding to the upper end of each interval of values z. Since the successive values of z differ by 0.4, the point halfway between each pair of values is found by adding 0.2 to the smaller one.

We note from the table that even for n as small as 25 and for only 100 observations, the mean does appear to be distributed approximately normally.

(m) Plot the last two columns against x on the same graph and connect each by a smooth curve.

A special case of (8), known as the *Laplace limit theorem* is obtained by considering a random variable that takes only the values 1 and 0 with probability p and q respectively. Here $\mu = 1 \cdot p + 0 \cdot q = p$, $\sigma^2 = (1 - p)^2 \cdot p + (0 - p)^2 \cdot q = q^2 p + p^2 q = pq(q + p) = pq$, and $\bar{x} = x/n$, where x is the number of successes in n trials. Then

$$(9) \qquad \frac{\bar{x} - \mu}{\sigma/\sqrt{n}} = \frac{x/n - p}{\sqrt{pq/n}} = \frac{x - np}{\sqrt{npq}}.$$

Applying (8), we have

> *If x is the number of successes in n independent trials with constant probability p of success and q of failure, then as n → ∞*

(10)
$$P\left[a \leq \frac{x - np}{\sqrt{npq}} \leq b \right] \to F(b) - F(a),$$

> *where F is the standard normal distribution function.*

That is, the normalized binomial distribution approaches the standard normal distribution as the number of trials increases. The word "normalized" refers to subtraction of the mean np and division by the standard deviation $\sqrt{npq}$.

The Laplace theorem enables us to use the right member of (10) as a convenient approximation to the left when n is large. For example, what is the probability of observing between 45 and 55 heads if an evenly balanced coin is tossed 100 times? Here $n = 100$, $p = q = 1/2$. It is convenient to rewrite the inequalities in (10) as $a\sqrt{npq} + np \leq x \leq b\sqrt{npq} + np$. Here $\sqrt{npq} = 5$, $np = 50$, $5a + 50 = 45$, $5b + 50 = 55$, $a = -1$, $b = 1$. From Table VI, $F(b) = 0.841$, $F(a) = 0.159$, and the desired probability is 0.682.

For the same situation, estimate the probability that the number of successes lies between: (n) 49 and 51,   (o) 45 and 50,   (p) 42 and 53.

PROBLEMS

1. Figure 8–11 was made by counting the number of successes (1's) in Table V starting at the left of the first row and continuing in the usual reading order. It was necessary to calculate 2500 relative frequencies and plot them. Calculate the first 50 relative frequencies starting on the right of the last row and reading backward. Plot your results.

2. Reading the table backward as in Problem 1, find x/n for $n = 10, 20, 30, 40, 50, 60, 70, 80, 90, 100$, and plot your results.

3. Derive (4) from (6) by considering the binomial distribution as generated in the following way. Let v be a random variable having just the values 1 (success) and 0 (failure), with probabilities p and q. Then take n observations of v and apply (6).

4. Show that with the same hypothesis as (10), $P[x_1 \leq x \leq x_2] \to F(b) - F(a)$, where $a = (x_1 - np)/\sqrt{npq}$ and $b = (x_2 - np)/\sqrt{npq}$.

5. A die is tossed 180 times. What is the probability that the face six appears exactly 30 times? (Suggestion: Let $x_1 = 29.5$ and $x_2 = 30.5$, and use the previous problem.)

6. Let x be the number of successes in a whole line of Table V. Make a table

of the observed frequency distribution of x and compare it with the normal distribution as we did in the table following (8).

7. Do the same for x the number of successes in each group of 10 digits in Table V.

8. A student pilot named Brom passed the time by tossing 10 nickels 40 times. Letting x be the number of heads appearing in a toss, the observed distribution was as follows:

x	Frequency x
2	4
3	1
4	6
5	7
6	14
7	4
8	4

Compare this with what might be expected by finding cumulative relative frequencies and comparing them with those for the normal distribution as we did in the section. Do you think that the nickels were biased toward heads?

9. A die was tossed 100 times. Six came up 25 times. What is the probability that for a balanced die, six would come up 25 or more times out of 100?

10. Two dice were tossed 50 times. Seven appeared only five times. What is the probability that seven would appear five or fewer times if the dice were fair?

11. In order to discover the average height of 1000 students, a random sample of 16 students was drawn. Suppose the true average height to be 5.4 ft and the standard deviation of the heights to be 0.4 ft. What is the probability that the mean of the sample will differ from the true mean by more than 0.1ft?

12. Answer the same question for a sample of 25.

13. For a sample of 100.

14. Answer the questions for Problems 11, 12, and 13 without the information about the true average height.

15. Referring again to Problem 11, how might one guess an approximation to the standard deviation?

★16. For an explanation of the fact that the graph in Fig. 8–11 remains on one side of the center line for long periods, see Feller, *An Introduction to Probability Theory and Its Applications*, Vol. 1, Second Edition, Chapter III.

ANSWERS TO EXERCISES

(a) The first few relative frequencies are 0, 1/2, 2/3, 3/4, 4/5, 5/6, 6/7, 7/8, 7/9, 8/10, 9/11, 9/12, 9/13, 9/14, 9/15, 9/16. Note that the observed relative frequency may not always move toward the expected. The last relative frequency is 29/50.　　(b) Each squared term is $\geq t$ since we limited the sum to such values. The sum of the probabilities over such values is just the prob-

ability that $|x - \mu| \geq t$.   (c) The sum in (2) is replaced by an integral from
$\mu - t$ to $\mu + t$. The rest of the proof is the same with summations replaced by
integrals.

(d) It must be less than or equal to $1/9$.   (e) For a normal distribution the
probability is 0.002.   The normal distribution is very special, whereas
Chebyshev's inequality applies to *any* distribution with mean and variance.
(f) $\forall \epsilon \; P[|x/n - p| \leq \epsilon] \to 1$ as $n \to \infty$.   (g) (8-2-6).   (h) No, it is likely
to be very large.   (i) $|x - np| \geq n\epsilon \leftrightarrow |x/n - p| \geq \epsilon$ since $n > 0$.   (j) If
we made many such charts and picked an ϵ, we would find that the proportion
in which x/n differed from μ by more than this ϵ would be very small for large n.

(k) The operations of taking the limit and the probability are interchanged.
The first tells us merely that the probability of any difference approaches 0,
the second that the probability that the difference approaches 0 is 1. The weak
law asserts merely that for all n sufficiently large, the probability of any given
deviation is arbitrarily small. However, larger deviations are still possible, no
matter how large n is, and hence it does not follow that $\bar{x} \to \mu$ as $n \to \infty$.
The strong law asserts that $\bar{x} \to \mu$ almost always, i.e., with the probability 1.
Its failure is less likely than that a monkey would type out Webster's unabridged
dictionary without a mistake if it were given a typewriter! The strong law
shows that the theory corresponds to experience and to our intuitive idea of
probability. In terms of the experiment with which this section begins, the
weak law asserts that if we draw horizontal lines at a distance ϵ above and below the 0.5 line and imagine the experiment continued indefinitely, we expect
the graph to cross these lines less and less often as n gets large. The strong law
asserts that in practically all such experiments, the graph approaches the 0.5
line asymptotically.

(l) If we made many such charts, in virtually all of them $\bar{x}$ would approach
μ as $n \to \infty$.   (m) The curves cross between 10 and 11.   (n) 0.158.   (o) 0.341.
(p) 0.671.

Answers to Problems

1. The last is $21/50$.   3. See derivation of Laplace limit theorem.   5. 0.08.
7. The frequencies of $x = 0, 1, \ldots, 10$ are 0, 0, 6, 36, 54, 60, 51, 35, 7, 1, 0.
9. About 0.01.   11. About 0.32.   13. 0.013.

★CHAPTER 9

STATISTICAL INFERENCE

9-1 The meaning of "statistics." When first introduced in the eighteenth century, "statistics" meant information about governments, and as late as 1885 the London Statistical Society proposed to define it as "the science which treats of the structure of human society." However, before the middle of the nineteenth century, "statistics" was used also as a new name for the systematic collection and study of numerical data on populations, which had been called "political arithmetic" in the seventeenth century and is now named "vital statistics." Gradually the meaning was broadened to include all numerical data, and today "statistics" as a plural refers to numerical information in any field. "Descriptive statistics" (singular) now means the science and art of collecting, interpreting, summarizing, and presenting numerical information.

Where it is possible to collect all the information about the phenomena in which we are interested, descriptive statistics is adequate. For example, if we are interested in the scholastic performance of the students in a certain college in a certain year, we may be able to get complete information from the registrar. However, in many situations complete information is denied us. This may be due merely to limitations of time and expense. For example, it would be possible to question each person in the United States at regular intervals to find out the number of unemployed, but the expense would be prohibitive. In other cases, gathering complete information is not even theoretically possible because it would require the observation of an infinite number of instances, often including some in the future. For example, we can never observe *all* freely falling objects to see whether they follow Newton's laws of motion, nor can we observe an infinite sequence of throws of a coin to see whether it follows the law of large numbers. In short, we often have a population of which only a sample can be observed. We nevertheless wish information about the population as a whole. Since our information is incomplete, we cannot make statements about the population with certainty. The best we can do is to work out procedures for making uncertain but useful inferences about the population. The process of drawing conclusions about a population from knowledge of a sample is called *statistical inference*, and the science and art of making such inferences is called *inferential statistics*.

Any statement about a population is either true or false. However, different procedures of statistical inference have different probabilities of yielding correct statements under various conditions. It is these probabilities that enable us to choose desirable procedures for statistical inference. Because of this the theory of statistical inference, also called *mathematical statistics*, becomes a special branch of probability theory. In its present form it is a product of the twentieth century, and especially of the last two decades.

Modern statistics, both descriptive and inferential, is an enormous field in terms of the number of people who are professional statisticians, the extent and rate of growth of the theory, and the increasing role that it plays in government, industry, and science. In this chapter we attempt nothing more than a very brief and incomplete survey of the most important theoretical ideas of statistical inference.

9–2 Testing statistical hypotheses. Imagine an urn containing 10 balls, of which some may be red and the rest green. We are not permitted to inspect the urn and count the balls. Our only means of gaining information is random sampling with replacement; i.e., we may remove one ball, note its color, replace it, and repeat in such a way that any ball is as likely to be picked as any other. This means that if the proportion of red balls is p, the probability of picking a red ball is p and of picking a green ball is $1 - p$. Now suppose that the claim is made that $p = 1/2$. We wish to test this hypothesis in order to be able with some justification to accept or reject it. Our problem is to decide on a good procedure for this purpose.

This may seem to be a rather unimportant problem, but consider the following "real life" situations:

A political scientist wishes to discover the proportion p of voters among a voting population of 10,000,000 favoring a certain candidate. He cannot ask all voters, so he must pick a sample and adopt a procedure for deciding about p on the basis of information provided by the sample.

A chemist wishes to determine the proportion p of a certain compound in a mixture. He cannot analyze the entire mixture, so he takes a sample and tries to determine p from the sample.

A biologist wishes to determine what proportion p of a certain type of animal have a given characteristic. He cannot observe all animals of this type, so he must resort to sampling.

Evidently, the problem of the urn is a simplified picture of a kind of problem that arises in practically all cases where we wish to secure information by observations.

To further simplify the urn problem let us agree first to consider only those procedures based on observing just four balls. Let x be the number of red balls observed. Then the possible outcomes of our experiment are $x = 0, 1, 2, 3,$ or 4. Since this is all our information, we must decide on a rule that tells us whether to accept or reject the hypothesis $p = 1/2$ on the basis of the value of x observed. Common sense suggests that $x = 2$ is favorable to our hypothesis and $x = 0$ or $x = 4$ are unfavorable. We may agree, then, to accept if $x = 2$, and reject if $x = 0$ or 4. But where shall we draw the line? What shall we say if $x = 1$ or $x = 3$? To see how one could answer this question, we adopt a rule and work out the consequences.

Suppose that we agree to accept if $x = 2$, and reject otherwise. Let us see what happens in all possible situations. Suppose, first, that our hypothesis is really true, that is, $p = 1/2$. Then the probability that $x = 2$ is 3/8 by (8–6–7)(n:4, x:2, p:1/2, q:1/2). This means that when our hypothesis is true, the probability of acceptance is 3/8 and of rejection is 5/8. In other words, if the hypothesis were true and we made many experiments of this kind to test it by this rule, we should accept it 3/8 of the time and reject it 5/8 of the time.

By considering the other values of p and calculating the probability of acceptance, $P[x = 2]$, in each case, we get the following table, in which probabilities are rounded to two decimal places.

(1)

True p	Probability of acceptance	Rejection
0.0	0.00	1.00
0.1	0.05	0.95
0.2	0.15	0.85
0.3	0.26	0.74
0.4	0.35	0.65
0.5	0.38	0.62
0.6	0.35	0.65
0.7	0.26	0.74
0.8	0.15	0.85
0.9	0.05	0.95
1.0	0.00	1.00

This test procedure has the obvious disadvantage that when the hypothesis is true we reject it more often than when we accept it. It has the advantage that when the hypothesis is false we are very likely to reject it. Figure 9–1 shows that the probability of rejection is smallest when the hypothesis is true and increases as the true p deviates more from $p = 1/2$.

(a) Suppose we adopt the following test of the hypothesis $p = 1/2$. Take a sample of four with replacement, accept if $x = 2$ or $x = 3$, and reject otherwise. Make a table like (1) and graph the probability of rejecting the hypothesis as a function of p. Do you think this is a better test?

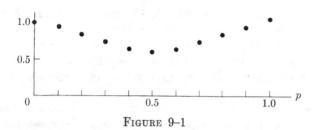

FIGURE 9–1

It is convenient to adopt some special terminology. First we note that in the above situation there is a random variable (the number of red balls observed) with a probability distribution about which we do not have complete information. We know that the distribution is binomial, but p is unknown. A statement about the distribution function of a random variable is called a *statistical hypothesis*. Above, our statistical hypothesis is $p = 1/2$.

Two things should be noted here. First, this use of "hypothesis" is different from that in logic, although a statistical hypothesis, being a sentence, may serve as the sufficient condition in an implication. Second, in scientific discourse any statement that is tentative may be called a hypothesis, but not all such statements are statistical hypotheses. A statistical hypothesis must refer to the distribution of a random variable. Since most quantitative observations in science involve random elements (for example, random errors of observation), many scientific hypotheses can be formulated as statistical hypotheses even when this is not obvious on the surface. For example, the hypothesis that the acceleration of gravity is 32 ft/sec/sec may be formulated as a statement about the distribution of the acceleration observed in a certain experiment. On the other hand, the statement "Mr. Jones' age is 28" can be verified or rejected by obtaining factual evidence not subject to random variation, and "All V-8's have eight cylinders" is a statement that can be established by logic without reference to observations. Neither are statistical hypotheses.

Give further examples of hypotheses that: (b) are statistical, (c) are not statistical.

A *test* of a statistical hypothesis is a procedure for deciding on acceptance or rejection of the hypothesis. In the above example our test is: accept if $x = 2$, reject otherwise.

In our example, there are a number of alternatives to the hypothesis tested. Each of these is also a statistical hypothesis. The rejection of the hypothesis tested amounts, then, to acceptance of the disjunction of the alternatives. In the example, we test the hypothesis that x is distributed binomially with $p = 1/2$ against the alternatives that it is distributed binomially with $p = 0, 0.1, 0.2$, etc. The claim that x is binomially distributed is not questioned. We call the set of hypotheses considered the *admissible hypotheses*. In our case they are specified by 11 values of a single parameter p.

Setting up a test of a statistical hypothesis about a random variable amounts to partitioning the sample space (set of values of the random variable) into two subsets such that if the observation falls in one, the hypothesis is accepted, and if it falls in the other, the hypothesis is rejected. The *critical region* of a test is that part of the sample space in which an observation calls for rejection. In our example, the sample space is $\{0, 1, 2, 3, 4\}$, and the critical region is $\{0, 1, 3, 4\}$. Evidently, the critical region determines the test.

We note above that there are two ways in which we may commit an error by using a certain test. First, we may reject the hypothesis when it is really true. This is called an *error of Type I*. The probability of the Type I error in our example is 0.62. On the other hand, we may accept the hypothesis when it is really false. This is called an *error of Type II*. In our example, the probability of Type II error depends on which one of the alternative hypotheses is true, and it is given by the entries in the acceptance column in (1), excluding the entry opposite $p = 0.5$. If the hypothesis tested consisted of more than one value of p, we should have more than one probability of error of Type I also.

Evidently we should like to use tests in which the probabilities of both types of errors were as small as possible. But a little reflection shows that such hopes cannot be satisfied. We can make the probability of Type I zero by adopting the test: accept no matter what the observation. But then the probability of Type II is 1! Similarly, we can make errors of Type II impossible at the cost of making errors of Type I certain. It appears that a compromise is necessary. Each compromise leads to a different test, and the most desirable compromise depends on circumstances.

One procedure is to decide on a satisfactory value for the probability of an error of Type I, to consider only those tests that keep the probability below this value, and among these to choose the test that makes the Type II error as small as possible. The probability of Type I error is called the *level of significance* of the test. Usually it is chosen on the order of 0.05 or even 0.01, so that the probability of rejecting a true hypothesis is very small.

For each of the possible critical regions indicated, find the probability of Type I error, i.e., the level of significance of the corresponding test.   (d) $\{0, 1, 2, 3, 4\}$.   (e) $\{1, 2, 3, 4\}$.   (f) $\{0, 2, 3, 4\}$.   (g) $\{0, 1, 3, 4\}$.   (h) $\{0, 1, 2, 4\}$.   (i) $\{0, 1, 2, 3\}$.   (j) $\{2, 3, 4\}$.   (k) $\{1, 3, 4\}$.   (l) $\{1, 2, 4\}$.   (m) $\{1, 2, 3\}$.   (n) $\{0, 3, 4\}$.   (o) $\{0, 2, 4\}$.   (p) $\{0, 2, 3\}$.   (q) $\{0, 1, 4\}$.   (r) $\{0, 1, 3\}$.   (s) $\{0, 1, 2\}$.   (t) $\{0, 1\}$.   (u) $\{0, 2\}$.   (v) $\{0, 3\}$.   (w) $\{0, 4\}$.   (x) $\{1, 2\}$.   (y) $\{1, 3\}$.   (z) $\{1, 4\}$.   (aa) $\{2, 3\}$.   (bb) $\{2, 4\}$.   (cc) $\{3, 4\}$.   (dd) $\{0\}$.   (ee) $\{1\}$.   (ff) $\{2\}$.   (gg) $\{3\}$.   (hh) $\{4\}$.   (ii) $\emptyset$.

Calculations such as those above tell us all we need to know about the errors of Type I associated with the possible tests. To consider errors of Type II, it is convenient to introduce the notion of power. The *power* of a test is the probability that it rejects the hypothesis. The power of the test of our example is given in the rejection column of (1). The *power function* of a test is the function that maps the admissible hypotheses into the corresponding probabilities of rejection of the hypothesis tested. The first and last columns of (1) are a table of a power function, and Fig. 9–1 is its graph.

Let P be the power function of a test, H an admissible hypothesis, and H_0 the hypothesis tested. Then $P[H]$ is the probability of rejecting H_0 when H is true. In particular, $P[H_0]$ is the level of significance. When $H \neq H_0$, $P[H]$ is 1 minus the probability of Type II error, i.e., it is the probability of correctly rejecting the hypothesis H_0. A very desirable power function would be one with $P[H_0]$ as small as possible and $P[H]$ otherwise as close to one as possible.

(jj) Graph the power functions of those tests in Exercises (d) through (ii) that have a level of significance less than or equal to 0.63. Begin by making a table of the probability of each value of x for each value of p, using your results from Exercises (g) through (o) in Section 8–6. Add appropriate entries to find the probabilities of an observation lying in each critical region for each value of p as we did in the last column of (1).

In relation to your results in (jj): (kk) of the three tests with level of significance 0.63, why is (g) the best? (ll) Can you answer the same question for those with level of significance 0.56? (mm) 0.50? (nn) 0.44? (oo) 0.38? (pp) 0.31? (qq) Comment on the others. (rr) What would you consider the best test, everything considered? (ss) What is your choice if it is very important to detect $p > 1/2$ but not $p < 1/2$? (tt) if it is important to detect p different from 1/2 by more than 0.2 but not important to detect smaller differences?

One of the principal tasks of the statistician is the choice of tests that are appropriate in practical situations. The previous example illustrates how this is done by studying the power functions of the possible tests. The proper choice requires a knowledge of the situation and of the actions that may be based on rejecting or accepting the hypothesis.

In the example of this section, our experiment consisted of four random drawings with replacement. This experiment limited our possible tests to those considered. It is evident that better tests could be devised if we took a larger sample. Indeed, common sense suggests that the larger the sample, the more information we may obtain. Perhaps, too, we might do better by sampling without replacement, if that were permitted. Indeed, it is obvious in the case considered that a sample of 10 without replacement would give us complete information! In more complex situations the variety of possible experiments would be very great.

It should be observed that changing the experiment may change the statistical hypothesis being tested. In the example of the balls in the urn, increasing the size of the sample introduces a different random variable and a different binomial distribution, even though we may still state our hypothesis in the form $p = 1/2$. It should be clear, then, that the design of experiments, the choice of tests, and the formulation of hypotheses are intimately interrelated. It is for this reason that statistical problems should be considered by scientists before rather than after they perform experiments. Often what appears to be a good experiment turns out to be useless because it was not designed in such a way as to permit a satisfactory test of a meaningful hypothesis. We cannot pursue this matter further here.

PROBLEMS

The first ten problems refer to the urn discussed in the section. The probabilities calculated in the exercises may be useful.

1. Suppose that we must judge whether or not $p = 1/2$ by observing just one ball. What is the sample space? What are the possible critical regions? What is the best test? Graph its power function.

2. Answer the same questions if a random sample of two balls is all that is permitted.

3. Design a good test of the hypothesis $p = 0.6$ with a sample of four members. Plot its power function.

4. Design what you consider the best test for the hypothesis $p = 0.8$ if it is particularly important to detect $p > 0.8$. (This might be the case, for example, if it were particularly important to reject a medicine with too high a concentration of a dangerous ingredient.) Graph the power function.

5. Design a good test for the hypothesis $0.4 \leq p \leq 0.6$.

6. Do the same for $p \leq 0.2$.

7. While increasing the sample size evidently improves the best tests, it is evident that the calculation of the power function would become more difficult. Illustrate by finding the level of significance and a few other values of the power function for a test of $p = 1/2$ based on a random sample with replacement of 25 balls, where the critical region is all points except $x = 11, 12,$ and 13.

8. When the sample size is large, we may use the central limit theorems of Section 8–9. Suppose, for example, that we test $p = 1/2$ by a sample of 100 observations. Experience with small samples suggests accepting the hypothesis if the observation lies in an interval whose center lies at $x = 50$, i.e., taking as a critical region $(\underline{50 - a \quad 50 + a})'$ or $(-\infty \underline{\quad} 50 - a) \cup (50 + a \underline{\quad} \infty)$. If e is our desired level of significance, we choose a so that for $p = 1/2$, $1 - P$ $[50 - a \leq x \leq 50 + a] = e$. Finding a is made easy by using the Laplace limit theorem (8–9–10), which tells us that $(x - 50)/5$ is approximately normally distributed. We have $50 - a \leq x \leq 50 + a \leftrightarrow -a/5 \leq (x - 50)/5 \leq a/5$. But by the Laplace theorem, $P[-a/5 \leq (x - 50)/5 \leq a/5] \doteq F(a/5) - F(-a/5) = 2H(a/5)$. For a level of significance of 0.05 we find $a/5$ from Table VI so that $2H(a/5) = 0.95$. Here $a/5 = 2$ or $a = 10$. What is

the critical region? What is the probability of error of Type I? Graph the power function of the above test by using the normal approximation and Table VI for each admissible value of p.

9. Repeat Problem 8 for a level of significance of 0.01, and graph the power function.

10. Same for level of significance of 10%.

★11. An election is to be predicted by a sample of 10 voters. Suppose that a level of significance of 0.01 is desired for a test of the hypothesis that the vote is even. Find a critical region. What will be the probability of detecting a sentiment of 54 or more percent in favor of one of the two candidates?

★12. Answer the same questions if we use a test with only 0.05 as level of significance.

13. A coin is tossed 100 times and comes up 55 times heads. Is the coin evenly balanced?

14. In the situation of Problem 9 in Section 8–9, is the die loaded?

15. In the situation of Problem 10 in Section 8–9, are the dice loaded?

★16. In the situation of Problem 13 in Section 8–9, suggest a test for the hypothesis that the true average height is 5.4 ft.

★17. Some people claim that after a run of several heads a coin is more likely to come up tails. Treating Table V as a record of 2500 throws of a balanced coin, test the hypothesis that following a run of five heads, the probability of heads on the next throw is 1/2.

★18. Test the hypothesis that $p = 1/2$ in Table V by considering the first line as a random sample and letting the level of significance be 0.68.

★19. Carry out a similar test for each line in Table V. In what proportion of the cases is the hypothesis rejected?

Answers to Exercises

(a) See the answer to (q) and to part (q) in Exercise (jj). (d) 1. (e) 0.94.
(f) 0.75. (g) 0.62. (h) 0.75. (i) 0.94. (j) 0.69. (k) 0.56. (l) 0.69.
(m) 0.86. (n) 0.38. (o) 0.50. (p) 0.69. (q) 0.38. (r) 0.56. (s) 0.69.
(t) 0.31. (u) 0.44. (v) 0.31. (w) 0.13. (x) 0.62. (y) 0.50. (z) 0.31.
(aa) 0.62. (bb) 0.44. (cc) 0.31. (dd) 0.06. (ee) 0.25. (ff) 0.38.
(gg) 0.25. (hh) 0.06. (ii) 0.

This entire paragraph relates to Exercise (jj). We give 11 powers corresponding to $p = 0, 0.1, \ldots, 1.0$ (g) See (9–2–1).

(k) 0, 0.30, 0.44, 0.49, 0.52, 0.56, 0.63, 0.73, 0.84, 0.95, 1.
(n) 1, 0.66, 0.44, 0.32, 0.31, 0.38, 0.50, 0.66, 0.82, 0.95, 1.
(o) 1, 0.71, 0.56, 0.51, 0.50, 0.50, 0.50, 0.51, 0.56, 0.70, 1.
(q) 1, 0.95, 0.82, 0.66, 0.50, 0.38, 0.31, 0.32, 0.44, 0.66, 1.
(r) 1, 0.95, 0.84, 0.73, 0.63, 0.56, 0.52, 0.49, 0.44, 0.30, 0.
(t) 1, 0.96, 0.82, 0.65, 0.47, 0.31, 0.18, 0.08, 0.03, 0.00, 0.
(u) 1, 0.70, 0.56, 0.50, 0.48, 0.44, 0.37, 0.27, 0.15, 0.05, 0.
(v) 1, 0.66, 0.44, 0.32, 0.28, 0.31, 0.37, 0.42, 0.42, 0.29, 0.
(w) 1, 0.66, 0.41, 0.25, 0.16, 0.12, 0.16, 0.25, 0.41, 0.66, 1.

(x)   0, 0.34, 0.56, 0.68, 0.69, 0.62, 0.50, 0.34, 0.18, 0.05, 0.
(y)   0, 0.30, 0.44, 0.49, 0.50, 0.50, 0.50, 0.49, 0.44, 0.30, 0.
(z)   0, 0.29, 0.41, 0.42, 0.37, 0.31, 0.28, 0.32, 0.44, 0.66, 1.
(aa) 0, 0.05, 0.18, 0.34, 0.50, 0.62, 0.69, 0.68, 0.56, 0.34, 0.
(bb) 0, 0.05, 0.15, 0.27, 0.37, 0.44, 0.48, 0.50, 0.56, 0.70, 1.
(cc) 0, 0.00, 0.03, 0.08, 0.18, 0.31, 0.48, 0.65, 0.82, 0.95, 1.
(dd) 1, 0.66, 0.41, 0.24, 0.13, 0.06, 0.03, 0.01, 0.00, 0.00, 0.
(ee) 0, 0.29, 0.41, 0.41, 0.35, 0.25, 0.15, 0.08, 0.03, 0.00, 0.
(ff) 0, 0.05, 0.15, 0.26, 0.35, 0.38, 0.35, 0.26, 0.15, 0.05, 0.
(gg) 0, 0.00, 0.03, 0.08, 0.15, 0.25, 0.35, 0.41, 0.41, 0.29, 0.
(hh) 0, 0.00, 0.00, 0.01, 0.03, 0.06, 0.13, 0.24, 0.41, 0.66, 1.
(ii) 0, 0, 0, 0, 0, 0, 0, 0, 0, 0, 0.

(kk) It has greatest power, except for one case in relation to each of (u) and (x), and there its power is almost as high.   (ll) No, one is better for $p > 1/2$, the other for $p < 1/2$.   (mm) Test (o) is better even though 2 belongs to critical region.   (nn) Like (ll).   (oo) Either (n) or (q) is better than (ff), but they are like (ll).   (pp) (t) is best for $p < 1/2$; (cc) for $p > 1/2$; other two might be good compromises.   (qq) All have low powers, except (w), which appears to be a good test except that it does not have high powers for p near $1/2$. (rr) No unique answer. Two candidates are (w), (g). A really satisfying test doesn't seem possible with such a small sample.   (ss) (cc) or (z).   (tt) (w).

Answers to Problems

1. $\{0, 1\}$; $\{0, 1\}$, $\{0\}$, $\{1\}$, $\emptyset$. Either $\{0\}$ or $\{1\}$ is better than others. See Fig. 9–2 for graph of power function of $\{0\}$, the best for testing $p = 1/2$ against $p < 1/2$. 3. Suggestion: $\{0, 1, 2, 4\}$ appears to be plausible. See Fig. 9–3 for the power function. To be sure of our choice we should investigate other tests.

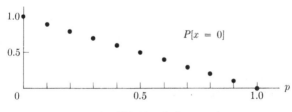

FIGURE 9–2

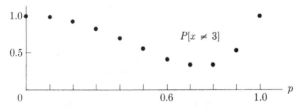

FIGURE 9–3

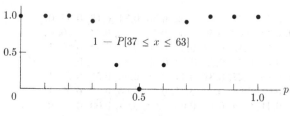

FIGURE 9–4

5. $\{0, 4\}$.   7. Level of significance is $1 - [C_{11}^{25} + C_{12}^{25} + C_{13}^{25}](1/2)^{25}$.   9. See Fig. 9–4. Critical region is $(\underline{37 \quad 63})'$.   13. Yes at 0.05 level of significance (see Problem 8). Yes at any level of significance ≤ 0.32.   15. If we test the hypothesis that the dice are fair, as above, the answer is yes at 0.05 significance level.

9–3 Estimation. In the previous section we approached the problem of incomplete information by formulating and testing specific hypotheses. An alternative approach is to design procedures for estimating the value of an unknown. Instead of asking whether the hypothesis $p = 1/2$ is true about the balls in the urn, we might adopt a formula to give an estimate of p in terms of our observations.

A method of estimation that seems very natural in this example is to observe the number x of red balls in a sample of n balls obtained by random sampling with replacement, then adopt x/n as an estimate of p. Why is this reasonable? For one thing, the expected value of x/n is p, since

(1) $E[x/n] = \sum (x/n)f(x)$ (8–6–6)

(2) $= (1/n) \sum xf(x)$ (6–7–18)

(3) $= (1/n)E[x]$ (8–6–6)

(4) $= (1/n)np = p$ (Problem 4 in Section 8–6).

Of course, in any particular sample x/n may differ considerably from the true value of p, but this shows that at least p is its average value over many samples. This suggests, of course, taking many samples and averaging the observed values of x/n. Such a procedure is equivalent to simply taking a larger value of n in a single sample.

To generalize, suppose a random variable v has a distribution function involving an unknown parameter θ. In the urn problem this parameter is p. Let $x_1, x_2, \ldots, x_n$ be n observed values of v obtained by a random sample with replacement. (This is not the only way of observing values of the random variable, but we consider this possibility for the sake of simplicity.) Let $\hat{\theta}$ be a function of $x_1, x_2, \ldots, x_n$. Then we call $\hat{\theta}$ an *estimator* of θ, and $\hat{\theta}(x_1, x_2, \ldots, x_n)$ an *estimate* of θ. In the example of the urn, v is a random variable that takes the values 1 and 0 with probability p and $1 - p$, 1 corresponding to an observation of red and 0 to green. If x is the number of red balls observed in a drawing of n balls with replacement, then $\hat{p}(x_1 \ldots x_n) = x/n = (x_1 + x_2 + \ldots x_n)/n$ is the estimate mentioned above.

When the expected value of an estimate is equal to the parameter being estimated we called the estimate *unbiased*. We showed above that x/n is an unbiased estimate of the proportion of red balls in the urn. More generally, suppose that a random variable has frequency function f with unknown mean μ. Let $\bar{x} = (x_1 + x_2 + \ldots + x_n)/n$ be the mean of n independent observations of v. Then it can be proved (though not without a little more theory about expectations than we have covered) that $E[\bar{x}] = \mu$, i.e., the mean of the sample is an unbiased estimate of the mean of the population.

Estimates of this type are called *point estimates*, for the following reason. We may consider the possible values of the unknown parameter as a set of points on a real axis. An estimator gives us a single value of the parameter, i.e., a single point.

Obviously, there are many different formulas for estimating a parameter. For example, to estimate the mean of a distribution, we might take the average of the largest and smallest values of the variable observed, or the value most frequently observed (the *mode* of the sample), or the value with an equal number of larger and smaller observed values (the *median* of the sample).

(a) Suppose one estimator is unbiased but frequently has values very much larger or very much smaller than the true value (these values balancing to give a mean equal to the true value). Suppose another were biased but had an expected value very near the true value and had no possible values differing widely from the true value. Which would be preferable?

From the definition of an estimate, it is evident that an estimate is a random variable with a distribution related to the distribution under consideration. Exercise (a) suggests that it is desirable for the variance of the distribution of an estimate to be small. Indeed, it may be better to use a biased estimate with small variance and expectation near the true value of the parameter than to use an unbiased estimate with large variance. Evidently we must choose between estimators on the basis of their properties and our needs just as was the case for tests. Lack of bias and small variance have been mentioned as desirable properties of point estimates. There are many others, some of which will be mentioned in the problems. It is usually impossible to find the best estimate from every point of view, and a compromise is required.

When we make a point estimate, we do not imagine that it gives the true value of the parameter exactly, except by chance. For example, in the problem of the ten balls in an urn, if the true value of p is $1/2$, the probability that $x/n = 1/2$ for a sample with $n = 4$ is 0.3750. On the other hand the probability that $1/4 \leq x/n \leq 3/4$ is 0.8750. If we were to take a sample of size 10, $P[x/n = 1/2] \doteq 0.25$, which is still smaller. But $P[0.4 \leq x/n \leq 0.6] \doteq 0.65$, and $P[0.3 \leq x/n \leq 0.7] \doteq 0.89$. In short, the estimate is likely to be near the true value; and as we increase the size of the interval around the true value, the probability that our estimate is within the interval increases.

(b) Check the above calculations.   (c) Find the same probabilities for a sample of size 100.

These considerations suggest that instead of determining a single value of the unknown parameter, we find an interval within which we may confidently assert that the true value lies. Of course, the longer we make the interval, the greater is our confidence and also the less the precision of our statement. Hence, as always in statistics, a compromise is necessary.

To make this more precise, let us suppose that we are estimating p in the problem of the ten balls in an urn by means of a sample of 100 observations. Now we know that $(x - np)/\sqrt{npq}$ is approximately normally distributed. Hence

$$(5) \qquad P[a \leq (x - np)/\sqrt{npq} \leq b] \doteq F(b) - F(a),$$

where F is the standard normal distribution function. This says that for a given p, $q = 1 - p$, and n, the probability that the observed x lies between the given limits, is as indicated. It seems natural to consider an interval with center at np, that is, to let $a = -b = z$. Then

$$(6) \qquad P[-z \leq (x - np)/\sqrt{npq} \leq z] = 2H(z),$$

where $H(z)$ is found from Table VI.

The inequalities in (6) may be written

$$(7) \qquad -z \sqrt{npq} \leq x - np \leq z\sqrt{npq},$$

$$(8) \qquad np - z\sqrt{npq} \leq \quad x \quad \leq np + z\sqrt{npq}.$$

Clearly, (6) gives the probability that the observed x lies within a certain number z of standard deviations $\sqrt{npq}$ from the mean np of the distribution.

Since we should like our interval to give a high degree of confidence, we begin by choosing z in (6) to accomplish this. For $z = 2$ we have

$$(9) \qquad P[-2 \leq (x - np)/\sqrt{npq} \leq 2] \doteq 0.95.$$

For a given n and an observed x there will be some values of p that satisfy the inequality and some that do not. It is evident that when $p \to 0$ or $p \to 1$, and $x \neq np$, $(x - np)/\sqrt{npq} \to \pm\infty$; and when $p = x/n$, $(x - np)/\sqrt{npq} = 0$. This suggests that there are values of p, say p_1 and p_2, between which the inequality is satisfied and that these values are just those that are given by

(10) $-2 = (x - np)/\sqrt{npq}$ and $(x - np)/\sqrt{npq} = 2$.

Substituting $q = 1 - p$ and squaring to eliminate radicals, we get from either equation

(11) $(x - np)^2 = 4np(1 - p)$,

which simplifies to

(12) $(n^2 + 4n)p^2 - (2nx + 4n)p + x^2 = 0$.

This can be solved for p to give us the two desired values p_1 and p_2. In our particular case, $n = 100$. If $x = 50$, the roots of the equation are approximately 0.4 and 0.6.

(d) Check that the solutions do satisfy (12) approximately.   (e) Carry through the algebra to find them.   (f) Repeat the above reasoning but with z chosen to make the probability in (9) approximately 0.68.   (g) Show that

(13) $[-z \leq (x - np)/\sqrt{npq} \leq z] \leftrightarrow [p_1 \leq p \leq p_2]$,

where p_1 and p_2 are the roots of

(14) $n(n + z^2)p^2 - n(2x + z^2)p + x^2 = 0$.

(h) Hence show that p_1 and p_2 so defined are such that

(15) $P[p_1 \leq p \leq p_2] = 2H(z)$.

We have shown that the interval $(p_1 \quad p_2)$, whose endpoints are roots of (14), has a probability $2H(z)$ of containing the true value of p. It is important to keep in mind that p is not a random variable, but is fixed (though unknown). The probability in (15) is the probability that p_1 and p_2 defined by (14) do enclose the mean. These endpoints are random variables, since they depend on x. By choosing z to get the desired probability in (15) we have a method of determining an interval within which the true value may be asserted to lie with a known probability that the statement will be true.

An interval within which an unknown parameter is asserted to lie is called a *confidence interval*. Its endpoints are called *confidence limits*, and the probability that the interval contains the true value is called the *confidence coefficient*. In the above example the confidence interval is (0.4 0.6), the confidence limits are 0.4 and 0.6, and the confidence coefficient is 95%.

The confidence coefficient gives the probability that the interval contains the true value in the following sense. If we make many such experiments, from each finding p_1 and p_2 in the same way, then the relative frequency of the occasions in which these limits include the true p between them will be nearly equal to the confidence coefficient. The parameter p is, of course, not a random variable. The random variable is x, and its different values determine different intervals. The meaning of (9) is that about 95% of the time the determined interval does include the true value of p. Hence, if we use this method of interval estimation we shall be making true statements 95% of the time. If the confidence coefficient is $a\%$, we call the endpoints of the confidence interval $a\%$ confidence limits.

With $n = 100$, find 95% confidence limits for p if: (i) $x = 60$, (j) $x = 80$, (k) $x = 20$.

(l) Repeat Exercise (i) for 68% confidence limits. (m) Repeat Exercise (i) for 99% confidence limits.

When parameters of other distributions are being estimated, the details of finding the confidence limits are different, but the essential idea is the same. We always adopt a procedure that results in true statements a certain desired percentage of the time. This type of estimation is known as *interval estimation* as opposed to point estimation discussed earlier in this section.

PROBLEMS

1. To estimate the total weight of 2000 men who are to board a ship, a sample of 100 men are weighed and their average weight found to be 165 lb. Make an unbiased estimate of the weight of the 2000 men.

2. Suppose that $p = 0.3$ in the urn problem. What is the probability that in a random sample with replacement of 10 balls that $x = 3$? that $2 \leq x \leq 4$? that $1 \leq x \leq 5$?

3. Suppose that $x = 10$ in a sample of size 25 in the urn problem. Use the method of this section to find 95% confidence limits for p.

4. Find 68% confidence limits in the same situation.

5. Could the same procedure be used to find confidence limits for a sample of size four?

6. Suppose that a sample of size four is drawn in the urn problem and that two red balls are observed. Find 95% confidence limits for p. (This requires finding z by trial so that the probability in the left of (6) is 0.95.)

7. In the situation of Problem 13 in Section 9–2, find 95% confidence limits for the probability that the coin comes up heads.

8. In trying to predict a presidential election, a polling group used a sample of 10,000. They found that 49% of the sample favored candidate A. With what confidence can they predict that A will lose?

9. Taking account of (8–9–8), find a 95% confidence interval for estimating the mean weight in Problem 1, assuming that the standard deviation of the distribution of the weights is 25 lb.

10. Do the same for 68% confidence limits.

11. Rework Problems 9 and 10 for a sample of size 25.

12. Could we apply the method for a sample of size 10?

★13. Let x_1, x_2, ..., x_n be observed values of a random variable v with frequency $f(x, \theta)$, where θ is to be estimated. Assuming independence of the observations, the probability of observing these n values is $f(x_1, \theta)f(x_2, \theta) \ldots f(x_n, \theta)$. The value of this expression is called the *likelihood*. A value of θ (if any) that makes the likelihood maximum is called a *maximum likelihood estimate* of θ. Show that the mean of a sample is a maximum likelihood estimate of the mean of a normal population with unknown mean μ and variance σ^2.

★14. Similarly show that the mean of a sample from a population following the Poisson distribution is a maximum likelihood estimate of the parameter λ.

★15. Show that $\sigma^2 = E[x^2] - (E[x])^2$.

★16. Show that $E[x^2] \geq (E[x])^2$.

★17. An insurance company wishes to be 99% sure that the mean age of the policy holders does not differ from 60 years by more than one year. How large a group of policy holders is required, assuming that the mean and standard deviation of all insurable people available as policy holders are 60 and five?

★18. Suppose that 10 fish in a lake are marked and put back. Then 10 more fish are caught and two fish are found to be marked. How many fish are there in the lake? (See *An Introduction to Probability Theory and Its Applications*, by William Feller, pp. 37 ff)

★19. Show that the midpoint of the interval for estimating p from (14) is not at x/n unless $x/n = 1/2$. Under what conditions is this midpoint less than x/n? Show that this midpoint approaches x/n as n approaches ∞ or as z approaches zero.

Answers to Exercises

(a) Perhaps the second, unless a very large number of estimates are to be made and averaged.   (c) $P[x/n = 1/2] \doteq P[49.5 \leq x \leq 50.5] \doteq 2H(1/10) = 0.08$; $P[0.4 \leq x/n \leq 0.6] \doteq 0.96$; and $P[0.3 \leq x/n \leq 0.7] \doteq 1$.   (f) 0.45 and 0.55.   (g) The inequalities are equivalent to $(x - np)/\sqrt{npq} \leq z$. Squaring both members yields (14). See (3–9–16), (3–5–41).   (h) See (6).   (i) 0.5 and 0.69.   (j) 0.71 and 0.87.   (k) 0.13 and 0.29.   (l) Same but with (12) replaced by $(n^2 + n)p^2 - (2nx + n)p + x^2 = 0$; 0.55 and 0.65.   (m) $z = 2.6$.

Answers to Problems

1. 165(2000).   3. (0.23   0.59) Note that $x/n = 10/25 = 0.4$.   5. No, since we are using the Laplace limit theorem (8–9–10) and 4 is hardly large enough to qualify.   7. (45   64).   9. Taking $-a = b = z$ in (8–9–8), $F(b) - F(a) = 2H(z) = 0.95$ when $z = 2$. Hence the interval is given by $-2 \leq (165 - \mu)/(25/10) \leq 2$ or $160 \leq \mu \leq 170$.   11. $155 \leq \mu \leq 175$.

9–4 Statistical decision functions. A decade ago modern statistical theory consisted essentially of the exploitation of the ideas and methods discussed in the last two sections. Most practical statistical work is still formulated in these terms, but the theory has been considerably generalized in its conceptions and broadened in its applications. The fundamental idea in this development is the concept of a statistical decision function. The purpose of this section is to introduce this idea, without, however, attempting to go very deeply into either the theory or applications, which are still in a state of flux and involve mathematical techniques more advanced than can be treated in this book.

The situations considered in the last two sections have several features in common. First, we are required to make a decision, e.g., to accept or reject an hypothesis, to decide on a method of testing a hypothesis, or to estimate a parameter. Second, we cannot obtain sufficient information to make a decision with certainty that it is correct, e.g., we could not observe the number of red balls in the urn. The basic problem is to decide on a satisfactory method of obtaining partial information and of basing a decision upon it.

These features are common to many other situations that appear in science, business, and government. Indeed, most of our decisions are made without complete information or certainty of the consequences. Nevertheless, we must make decisions and we need satisfactory procedures for doing so. A procedure for decision must tell us what experiments to perform in order to get information and what action to take on the basis of the information obtained. To specify such a procedure is equivalent to specifying a function whose domain is the set of possible outcomes of some experiment and whose range is the set of possible actions. The image of each experimental outcome is the action we are to take, so that the function gives a complete guide for decision. We call such a function a *decision function*. If the action taken depends on observed values of a random variable, so that the independent variable and hence the dependent variable in the decision function are random variables, the function is called a *statistical decision function*. A test of a hypothesis is a statistical decision function since it gives a correspondence between observed values of a random variable and two possible actions, acceptance or rejection. A method of point estimation is a statistical decision function that maps the experimental outcomes into the set of all possible values of the parameter to be estimated. Here the set of possible actions may be infinite, each action consisting of asserting that the parameter has one of its possible values. Similar remarks apply to methods of interval estimation.

(a) Verify this by indicating the domain and range of the statistical decision function corresponding to the method of interval estimation.

The basic problem in statistics is to choose statistical decision functions appropriate to specific situations. To do this we wish to have criteria for choosing such functions. To decide on satisfactory decision functions we must consider all the consequences of adopting them. This means considering the losses that may result from wrong decisions as well as the cost of gathering information. In the previous sections the cost of obtaining information did not enter since we specified an experiment and left open only the question of how the resulting information was to be utilized.

To indicate in a very simple context the factors that are involved in choosing a decision function, we imagine a manufacturer of a product (say light bulbs) who sells the product in lots of 100. His customers expect an occasional faulty bulb, but there will be complaints and loss of business if there are too many. For simplicity we assume that there are just two possibilities, 1 faulty bulb in a lot or 10 faulty bulbs. In the former case the customer is satisfied, in the latter case we assume that his dissatisfaction costs the manufacturer $500. We assume that it costs $1 to test a bulb, and $100 to reject a lot and refrain from selling it. Suppose also that from quality control records it is known that the probability of 1 faulty bulb is 9/10, and that of 10 faulty bulbs is 1/10. What procedure should be followed?

Letting x be the number of faulty bulbs in a lot, we may summarize the situation as follows.

Possible states:

G: the lot is good, i.e., $x = 1$; $P[G] = 9/10$.

B: the lot is bad, i.e., $x = 10$; $P[B] = 1/10$.

Possible terminal actions:

A: accept the lot and sell it.

R: reject the lot and do not sell it.

Either situation may be combined with either terminal action. The following table summarizes the costs associated with each combination.

(1)

	G	B
A	0	500
R	100	100

One procedure is simply to reject all lots. Then the cost per lot is $100. There is no danger of losing $500, but there is also no possibility of doing any business! Another possibility is to accept every lot. Now the cost C under any procedure is a random variable. In this case it has two values,

0 and 500, with probabilities of 9/10 and 1/10 respectively. Hence, if we accept all lots, the expected cost per lot is given by $E[C] = 0 \cdot (9/10) + 500 \cdot (1/10) = 50$. Apparently, in this situation, it costs less to accept all lots than to reject all lots.

Make a table such as (1) and find expected cost of accepting all lots and rejecting all lots for: (b) Cost of accepting a bad lot is \$1000 instead of \$500. (c) Cost of accepting a bad lot is \$2000. (d) Probabilities are 8/10 and 2/10 instead of 9/10 and 1/10. (e) Probabilities are 99/100 and 1/100 instead of 9/10 and 1/10.

Evidently if we are required to accept or reject all lots, it may pay to accept all of them or reject all of them (and go out of business on this item) depending on the underlying costs and probabilities. However, it seems obvious that we might reduce costs by testing some bulbs from each lot. Suppose we adopt the decision function D_1 consisting of testing one bulb and rejecting or accepting the lot according as this one bulb is bad or good. The expected cost C is \$1 plus the sum of the entries in (1) each multiplied by the probability that it is the outcome. For example, letting y be the number of faulty bulbs observed, the probability of rejection when the lot is good is $P[G]P[y = 1 \mid G] = (0.9)(0.01)$. Hence for D_1, $E[C] = 1 + 0(0.9)(0.99) + 500(0.1)(0.9) + 100(0.9)(0.01) + 100(0.1)(0.1) = 47.90$.

(f) Let D_2 consist of testing two bulbs and rejecting the lot if either is faulty. Show that the expected cost is \$46.20. (g) Find the expected cost if three bulbs are tested and the lot rejected if one or more is faulty. (h) Find the expected cost for D_{10} consisting of testing 10 bulbs and rejecting if one or more is faulty.

We have considered here only a few possible decision functions. It seems reasonable to prefer one decision function to another if it yields a lower expected cost. (Why?) Considering cost as negative income (and income as negative cost), we may rephrase this by saying that a decision function is preferable if it yields higher expected income. This suggests the following *maximization principle*:

(2)   *Choose a decision function so as to maximize expected income (i.e., minimize expected cost).*

(i) Argue that among the decision functions D_n consisting in testing n bulbs and rejecting if one or more is bad, there is a best one according to criterion (2). ★(j) Find it.

PROBLEMS

★1. Rework Exercises (f) through (h) with $P[G] = 8/10$, $P[B] = 2/10$.
★2. Do the same for $P[G] = 0.99$, $P[B] = 0.01$.

★3. Suppose we know that the urn with ten balls has been filled by someone who chose at random a number between 0 and 10 inclusive, placed that number of red balls in the urn and completed with green balls. We are asked to estimate the number of red balls by sampling four balls with replacement. If our estimate is correct, we get $500. For each 1/10 we miss, we get $50 less, so that we gain nothing if we estimate 0 and there are really 10. How should we estimate? What is our expected pay?

★4. Reconsider Problem 3 for a sample of 100 with replacement.

★5. Verify the statement in the footnote on page 58 of "The Theory of Statistical Decision," by L. J. Savage, in the *Journal of the American Statistical Association*, March, 1951.

★6. Read the article referred to in Problem 5, and formulate the minimax principle in your own words.

★7. Read "Statistical Decisions," by E. S. Keeping, in the *American Mathematical Monthly*, March, 1956. Check the author's calculations.

★8. In the ancient method of trial by ordeal, the accused was subject to an ordeal, such as walking over hot coals, and acquitted if he survived. Suppose that the probability of survival for an innocent man was 80% and for a guilty man 50% (because of superstitious fear of the ordeal). Discuss the effectiveness of this method in terms of the probabilities of the accused being guilty, a probability that depends on the care exercised by officials and the community.

Answers to Exercises

(a) Domain is set of possible observed values of some random variable. Range is set of corresponding intervals.   (b) 100, 100.   (c) 200, 100.   (d) 100, 100.   (e) 5, 100.   (g) 44.9.   (h) 42.5.   (i) There are only a finite number of such decision functions, since $n \leq 100$.

Further Reading

The modern theory of statistical inference is closely related to the theory of games. The following references are suitable for the student who wishes to follow up the meager introduction given in this chapter.

Statistics, A New Approach, by W. Allen Wallis and Harry V. Roberts, The Free Press, Glencoe, Illinois, 1956. An elementary book with many interesting examples.

First Course in Probability and Statistics, by Jerzy Neyman, Henry Holt and Co., New York, 1950. A careful discussion of basic ideas by one of the architects of modern statistics.

The Compleat Strategyst, by John D. Williams, McGraw-Hill, New York, 1954. Very easy introduction to game theory.

Games and Decisions, by R. D. Luce and H. Raiffa, Wiley, New York, 1957.

Theory of Games and Statistical Decisions, by D. H. Blackwell and M. A. Girshick, Wiley, New York, 1954.

Introduction to the Theory of Games, by J. McKinsey, McGraw-Hill, New York, 1952.

★CHAPTER 10

ABSTRACT MATHEMATICAL THEORIES

10-1 The nature of abstract theories. The dictionary descriptions of "abstract" that apply here are "considered apart from any application to a particular object," "without reference to a thing or things," and "general, as opposed to particular." All mathematical theories are abstract to some extent. For example, the statement $2 + 3 = 5$ says in effect that two things and three things are five things, without reference to any particular things. Of course, it is also a statement about the particular numbers 2, 3, and 5. The law $a + b = b + a$ is more abstract, since it does not refer to any particular numbers. Still, it does refer to numbers rather than some other objects. In the form $\forall a\, \forall b\; a \in Re \wedge b \in Re \to a + b = b + a$, it is clearly a statement about a particular thing, namely the real numbers Re. We see from these examples that there are various levels of abstraction and that a theory may be abstract from one point of view and concrete from another.

(a) To what extent is $2 + 3 = 5$ "apart from any application to a particular object," and to what extent does it apply to a particular object? (b) Discuss the physical law, "A freely falling body moves with constant acceleration," from the point of view of the preceding paragraph. (c) Is it correct to say that all scientific theories are abstract to some extent? (d) Does abstractness imply lack of applicability?

The term "abstract mathematical theories" designates mathematical theories on a considerably higher level of abstraction than the examples mentioned. To make clear how the term is used, we give a series of progressively more abstract statements. "Two ducks and three ducks are the same as three ducks and two ducks" is a statement of fact about ducks. It is already abstract to some extent, since no particular ducks are mentioned. "Two plus three is the same as three plus two" is more abstract, since we do not specify what is being counted. On the other hand, we are referring specifically to the numbers 2 and 3. The law $a + b = b + a$ in the context of the real numbers, is much more abstract, as we suggested above. Now suppose we simply write the law $\forall a\, \forall b\; a + b = b + a$, without indicating whether we are talking about integers, rational numbers, real numbers or complex numbers. We may investigate the consequences of this law without concerning ourselves with what numbers are involved. If we do so, we are considering the commutative law in abstraction from the particular numbers to which it applies. However, we are still thinking of numbers and the operation of addition. But we have seen that the commutative property applies to things other than numbers. Thus we have such laws as $p \lor q = q \lor p$, $p \land q = q \land p$, $A \cup B = B \cup A$, and $A \cap B = B \cap A$. We may abstract from the particular operations by considering an unspecified binary operation (indicated by $*$) on an unspecified set of objects S such that the law $a * b = b * a$ holds for all members of S.

In the last example we have abstracted to the point where the objects under discussion are not specified (the name S is adopted merely as a convenience and tells us nothing about its members) and the operation is unspecified except to indicate that it serves to yield an object when it is applied to an ordered pair of members of S. After all this abstraction, what do we have left? Merely the statement that the binary operation $*$ (unspecified) is commutative in S (unspecified). We can then investigate the consequences of this law. Of course, a theory based on only this one law would not be of much interest. The typical *abstract mathematical theory* involves a number of laws about unspecified objects, relations, and operations.

We have presented some of the theory of this book in axiomatic form. That is, we have begun each theory with undefined terms and axioms, introduced definitions, and derived theorems from the axioms on the basis of rules of proof. This is not the only way in which mathematical theories can be developed. But it has many advantages, and it is particularly useful in abstract theories, in which we cannot rely on any intuitive knowledge of the objects under consideration. Of course, in developing any axiomatic theory we can base our proofs only on the stated axioms and definitions, even though we may "know" a good deal about the objects of the theory. But in an abstract theory, the objects are unspecified and we know nothing about them except that they satisfy whatever axioms we are using as the basis for our theoretical construction.

(e) Would it be correct to say that we make no use of our intuitive knowledge in constructing mathematical theories?

Of what use is a theory about unspecified relations and operations in the context of unspecified objects? Very useful indeed just *because* of the lack of specification! An abstract mathematical theory is applicable to any particular situation in which the objects and operations "fit" the theory, that is, satisfy the axioms of the theory. If the objects in any particular situation satisfy the axioms, then they must satisfy all the laws of the theory. Hence, an appropriately chosen abstract theory is "tailor made" to give us answers in particular situations. It frequently happens that a single abstract theory is applicable in many entirely different situations. (A trivial example is the commutative law mentioned above.) Hence, by developing an abstract theory we may be able at one stroke to solve problems in many different areas. This is precisely the value of any abstract theory, from the simplest generalizations in science to the most abstract mathematics. The more abstract the theory, the wider its potential applicability.

(f) Which is of wider applicability, $2 + 3 = 3 + 2$ or $a + b = b + a$? (g) Which is more abstract and which is more widely applicable, the theory of freely falling bodies or the theory of bodies moving with constant acceleration? (h) Bertrand Russell is reported to have said in an after-dinner speech that mathematics is the science in which we never know what we are talking about or whether what we say is true. Explain.

The purposes of this chapter are (1) to introduce the reader to several of the most interesting theories of abstract mathematics, (2) to give him some insight into the way in which abstract theories are constructed, and (3) to review some of the ideas of previous chapters in an abstract setting.

ANSWERS TO EXERCISES

(c) Yes.   (d) On the contrary!   (e) No.   (f) The latter.   (g) The latter.

10–2 Fields. The notion of a field may be obtained by abstraction from the properties of real numbers embodied in the axioms (1–16–1) through (1–16–10). Suppose we have a set S that is closed under two binary operations (1–16–1) which are associative (1–16–4 and 1–16–5) and commutative (1–16–2 and 1–16–3); that contains an identity element for each operation (1–16–7 and 1–16–8); that contains an inverse of each of its elements with respect to the first operation (1–16–9); that contains an inverse of each of its elements except the first identity element with respect to the second operation (1–16–10); and such that the second operation is distributive over the first (1–16–6). We call the operations plus (+) and times (·) and call the corresponding identity elements zero (0) and one (1), though there is no requirement that they be the familiar operations or numbers formerly so designated. A system $\{S, +, \cdot, 0, 1\}$ satisfying these axioms is called a *field*.

(a) Name three sets of numbers that are fields.   (b) If we know that a system is a field, can we apply to it any of the algebraic identities proved for real numbers? Why?

Even if the examples in Exercise (a) were the only possible fields, the concept would be of some value because of Exercise (b). However, there are many other (indeed an infinite number) of different fields. Consider, for example, the set S consisting of the numbers 0, 1, and 2. We define addition and multiplication as follows: Add or multiply as usual and then subtract multiples of 3 until the result is one of the members of S. Take 0 and 1 as the identity elements. With these definitions $\{S, +, \cdot, 0, 1\}$ is a field. To justify this statement we must verify that all the axioms hold. The following addition and multiplication tables help "see" the situation.

	+	0	1	2
		0	1	2
	0	0	1	2
(1)				
	1	1	2	0
	2	2	0	1

	·	0	1	2
		0	1	2
	0	0	0	0
(2)				
	1	0	1	2
	2	0	2	1

(c) Verify the tables by doing the arithmetic.

From the tables or the definition of addition and multiplication it is obvious that S is closed. Associativity follows from the definition of the operations, since we begin by applying the usual operations (which we know are associative) and then subtract multiples of three (which is independent of the way in which we have associated the numbers in the first calculation). Commutativity and the distributive law follow in the same way. We can also verify the commutative law by noting that the tables are symmetric about the *main diagonal* from upper left to lower right. From the tables also we see immediately that 0 and 1 have the desired properties. The existence of an inverse is shown in the tables by the fact that 0 appears in each row and column of (1) and 1 appears in each row and column of (2) except the row and column corresponding to 0.

(d) Show that {0, 1} together with addition and multiplication defined as above but subtracting multiples of 2 instead of multiples of 3, is a field.   (e) Show that {0, 1, 2, 3}, with the operations defined as above but subtracting multiples of 4 is *not* a field.   (f) Show without reference to the tables (1) and (2) that 0 and 1 have the desired properties and that the required inverses exist.

We have given two examples of *finite fields*, that is, fields with a finite number of elements. There are an infinite number of finite fields constructible similarly. Indeed, consider the set {0, 1, 2, ... $n - 1$}, where n is a prime number. Define the sum and product as obtained by the usual operations followed by the subtraction of multiples of n until one of the elements is obtained. Then an argument similar to those above shows that the system is a field.

(g) Why must n be prime?

Are there infinite fields different from the rationals, reals, and complex numbers? Yes, an infinite number. Consider, for example, the set of all real numbers of the form $a + b\sqrt{2}$, where a and b are rationals, together with the usual operations of addition and multiplication and the two identity elements $0 + 0\sqrt{2}$ and $1 + 0\sqrt{2}$. It is obvious that all axioms are satisfied, except possibly for the existence of the inverses. However, the identities

(3) $(a + b\sqrt{2}) + (-a - b\sqrt{2}) = 0 + 0\sqrt{2}$

(4) $(a + b\sqrt{2}) \left[\dfrac{a}{a^2 - 2b^2} - \dfrac{b}{a^2 - 2b^2} \sqrt{2} \right] = 1 + 0\sqrt{2}$

exhibit the existence.

(h) How does (4) show lack of a multiplicative inverse for $0 + 0\sqrt{2}$? (i) Show that the above system is closed under multiplication.   (j) What about the possibility of $a^2 - 2b^2 = 0$ in (4)?   (k) Show that the numbers of the form $a + b\sqrt{c}$, for fixed c an integer not a perfect square and a and b rational, form a field.   (l) Show that if we modify Exercise (k) to require that a and b be integers, the result is not a field.   (m) Show that we do get a field if we allow a and b to be any reals in Exercise (k).

We defined subtraction and division in (1–14–19) and (1–14–1) so that (as proved in Section 1–16)

$$(5) \qquad\qquad a - b = a + (-b),$$

$$(6) \qquad\qquad a \div b = a \cdot b^{-1}.$$

Evidently this can be done in any field because of the existence of the inverses $-b$ and b^{-1}.

(n) Do subtraction and division in any field have all the properties that we derived for real numbers?   (o) In the system of Exercise (e) find 2/3, 3/2, (2/3)/(1/2), and (1/3)(1/2).   (p) Why is division by zero undefined in a field?

Problems

1. Show that the complex numbers with rational components form a field under the usual operations and with the usual identity elements.

2. How do the tables (1) and (2) illustrate the fact that in a field there is only one element with the properties of 0 and only one with the properties of 1? (We describe this by saying that the identities are unique.)

3. Can a field have no elements? Can a field have just one element?

4. Show that the real numbers of the form $a\sqrt{2} + b\sqrt{3}$ with a and b rational do not form a field.

5. Do all numbers of the form $a/2^n$, where a is an integer, form a field?

6. Answer the same question if a may be any rational.

7. In the definition of a field, we used equality ($=$) as the one relation of our system. A slightly more abstract theory results from using an unspecified relation R. We require that this unspecified relation be reflexive, symmetric, and transitive, i.e.,

$$(7) \qquad\qquad x \, R \, x \qquad\qquad \text{(Reflexivity)},$$

$$(8) \qquad\qquad x \, R \, y \to y \, R \, x \qquad\qquad \text{(Symmetry)},$$

$$(9) \qquad\qquad (x \, R \, y \land y \, R \, z) \to (x \, R \, z) \qquad \text{(Transitivity)}.$$

Such a relation is called an *equivalence relation*. We need also to be able to prove or to assume that if $a \, R \, b$ and $c \, R \, d$, then $a + c \, R \, b + d$ and $ac \, R \, bd$. Then we may define a field as a set $\{S, +, \cdot, 0, 1, \equiv\}$, where $\equiv$ is some equivalence relation. Let a relation called "congruence modulus 3" be defined as follows:

(10)   **Def.** $[a \equiv b \pmod 3] = [a - b \text{ is divisible by 3}]$.

We read $a \equiv b$ (mod 3) as "a is congruent to b mod 3." Show that congruence mod 3 is an equivalence relation. Show that the integers form a field with this equivalence relation.

8. Is $=$ an equivalence relation? Name as many equivalence relations as you can. What property does $=$ have that is not possessed by other equivalence relations?

9. Define congruence modulo m. Show that the integers with congruence mod m form a field if and only if m is a prime.

10. Solve $3x - 2 \equiv x + 7$ (mod 11). How many solutions are there?

11. Do the procedures for solving equations apply to solving congruences?

12. Does $x^2 + 2x + 1 \equiv 0$ (mod 3) have a solution? Does $x^2 + 2x + 2 \equiv 0$ (mod 3)?

13. Solve the following simultaneous congruences.

$$3x + y \equiv 1,$$
$$(\text{mod } 5)$$
$$x - y \equiv 2.$$

14. Solve

$$x + 2y \equiv 6,$$
$$(\text{mod } 7)$$
$$2x - y \equiv 1.$$

15. Why does

$$-2x - y \equiv 1,$$
$$(\text{mod } 5)$$
$$4x + 2y \equiv 1,$$

have no solution?

16. The field of integers with congruence mod m is called the *field of integers mod m*. What numbers in the field of integers mod 3 are perfect squares (i.e., squares of elements of the field)?

17. Answer the same question for the integers mod 7; mod 11; mod 13; and mod 17.

18. A system that satisfies all the requirements for a field, except for the deletion of the requirement of the existence of multiplicative inverses and the addition of the cancellation law, $ac = bc \rightarrow a = b$, is called an *integral domain*. Give a familiar example.

19. Argue that any field is an integral domain.

20. An *ordered field* is a field with a subset (called the positives) closed under the two operations and such that any member of the field belongs to this subset, or else is zero, or else its additive inverse belongs. State this definition using logical symbols. Are the reals an ordered field? are the rationals? How is this definition tied in with the concept of "less than"?

21. Show that the integers mod m (prime) are not an ordered field.

22. Show that the field $0, 1, 2, \ldots, n - 1$ defined in this section is not an ordered field.

23. Show that the complex numbers are not an ordered field.

24. Show that the set of all numbers formed from $\sqrt{2}$ by the operations $+$, $-$, $\cdot$, and $\div$ form a field.

Answers to Exercises

(a) Rationals, reals, complex numbers.   (b) Yes, since all the identities were derived from the axioms alone. Hence we could prove them for any field with the same proofs.   (c) For example, $2 + 2 = 4$. Subtracting 3, we find 1 as in the table.   (d) Write tables and verify axioms.   (e) $2 \cdot 2 = 0$ in contradiction with (2–8–5).   (f) From the definition of the operations it is immediate. (g) Otherwise we would have $a \cdot b = 0$ with neither a nor b zero, in contradiction to (2–8–5).   (h) Since $a^2 - 2b^2$ is then zero and the inverse is undefined. (i) $(a + b\sqrt{2})(c + d\sqrt{2}) = (ac + 2bd) + (ad + bc)\sqrt{2}$.   (j) Impossible since $\sqrt{2}$ is not rational.   (k) Verify axioms and find inverses.   (l) Multiplicative inverse fails.   (m) We get the field of real or complex numbers according as $c \geq 0$ or $c < 0$.   (n) Yes, since all the derived properties were proved from the axioms and definitions.   (o) $2/3 = 2$; $3/2$ does not exist; same for other two.   (p) See Exercise (n).

Answers to Problems

1. Closure is the only nonobvious property and can be shown by calculating a typical product. The essential thing is that the components of a sum, difference, product, and quotient are rational.   3. No, because we require the elements 0 and 1. Possibly, since we might have just one element with $0 = 1$. Fields with less than two elements are called *trivial*.   5. No. Multiplicative inverse fails. 7. It has the tables (1) and (2). By use of the definition of congruence, each axiom can be verified.   9. a is congruent to b mod m if and only if $b - a$ is divisible by m. If m is not prime, we have the difficulty indicated in Exercise (g). 11. Yes. The proofs require some modification, since the Rule of Replacement does not apply to all equivalence relations, but the properties (7) through (9) together with the axioms are sufficient.   13. (2, 0).   15. If the first equation be multiplied by -2, it becomes $4x + 2y = 2$, which is inconsistent with the second equation.   17. 0, 1, 4, 2; 0, 1, 4, 9, 5, 3; 0, 1, 4, 9, 3, 12, 10; 0, 1, 4, 9, 16, 8, 2, 15, 13.   19. The cancellation law holds in any field; hence a field has all the integral domain properties.   20. See Section 3–5.   21. Zero cannot be in the positive subset, but if we add 1 to itself m times, we get zero. Hence 1 cannot be in the subset. Then -1 must be in the subset, and $(-1)(-1) = 1$ must be in the subset, which is a contradiction. 23. Since in any ordered field the square of any nonzero element is positive (the proof of 3–5–21 applies), and since $i^2 = -1$ is not positive.

10–3 Groups. The axioms in Section 1–16 and the various theorems derived from them suggest that the operations of addition and multiplication are similar in many ways. In Section 1–9 we have paired laws to bring out the analogy. Suppose we were to put $*$ in place of both the plus sign and the dot in (1–16–2) and (1–16–3). The result would be the same, namely $a * b = b * a$. Suppose, in addition, we put e in place of both 0 and 1 in (1–16–7) and (1–16–8). In either case we find $a * e = a$.

Finally, suppose we put a' for both $-a$ and $1/a$ in (1–16–9) and (1–16–10). We get $a * a' = e$. These observations suggest the possibility of constructing an abstract theory about an unspecified operation of which the familiar addition and multiplication would be special cases.

We take as undefined a set S, a binary operation $*$, and an element e of S. We are going to require that S be closed under $*$, that $*$ be associative, that e have the familiar property of an identity element, and that every element have an inverse with respect to $*$. Such a system $\{S, *, e\}$ we call a *group*. If, in addition, $*$ is commutative, we call the system a *commutative group*.

Which of the following are groups?   (a) The integers under addition. (b) The integers under multiplication.   (c) The rationals under division. (d) The nonzero real numbers under division.

The defining axioms of a group may be stated as follows:

(1)   *For every a and b in S, $a * b$ is an element of S.* (Closure)

(2)   *For every a, b, c in S, $a * (b * c) = (a * b) * c$.* (Associativity)

(3)   *There is an element of S (call it e) such that for every member a of S, $a * e = e * a = a$.* (Identity element)

(4)   *For every element a of S there is an element (call it a') of S such that $a * a' = a' * a = e$.* (Inverse)

(e) Replace words in (1) through (4) by quantification symbols.   (f) In the above formulation, what are the undefined constants, variables, and formulas?   (g) Sometimes in defining a group it is specified that $a * b$ is "unique," meaning that for any value of a and b in S there is just one value of $a * b$. Why is this implicit in the above definition?   (h) Construct an example of a noncommutative group.   (i) Suppose we make the definition

(5) **Def.** $a \circ b = a * b'$.

To what operation does $\circ$ correspond if $*$ is multiplication? addition?   (j) Addition mod m is the process (described in Section 10–2) of adding in the usual way and subtracting multiples of m until a number from the set $\{0, 1, \ldots, m - 1\}$ is obtained. Similar meaning is assigned to other operations mod m. Show that $\{0, 1, \ldots, m - 1\}$ form a group under addition mod m for any integer m. (k) Under what condition does $\{1, 2, \ldots, m - 1\}$ form a group under multiplication mod m?   (l) In what sense is a clock a device for addition mod 12?

Some very interesting groups result from the study of the symmetries of geometric figures, and these groups have applications in chemistry, physics, crystallography, and art. Consider, for example, the square pictured in Fig. 10–1. The figure is obviously symmetric about the dotted

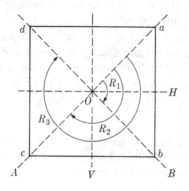

FIGURE 10–1

lines A, B, H, and V. One way of expressing this fact is to say that the figure can be rotated about these lines (up out of the plane and back into it) through an angle of 180° without changing its appearance. Or we may think of the figure as being reflected in one of the lines without altering its appearance. The figure is also symmetric about the point O. Moreover, this central symmetry has the special feature that if we rotate the square in the plane about O through an angle of $\pi/2$, π, $3\pi/2$, and of course 2π, the appearance of the figure remains unchanged.

(m) Check the definitions of symmetry in Section 5–15.

To analyze these symmetries more carefully, we label the vertices of the square as indicated in the figure. We imagine the point O and the lines A, B, V, and H as remaining fixed. We let the letters A, B, V, and H serve also to name the action of reflecting the figure in the corresponding lines (or rotating it through 180° about these lines). We let R_1, R_2, and R_3 be the rotations of the figure about O through the indicated angles. Finally, we introduce I to represent the action of no motion at all. Now we let $X * Y$ mean the motion that achieves the same result as X followed by Y.

 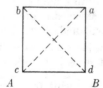 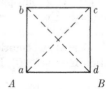

FIGURE 10–2

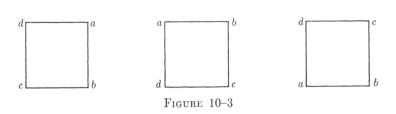

FIGURE 10–3

For example, $R_1 * R_1 = R_2$, since a rotation of 90° followed by another of 90° is the same as a rotation of 180°. Similarly, $R_1 * R_2 = R_3$, and so on. We see also that $A * A = I$, since reflecting twice about the same line re-establishes the status quo. It is easy to calculate these "products," but for a case such as $A * B$ or $R_3 * H$, it is convenient to keep track of what happens by drawing diagrams. For example, to find $A * B$, we first draw the square in the original position, then as it appears after reflection in the line A, and finally as it appears after reflecting *this* square in line B. (See Fig. 10–2.) Note that the axes do not move, and that B means reflection in the fixed line B of the square in whatever position the square is as the result of previous operations. The final position shows that $A * B = R_2$.(!) Similarly, from Fig. 10–3 we see that $R_3 * H = B$.

Draw the squares and find: (n) $B * A$,   (o) $H * R_3$.

For reasons that will soon appear, we are interested at the moment particularly in the rotations. The reader can easily verify the following "multiplication table" of the rotations of the square.

*	I	R_1	R_2	R_3
I	I	R_1	R_2	R_3
R_1	R_1	R_2	R_3	I
R_2	R_2	R_3	I	R_1
R_3	R_3	I	R_1	R_2

(6)

It is not hard to see that these rotations form a group, with I the identity element.

(p) How does the table reflect that I has the property (3)?   (q) How is property (4) shown?   (r) How is closure indicated?   (s) How could you check (2) from the table?   (t) Can you think of an easier way to demonstrate associativity by an argument?

Now we consider the group of integers $\{0, 1, 2, 3\}$ under addition mod 4. We have the following table.

$$
\begin{array}{c|cccc}
+ & 0 & 1 & 2 & 3 \\
\hline
0 & 0 & 1 & 2 & 3 \\
1 & 1 & 2 & 3 & 0 \\
2 & 2 & 3 & 0 & 1 \\
3 & 3 & 0 & 1 & 2 \\
\end{array}
$$

(7)

If the reader compares (7) with (6) term for term, he will find that they have the same structure. Indeed, the correspondence $\{(0, I), (1, R_1), (2, R_2), (3, R_3)\}$ carries (7) into (6). This correspondence is one to one. Calling it F, it is evident that $F(a + b) = F(a) * F(b)$ for any a and b in $\{0, 1, 2, 3\}$. Hence the two sets $\{0, 1, 2, 3\}$ and $\{I, R_1, R_2, R_3\}$ are isomorphic with respect to the relation of equality and the operations $+$ and $*$. When two groups are isomorphic with respect to the group operations, we say simply that they are *isomorphic groups*. Isomorphisms are of considerable use. If we know a property of one isomorphic group, we know immediately that it applies to the other. In this example, we can answer any question about the rotations of a square by reference to addition mod 4!

(u) In Fig. 10–4 we show an equilateral triangle with the indicated symmetries. Make a table of the group of rotations I, R_1, R_2. (v) Show that this group is isomorphic to the group of integers under addition mod 3. (w) Show by writing down the multiplication table that the four fourth roots of 1, $\{1, i, -1, -i\}$ form a group under multiplication. (x) Is this group isomorphic to the groups in (6) and (7)? (y) What can you say about the square roots of 1 under multiplication?

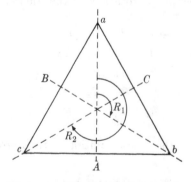

FIGURE 10–4

PROBLEMS

1. Complete the multiplication table of the group of symmetries of the square.

2. Complete the table for the group of symmetries of the equilateral triangle.

3. Show that the cube roots of 1 under multiplication is a group isomorphic to the groups in Exercises (u) and (v).

4. Can you suggest an underlying reason for the isomorphisms indicated in Problem 3?

In Problems 5 through 32 show that the set under the given operation is a group (cite the identity element and inverse) or show that it is not a group and indicate why.

5. The even numbers under addition.

6. The odd numbers under multiplication.

7. The odd numbers under addition.

8. The even numbers under multiplication.

9. Numbers of the form 2^n under multiplication, where n may be any integer.

10. Numbers of the form 10^n under multiplication, where n is any real number.

11. Numbers of the form 10^n under addition.

12. The integers under subtraction.

13. The positive reals under multiplication.

14. The negative reals under multiplication.

15. The nonconstant linear functions under composition.

16. The quadratic functions under composition.

17. The exponential functions under composition.

18. The exponential functions under multiplication.

19. The power functions I^r with domain the positive reals and with exponent any rational number except 0, under composition.

20. The debit and credit entries in a charge account, where $a * b$ means the entry that has the same effect as a followed by b.

21. The set of all valid contracts between two individuals, where $a * b$ means the contract equivalent to contract a followed by contract b.

22. Numbers of the form $n/\sqrt{2}$, where n is an integer, under addition.

23. The five fifth roots of 1 under multiplication.

24. Numbers of the form $1/5^n$, where n is an integer, under addition.

25. The power set of a given set under the operation of union.

26. Operations of the form D^n (where n is any integer, $D^0 f = f$, where D is the derivative, D^{-1} is the antiderivative, and these operators all have as domain and range the set of all functions that can be differentiated any number of times) under composition.

27. Sentences under the operation of conjunction; disjunction.

28. Sentences under the operation $p * q = (p \lor \sim q) \lor (q \lor \sim p)$.

29. Two-place decimals under multiplication, where the product is obtained in the usual way followed by rounding off to two places.

30. The odd functions under multiplication.

31. Plane vectors under the dot product.

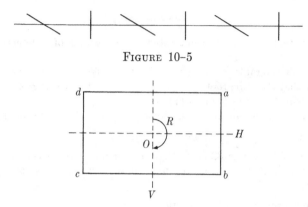

FIGURE 10–5

FIGURE 10–6

32. The motions that leave the pattern of Fig. 10–5 unchanged. (We imagine the pattern continuing indefinitely in either direction.)

33. Show that the group of positive real numbers under multiplication is isomorphic to the group of real numbers under addition. (*Hint*: This isomorphism is the reason that logarithms are so useful!)

34. Sketch an isosceles triangle and write the multiplication table of its symmetries. To what group of integers under addition is this symmetry group isomorphic?

35. Answer the same question for the rectangle sketched in Fig. 10–6.

36. Show that the group of symmetries of an ellipse or hyperbola is isomorphic to that of a rectangle (except when the ellipse is a circle).

37. Construct and justify a similar statement for the symmetries of a parabola.

38. What is the group of symmetries of a figure that has "no" symmetries?

★39. Sketch a regular pentagon and write the multiplication table of its symmetries.

40. Describe the group of symmetries of a regular polygon of an even number of sides.

41. Do the same for one with an odd number of sides.

42. Suppose we have a certain geometric figure G. We call a movement of the figure that leaves its appearance unchanged a *symmetry*. Let G be the set of all symmetries of a figure, and let $a * b$ be the movement equivalent to a followed by b. Show that the symmetries form a group.

43. Show that the n nth roots of 1 form a group under multiplication. To what group of symmetries is this group isomorphic? To what group of integers under addition is it isomorphic?

44. Describe the group of all symmetries of a circle. To what group of numbers under addition is the group of rotations isomorphic?

45. Consider the set of all angles under the operation of addition and with the relation of congruence as defined in Section 4–9. Do they form a group? What is the relation between congruence in the geometric sense of Section 4–9 and congruence modulo some number m?

★46. A pattern such as that of Fig. 10–5 laid out on an infinite straight line is called a *border*. Any border has a characteristic group of motions that leave it unchanged in appearance. Construct some borders and characterize their groups.

★47. A pattern that covers the entire plane in some repetitive fashion is called a *tesselation*. (Imagine an infinitely large wall covered with wallpaper or an infinite floor inlaid with some pattern of tiles.) Each tesselation has a characteristic group of motions that leave its appearance unchanged. Draw some tesselations and investigate their groups.

★48. Solid figures have symmetries that may be studied as we have those of plane figures above. Write the multiplication table of the symmetries of a cube. Investigate other solid figures. (*Hint:* Consider reflections in a plane of symmetry, reflections in a point of symmetry, and rotations about a line of symmetry.)

★49. Consider the set of all possible colors and let $xa * yb$ be the color resulting from mixing x parts of a and y parts of b. Do the colors form a group under $*$?

★50. Consider the substitutions that can be made for variables in formulas. Do they form a group?

★51. Do the productive operations of a factory form a group?

★52. Let $a * b = a + b - ab$, for a and b real numbers. Is this a group?

ANSWERS TO EXERCISES

(a) Yes.   (b) No, inverse fails.   (c) No, since 0 has no inverse.   (d) No, not associative.   (e) $\forall a \, \forall b \, a \in S \wedge b \in S \to a * b \in S$; $\forall a \, \forall b \, \forall c \, a \in S \wedge b \in S \wedge c \in S \to a * (b * c) = (a * b) * c$; $\exists x \, x \in S \wedge \forall a \, a \in S \to a * x = x * a = a$ (then we can use 2–10–9 to name this element e); $\forall a \, a \in S \to \exists x \, a * x = x * a = e$ (again we use 2–10–9 to name this element a').   (f) Constants are "S," "$*$," and "e"; variables are small Latin letters, and formula is "$a * b$." (g) In our treatment we implicitly took "$a * b$" as an undefined formula. But by definition a formula has only one value for each set of values of its variables. (i) Division; subtraction. (j) Verify axioms. (k) m must be prime. (l) Two times that differ by a multiple of 12 yield the same clock position.

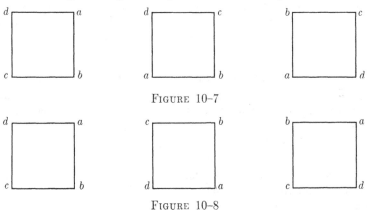

FIGURE 10–7

FIGURE 10–8

(n) Fig. 10–7.   (o) Fig. 10–8.   (p) The row and column of I repeat the borders.   (q) I appears once in each row and column.   (r) All entries are members of the set.   (s) By trying all cases, all 24 of them!   (t) Since in this case $a * (b * c)$ means the result of the motion a followed by the result of the motion b followed by c, and $(a * b) * c$ means the result of motion a followed by b and then followed by c, each mean the result of performing a, b, and c in that order.   (u) See Exercise (v).   (v) Under the correspondence $\{(0, I),$ $(1, R_1), (2, R_2)\}$.   (w) See Exercise (x).   (x) Under the correspondence $\{(0, 1), (1, i), (2, -1), (3, -i)\}$.   (y) Group isomorphic to the integers mod 2.

ANSWERS TO PROBLEMS

1.

$*$	I	R_1	R_2	R_3	H	V	A	B
I	I	R_1	R_2	R_3	H	V	A	B
R_1	R_1	R_2	R_3	I	A	B	V	H
R_2	R_2	R_3	I	R_1	V	H	B	A
R_3	R_3	I	R_1	R_2	B	A	H	V
H	H	B	V	A	I	R_2	R_3	R_1
V	V	A	H	B	R_2	I	R_1	R_3
A	A	H	B	V	R_1	R_3	I	R_2
B	B	V	A	H	R_3	R_1	R_2	I

3. Show that if $w = (-1 + \sqrt{3}i)/2$, then the three cube roots of 1 are w, w^2, and $w^3 = 1$. Then make the table. 5. Yes, 0 is identity, and $-a$ is inverse. 7. No, not closed. 9. Yes, 2^0 is identity, 2^{-n} is inverse. 11. No, not closed. 13. Yes, 0, $1/x$. 15. Yes, I, $(I - b)/m$ is inverse of $mI + b$. 17. No, not closed. 19. Yes, I, $I^{1/r}$ is inverse of I^r. 21. No, inverse fails. 23. Yes, isomorphic to addition mod 5 of integers. 25. No, inverse fails. 27. No, inverse fails. 29. No, identity not unique. 31. No, closure fails. 33. A correspondence is given by 10^I, e^I, or any other exponential with positive base. 35. None. 37. Integers mod 2. 41. There are $n - 1$ rotations through multiples of $2\pi/n$ and n reflections through lines joining vertices to midpoints of the opposite sides. 43. Rotational symmetries of regular n-sided polygon. Integers mod n. 45. Yes. Congruence of angles is isomorphic to congruence of numbers mod 2π.

10–4 Transformation groups. Since composition of functions is associative by (5–9–6), it seems possible that some sets of functions might be groups under this operation. Indeed, we see from (5–9–8) that the identity function I_A might play the role of the group identity, and from (5–10–19) and (5–10–20), that the inverse of a function might play the role of the group inverse. However, for this to work out it is evidently necessary that the set of functions have inverses (i.e., that their converses be functions) and that a fixed set serve as both range and domain for all of them.

(a) Explain why this is necessary in terms of (5–9–8), (5–10–19), and (5–10–20).

Suppose that we have a set of one-to-one correspondences (functions, transformations) that map a set A onto itself (i.e., A is both domain and range of each function) and such that the set is closed under composition, contains I_A, and contains the inverse of each of its members. Then the set is a group under composition, with I_A the identity element and T^* the group inverse of T. We write I for I_A and $T_1 T_2$ for $T_1 \circ T_2$ unless ambiguity seems likely. Note that $T_1 T_2$ is the operation of applying T_1 and then T_2, that is, $(T_1 T_2)(x) = T_2(T_1(x))$. So as to always be working from left to right, we write $x T_1 T_2$ for the image of x under $T_1 \circ T_2$.

As a simple example of a group of transformations, consider the linear functions of the form $I + a$ and let $T_a = I + a$ with domain and range the real numbers. The identity is I on the reals, the inverse of T_a is T_{-a}, and $T_a T_b = T_{a+b}$. The geometric interpretation in terms of an axis is interesting. We may think of T_a as translating each point of the axis a directed distance a. (We could alternatively think of T_a as moving the origin a directed distance $-a$.) Then I is the act of translation through O, T_{-a}, reverses T_a, and T_{a+b} is a translation equivalent to translating through a and then through b.

(b) Show that the group of translations T_a is isomorphic to the real numbers under addition.   (c) Show that the distance between two points on an axis is invariant (unchanged) under the group of translations; i.e., show that $|x_1 - x_2| = |x_1 T_a - x_2 T_a|$.

The above example is easily generalized to the plane. Let $T_{a,b}$ be the transformation that carries (x, y) into $(x + a, y + b)$, that is, $(x, y) T_{a,b} = (x + a, y + b)$. Let the domain and range be $Re \times Re$. Then the set of all such translations is a group. The geometric interpretation is that $T_{a,b}$ moves all points of the plane through the vector (a, b).

(d) Show that $(x, y) T_{a,b} = (x, y) + (a, b)$.   (e) Find $T_{a,b} T_{c,d}$.   (f) Is this group commutative?   (g) What is the inverse of $T_{a,b}$? the identity element?   (h) To what group of numbers is this group of translations isomorphic? (i) To what group of transformations of coordinates is it isomorphic?

Because of the convenience of geometric imagery, we often speak of the domain and range of a transformation as a "space" and of its members as "points," even when geometry is not involved.

The above examples referred to spaces with an infinity of points. By way of contrast consider the set of all one-to-one transformations of $\{a, b, c, d\}$ onto itself. The identity is $I = \{(a, a), (b, b), (c, c), (d, d)\}$ and another transformation is $\{(a, b), (b, c), (c, d), (d, a)\}$. A more convenient notation for such transformations is to list the points on one line and the corresponding images below them. The transformations just named are given by

(1) $\qquad \begin{pmatrix} a & b & c & d \\ a & b & c & d \end{pmatrix} \quad \text{and} \quad \begin{pmatrix} a & b & c & d \\ b & c & d & a \end{pmatrix}.$

To find the product of two transformations we simply trace each point. Thus we have

(2) $\qquad \begin{pmatrix} a & b & c & d \\ b & c & d & a \end{pmatrix} \begin{pmatrix} a & b & c & d \\ c & a & d & b \end{pmatrix} = \begin{pmatrix} a & b & c & d \\ a & d & b & c \end{pmatrix},$

since on the left, working from left to right, a goes into b and then b goes into a, so that the net result is a into a. Similarly, b goes into c and then c goes into d, so that the net result is b into d.

(j) Check the other entries in (2).   (k) Calculate

$$\begin{pmatrix} a & b & c & d \\ b & d & a & c \end{pmatrix} \begin{pmatrix} a & b & c & d \\ d & c & b & a \end{pmatrix}.$$

(l) Find

$$\begin{pmatrix} a & b & c & d \\ c & a & d & b \end{pmatrix} \begin{pmatrix} a & b & c & d \\ b & d & a & c \end{pmatrix}.$$

One-to-one transformations on finite sets are often called permutations, and the corresponding groups are called *permutation groups*.

(m) What is the connection with the use of this term in Section 8–3? (n) Conjecture a group of permutations isomorphic to the symmetries of the equilateral triangle and justify your result.

The isomorphism between the group of permutations on three objects and the symmetries of the equilateral triangle is not accidental. Whenever we move the triangle by one of its symmetries we permute the vertices in some way, and conversely. Indeed the symmetries of the triangle is just a geometric interpretation of the corresponding permutation group. The group of all permutations on n objects is known as the *symmetric group of degree n*.

(o) Show that the permutations on $\{a, b, c, d\}$ and the permutations on $\{1, 2, 3, 4\}$ are isomorphic. How does this justify the use of the name "*the* symmetric group"?   (p) Is the symmetric group of degree four isomorphic to the symmetries of the square?   (q) How many members does the symmetric group of degree n have?   (r) What property of the regular n-sided polygon remains invariant under all transformations of the symmetric group applied to its vertices?   (s) under those of the group of symmetries of the regular n-sided polygon?   (t) A formula is said to be *symmetric* with respect to a set of variables if it is invariant under the symmetric group on these variables. List several formulas defining polynomials that are invariant under the symmetric group on $\{a, b, c, d\}$.

We have seen that transformation groups seem to be associated with properties that remain invariant. This is no accident. Let $f(x_1, x_2, \ldots, x_n)$ be a formula depending on the n variables indicated. Consider the set of all one-to-one transformations mapping $(x_1, x_2, \ldots, x_n)$ into $(x_1', x_2', \ldots, x_n')$ so that $f(x_1, x_2, \ldots, x_n) = f(x_1', x_2', \ldots, x_n')$ for all values of the variables, i.e., so that the expression remains invariant. Then the set of transformations is a group under composition.

(u) Prove this.

This idea has been used to classify geometries of different types. Recall that in Euclidean geometry we are permitted to move figures about (transformations) provided their shape or size is not distorted, i.e., provided distances remain invariant. Such motions are required, for example, in proving congruence theorems by superposition. The set of all such motions form a group. What are they and what is the nature of this group? They are the translations (considered at the beginning of this section), the rotations (see Section 5–17), and the reflections in any line. For example, to prove the congruence of the two triangles ABC and $A'B'C'$ in Fig. 10–9, we need to be able to "flip one of them over," i.e., to reflect $A'B'C'$ in the line L to get $A''B''C''$. In Euclidean geometry these two triangles are congruent (i.e., can be superposed), but this is only possible if reflection is permitted. The translations, rotations, and reflections are called the *rigid motions*, and their group is sometimes called the *Euclidean group*.

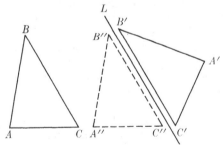

FIGURE 10–9

There are many types of geometry, in each of which different types of transformations are allowed. The most general geometry so far considered by mathematicians is the geometry that treats properties invariant under any continuous one-to-one transformations with continuous inverses. Geometrically, this means that we can do any thing we wish with the plane in the way of stretching, moving, or distorting, *except* that we must not cut or tear. This geometry, which studies the properties invariant under the group of all one-to-one continuous transformations is called *topology*. In topology, a square and a circle are "congruent"! In the problems we consider these and other transformation groups in further detail.

(v) Why does the continuity of the transformations preclude tearing or cutting? (w) Why would a square and a circle be "congruent"? (x) Draw a figure *not* congruent to a circle in plane topology.

PROBLEMS

1. In Section 5–17 we derived equations (5–17–8) giving the old coordinates of a point in terms of the new coordinates when the axes are rotated through an angle ϕ. Here we wish the formulas giving the new coordinates in terms of the old when the plane (axes fixed) is rotated around the origin through an angle ϕ. Show that the required formulas for this transformation T_ϕ are

$$
\begin{aligned}
x' &= x \cos \phi - y \sin \phi, \\
y' &= x \sin \phi + y \cos \phi.
\end{aligned}
\tag{3}
$$

2. Letting T_ϕ be the rotation of the plane about the origin, what is T_0?

3. Find $T_\phi T_\theta$ by noting that $(x, y)T_\phi = (x \cos \phi - y \sin \phi,\ x \sin \phi + y \cos \phi)$.

4. Find the inverse of T_ϕ by "looking at the geometry."

5. Find the inverse by algebra from the requirement that $T_\phi^* T_\phi = T_0$.

6. Both translations and rotations leave distance invariant. What additional invariance is related to translation? to rotation?

7. Consider the group of all translations and rotations where $T_\phi T_{a,b}$ means a rotation through ϕ followed by a translation through (a, b). Is $T_{a,b}T_\phi$ the same transformation? What is the identity element?

8. Find the equations defining a reflection in the x-axis; the y-axis.

9. Do the same for any line through the origin.

10. Does the set of all reflections in all lines of the plane form a group?

11. What properties of the conics remain invariant under rigid motions? How does this appear algebraically?

12. A transformation T for which $xT = mx + b$ is called a linear transformation. What is the geometric interpretation in terms of points on a line when $m > 1$ and $b = 0$?

13. When $m > 1$ and $b \neq 0$?

14. When $0 < m < 1$ and $b \neq 0$?

15. When $m < 0$?

16. In general?

17. Show that the linear transformations with $m \neq 0$ form a group.

18. Can you think of anything that remains invariant under this group?

19. Make a multiplication table of the symmetric group on two objects. Let

$$
I = \begin{pmatrix} 1 & 2 \\ 1 & 2 \end{pmatrix}, \qquad a = \begin{pmatrix} 1 & 2 \\ 2 & 1 \end{pmatrix}.
$$

20. What group of symmetries of a geometric figure is isomorphic to this group?

21. Write down the permutations on the vertices $\{a, b, c, d\}$ that result from each symmetry of the rectangle (Fig. 10-6). Treating them as permutations, find their multiplication table.

22. Do the same for the permutations of the vertices of the square corresponding to the symmetries.

23. Write the multiplication table of the symmetric group of degree 3.

★24. We have shown that vectors in the plane form a group under addition. Could you generalize this by defining vectors in space of three dimensions in terms of three components?

★25. Let $(x_1, x_2, \ldots, x_n)$ be an n-dimensional vector. Define addition so that all such vectors with real components form a group. Indicate some subgroups.

★26. The *linear* (homogeneous) *transformations* in the plane are those such that $(x, y)T = (ax + by, cx + dy)$. Those for which $ad - bc \neq 0$ are called *nonsingular*. Show that the set of all nonsingular linear transformations forms a group.

★27. An *affine transformation* in the plane is one for which $(x, y)T = (ax + by + m, cx + dy + n)$. Show that an affine transformation is the composition of a linear homogeneous transformation with a translation in that order. Show that the nonsingular affine transformations form a group. (The corresponding geometry is called *affine geometry*. The thing that remains invariant is the parallelism of lines, i.e., two parallel lines remain parallel after any such transformation.)

★28. Construct a multiplication table for the symmetric group of degree 4.

ANSWERS TO EXERCISES

(a) Otherwise the identity would not be unique, and the composite might not always be defined.   (b) This was the essential idea behind the discussion in Sections 1-2 and 1-4.   (c) Replace $x_i T_a$ by the equivalent $x_i + a$.   (d) See Section 4-3.   (e) $T_{a+c, b+d}$.   (f) Yes.   (g) $T_{-a, -b}$.   (h) The complex numbers under addition.   (i) The translation of the axes, where the translation of the origin to $(-a, -b)$ corresponds to $T_{a,b}$.   (k) The second line is $(cadb)$.   (l) The identity.   (m) There is a one-to-one correspondence between arrangements of the letters (permutations in the sense of Section 8-3) and transformations (permutations in the present sense). Indeed, the corresponding arrangement is just that of the second line in the symbol for the transformation. (n) The permutations on $\{a, b, c\}$. See the paragraph that follows the exercise.

(o) Just a change of names. "The" is justified, because all of them are isomorphic, i.e., of the same structure.   (p) No. There are only 8 symmetries, but 24 permutations.   (q) $n!$ (r) Its appearance.   (s) Distance between vertices.   (t) $abcd$; $abc + abd + acd + bcd$; $ab + ac + ad + bc + bd + cd$; $a + b + c + d$. (These are called the *elementary symmetric polynomials*.)   (u) If f is invariant under T_1 and under T_2, it will certainly be so under $T_1 T_2$. Hence the set is closed. The identity transformation does not change f, so it belongs to the group. The inverse of a transformation that leaves f invariant will certainly leave f invariant. Finally, all transformations are associative under composition.

(v) Roughly speaking, nearby points would not then be transferred into nearby points, which is essential for continuity. (See Section 7–3, which you will need to generalize from one to two dimensions. The essential thing is that the continuity of F means that as $x \to y$, $F(x) \to F(y)$, i.e., when x is close to y, $F(x)$ is close to $F(y)$. In other words, nearby points go into nearby points.)

(w) Since one can be superposed on the other by distortion short of cutting.

(x) A figure eight.

Answers to Problems

1. To think of the points instead of the axes moving, reverses the relation between old and new. 3. $T_\phi T_\theta = T_{\phi+\theta}$. 5. $T_{\phi+\theta} = T_0 \to \theta = -\phi + 2k\pi$. But $T_{\alpha+2k\pi} = T_\alpha$. 7. No. Draw a figure and trace a point through a translation followed by a rotation, and through the same rotation following by the translation. 9. Suppose that the line has equation $y = (\tan\phi)x$. Rotate plane through $-\theta$ to bring line on x-axis, reflect in x-axis, rotate back through θ. This yields, $x' = x\cos 2\theta + y\sin 2\theta$, $y' = x\sin 2\theta - y\cos 2\theta$. 10. No. 11. All properties of curve itself. Invariance of certain functions of the coefficients of the equations. See Section 5–17. 13. Stretches the line and shifts it a directed distance b. 15. As before, but reverses directions. 17. Check axioms. 19. Isomorphic to addition mod 2. 21. Should be isomorphic to symmetries of the rectangle. 23. Should be same as that of symmetries of equilateral triangle if you use the same labels.

10–5 Group theory. The last two sections defined the group concept and illustrated it in various ways. Group theory itself is concerned with developing laws that apply to abstract groups, independent of such illustrations. The theory is built solely upon the axioms (10–3–1) through (10–3–4) and well-chosen definitions. It is much too vast to be even sketched here. One could easily spend a lifetime exploring and adding to it. Our purpose here is merely to hint at this great structure by a few examples.

First of all, it is obvious that any law that holds for addition of real numbers or for multiplication of nonzero real numbers, where just one of these operations is involved, must hold for any commutative group.

(a) Why? (b) Why nonzero real numbers? (c) Why commutative group? (d) Generalize by referring to any field instead of to just the real numbers.

Furthermore, any law of this kind whose proof does not require the commutative law, will hold in any group. Of course, it is naturally most interesting to derive laws that hold in any group to replace laws that were derived for commutative groups but do not hold more generally. For example, for real numbers (and in any field) we have $-(a+b) = -a +$

$(-b)$ and $1/ab = (1/a)(1/b)$. The corresponding law for commutative groups is $(a * b)' = a' * b'$, but this does not hold for all groups. However, the following law does hold in any group.

$$(1) \qquad\qquad (a * b)' = b' * a'.$$

The proof is not difficult: Using associativity and the property of the inverse, we have $(ab)(b'a') = a(bb')a' = (ae)a' = aa' = e$. Since the inverse of an element is the only element whose product with the given element yields e, we must have $(ab)' = b'a'$.

Of course we have not yet proved here that there is one and only one element x such that $ax = e$, that is,

$$(2) \qquad\qquad \forall a\ \exists! x\ a * x = e.$$

But this is not difficult. For suppose that $a * x = e$ and $a * y = e$. Then $a * x = a * y$. Hence $a' * (a * x) = a' * (a * y)$, $e * x = e * y$, and $x = y$. Hence there can be *at most* one x such that $a * x = e$. But axiom (10–3–4) guarantees that there is *at least* one.

(e) Prove (3).   (f) Prove (4).   (g) Prove (6) and (7).

$$(3) \qquad\qquad \forall a\ \exists! x\ x * a = e.$$

$$(4) \qquad\qquad \exists! x\ \forall a\ x * a = a.$$

$$(5) \qquad\qquad \exists! x\ \forall a\ a * x = a.$$

$$(6) \qquad\qquad (a * x = a * y) \rightarrow (x = y).$$

$$(7) \qquad\qquad (x * a = y * a) \rightarrow (x = y).$$

(h) Why do we need two such cancellation laws? (They are called the *left* and *right cancellation laws*).

We spoke above of "well-chosen definitions." As an example, we introduce the notion of subgroup, upon which rests many of the most fascinating results of group theory. If S is a subset of G and if S and G are groups with respect to the same operation, then S is called a *subgroup* of G.

(i) Show that every group is a subgroup of itself.   (j) Show that any nontrivial group has at least one other subgroup.   (k) What element of a group belongs to every subgroup?   (l) Mention several subgroups of the group of integers under addition.   (m) Find all the subgroups of the group of symmetries of the equilateral triangle.   (n) Show that if a group is commutative, so must be all of its subgroups.

One of the most remarkable elementary results in group theory is *Lagrange's theorem*:

(8)   *The number of members of any subgroup of a finite group is a divisor of the number of members in the group.*

(o) What conclusion can you draw about groups with a prime number of members?   (p) Verify (8) in the situation of Exercise (m).

To prove (8), let us consider any subgroup S of a group G. Let Sa be the set of elements of G obtained by multiplying each member of S on the right by a, where $a \in G$. Of course, if $a \in S$, then $Sa = S$, but otherwise Sa is not a group.

(q) Why not?

For any a, Sa has exactly as many members as S. For obviously it cannot have more members, since each of its elements is obtained from one of S by multiplication by a. It could have less only if $b * a = c * a$ for two different elements b and c in S. But this is impossible because of (6).

Also, for any a and b we have either $Sa = Sb$ or else $Sa // Sb$, that is, two different sets of this kind are always disjoint. For suppose that Sa and Sb have a common element c, so that $c = ma = nb$. Then for any element sa of Sa, $sa = sm'ma = sm'nb$. That is, any member of Sa is also a member of Sb. Hence if Sa and Sb are not disjoint, they are identical.

Finally, we note that the union of all the sets of the form Sa, for a running over the whole membership of G, is just G itself, i.e., every member of G belongs to one of these sets, for the element g of G clearly belongs to Sg, since e belongs to S and $eg = g$.

What we have shown is that the set G can be partitioned into subsets each of which has the same number of members as S. It follows that the number of members in G is a multiple of the number of members in S, as we wished to prove.

(r) Restate the last paragraph without using the word "partitioned." (s) Write a similar proof based on using sets of the form aS consisting of the left multiples of members of S.

PROBLEMS

1. Show that in any group $(a')' = a$.
2. Generalize (10–5–1) to the inverse of the product of any number of elements.
3. How can we be sure that it is impossible from the group axioms to prove the commutative law?
4. Show that in any group for any a and b in the group there exists one and only one element c of the group such that $a * c = b$.

5. Do the same but with the final equation $a * b = c$.

6. What is the relation between the group of symmetries of the square and the symmetric group of order 4?

7. Mention some subgroups of the group of rigid motions of the plane.

8. Do the reflections in the plane form a subgroup of the rigid motions?

9. Prove that in any group $e' = e$.

10. Let $a^1 = a$ and $a^2 = a * a$. Give a recursive definition of a^n for n a positive integer. Define a^0 and a^{-n} for n positive.

11. Prove that with these definitions the laws of exponents hold.

12. A group that consists entirely of powers of one of its elements is called *cyclic*. Cite some examples.

13. List all the subgroups of the group of the 15 fifteenth roots of 1 under multiplication.

14. State a general result of which your result in Problem 13 is a special case.

15. Show that every group with three or fewer members must be commutative.

16. Show that a set satisfying axioms (10-3-1), (10-3-2), and the following alternatives for (10-3-3) and (10-3-4) is a group: There is an element e such that $e * a = a$. For any a, there is an element a' such that $a' * a = e$.

17. Make up another weaker set of axioms similar to those given in Problem 16 and prove that they are sufficient to define a group.

★18. Show that the following axioms define a group. 1. If a and b are in S, then $a * b$ is in S. 2. $*$ is associative. 3. For every a in S there is an a' in S such that for any b in S, $(b * a) * a' = b$. 4. $a * b = a * c \rightarrow b = c$.

★19. Make up and justify another set of axioms.

★20. Prove that every finite group is isomorphic to some group of permutations! (*Suggestion*: Let $a_1, a_2, \ldots, a_n$ be the elements of the group. Then consider the set of permutations $A_1, A_2, \ldots, A_n$ on these elements, where A_i is the permutation that carries a_1 into $a_1 a_i$, a_2 into $a_2 a_i$, a_3 into $a_3 a_i$, $\ldots$, and a_n into $a_n a_i$.)

ANSWERS TO EXERCISES

(a) Since such laws are derived only from the axioms for reals that are the same as the group axioms.   (b) The reals including zero do not form a group.   (c) Reals are commutative.   (d) Substitute "any field" for "the real numbers."   (e) Like (2).   (f) $xa = a$ and $ya = a$ imply $xa = ya$ implies $x = y$. Similarly for other.   (g) Multiply on left by a' in (6) and on the right by a' in (7).   (h) We do not have commutativity.   (i) By the definition, since every set is a subset of itself.   (j) The group consisting of only the identity element.   (k) The identity.   (l) Evens, multiples of 3, multiples of 4.   (m) The rotations; the groups consisting of the identity and just one reflection; all symmetries; the identity.   (n) Since all elements of the subgroup belong to the group.   (o) They have no subgroups but the trivial one and the group itself.   (p) The divisors of 6 are just 2 and 3.   (q) No identity element.   (r) The sets of the form Sa are such that each member of G belongs to one and only one and each such set has exactly as many members as S. Hence . . . .   (s) Everything goes exactly the same with reversal of order of multiplication.

Answers to Problems

1. Since $aa' = e$ and by (3), there is just one x, namely $(a')'$, for which $xa' = e$. 2. Induction! 3. Since we have examples of groups that are not commutative. 5. There exists at least one, since $(b:a'c)$ yields a true statement. There is at most one, since $ax = c$ and $ay = c$ imply $x = y$. 7. The rotations, the translations, the translations parallel to the x-axis, the rotations through an even number of degrees, etc. 9. Since $ee = e$, and use (3). 11. The definitions and proofs are exactly the same as in Section 6–6. 13. The five roots with angles multiples of 72° (i.e., the five fifth roots of unity), the three roots with angles multiples of 120° (i.e., the three cube roots), the single number 1, the entire set of 15 roots. 15. Construct multiplication tables for one, two, and three members. 17. There is an element e such that $ae = a$, and for any a there is an element a' such that $aa' = e$. 19. See "Definitions of Group Involving Quasi-inverse Elements," by Rafael Sanchez-Diaz, *Proceedings of the American Mathematical Society*, Vol. 4, 1953, pp. 424–428.

10–6 Boolean algebra. Let B be a set of elements on which are defined two binary operations $\cup$ (cup) and $\cap$ (cap) and a unary operation $'$ (prime) and an equivalence relation $\equiv$ such that

$$(1) \qquad (a, b \in B) \to (a \cup b \in B \wedge a \cap b \in B \wedge a' \in B).$$

For any a, b, and c in B

(2)	$a \cup b \equiv b \cup a$,	(2′)	$a \cap b \equiv b \cap a$,
(3)	$a \cup (b \cup c) \equiv (a \cup b) \cup c$,	(3′)	$a \cap (b \cap c) \equiv (a \cap b) \cap c$,
(4)	$a \cap (b \cup c) \equiv (a \cap b) \cup (a \cap c)$,		
		(4′)	$a \cup (b \cap c) \equiv (a \cup b) \cap (a \cup c)$,
(5)	$a \cup a \equiv a$,	(5′)	$a \cap a \equiv a$,
(6)	$(a \cup b)' \equiv a' \cap b'$,	(6′)	$(a \cap b)' \equiv a' \cup b'$.
(7)	$(a)' \equiv a$		

Let there be e_0 and e_1, belonging to B, such that

(8)	$a \cup e_0 \equiv a$,	(8′)	$a \cap e_0 \equiv e_0$,
(9)	$a \cup e_1 \equiv e_1$,	(9′)	$a \cap e_1 \equiv a$,
(10)	$a \cup a' \equiv e_1$,	(10′)	$a \cap a' \equiv e_0$.

We assume also that if $a \equiv b$ and $c \equiv d$, then

$$a \cup c \equiv b \cup d \quad \text{and} \quad a \cap c \equiv b \cap d.$$

Then we call the set $\{B, \cup, \cap, ', e_0, e_1\}$ a *Boolean algebra*.

(a) Show that the set B of all subsets of a universe U is a Boolean algebra under the operations of union, intersection, and complimentation.   (b) Suggest names for the axioms (1) through (7).   (c) Is a Boolean algebra a group with respect to either operation?

In the algebra of sets we also had binary relations, in particular the relation of inclusion $\subseteq$. In any Boolean algebra we may define a similar binary relation, as follows:

(11)   **Def.** $a \, R \, b = [a \cap b' \equiv e_0]$.

(d) To what law of sets does (11) correspond?

With this definition it is not hard to prove

(12) $e_0 \, R \, a$,   (12′) $a \, R \, e_1$,

(13) $a \, R \, a$,

(14) $[a \, R \, b \,\wedge\, b \, R \, a] \to (a \equiv b)$,

(15) $[a \, R \, b \,\wedge\, b \, R \, c] \to (a \, R \, c)$,

(16) $(a \, R \, b) \leftrightarrow (a \cup b \equiv b) \leftrightarrow (a \cap b \equiv a)$.

(e) To what laws of sets do (12) through (16) correspond?

As a simple example, quite different from sets, let $B = \{0, 1\}$, $e_0 = 0$, $e_1 = 1$, $x' = 1 - x$, $x \cup y = x + y - xy$, and $x \cap y = xy$. Then it is easy to verify that all the axioms hold.

(f) Write the tables giving the values of x', $x \cup y$, and $x \cap y$ for the possible values of the variables.   (g) What is R in this case?   (h) Prove that (4′) holds by applying the definitions of the operations to the formulas in both members.
(i) Let B be a nonempty set of propositions that contains the negation of each of its member and conjunction and disjunction of each pair of its members. Let e_0 be a false proposition f and e_1 a true proposition t. Let $x' = \sim x$, $x \cup y = x \vee y$, $x \cap y = x \wedge y$, and the equivalence relation be $\leftrightarrow$. Show that this is a Boolean algebra.   (j) What is R in this algebra?   (k) Why does it happen that sets and sentences satisfy formally identical laws?

One of the most interesting examples of a Boolean algebra is the algebra of switching circuits. Let us imagine an electric circuit between two terminals T_1 and T_2. Figure 10–10 shows the simplest possible circuit, a single wire joining the terminals with one switch. Suppose we indicate the state of this switch by a variable x that has the value 0 when the switch is open and 1 when it is closed.

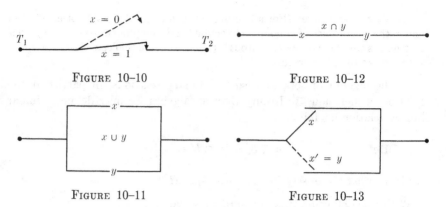

FIGURE 10-10   FIGURE 10-12

FIGURE 10-11   FIGURE 10-13

If a circuit contains several switches, we use a different variable for each one. For example, Fig. 10-11 shows a circuit containing two switches in parallel. We say that the entire circuit is closed if current can flow from one terminal to the other, and open otherwise. We let $x \cup y$ stand for the state of the circuit. It takes the value 0 if both x and y are 0 and the value of 1 if either one is 1.

Figure 10-12 shows a circuit containing two switches in series. We let $x \cap y$ stand for the state of the circuit. Evidently $x \cap y$ is 1 if both x and y are 1, otherwise it is 0.

Finally, we let x' be the state opposite to x, so that if $y = x'$, then y is open and x closed, or else y closed and x open. Such a situation exists when one switch is linked to another so that when one closes the other opens. A simple possibility is sketched in Fig. 10-13. In practice, switches are usually linked by relay mechanisms consisting of coils and magnets, or by electronic devices.

(l) Write simple formulas for a', $a \cup b$, $a \cap b$, in terms of the usual arithmetic of 0 and 1.   (m) Show that the above is a Boolean algebra.   (n) Do you think it might be possible to construct machines to solve logical problems? Why?

It should be noted that $=$ in switching algebra is the equivalence relation. It does not mean that the circuits are the same, but merely that their states are the same, since the variables stand for the states of the circuits. Thus (6) means that a circuit consisting of two switches a and b in parallel is always in the state opposite to that of a circuit consisting of two switches c and d in series that are respectively in the opposite state from the originals, i.e., for which $c = a'$ and $d = b'$. The two circuits are sketched in Fig. 10-14. In the figure, a and b are shown both open, which implies that a' and b' are closed. As is evident in the figure, the circuit $a \cup b$ is open and the circuit $a' \cap b'$ is closed. If we make any

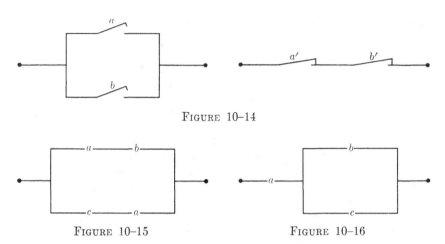

Figure 10-14

Figure 10-15   Figure 10-16

change that closes the first circuit, the second circuit opens. This is the meaning of (6) in the present interpretation.

(o) Draw diagrams to illustrate other laws of Boolean algebra as applied to switching circuits.   (p) What is R here?

Because the algebra of switching circuits is Boolean, this type of abstract system plays a very important role in the design of telephone networks, electronic "brains," and other circuits.   To give some idea of the applications, we consider a very trivial example. The circuit in Fig. 10-15 has the formula $(a \cap b) \cup (c \cap a)$. But by the laws of Boolean algebra, this formula is equivalent to $a \cap (b \cup c)$, whose corresponding circuit is given in Fig. 10-16. But this second circuit involves less wiring and one fewer switches.   In a similar way, very complicated circuits may often be replaced by simpler ones with very substantial savings.

(q) Prove that in any Boolean algebra

(17') $a \cup (a \cap b) \equiv a.$

by giving reasons for the following; $a \cup (a \cap b) \equiv (a \cap e_1) \cup (a \cap b) \equiv a \cap (e_1 \cup b) \equiv a \cap e_1 \equiv a.$ (r) Draw diagrams of the two members interpreted as circuits.

PROBLEMS

1. Prove the following laws of Boolean algebra.

(17) $a \cap (a \cup b) \equiv a,$ (17') $a \cup (a \cap b) \equiv a,$

(18) $a \cap (a' \cup b) \equiv a \cap b,$ (18') $a \cup (a' \cap b) \equiv a \cup b,$

(19) $a \; R \; (a \cup b),$ (19') $(a \cap b) \; R \; a,$

(20) $(a \cap b) \cup (a' \cap b) \equiv (a \cup b) \cap (a' \cup b)$,

(21) $(a \cap c) \cup (a' \cap b) \cup (b \cap c) \equiv (a \cap c) \cup (a' \cap b)$.

2. Show that e_0 is the only element x such that $x \, R \, e_0$.

3. Show that $(a \cap b) \cup (a' \cap b') \equiv (a \cup b') \cap (a' \cup b)$.

4. Draw the circuits corresponding to the two members of Problem 3.

5. Argue that if a light is placed in either of the circuits of Problem 3 appropriately, it can be turned on or off from either of the switches a and b.

6. State some other identities of Boolean algebra.

★7. Show that the complement of any expression built up from primed and unprimed letters, cups, caps, and parentheses (but no primed parentheses) is obtained by interchanging cups and caps, priming each unprimed letter, and unpriming each primed letter.

8. Show that if $A \cap X \equiv A \cap Y$ and $A \cup X \equiv A \cup Y$, then $X \equiv Y$.

9. Show that neither one of these hypotheses is sufficient for the conclusion.

10. What are the undefined constants, formulas, and sentences in the presentation of Boolean algebra in this section? What formulas are defined? What are the variables?

11. Could the number of undefined formulas be reduced?

★12. A farmer wants to be able to turn on or off his yard light from three different locations. Design the circuit for the three switches.

★13. A man has to get a wolf, goat, and a basket of cabbages across a river by using a boat that will hold only himself and one of the three. He must not leave the wolf and goat, or the goat and cabbages, together out of his presence. How does he get them across? Do you see any connection with Boolean algebra?

★14. Argue that any law of Boolean algebra involving cup, cap, and prime only, remains true when cup and cap, and e_0 and e_1 are interchanged.

★15. Prove (22).

(22) $(a \cap b) \cup (a \cap b') \cup (a' \cap b) \cup (a' \cap b') \equiv e_1$.

★16. Prove (23).

(23) $(a \equiv e_0) \leftrightarrow [\forall b \, (a \cap b') \cup (a' \cap b) \equiv b]$.

★17. Show that $(a \, R \, b) \leftrightarrow (a' \cup b \equiv e_1)$.

★18. Show that every Boolean algebra has a subalgebra of just two elements. (First you must define a subalgebra!)

★19. Show that the elements of a Boolean algebra form a group under $*$ defined by $a * b = (a \cap b') \cup (a' \cap b)$.

ANSWERS TO EXERCISES

(a) Each of the axioms appears in Section 3–7.   (b) See Section 3–7
(c) No, inverses do not always exist.   (d) (3–8–16).   (e) See Section 3–8
(g) $\leq$.   (h) $a \cup (b \cap c) = a + bc - abc$. $(a \cup b) \cap (a \cup c) = (a + b - ab)(a + c - ac) = a^2 + ac - a^2c + ab + bc - abc - a^2b - abc + a^2bc = a + bc - abc$.   (i) Again, the axioms appear in Chapter 2 as laws of logic.

Note that (9) becomes $p \lor t \leftrightarrow t$. Recalling that t is a true proposition, this is equivalent to (2–7–32). Similar interpretations have to be given to other identities involving t and f.
(j) $\to$. There is a double use of $\to$ here, both as the relation of the Boolean algebra and as the implication sign for stating laws of the algebra as, for example, (14).   (k) Each operation and relation on sets was defined in terms of the corresponding operation and relation on sentences.   (l) They are just those of the Boolean algebra of $\{0, 1\}$.   (m) Obvious from (l).   (n) Yes, since electrical switches being open and closed could corresponding to sentences being false or true, and a circuit would then correspond to a logical formula.   (p) $a \, R \, b$ means that if a is closed, so is b.   (q) (9′), (4), (9), (9′).

ANSWERS TO PROBLEMS

1. Just like corresponding proofs for set theory. 3. Left member $= (a \cup a') \cap (a \cup b') \cap (b \cup a') \cap (b \cup b') = e_1 \cap (a \cup b') \cap (b \cup a') \cap e_1 =$ right member. 5. The circuit on the left is closed if either both switches are on or both are off. Hence any change at either switch changes the light. Same statement applies to right member. 9. Counterexamples: $A = e_0$ and $A = e_1$. 11. Yes, just as we did in Chapter 2 by defining cup (or cap) in terms of cap (or cup) and prime. 17. Prime both sides of the defining term in (11).   19. Verify each axiom.

10–7 Axiomatics.   In this chapter we have described three abstract mathematical theories. In each case the theory was constructed by specifying certain undefined constants, variables, and formulas, then formulating axioms, definitions, and theorems derived with the aid of logic and rules of proof. These three theories were natural abstractions from theories of a less abstract nature. It would appear that we could construct many such abstract theories by incorporating in an abstract theory certain features of familiar theories. For example, the theory of groups takes the properties of closure, associativity, identity, and inverse. Could we not omit associativity and include commutativity instead in order to get a different theory? Yes, of course. Moreover, we are not required to use properties from familiar theories. Indeed we are free to choose our terms and axioms as we wish and even to specify our rules of proof and logical procedures at pleasure. This notion of freedom in constructing theories is one of the essential features of modern mathematics.

Of course, the fact that we may construct theories as we wish does not mean that the theories we construct will inevitably be interesting, useful, or aesthetically appealing to others or even to ourselves. With the freedom of mathematical construction comes inevitably the necessity of a critical examination of the results. Since a theory is completely determined by its axioms, a critical examination of a theory naturally centers on its axioms. *Axiomatics* is the theory of axioms systems.

One of the first things we might ask about an axiom system is whether or not it is consistent. A set of axioms is called *inconsistent* if we can derive from it a contradiction, that is, some proposition of the form $p \wedge \sim p$. An axiom system is called *consistent* if this is not possible. If we discover a contradiction, inconsistency is proved, but our failure to discover one is not conclusive proof that none exists. The accepted method of proving consistency is to exhibit a set of things that satisfy the axioms, that is, to point to an example or realization of the theory.

For example, we may establish the consistency of the group axioms by showing that the three cube roots of one form a group under ordinary multiplication. The idea behind this method of demonstrating consistency is the faith that the world is not inconsistent, and hence that if the axiom set is satisfied by some real objects, then the axioms must be consistent. However, when we show that a theory is realized in an example, we merely show that it is consistent with our theory of this example. For example, when we exhibit the integers under addition as an example of a group, we show that group theory is consistent if the theory of the integers under addition is consistent. Hence, such a proof of consistency is *relative* to some assumptions and has to go outside mathematics for its final justification. In practice, consistency means that the theory applies to some other theory that is accepted because of its connection with experience.

(a) Give a definition of "example" as we have used this word above. (b) Show that the axioms of Boolean algebra are consistent if the theory of the arithmetic of integers is consistent. (c) Do you think the arithmetic is consistent? Why?

Another question we might ask is whether a set of axioms is as simple as it might be. Could some be derived from the others? If so, the former could be omitted. It would be worth knowing whether some axioms could be proved as theorems, even if we preferred (as is often the case) to use an axiom system larger than absolutely necessary. An axiom system no one of whose members can be derived from the others is called *independent*.

The independence of a particular axiom may be shown by finding an example for which the axiom is false and the other axioms are true. If this can be done, it is evident that the axiom could not be derived from the others, for such a derivation would amount to proving that the axiom must hold if the others do.

(d) Cite an example of a set of numbers that satisfy the group axioms except for closure. (e) Cite an example in which all group axioms are satisfied except the existence of inverses. (f) Can you cite one in which only the existence of the identity fails? (g) Explain: An axiom system is independent if and only if every system obtained by replacing one axiom by its negation is a consistent system.

There is a stronger kind of independence called *complete independence*. This is said to hold for a set of axioms if no subset of the axioms implies the disjunction of the others. For example, if the axiom set is $\{p, q, r, s\}$, then $p \wedge q$ does not imply $r \vee s$, nor does any other conjunction of axioms imply the disjunction of the rest. This means that every axiom system made up by picking certain of the axioms and the negations of the remainder is a consistent system. For example, if there are three axioms $\{p, q, r\}$, then complete independence means that $\{p, q, r\}$, $\{\sim p, q, r\}$, $\{p, \sim q, r\}$, $\{p, q, \sim r\}$, $\{\sim p, \sim q, r\}$, $\{\sim p, q, \sim r\}$, $\{p, \sim q, \sim r\}$, and $\{\sim p, \sim q, \sim r\}$ are all consistent systems.

(h) Show that the axiom system consisting of the commutative law, the associative law, and the closure law is completely independent.

Another quality that may be possessed by an axiom system is completeness. A system is said to be *complete* when it is impossible to introduce an additional axiom that is independent of the original axioms. In other words, any further true propositions must be consequences of the given axioms. Of course, by further axioms and propositions we mean those involving only terms of the given theory.

Just as for the notion of consistency, there seems no way to establish completeness directly. To deal with consistency, we introduced the idea of the existence of an example, which we took as a sufficient condition for consistency. For completeness we use a similar strategy. An axiom system is said to be *categorical* if all its examples are isomorphic. Since any two theories that satisfy a set of categorical axioms are isomorphic, they differ only in the symbols used or in their interpretation. Every true sentence in one has a twin in the other. Two such theories are "essentially the same" in the same sense that they do not differ in any way as far as the theory is concerned.

It is not hard to see that categoricalness implies completeness, for if an independent axiom could be added, the systems obtained by adding it and by adding its negation would both be consistent. It would therefore be possible to construct two examples satisfying the original axioms, one of which included the new axiom and the other its contradiction. These could certainly not be isomorphic without a contradiction.

(i) The previous argument proves actually that incompleteness implies noncategoricalness. Why does this prove that categoricalness implies completeness? (j) Show that the group axioms are not categorical. (k) Show that the group axioms are not complete. (l) Do you think it is true that completeness implies categoricalness? (m) Show the completeness of the axiom system consisting of the group axioms and the following additional axiom: G contains exactly three members.

The qualities mentioned above are certainly important ones, but there is no reason to think that they are the only important aspects of axiomatic systems. Nor should it be assumed that the named qualities are desirable under all circumstances. If consistency is interpreted as absence of contradictions in the broadest sense, it is not a desirable feature of a theory that is intended to describe a situation involving contradictions! If consistency is interpreted as applicability, it may not be desirable for a theory that is intended as an object of amusement. An independent axiom system may be less useful in teaching a subject than a dependent system that is more appealing and easier to use. An incomplete system is more powerful than a complete one in the sense that it has a larger variety of applications. A categorical system is desirable only if we wish to construct a precise and elegant theory of a unique kind of structure. In short, the criteria discussed above are some of the important ones to apply to a theory in order to understand its nature and to see how it is adapted to the purposes that motivated its creation.

PROBLEMS

1. Show that the axioms of logic are consistent if elementary arithmetic is consistent.

2. Show that the axioms of logic are neither categorical nor complete. Would it be desirable for them to be so?

3. Show that any inconsistent set of axioms is complete.

4. It may be objected that since basic terms are undefined and axioms unproved, a mathematical theory gives no final proof or explanation of anything. Discuss this.

5. Construct an abstract theory generalizing the notion of distance.

6. Consider a field with the additional operation $\circ$ defined by $a \circ b = a + b - ab$. Is there an identity? inverse? associativity? Derive as many theorems as you can.

7. Generalize the notion of an ordered pair to an ordered n-tuple (ordered set of n members). Include a definition of the cartesian product of n sets.

8. Let F be a function satisfying the following axioms: 1. The domain of F is the n-factor cartesian product $U \times U \times U \times \ldots \times U$, and its range is U, where $U = \{-1, 0, 1\}$. 2. If x' is obtained by permuting the elements of x, then $F(x') = F(x)$. 3. If x' is obtained from x by negating every element of x, then $F(x') = -F(x)$. 4. If $F(x) \geq 0$, and x' is obtained from x by increasing only one element of x, then $F(x') = 1$. Prove that for a fixed n, these axioms are consistent, categorical, and completely independent. (For an interpretation of these axioms, see "A Set of Independent Necessary and Sufficient Conditions for Simple Majority Decision," in *Econometrica*, October, 1952, and "A Note on the Complete Independence of the Conditions for Simple Majority Decision," *Econometrica*, January, 1953.)

ANSWERS TO EXERCISES

(a) An interpretation satisfying the axioms.   (b) The Boolean algebra of $\{0, 1\}$ is an example.   (c) Experience.   (d) Integers between -3 and 3

under addition.   (e) Reals under multiplication.   (f) In the form in which we stated the axioms, the statement of the existence of an inverse depends on the existence of the identity.   (g) Since we find examples to prove the independence.   (h) Cite examples.   (i) $p \to q = \sim q \to \sim q$.   (j) We know of groups with different numbers of members.   (k) We can add an axiom of commutativity.   (m) Only one possible multiplication table.

10–8 What is mathematics?   Mathematics appears in our culture in so many varied ways that it is difficult to describe it as a whole. We wish here merely to suggest a tentative answer to the title of the section.

The child first meets mathematics when he begins to count and to do simple problems in arithmetic. He thinks of mathematics as having to do with calculations which become more complicated as he advances in school. In high school he finds that geometry is part of mathematics, too, and at this stage he might be satisfied with the old definition that mathematics is the science that deals with numbers and space. Indeed, this is a generally accepted view of mathematics. Certainly arithmetic and geometry are part of mathematics, but the reader of this book can see that they are merely excerpts from certain parts of mathematics, which is far too broad to be characterized as dealing with merely numbers and space.

One of the characteristic features of mathematics is its special symbolism. This had led some people to say that mathematics *is* a language. Certainly mathematical language has some of the characteristics of everyday languages. For example, people can communicate mathematically almost without using the usual verbal languages. However, every mathematical theory could be restated without the familiar special symbolism. The results would be most awkward and inconvenient, but the possibility shows that the symbolism of mathematics is not mathematics itself. Indeed the view that mathematics is a language involves the confusion of symbols and their use. Mathematics, like any other discipline, has special terms and symbols, but mathematics is the subject that is expressed by these symbols and not the symbols themselves.

Another feature of mathematics is its wide applicability. It is used by everyone, from the clerk making change in the store to the physicist setting up a differential equation. This has led to mathematics being described as a tool. Of course, it is true that mathematical theories are useful. Many are developed in answer to specific scientific needs, and experience suggests that even those theories that grow without any practial motivation eventually are found useful by someone. However, to define mathematics in terms of its applications would be to exclude much mathematics that is developed without regard to applications and has not as yet found any.

The reader of this book has seen that mathematicians feel free to choose basic terms, axioms, and definitions at will. On the basis of any such choice, one can build a theory by the discovery and proof of theorems. Evidently mathematical work has a creative aspect. This has led to mathematics being described as a creative art. It is certainly true that the creative mathematician has much in common with the creative person in music and the visual arts. Both have a freedom to choose their forms and structures and both must accept the logical and aesthetic consequences of such choices. But this characterization of mathematics is also biased since it ignores the important link with applications.

Perhaps one might give up the attempt to define mathematics in any simple way and simply say that it is what mathematicians do and that mathematicians are those people in our society with certain kinds of education and professional memberships. Such a definition could hardly be wrong, but it is not satisfying if one wants to understand what it is that mathematicians do!

Let us look at mathematics as seen in this book and in the reader's previous studies. It appears as a collection of abstract theories. These theories differ from the abstract theories of other sciences (such as those of physics, biology, or economics) by being on a higher level of abstraction, that is, they do not include any concrete interpretation but are capable of many different interpretations. When mathematical theories are expressed formally in terms of axioms and theorems, this abstractness is reflected by the fact that the undefined terms remain ambiguous in their meaning. Some mathematical theories have many applications outside of mathematics (for example, the theory of growth and decay) and other theories apply to various theories within mathematics itself (for example, group theory). Theories of the latter sort are considered to be more abstract, and some mathematicians use the term "abstract mathematics" only in connection with them. We might be tempted, then, to define mathematics as a collection of abstract theories, but there are certainly many abstract theories that are not mathematical.

A common characteristic of the mathematical theories familiar to the reader is their logical character, their organization in terms of axioms, definitions, and proofs. Of course, this feature is common to theories outside mathematics also, for example, in theology and physics.

The above discussion suggests that it may not be possible to give a definition of mathematics in terms of any single one of its characteristics. However, if we include several of them we may find one that is neither too restrictive nor too inclusive. Here is such an attempt: Mathematics is the collection of known abstract, axiomatic, consistent theories. That is, mathematics is the set to which an object belongs if and only if it is a theory that is abstract (in the sense of having multiple interpretations), axiomatic (logically organized), and consistent (having at least one application or example, as indicated in Section 10–7). This characterization seems to fit the criteria used by mathematicians. An alleged contribution to mathematics is rejected as mathematics if it is limited to a particular application outside mathematics (not sufficiently abstract), fails to be, or to be a part of, an axiomatic theory, or fails to have any realizations (within or outside of mathematics).

Other aspects of mathematics, such as its special symbolism, its aesthetic values, the creative activity associated with it, its numerous inter-relations with other disciplines, and its role in various other aspects of our culture are not part of mathematics itself but rather related to its creation and use by human beings.

FURTHER READING

In place of a set of problems, we follow this section by a list of books in which the reader will find stimulating ideas on the nature of mathematics and further examples extending his knowledge of it.

BELL, E. T., *Mathematics, Queen and Servant of Science*. McGraw-Hill Book Co., New York, 1951.

BELL, E. T., *Men of Mathematics*. Simon and Schuster, New York, 1937.

COURANT, RICHARD, and ROBBINS, HERBERT, *What is Mathematics?* Oxford University Press, New York, 1941.

EVES, HOWARD, *An Introduction to the History of Mathematics*. Rinehart and Co., New York, 1953.

EVES, HOWARD, and NEWSOM, CARROLL V., *An Introduction to the Foundations and Fundamental Concepts of Mathematics*. Rinehart and Co., New York, 1958.

KLINE, MORRIS, *Mathematics in Western Culture*. Oxford University Press, New York, 1953.

LIEBER, LILLIAN R., *The Education of T. C. Mits*. W. W. Norton and Co., New York, 1944. (There are several other delightful books by the same author.)

POLYA, G., *How to Solve It*. Princeton University Press, Princeton, 1945.

STABLER, E. R., *An Introduction to Mathematical Thought*. Addison-Wesley Publishing Co., Reading, Mass., 1953.

WHITEHEAD, A. N., *An Introduction to Mathematics*. Oxford University Press, New York, 1948.

WILDER, R. L., *The Foundations of Mathematics*. John Wiley and Sons, New York, 1952.

APPENDIX

TABLE I

ROOTS AND RECIPROCALS

x	$x^{1/2}$	$x^{1/3}$	$1000/x$	x	$x^{1/2}$	$x^{1/3}$	$1000/x$
1	1.000	1.000	1000.00	36	6.000	3.302	27.76
2	1.414	1.260	500.00	37	6.083	3.332	27.03
3	1.732	1.442	333.33	38	6.164	3.362	26.32
4	2.000	1.587	250.00	39	6.245	3.391	25.64
5	2.236	1.710	200.00	40	6.325	3.420	25.00
6	2.449	1.817	166.67	41	6.403	3.448	24.39
7	2.646	1.913	142.86	42	6.481	3.476	23.81
8	2.828	2.000	125.00	43	6.557	3.503	23.26
9	3.000	2.080	111.11	44	6.633	3.530	22.73
10	3.162	2.154	100.00	45	6.708	3.557	22.22
11	3.317	2.224	90.91	46	6.782	3.583	21.74
12	3.464	2.289	83.33	47	6.856	3.609	21.28
13	3.606	2.351	76.92	48	6.928	3.634	20.83
14	3.742	2.410	71.43	49	7.000	3.659	20.41
15	3.873	2.466	66.67	50	7.071	3.684	20.00
16	4.000	2.520	62.50	51	7.141	3.708	19.61
17	4.123	2.571	58.82	52	7.211	3.733	19.23
18	4.243	2.621	55.56	53	7.280	3.756	18.87
19	4.359	2.668	52.63	54	7.348	3.780	18.52
20	4.472	2.714	50.00	55	7.416	3.803	18.18
21	4.583	2.759	47.62	56	7.483	3.826	17.86
22	4.690	2.802	45.45	57	7.550	3.849	17.54
23	4.796	2.844	43.48	58	7.616	3.871	17.24
24	4.899	2.884	41.67	59	7.681	3.893	16.95
25	5.000	2.924	40.00	60	7.746	3.915	16.67
26	5.099	2.962	38.46	61	7.810	3.936	16.39
27	5.196	3.000	37.04	62	7.874	3.958	16.13
28	5.290	3.037	35.71	63	7.937	3.979	15.87
29	5.385	3.072	34.48	64	8.000	4.000	15.62
30	5.477	3.107	33.33	65	8.062	4.021	15.38
31	5.568	3.141	32.26	66	8.124	4.041	15.15
32	5.657	3.175	31.25	67	8.185	4.062	14.93
33	5.745	3.208	30.30	68	8.246	4.082	14.71
34	5.831	3.240	29.41	69	8.307	4.102	14.49
35	5.916	3.271	28.57	70	8.367	4.121	14.29

TABLE I (*Continued*)

ROOTS AND RECIPROCALS

x	$x^{1/2}$	$x^{1/3}$	$1000/x$	x	$x^{1/2}$	$x^{1/3}$	$1000/x$
71	8.426	4.141	14.08	86	9.274	4.414	11.63
72	8.485	4.160	13.89	87	9.327	4.431	11.49
73	8.544	4.179	13.70	88	9.381	4.448	11.36
74	8.602	4.198	13.51	89	9.434	4.465	11.24
75	8.660	4.217	13.33	90	9.487	4.481	11.11
76	8.718	4.236	13.16	91	9.539	4.498	10.99
77	8.775	4.254	12.99	92	9.592	4.514	10.87
78	8.832	4.273	12.82	93	9.644	4.531	10.75
79	8.888	4.291	12.66	94	9.695	4.547	10.64
80	8.944	4.309	12.50	95	9.747	4.563	10.53
81	9.000	4.327	12.35	96	9.798	4.579	10.42
82	9.055	4.344	12.20	97	9.849	4.595	10.31
83	9.110	4.362	12.05	98	9.899	4.610	10.20
84	9.165	4.380	11.90	99	9.950	4.626	10.10
85	9.220	3.397	11.76	100	10.000	4.642	10.00

Table II

Trigonometric Functions

Angle		Sine	Cosine	Tangent	Cotangent
Radians	Degrees				
0.00	0.0	0.00	1.00	0.00	*
0.09	5.0	0.087	0.996	0.087	11.4
0.10	5.7	0.10	0.995	0.10	10.0
0.17	10.0	0.17	0.98	0.18	5.7
0.20	11.5	0.20	0.98	0.20	4.9
0.26	15.0	0.26	0.97	0.27	3.7
0.30	17.2	0.30	0.96	0.31	3.2
0.35	20.0	0.34	0.94	0.36	2.7
0.40	22.9	0.39	0.92	0.42	2.4
0.44	25.0	0.42	0.91	0.47	2.1
0.50	28.6	0.48	0.88	0.55	1.8
0.52 ($\pi/6$)	30.0	0.50	0.87	0.58	1.7
0.60	34.4	0.56	0.83	0.68	1.5
0.61	35.0	0.57	0.82	0.70	1.4
0.70	40.1	0.64	0.76	0.84	1.2
0.79 ($\pi/4$)	45.0	0.71	0.71	1.00	1.00
0.80	45.8	0.72	0.70	1.0	0.97
0.87	50.0	0.77	0.64	1.2	0.84
0.90	51.6	0.78	0.62	1.3	0.79
0.96	55.0	0.82	0.57	1.4	0.70
1.00	57.3	0.84	0.54	1.6	0.64
1.05 ($\pi/3$)	60.0	0.87	0.50	1.7	0.58
1.10	63.0	0.89	0.45	2.0	0.51
1.13	65.0	0.91	0.42	2.1	0.47
1.20	68.7	0.93	0.36	2.6	0.39
1.22	70.0	0.94	0.34	2.7	0.37
1.30	74.5	0.96	0.27	3.6	0.28
1.40	80.2	0.985	0.17	5.8	0.17
1.48	85.0	0.996	0.09	11.4	0.09
1.50	85.9	0.998	0.07	14.1	0.07
1.57 ($\pi/2$)	90.0	1.00	0.00	*	0.00

* Undefined.

Table III

Common Logarithms

N	0	1	2	3	4	5	6	7	8	9
1	000	041	079	114	146	176	204	230	255	279
2	301	322	342	362	380	398	415	431	447	462
3	477	491	505	519	531	544	556	568	580	591
4	602	613	623	633	643	653	663	672	681	690
5	699	708	716	724	732	740	748	756	763	771
6	778	785	792	799	806	813	820	826	833	839
7	845	851	857	863	869	875	881	886	892	898
8	903	908	914	919	924	929	934	940	944	949
9	954	959	964	968	973	978	982	987	991	996

Table III(a)

N	0	1	2	3	4	5	6	7	8	9
10	000	004	009	013	017	021	025	029	033	037
11	041	045	049	053	057	061	064	068	072	076
12	079	083	086	090	093	097	100	104	107	111
13	114	117	121	124	127	130	134	137	140	143
14	146	149	152	155	158	161	164	167	170	173
15	176	179	182	185	188	190	193	196	199	201
16	204	207	210	212	215	217	220	223	225	228
17	230	233	236	238	241	243	246	248	250	253
18	255	258	260	262	265	267	270	272	274	276
19	279	281	283	286	288	290	292	294	297	299

TABLE IV

NATURAL LOGARITHMS

n	$\log_e n$	n	$\log_e n$	n	$\log_e n$
0.1	−2.303	1.6	0.470	3.5	1.253
0.2	−1.609	1.7	0.531	4.0	1.386
0.3	−1.204	1.8	0.588	4.5	1.504
0.4	−0.916	1.9	0.642	5.0	1.609
0.5	−0.693	2.0	0.693	5.5	1.705
0.6	−0.511	2.1	0.742	6.0	1.792
0.7	−0.357	2.2	0.788	6.5	1.872
0.8	−0.223	2.3	0.833	7.0	1.946
0.9	−0.105	2.4	0.876	8.0	2.079
1.0	0.000	2.5	0.916	9.0	2.197
1.1	0.095	2.6	0.956	10	2.303
1.2	0.182	2.7	0.993	20	2.996
1.3	0.262	2.8	1.030	40	3.689
1.4	0.336	2.9	1.065	100	4.605
1.5	0.405	3.0	1.099	500	6.215

TABLE V

RANDOM SEQUENCE $p = 1/2$

1.	1100100111	0101000100	1011111100	0110000011	0111101100
2.	0111111101	1000001110	1111110001	0110101010	0011010110
3.	1001010011	1110110101	1100011000	1111000111	1001111111
4.	1001011000	1101000001	1100011111	0101000001	0111110011
5.	1101100101	0111110000	1111001110	1101001001	0111000100
6.	0001100001	0110100011	1011100101	0101000101	0010011001
7.	0011110011	0010000110	1001001010	0101001000	1110010000
8.	1110010011	1000101000	1100011110	1001001010	0011011011
9.	1001010110	0100000111	1000101101	1010000110	0110101110
10.	0010100001	1001010001	1111001110	1011000011	1000000100
11.	1111101010	0011010010	1000000101	1001110101	1111011011
12.	0100111011	1100001110	1101011100	1011101101	1110011101
13.	0000110011	1010101110	0111010101	1010110001	1011010010
14.	0001010001	0101010110	1100110010	0000100110	0011100111
15.	1110100100	0100101011	0000000111	1001001001	1011101011
16.	0011101000	0010100000	0111000011	1000011100	0111111100
17.	1011100100	0011100000	0011101011	1011010110	1000011001
18.	0000110111	1010101000	0111001110	0011000101	1110100011
19.	1000110010	1001111011	0010011101	1110010101	0101000010
20.	1110110110	0011011110	0100011110	1011010110	0010110111
21.	1111100010	1110011011	0100010100	0011100110	0111001111
22.	1110101110	0111011100	1001011100	1010100001	0100100101
23.	0001111111	0000010010	1111001101	0101000100	1100010010
24.	1000110011	1110101110	0110101000	1101011101	1000110100
25.	0101011101	0100101100	1001101001	1100100101	0100101001
26.	0110100010	1010001000	0110111100	1000001111	0001011000
27.	0001111010	1100100101	1110100110	0001111111	0000000111
28.	0110010011	0111100001	0001000011	0100000111	1000100110
29.	0101011100	0000111101	1010001000	0011010000	1111110000
30.	0011011010	0100101101	1100100101	1011010000	0011101000
31.	1100000100	0111011100	0011011000	1010011100	0010111101
32.	0001001010	0000100001	0011010100	1001000111	1100010010
33.	1010000101	1010001001	1110101101	0000110000	0010111011
34.	0011010000	1000011110	0001110010	0010000111	1011000000
35.	1010010001	1011111100	1101001010	0000111001	0010110000
36.	1011010111	0001110000	0110010100	0011010111	0101001000
37.	0101000101	0101001011	0100001010	0011100101	0010111101
38.	1101100001	0110100111	1110011110	1011110010	1011010110
39.	1011001110	1000111110	1001101111	0101111110	1000011110
40.	0010111000	0110110110	0011111000	1000000001	1101111011
41.	0100011001	1100000101	0010101101	1011111001	1111110111
42.	1110011001	0001111100	0011000111	0010110110	0101100111
43.	1011001100	1001001011	0010010100	1000111001	0001011000
44.	1001111110	1010001000	0000111101	1000100001	0010111101
45.	1101111000	1111010001	1101101001	1110111110	1111101011
46.	1001100100	1111011100	1000101001	1101010111	0110001000
47.	0010111000	1110100010	1111111100	0100111100	0111100100
48.	1100110001	1000101000	1111010001	1101100111	0011111011
49.	1110001100	0110010101	1101101001	1010110001	0100000111
50.	1011000100	1001001000	0010011111	0000010001	1111000110

Table VI

Normal Distribution

$$f(z) = (2\pi)^{-1/2}e^{-z^2/2} \qquad H(z) = \int_0^z f$$

z	$f(z)$	$H(z)$	z	$f(z)$	$H(z)$
0.0	0.399	0.000	1.3	0.171	0.403
0.1	0.397	0.040	1.4	0.150	0.419
0.2	0.391	0.079	1.5	0.130	0.433
0.3	0.381	0.118	1.6	0.111	0.445
0.4	0.368	0.155	1.7	0.094	0.455
0.5	0.352	0.191	1.8	0.079	0.464
0.6	0.333	0.226	1.9	0.066	0.471
0.7	0.312	0.258	2.0	0.054	0.477
0.8	0.290	0.288	2.2	0.035	0.486
0.9	0.266	0.316	2.4	0.022	0.492
1.0	0.242	0.341	2.6	0.014	0.495
1.1	0.218	0.364	2.8	0.008	0.497
1.2	0.194	0.385	3.0	0.004	0.499

TABLE OF SPECIAL SYMBOLS

After each symbol is given a brief indication of meaning and references to first use and definition.

ARITHMETIC AND ALGEBRA

$-$	operation of negation, 1–2; operation of subtraction, 1–4.
$1/a$	reciprocal of a, 1–2.
$+$	operation of addition, 1–4, 6–3, 6–8, 6–9, 6–13.
$\cdot$	operation of multiplication, 1–5, 6–3, 6–8, 6–9, 6–13. $(a \cdot b = ab)$
$\div$	operation of division, 1–4. $(a \div b = a/b)$
$a^2, a^3, \ldots$	powers of a, 1–5, 1–13.
a^n	nth power of a, 6–6, 6–8, 6–9.
$\lvert a \rvert$	absolute value of a, 1–4, 1–13, 3–9.
$\sqrt{a}$	non-negative square root of a, 1–13.
$\sqrt[3]{a}$	real cube root of a, 5–5.
$\sqrt[4]{a}$	principal fourth root of a, 5–6.
$<$	less than, 1–7, 3–5.
$>$	greater than, 1–7, 3–5.
$\leq$	less than or equal to, 2–2, 3–5.
$\geq$	greater than or equal to, 2–2, 3–5.
$\%$	percent, 1–13.
$a < x < b$	a less than x and x less than b, 3–4.
$(a\_b), (a\_b), (a\_b), (a\_b)$	intervals, 3–4.
$3.21\overline{4}$	repeating decimal, 3.214141414..., 1–15.
N	natural numbers $\{1, 2, 3, \ldots\}$, 6–2.
G	the digits, 5–2.
J	integers, 3–2.
J^+	positive integers, 3–2, 6–8.
E	even numbers, 3–2.
D	odd numbers, 3–2.
Ra	rational numbers, 3–2.
Ra^+	positive rationals, 6–8.
Re	real numbers, 3–5.
Re^+	positive reals, 3–5.
P	prime numbers, 3–2.

$a \mid b$ a divides b, 5–3.

max S largest member of S, 3–9.

min S smallest member of S, 3–9.

$\mathfrak{N}(S)$ cardinal number of the set S, 3–4, 6–2.

$\aleph_0$ aleph null, 6–2.

$S(x)$ the successor of x, 6–4.

$n!$ n-factorial, 6–6.

$\sum$ summation, 6–7.

$\prod$ continued product, 6–7.

(a, b) complex number, 6–13. $[(a,b) = a + ib]$

$(r, \theta)_p$ complex number in polar form, 4–10, 6–13.

$\bar{z}$ complex conjugate, 6–13.

i $(0,1)$, 6–13.

e base of natural logarithms, 7–14. $(e \doteq 2.71828)$.

π ratio of circumference to diameter of a circle, 1–2.

$\doteq$ approximately equal to, 1–4, 6–11.

Logic and Sets

$=$ relation of identity, 1–6.

$\neq$ relation of nonidentity, 1–6.

$\sim$ not, logical negation, 2–2.

$\vee$ or $(=$ and/or$)$, 2–2.

$\wedge$ and, 2–2.

$\underline{\vee}$ or else, 2–2.

$\rightarrow$ implies, 2–3.

$\leftrightarrow$ if and only if, 2–3.

$\therefore$ therefore, 2–6.

$\forall x$ for all x, 2–10.

$\exists x$ for some x, 2–10.

$\exists ! x$ for one and only one x, 2–11.

a_i a-sub-i, 1–3.

$(a{:}b), (a{:}b, c{:}d)$ substitution, 1–11.

$\in$ is a member of, 3–2.

$\notin$ is not a member of, 3–2.

$\{a, b, \ldots\}$ set whose members are a, b, $\ldots$, 3–2.

$\{x \mid \ldots\}$ set of x's such that $\ldots$, 3–2.

∅   empty set, 3–3.

U   universal set, 3–3.

⊆   is a subset of, 3–4.

𝒫(A)   power set of A, 3–4.

A′   complement of A, 3–6.

A ∩ B   intersection of A and B, 3–6.

A ∪ B   union of A and B, 3–6.

A ⊻ B   exclusive union of A and B, 3–6.

A − B   relative complement of B, 3–6.

A // B   A and B are disjoint, 3–8.

⊂   is a proper subset of, 3–8.

⟩⟨   overlaps, 3–8.

x ϱ . . .   the x such that . . . , 3–9.

A ≈ B   the set A is similar to the set B, 6–2.

ANALYTIC GEOMETRY

(x, y)   ordered pair, vector, point, 4–2, 4–3, 6–13.

{(x, y)|f(x, y)}   the set of ordered pairs (x,y) such that f(x,y), 4–2.

A × B   cartesian product, 5–2.

u   vector, 4–3.

0   null vector (0, 0), 4–4.

|(x, y)|, |**u**| length of vector, 4–4.

(r, θ)$_p$ point with polar coordinates r and θ, 4–10, 6–13.

e   eccentricity of a conic, 5–13 through 5–17.

m   slope of straight line, 4–5.

$\overline{AB}$ distance between A and B, 4–4.

FUNCTIONS AND RELATIONS

f(x)   f-of-x, image of x under f, 2–9, 5–4.

f(x,y)   f-of-x-and-y, 2–9.

Dom (ρ)   domain of ρ, 5–3.

Rge (ρ)   range of ρ, 5–3.

Fun   the set of functions, 5–4.

I   identity function on the reals, 5–7.

I_A identity function on A, 5–9.

sin, cos, tan, cot, sec, csc   circular functions, 4–11, 4–12.

LIMITS AND CALCULUS

PROBABILITY AND STATISTICS

$C^n_{r_1, r_2, \ldots, r_k}$ number of ordered partitions, 8–4.

$P[A|B]$ probability of A given B, 8–5.

$E[g(x)]$ expectation of $g(x)$, 8–6.

μ mean, 8–6, 8–7.

σ standard deviation, 8–6, 8–7.

$\hat{\theta}$ estimator of θ, 9–3.

ABSTRACT THEORIES

e identity element, 10–2.

$*$ unspecified binary operation, 10–1.

$\cup, \cap, {}'$ operations of Boolean algebra, 10–6.

e_1, e_0 special elements of a Boolean algebra, 10–6.

GREEK ALPHABET

A	α	Alpha	N	ν	Nu
B	β	Beta	Ξ	ξ	Xi
Γ	γ	Gamma	O	o	Omicron
Δ	δ	Delta	Π	π	Pi
E	ϵ	Epsilon	P	ρ	Rho
Z	ζ	Zeta	Σ	σ	Sigma
H	η	Eta	T	τ	Tau
Θ	θ	Theta	Υ	υ	Upsilon
I	ι	Iota	Φ	ϕ	Phi
K	κ	Kappa	X	χ	Chi
Λ	λ	Lambda	Ψ	ψ	Psi
M	μ	Mu	Ω	ω	Omega

INDEX

Absolute maximum, 411
Absolute probability, 502
Absolute value, 11, 52, 53, 165, 269
Abstract theories, 548 ff.
Acceleration, 388, 462
of gravity, 428
Accuracy, 361, 366
Addition, 10, 27 ff., 327, 332, 348
of ordinates, 295
Adjacent members of a set, 353
Admissible hypothesis, 533
Affine transformation, 567
Aleph null, 324
Algebra
elementary, 1 ff., 25, 27 ff., 41 ff., 65, 94, 104 ff., 146 ff., 321 ff.
of sets, 155 ff.
All, 117, 121, 169
Amplitude, 290
Analytic geometry
on a line, 2 ff., 22–23, 143 ff.
in the plane, 174 ff., 245
And, 71
Angle, 214 ff., 238 ff.
between two lines, 238, 240
Angular velocity, 282
Antiderivative, 429, 479, 559 ff.
Approaches, 387, 390
Approximation, 14, 360, 365, 368, 377, 405, 436, 471
Arc, 214 ff.
Arc length, 217, 468, 477
Area, 377, 434, 443
Area of a circular sector, 218
Arithmetic, 10 ff., 27 ff., 44–46, 298, 327, 357, 360, 365
Arithmetic mean, 150
Arithmetic operations on functions, 294
Arithmetic progression, 339, 346
Associative law
of addition, 28, 65
of multiplication, 29, 65
Associativity, 155, 281, 555
Asymptote of a hyperbola, 311
Average, 189, 539
Average rate of change, 379

Average velocity, 378
Axial symmetry, 310
Axiomatics, 577 ff.
Axioms, 64, 321, 571
for the natural numbers, 330
of probability, 488
for the real numbers, 64 ff.
Axis, 2, 175, 306
of symmetry, 277, 307

Bar, 34
Base, 357, 453
Basic assumption, 64
Basic term, 47
Belong, 134
Biconditional, 77
Binary operation, 261, 549, 550, 572, 573
Binary relation, 248
Binary system, 358
Binomial coefficient, 500
Binomial distribution, 509
Binomial theorem, 419, 500
Bisector of a segment, 188
Boolean algebra, 572 ff.
Border, 561
Bounded set, 355
Brace, 34, 135
Brackets, 34
Break-even chart, 203
Buffon's needle problem, 516

Calculus, 377 ff.
Calculation, 365 ff., 455, 462
Cancellation, 31, 32, 569
Cap, 152
Cardinal number, 321 ff.
Cartesian graph of a sentence, 246
Cartesian plane, 245
Cartesian product, 244, 328
Categorical theory, 579
Causality, 261
Center
of ellipse, 306
of the hyperbola, 310
of symmetry, 310

Integers, 1 ff., 137, 350
Integral, 478
Integral domain, 553
Integrand, 446
Intercept, 195, 258
Intercept form, 197
Intercepted arc, 217
Interest, 342, 389, 421, 461
Interpolation, 227 ff., 456 ff.
Intersection, 199
  of relations, 265
  of sets, 151
Interval, 144, 156
Interval estimation, 542
Invalid argument, 96
Invariance, 565
Invariants of a conic, 320
Inverse, 296, 348, 550, 555
Inverse circular function, 291
Inverse element of group, 569
Inverse function, 285
Inverse variation, 426
Inversely proportional, 426
Involution laws, 155
Irrationality of the square root of 2, 351
Irrational number, 4, 61 ff., 351 ff.
Irrational point, 356
Isomorphic group, 558
Isomorphism, 349, 370, 571

Just, 125

Lagrange's theorem, 570
Large numbers, laws of, 522 ff.
Law, 24 ff., 77, 119
  of cosines, 239
  of large numbers, 521
  of mean, 430
  of sines, 239
Least upper bound, 355, 438
Lemma, 99
Length, 378
  of an arc, 217, 468, 477
  of a vector, 11, 52, 186
Less than, see Inequalities
Level of significance, 533
Likelihood, 543
Limit, 390 ff., 442, 514
  of integration, 446
  of summation, 344
Limiting case, 303
Linear equation, 107, 193

Linear function, 268 ff., 282
Linear graph, 144
Linear (homogeneous) transformation, 567
Linear interpolation, 227, 456
Linear transformation, 567
Linear velocity, 282
Local maximum, 411
Locus, 188
Logarithm, 453 ff.
Logic, 70 ff., 252, 337, 559, 573
Logical relation, 267
Logically equivalent, 78, 106, 139
Logistic law, 462
Lower bound, 355
Lower limit of integration, 446
Lower sum, 440

Major axis of ellipse, 306
Malthus, 342
Mantissa, 455
Many-many correspondence, 260
Many-one correspondence, 260
Mapping, 260, 285
Marginal cost, 399, 414
Marginal revenue, 414
Marginal unit cost, 427
Marginal utility, 414
Material implication, 76
Mathematical statistics, 530
Matrix, 202
Maxima and minima, 164, 276, 409 ff.
Maximization principle, 546
Maximum likelihood estimate, 543
Maximum of a function, 278
Mean, 150, 414, 509
  law of, 430
Mean deviation, 510
Measure, 434, 488
  of central tendency, 414
Median, 539
Member, 134
  of equation, 18
  of inequality, 22
Metatheory, 96
Midpoint of a segment, 188
Mil, 218
Minimum, 164, 276
  of a function, 278
Minor axis of ellipse, 306
Minus sign (see also Negative), 30, 56 ff.